EARTH OBSERVATION
OF ECOSYSTEM SERVICES

Earth Observation of Global Changes

Series Editor
Chuvieco Emilio

Earth Observation of Ecosystem Services
edited by Domingo Alcaraz-Segura, Carlos Marcelo Di Bella,
and Julieta Veronica Straschnoy

Global Forest Monitoring from Earth Observation
edited by Frédéric Achard and Matthew C. Hansen

EARTH OBSERVATION
OF ECOSYSTEM SERVICES

EDITED BY

Domingo Alcaraz-Segura
Carlos Marcelo Di Bella
Julieta Veronica Straschnoy

CRC Press
Taylor & Francis Group
Boca Raton London New York

CRC Press is an imprint of the
Taylor & Francis Group, an **informa** business

CRC Press
Taylor & Francis Group
6000 Broken Sound Parkway NW, Suite 300
Boca Raton, FL 33487-2742

First issued in paperback 2017

© 2014 by Taylor & Francis Group, LLC
CRC Press is an imprint of Taylor & Francis Group, an Informa business

ISBN-13: 978-1-4665-0588-9 (hbk)
ISBN-13: 978-1-138-07392-0 (pbk)

Library of Congress Cataloging-in-Publication Data

Earth observation of ecosystem services / [edited by] Domingo Alcaraz-Segura, Carlos
    Marcelo Di Bella, Julieta Veronica Straschnoy.
        pages cm. -- (Earth observation of global changes)
    Includes bibliographical references and index.
    ISBN 978-1-4665-0588-9 (hardback)
    1. Biodiversity--Monitoring. 2. Biodiversity--Remote sensing. 3. Environmental
monitoring. 4. Global environmental change. I. Alcaraz-Segura, Domingo.

QH541.15.B56E34 2013
577.2'2--dc23   2013036753

Visit the Taylor & Francis Web site at
http://www.taylorandfrancis.com

and the CRC Press Web site at
http://www.crcpress.com

Contents

Foreword ix

Editors..xi

Contributors... xiii

Reviewers xix

Section I   Introduction

1. **A Global Vision for Monitoring Ecosystem Services with Satellite Sensors**..3

   D. Alcaraz-Segura and C. M. Di Bella

Section II   Ecosystem Services Related to the Carbon Cycle

2. **Ecosystem Services Related to Carbon Dynamics: Its Evaluation Using Remote Sensing Techniques** 17

   J. M. Paruelo and M. Vallejos

3. **Recent Advances in the Estimation of Photosynthetic Stress for Terrestrial Ecosystem Services Related to Carbon Uptake**.................39

   M. F. Garbulsky, I. Filella, and J. Peñuelas

4. **Earth Observation of Carbon Cycling Pools and Processes in Northern High-Latitude Systems**...63

   H. E. Epstein

5. **Monitoring the Ecosystem Service of Forage Production**....................87

   J. G. N. Irisarri, M. Oesterheld, M. Oyarzabal, J. M. Paruelo, and M. Durante

6. **Missing Gaps in the Estimation of the Carbon Gains Service from Light Use Efficiency Models**... 105

   A. J. Castro Martínez, J. M. Paruelo, D. Alcaraz-Segura, J. Cabello, M. Oyarzabal, and E. López-Carrique

7. **Biomass Burning Emission Estimation in Amazon Tropical Forest** ... 125

   Y. E. Shimabukuro, G. Pereira, F. S. Cardozo, R. Stockler, S. R. Freitas, and S. M. C. Coura

Section III   Ecosystem Services Related to Biodiversity

8. Earth Observation for Species Diversity Assessment
   and Monitoring 151
   N. Fernández

9. Ecosystem Services Assessment of National Parks Networks for
   Functional Diversity and Carbon Conservation Strategies Using
   Remote Sensing 179
   J. Cabello, P. Lourenço, A. Reyes, and D. Alcaraz-Segura

10. Catchment Scale Analysis of the Influence of Riparian Vegetation
    on River Ecological Integrity Using Earth Observation Data 201
    T. Tormos, K. Van Looy, P. Kosuth, B. Villeneuve, and Y. Souchon

Section IV   Ecosystem Services Related to the Water Cycle

11. Evaluation of Hydrological Ecosystem Services through Remote
    Sensing 229
    *C. Carvalho-Santos, B. Marcos, J. Espinha Marques, D. Alcaraz-Segura,
    L. Hein, and J. Honrado*

12. Assimilation of Remotely Sensed Data into Hydrologic
    Modeling for Ecosystem Services Assessment 261
    J. Herrero, A. Millares, C. Aguilar, F. J. Bonet, and M. J. Polo

13. Detecting Ecosystem Reliance on Groundwater Based on
    Satellite-Derived Greenness Anomalies and Temporal Dynamics 283
    S. Contreras, D. Alcaraz-Segura, B. Scanlon, and E. G. Jobbágy

14. Surface Soil Moisture Monitoring by Remote Sensing:
    Applications to Ecosystem Processes and Scale Effects 303
    M. J. Polo, M. P. González-Dugo, C. Aguilar, and A. Andreu

15. Snowpack as a Key Element in Mountain Ecosystem Services:
    Some Clues for Designing Useful Monitoring Programs 329
    F. J. Bonet, A. Millares, and J. Herrero

Section V   Ecosystem Services Related to the Land-Surface Energy Balance

16. Characterizing and Monitoring Climate Regulation Services 351
    D. Alcaraz-Segura, E. H. Berbery, O. V. Müller, and J. M. Paruelo

17. **Ecosystem Services Related to Energy Balance: A Case Study of Wetlands Reflected Energy**..379
C. M. Di Bella and M. E. Beget

18. **Energy Balance and Evapotranspiration: A Remote Sensing Approach to Assess Ecosystem Services**..399
V. A. Marchesini, J. P. Guerschman, and J. A. Sobrino

19. **Urban Heat Island Effect**..417
J. A. Sobrino, R. Oltra-Carrió, and G. Sòria

Section VI   Other Dimensions of Ecosystem Services

20. **Multidimensional Approaches in Ecosystem Services Assessment** 441
A. J. Castro Martínez, M. García-Llorente, B. Martín-López, I. Palomo, and I. Iniesta-Arandia

Index..469

Foreword

Trend monitoring is considered increasingly critical for a better understanding of environmental changes. Our planet is a living system with multiple interactions between physical and biological processes that are continually modifying the Earth's landscapes. In addition to these natural processes, human activities play a very relevant role in explaining the environmental processes, as humans interact with a wide variety of atmospheric, ocean, and land flows. This interaction has two complementary heads—how humans affect and are affected by ecological processes. Planet Earth primarily serves as our home. We rely on natural resources to find food and shelter and even to find our spiritual guidance. For millennia, we have used those resources in diverse ways, perceiving them as infinite. This planet was considered vast and diverse enough to serve human needs unconditionally. However, we now realize that the human footprint is almost everywhere, and we have gradually begun to consider natural resources as precious and limited goods. Our house is becoming too small, or at least this is how we perceive it. We can react to this perception by ignoring the potential collapse to which our current way of living may bring us or we can limit our growth. But the first effort should be to better understand the problem: whether this perception is true and what are the trends to estimate near-term scenarios.

Earth observation by satellite has become an indispensable tool for obtaining a global view of many natural processes as well as for monitoring their trends. Even though the historical archives of satellite images are small (the first reliable satellites were launched only 25 to 30 years ago), they provide critical information on tropical deforestation, land use trends, water quality, crop yields, snow and ice extents, coastal processes, oceans, cloud and aerosol distribution, and many other variables that are essential to describe the global system.

Ecosystem services refer to all natural processes that have a significant impact on human societies. In recent decades, these services have been as inputs to be accounted for in any economical evaluation. Nature provides a wide range of services to humanity, from water quality to wood or pasture, from hunting to fishing, from biodiversity conservation to snow cover, from carbon stocks to soil erosion protection, from cork to nuts to mushrooms. Many of these aspects are covered in this book, which serves as a very relevant method of facilitating updated material for better appreciating how satellite images can be used operationally in monitoring ecosystem services. This is probably the first book to cover most of those topics, providing a comprehensive analysis of a very innovative field of research that should be promoted in the near future. The editors' efforts to cover such a wide range of topics with such a diverse list of authors should be greatly acknowledged. The resulting text will facilitate extensive material for ecologists, hydrologists, biologists, geographers, and many other environmental scientists, who can further rely on the growing availability of satellite images for better understanding and monitoring our fragile environment.

Emilio Chuvieco
Professor of Geography
Universidad de Alcalá

Editors

Domingo Alcaraz-Segura was born in Alhama de Almería, Spain, in 1978. He received his bachelor degree in environmental sciences in 2000 and his PhD degree in 2005, both at the University of Almería. He has enjoyed post-doctoral positions at the University of Virginia, University of Buenos Aires, University of Texas at Austin, University of Maryland, and Spanish Council for Research (Doñana Biological  Station). Currently, he is a professor at the University of Granada (Spain) and an associate researcher of the Andalusian Center for the Assessment and Monitoring of Global Change. He teaches courses in botany, geobotany, global change, biodiversity conservation and human well-being, and time series analysis of satellite images. His current research interests are the environmental controls of biodiversity, the impact of land cover and land use changes on ecosystem functioning and services and on hydroclimate, and the development of monitoring and alert systems of global change effects on protected areas. His research is based on fieldwork, remote sensing techniques, time series analysis, and geographical information systems.

Carlos Marcelo Di Bella was born in Buenos Aires, Argentina, in 1969. He graduated as an agronomist (Faculty of Agronomy, University of Buenos Aires, Argentina) in 1994 and received a PhD degree from the Institut National Agronomique Paris-Grignon, France, in 2002. Since 1998, he has been a staff researcher at the Institute of Climate and Water (INTA, National Institute of Agricultural Technology) and since 2006 at the National Scientific and Technical Research  Council (CONICET). He is also a director of postgraduate career: remote sensing and geographic information systems (GISs) applied to the study of natural resources and agricultural production (Alberto Soriano Graduate School, Faculty of Agronomy, University of Buenos Aires, Argentina). His research focuses on the application and development of remote sensing and the application of GISs to natural resources and agroecosystems study, management, and monitoring.

Julieta Verónica Straschnoy was born in Buenos Aires, Argentina, in 1974. She graduated as a teacher of mathematics and physics in 1997. She received her bachelor degree in environmental management in 2002 and currently is finalizing her masters in environmental studies, both at the University of Business and Social Sciences (Buenos Aires). Since 2003, she has been a staff researcher at the Institute of Climate and Water (INTA, National Institute of Agricultural Technology). She participates in the development of different national and international projects in the area of permanent observation of agroecosystems.

Contributors

C. Aguilar
Fluvial Dynamics and Hydrology
  Research Group
Andalusian Institute of Earth
  System Research
Agrifood Campus of International
  Excellence (ceiA3)
University of Córdoba
Córdoba, Spain

D. Alcaraz-Segura
Botany Department
Faculty of Sciences
University of Granada
Granada, Spain

and

Andalusian Center for the
  Assessment and Monitoring of
  Global Change (CAESCG)
University of Almería
Almería, Spain

A. Andreu
Andalusian Institute for
  Agricultural and Fisheries
  Research and Training (IFAPA)
Cordoba, Spain

M. E. Beget
Climate and Water Institute
Research Center of Natural
  Resources (CIRN)
National Institute of Agricultural
  Technology (INTA)
Hurlingham, Argentina

E. H. Berbery
Cooperative Institute for Climate
  and Satellites
Earth System Science
  Interdisciplinary Center
University of Maryland
College Park, Maryland

F. J. Bonet
Terrestrial Ecology Research
  Group
Andalusian Institute for Earth
  System Research
University of Granada
Granada, Spain

J. Cabello
Department of Biology and
  Geology
Andalusian Center for
  the Assessment and
  Monitoring of Global Change
  (CAESCG)
University of Almería
Almería, Spain

F. S. Cardozo
National Institute for Space
  Research (INPE)
São José dos Campos, Brazil

C. Carvalho-Santos
Department of Biology
Faculty of Sciences and
    Research Centre in
    Biodiversity and Genetic
    Resources (CIBIO)
University of Porto
Porto, Portugal

and

Environmental Systems Analysis
    Group
Wageningen University
Wageningen, The Netherlands

A. J. Castro Martínez
Oklahoma Biological Survey (OBS)
University of Oklahoma
Norman, Oklahoma

and

Andalusian Center for
    the Assessment and
    Monitoring of Global Change
    (CAESCG)
Department of Plant Biology and
    Ecology
University of Almería
Almería, Spain

S. Contreras
Centre of Pedology and Applied
    Biology of Segura
Spanish Council for Scientific
    Research (CSIC)
Murcia, Spain

S. M. C. Coura
National Institute for Space
    Research (INPE)
São José dos Campos, Brazil

C. M. Di Bella
Climate and Water Institute
Research Center of Natural
    Resources (CIRN)
National Institute of Agricultural
    Technology (INTA)
National Scientific and
    Technical Research Council
    (CONICET)
Hurlingham, Argentina

M. Durante
National Institute of Agricultural
    Technology (INTA)
Concepción del Uruguay
Argentina

H. E. Epstein
Department of Environmental
    Sciences
University of Virginia
Charlottesville, Virginia

J. Espinha Marques
Geology Centre and Department of
    Geosciences
Environment and Spatial Planning
Faculty of Sciences
University of Porto
Porto, Portugal

N. Fernández
Doñana Biological Station
Spanish National Research Council
    EBD-CSIC
Sevilla, Spain

I. Filella
National Research Council (CSIC)
Center for Ecological Research
    and Forestry Applications
    (CREAF)
Catalonia, Spain

S. R. Freitas
National Institute for Space
    Research (INPE)
São José dos Campos, Brazil

M. F. Garbulsky
School of Agriculture
University of Buenos Aires
Institute for Agricultural
Plant Physiology and Ecology
    (IFEVA)
National Scientific and Technical
    Research Council (CONICET)
Buenos Aires, Argentina

M. García-Llorente
Sociology and the Environment
    Research Area
Social Analysis Department
Carlos III University of Madrid
and
Social-Ecological Systems
    Laboratory
Department of Ecology
Autonomous University of Madrid
Madrid, Spain

M. P. González-Dugo
Andalusian Institute for
    Agricultural and Fisheries
    Research and Training (IFAPA)
Cordoba, Spain

J. P. Guerschman
Commonwealth Scientific and
    Industrial Research Organisation
    (CSIRO) Land and Water
Canberra, Australia

L. Hein
Environmental Systems Analysis
    Group
Wageningen University
Wageningen, The Netherlands

J. Herrero
Fluvial Dynamics and Hydrology
    Research Group
Andalusian Institute for Earth
    System Research
University of Granada
Granada, Spain

J. Honrado
Department of Biology
Faculty of Sciences and Research
    Centre in Biodiversity and
    Genetic Resources (CIBIO)
University of Porto
Porto, Portugal

I. Iniesta-Arandia
Social-Ecological Systems
    Laboratory
Department of Ecology
Autonomous University of Madrid
Madrid, Spain

J. G. N. Irisarri
School of Agriculture
University of Buenos Aires
Regional Analysis
    Laboratory and Remote
    Sensing (LART)
Institute for Agricultural
    Plant Physiology and Ecology
    (IFEVA)
National Scientific and
    Technical Research Council
    (CONICET)
Buenos Aires, Argentina

E. G. Jobbágy
Environmental Research Group
    (GEA)
San Luis Institute of Applied
    Mathematics (IMASL)
National Scientific and Technical
    Research Council (CONICET)
San Luis, Argentina

P. Kosuth
National Research Institute of
    Science and Technology for
    Environment and Agriculture
    (IRSTEA)
Earth Observation and
    Geo-Information for Environment
    and Land Management Unit
    (UMR TETIS)
Montpellier, France

E. López-Carrique
Andalusian Center for the
    Assessment and Monitoring of
    Global Change (CAESCG)
University of Almería
Almería, Spain

P. Lourenço
Department of Biology and
    Geology
Andalusian Center for the
    Assessment and Monitoring
    of Global Change (CAESCG)
University of Almería
Almería, Spain

V. A. Marchesini
School of Agriculture
University of Buenos Aires
Regional Analysis Laboratory and
    Remote Sensing (LART)
Institute for Agricultural
    Plant Physiology and Ecology
    (IFEVA)
National Scientific and Technical
    Research Council (CONICET)
Buenos Aires, Argentina

and

School of Plant Biology
The University of Western
    Australia
Perth, Australia

B. Marcos
Department of Biology
Faculty of Sciences and Research
    Centre in Biodiversity and
    Genetic Resources (CIBIO)
University of Porto
Porto, Portugal

B. Martín-López
Social-Ecological Systems
    Laboratory
Department of Ecology
Autonomous University of Madrid
Madrid, Spain

A. Millares
Fluvial Dynamics and Hydrology
    Research Group
Andalusian Institute for Earth
    System Research
University of Granada
Granada, Spain

O. V. Müller
CEVARCAM
Faculty of Engineering and Water
    Resources
National University of Litoral
National Scientific and Technical
    Research Council (CONICET)
Santa Fe, Argentina

M. Oesterheld
School of Agriculture
University of Buenos Aires
Regional Analysis
Laboratory and Remote Sensing
    (LART)
Institute for Agricultural Plant
    Physiology and Ecology (IFEVA)
National Scientific and
    Technical Research Council
    (CONICET)
Buenos Aires, Argentina

R. Oltra-Carrió
Global Change Unit
Image Processing Laboratory
University of València
València, Spain

M. Oyarzabal
Department of Quantitative
  Methods and Information
  Systems
School of Agriculture
University of Buenos Aires
Regional Analysis Laboratory and
  Remote Sensing (LART)
Institute for Agricultural
  Plant Physiology and
  Ecology (IFEVA)
National Scientific and
  Technical Research Council
  (CONICET)
Buenos Aires, Argentina

I. Palomo
Social-Ecological Systems
  Laboratory
Department of Ecology
Autonomous University of Madrid
Madrid, Spain

J. M. Paruelo
Department of Quantitative
  Methods and Information
  Systems
School of Agriculture
University of Buenos Aires
Regional Analysis Laboratory and
  Remote Sensing (LART)
Institute for Agricultural
  Plant Physiology and
  Ecology (IFEVA)
National Scientific and
Technical Research Council
  (CONICET)
Buenos Aires, Argentina

J. Peñuelas
National Research Council (CSIC)
Center for Ecological Research
  and Forestry Applications
  (CREAF)
Catalonia, Spain

G. Pereira
National Institute for Space
  Research (INPE)
São Paulo, Brazil

and

Federal University of São João
  del-Rei (UFSJ)
São João del-Rei, Brazil

M. J. Polo
Fluvial Dynamics and
  Hydrology Research Group
Andalusian Institute of Earth
  System Research
Agrifood Campus of International
Excellence (ceiA3)
University of Córdoba
Córdoba, Spain

A. Reyes
Department of Biology and
  Geology
Andalusian Center for the
  Assessment and Monitoring of
  Global Change (CAESCG)
University of Almería
Almería, Spain

B. Scanlon
Bureau of Economic Geology
Jackson School of Geosciences
The University of Texas at Austin
Austin, Texas

Y. E. Shimabukuro
National Institute for Space
  Research (INPE)
São José dos Campos, Brazil

J. A. Sobrino
Global Change Unit
Image Processing Laboratory
University of València
València, Spain

G. Sòria
Global Change Unit
Image Processing Laboratory
University of València
València, Spain

Y. Souchon
National Research Institute of Science
    and Technology for Environment
    and Agriculture (IRSTEA)
Aquatic environments, ecology and
    pollution (UR MALY)
River Hydro-Ecology Unit
    (ONEMA–IRSTEA)
Lyon, France

R. Stockler
National Institute for Space
    Research (INPE)
São José dos Campos, Brazil

T. Tormos
National Research Institute of Science
    and Technology for Environment
    and Agriculture (IRSTEA)
Aquatic environments, ecology and
    pollution (UR MALY)
French National Agency for Water
    and Aquatic Environments
    (ONEMA)
River Hydro-Ecology Unit
    (ONEMA–IRSTEA)
Lyon, France

M. Vallejos
Department of Quantitative
    Methods and Information
    Systems
School of Agriculture
University of Buenos Aires
Regional Analysis Laboratory and
    Remote Sensing (LART)
Institute for Agricultural
    Plant Physiology and Ecology
    (IFEVA)
National Scientific and
    Technical Research Council
    (CONICET)
Buenos Aires, Argentina

K. Van Looy
National Research Institute of Science
    and Technology for Environment
    and Agriculture (IRSTEA)
Aquatic environments, ecology and
    pollution (UR MALY)
River Hydro-Ecology Unit
    (Onema–Irstea)
Lyon, France

B. Villeneuve
National Research Institute of
    Science and Technology for
    Environment and Agriculture
    (IRSTEA)
Aquatic environments, ecology and
    pollution (UR MALY)
River Hydro-Ecology Unit
    (ONEMA–IRSTEA)
Lyon, France

Reviewers

We sincerely thank the many reviewers that contributed to this book; their thoughtful evaluations and constructive comments substantially improved the content and presentation of the chapters. We are also grateful to the numerous universities, research centers, public agencies, and taxpayers that funded their works.

Flor Álvarez-Taboada, University of León, Spain
Roxana Aragón, National University of Tucumán, Argentina
Olga Barron, CSIRO Land and Water, Australia
J. Jesús Casas, University of Almería, Spain
Antonio Castro, University of Oklahoma, USA
Sérgio Bruno Costa, Simbiente, Portugal
Piedad Cristiano, University of Buenos Aires, Argentina
Miguel Delibes, CSIC, Spain
Koen De Ridder, VITO, Belgium
Heriberto Díaz-Solís, Antonio Narro Agrarian Autonomous University, Mexico
Martial Duguay, EURAC, Italy
Michael Ek, NOAA Center for Weather and Climate Prediction, USA
Martín Garbulsky, University of Buenos Aires, Argentina
Monica García, Columbia University, USA
Gregorio Gavier-Pizarro, National Institute of Agricultural Technology, Argentina
Artur Gil, University of the Açores, Portugal
Anatoly Gitelson, University of Nebraska, USA
Silvana Goirán, National University of Cuyo, Argentina
Alexander Graf, Forschungszentrum Jülich, Germany
Diego Gurvich, National University of Córdoba, Argentina
Michael Heinl, University of Innsbruck, Austria
Robert Höft, Secretariat of the Convention on Biological Diversity, Canada
João Honrado, University of Porto, Portugal
Ned Horning, American Museum of Natural History, USA
Charles Ichoku, NASA Goddard Space Flight Center, USA
Akihiko Ito, National Institute for Environmental Studies, Japan
Eva Ivits, Institute for Environment and Sustainability, Italy
Gensuo Jia, Chinese Academy of Sciences, China
Juan Carlos Jiménez-Muñoz, University of Valencia, Spain
Eric Kasischke, University of Maryland, USA
William Lauenroth, University of Wyoming, USA
Feliciana Licciardello, University of Catania, Italy

César Agustín López Santiago, University Autonomous of Madrid, Spain
Néstor Oscar Maceira, National Institute of Agricultural Technology, Argentina
Priscilla Minotti, National University of San Martín, Argentina
Claudia Notarnicola, EURAC, Italy
Miren Onaindia, University of the Basque Country, Spain
Pedro Peña Garcillán, Biological Research Center of the Northwest, Mexico
César Pérez-Cruzado, University of Göttingen, Germany
Gabriela Posse, National Institute of Agricultural Technology, Argentina
Serge Rambal, CEFE CNRS, Canada
Duccio Rocchini, Edmund Mach Foundation, Italy
Nilda Sánchez Martín, CIALE, University of Salamanca, Spain
Fernando Santos-Martín, Autonomous University of Madrid, Spain
David Sheeren, National Polytechnic Institute of Toulouse, France
Bob Su, University of Twente, Netherlands
Anke Tetzlaff, EURAC, Italy
Santiago Verón, National Institute of Agricultural Technology, Argentina
Donald Young, Virginia Commonwealth University, USA
Julie Zinnert, Virginia Commonwealth University, USA

Section I

Introduction

1

A Global Vision for Monitoring Ecosystem Services with Satellite Sensors

D. Alcaraz-Segura

University of Granada, Spain; University of Almería, Spain

C. M. Di Bella

National Institute of Agricultural Technology (INTA), Argentina

CONTENTS

1.1 General Overview of Remote Sensing of Ecosystem Services 3
1.2 Overview of Book Sections and Chapters 7
    1.2.1 Ecosystem Services Related to the Carbon Cycle 7
    1.2.2 Ecosystem Services Related to Biodiversity 8
    1.2.3 Ecosystem Services Related to the Water Cycle 8
    1.2.4 Ecosystem Services Related to Energy Balance 9
    1.2.5 Other Dimensions of Ecosystem Services 9
Acknowledgments 9
References 10

1.1 General Overview of Remote Sensing of Ecosystem Services

Ecosystem services can be defined as "an activity or function of an ecosystem that provides benefit (or occasionally detriment) to humans" (Mace et al. 2012; see also Burkhard et al. 2012; Crossman et al. 2012). Repeated efforts have been made to quantify, value, map, monitor, and analyze the various ecosystem service components that sustain human well-being: from the early attempts of Costanza et al. (1997) and the Millennium Ecosystem Assessment (MA 2005) to the more recent integrative initiatives of the Intergovernmental Science-Policy Platform on Biodiversity and Ecosystem Services (IPBES 2013) and the Global System for Monitoring

Ecosystem Service Change (Tallis et al. 2012; GEO BON ES 2013). The latter initiatives join more than 50 international organizations and 80 governmental representations under the auspices of the Group on Earth Observations in order to develop the Global Earth Observation System of Systems (GEOSS). Their aim is to compile extensive and standard monitoring of multiple services to allow policymakers and scientists to explore, better understand, and prioritize trade-offs across socioecological settings and scales (from the national to the global level). The Group on Earth Observations Biodiversity Observation Network (GEO BON) identified four main sources of data: national statistics, field-based observations, remote sensing, and numerical simulation models (Tallis et al. 2012). Of these, remote sensing offers the potential to use the same standard protocols from local to global scales and through time, which is essential for long-term monitoring and for trade-off assessment across regions. Satellite-based earth observation is probably the most economically feasible means to systematically retrieve global information with high temporal, spatial, and spectral resolution over large areas (Ayanu et al. 2012). Decline in the cost involved in obtaining such data and increase in the computational power of higher resolution sensors to cope with larger data sets will maximize this advantage (Kreuter et al. 2001).

Remote sensing of ecosystem services has been a fast-growing research field in recent years. On searching the Scopus database (http://www.scopus.com) for documents that contain the terms "remote sensing," "earth observation," and/or "ecosystem services" in the title, abstract, or keywords, we found a total of 270 documents (Table 1.1). This search revealed that this is a relatively young research field (the first article dates back to 2001) that has been growing fast recently (50% of the articles have been published in the past two-and-half years, 2011–present). It is interesting to note the relatively active role that Chinese research is playing in the explicit assessment of ecosystem services through remote sensing tools, with 50% of articles being contributed by them. Of the remaining publications, 25% correspond to U.S. affiliations and the rest are affiliated with various countries. From the first 15 institutions (those who published five or more documents in this field), only five of them were not from China (i.e., the University of Buenos Aires, the German Aerospace Center, the Virginia Polytechnic Institute and State University, and Columbia University in New York). The journals that attracted more articles (five or more) were *Remote Sensing of Environment* (impact factor [IF] 4.574), *Acta Ecologica Sinica*, *Chinese Journal of Ecology*, *Ecological Indicators* (IF 2.695), *Chinese Journal of Applied Ecology*, *Acta Geographica Sinica*, and *Agriculture Ecosystems and Environment* (IF 3.004).

Some recent works have already compiled and highlighted the many possibilities that remote sensing offers for ecosystem services mapping (e.g., Naidoo et al. 2008; Feld et al. 2010; Feng et al. 2010; Ayanu et al. 2012; Crossman et al. 2012; Martínez-Harms and Balvanera 2012; Tallis et al. 2012).

TABLE 1.1

Number of Publications Per Year, and Ranking of the Top Countries, Authors, and Journals Publishing Articles on Earth Observation of Ecosystem Services

Year	#	Country	#	Authors	#	Journal Title	#
2013	18	China	134	Wang, K.	5	*Acta Ecologica Sinica*	17
2012	53	U.S.	66	Paruelo, J.	4	*Remote Sensing of Environment*	11
2011	61	Germany	19	Kuenzer, C.	4	*Chinese Journal of Ecology*	9
2010	43	U.K.	15	Pan, Y.	4	*Chinese Journal of Applied Ecology*	6
2009	35	Australia	11	Gu, X.	4	*Ecological Indicators*	6
2008	17	Netherlands	8	Zhu, W.	4	*Acta Geographica Sinica*	6
2007	21	Argentina	7	DeFries, R.	3	*Agriculture Ecosystems and Environment*	5
2006	6	Spain	7	Chong, J.	3	*Applied Geography*	4
2005	5	Italy	7	He, H.	3	*International Journal of Applied Earth Observation and Geoinformation*	4
2004	2	India	6	Li, J.	3	*Environmental Earth Sciences*	4
2003	3	Canada	6	Shi, P.	3	*Ecological Economics*	4
2002	3	Finland	6	Zhang, C.	3	*Environmental Monitoring and Assessment*	4
2001	3	Switzerland	5	Gond, V.	2	*International Journal of Remote Sensing*	4
Total	270	France	5	Dech, S.	2	*Remote Sensing*	4

Note: Publications in Scopus, from January 1, 2001 to April 1, 2013, containing the terms "remote sensing," "earth observation," and/or "ecosystem services" in the title, abstract, or keywords. # = number of articles.

Nevertheless, there is still an urgent need to develop a standardized and consistent methodological approach, or even a blueprint, for mapping the stocks and flows and the supply and demand of a fuller suite of ecosystem services (Crossman et al. 2012; Martínez-Harms and Balvanera 2012; Palomo et al. 2012). During the past 12 years, remotely sensed information has been mainly used to estimate provisioning and regulating ecosystem services due to the biophysical dimension of their supply but has scarcely been used to retrieve cultural services due to the inherent socioeconomic characteristics of their demand. Satellite images have provided relevant information for the assessment of ecosystem services from a multidimensional perspective (see Chapter 20), from the "supply side" (what does nature supply?) to the "demand side" (what do humans demand?). Only a few works have used remote sensing tools as accompanying information for the assessment of the demand side of ecosystem services (e.g., Sutton and Costanza 2002; Scullion et al. 2011).

Up until now, the most frequent use of satellite images in the supply side has been in the production of land use/cover maps that are later utilized in models to simulate the delivery of ecosystem services and their changes through time and space (Kreuter et al. 2001; Konarska et al. 2002; Zhao et al. 2004; De-Yong et al. 2005; Wang et al. 2006, 2009; Li et al. 2007, 2011, 2012; Hu et al. 2008; Du et al. 2009; Huang et al. 2009, 2011; Liu et al. 2009, 2012; Feng et al. 2010; McNally et al. 2011; Burkhard et al. 2012; Estoque and Murayama 2012; Hao et al. 2012; Bian and Lu 2013; Duan et al. 2013; Verburg et al. 2013; Zhao and Tong 2013). The former attempts aim to map ecosystem services by linking ecosystem service values to the different land use/cover types present in image classification. Then, changes in the ecosystem services delivery are estimated from changes in land use/cover types. However, there are spatial variabilities within land cover types; the same land cover may offer different landscape functions and, therefore, the delivery of multiple ecosystem services (Verburg et al. 2009). For example, the grazing capacity of semiarid grasslands varies greatly between the northern and southern exposures and between dry and humid years. Likewise, the same cropland category (e.g., soybean plantations) may have very different ecosystem functioning and may provide different ecosystem services depending on the land management (e.g., tillage versus no tillage) (Jayawickreme et al. 2011; Viglizzo et al. 2011). This is one of the reasons that Verburg et al. (2009) give for using land function dynamics assessments (which are continuous) rather than land cover change characterizations in ecosystem service assessments (which are categorical). Similarly, Euliss et al. (2010) point out that a more holistic and integrative approach to monitoring ecosystem processes and related services is necessary. For example, Schneider et al. (2012) combined the retrieval of direct estimates of ecosystem services related to land surface albedo and energy balance dynamics, derived from time series of Moderate Resolution Imaging Spectroradiometer (MODIS) satellite images, with other simulated ecosystem services, derived from running a model (the Agro-IBIS dynamic global vegetation model) on a land cover map (Kucharik 2003), covering the effects on the carbon and water cycles and on the energy balance. In general, all models used in ecosystem services modeling and mapping utilize spatially explicit information regarding land cover, climate variables, and topography. In addition, they may incorporate other biophysical variables, such as evapotranspiration (ET), that can be derived from remote sensing data. Currently, the most widespread ecosystem service models are InVEST, ARIES, and POLYSCAPE (for a review and further details, see Chapter 20).

Satellite images can also be used to directly quantify one or a set of particular ecosystem services, that is, without modeling them through their link to particular land cover types. The remote sensing literature has plenty of conceptual and empirical models that link the spectral information to critical biophysical variables and ecosystem processes such as primary production, biomass, surface temperature, albedo, ET, soil moisture, surface roughness, species richness, and others (Verstraete et al. 2000; Jensen 2007; Chuvieco 2008). These biophysical variables are tightly linked to ecosystem functions and their associated ecosystem services; hence, they have been widely used in ecosystem service mapping and assessments (Jin et al. 2009; Krishnaswamy et al. 2009; Malmstrom et al. 2009; Newton et al. 2009; Tenhunen et al. 2009; Feng et al. 2010; Lane and D'Amico 2010; Porfirio et al. 2010; Rocchini et al. 2010; Woodcock et al. 2010; McPherson et al. 2011; Turner et al. 2011; Caride et al. 2012; Frazier et al. 2012; Ivits et al. 2012; Martínez-Harms and Balvanera 2012; Politi et al. 2012; Shi et al. 2012; Tallis et al. 2012; Volante et al. 2012; Forsius et al. 2013; Oki et al. 2013).

1.2 Overview of Book Sections and Chapters

The remaining 19 chapters in this book are presented in five sections: carbon cycle, biodiversity, water cycle, energy balance, and other components in ecosystem services. Each section provides a review of conceptual and empirical methods, techniques, and case studies linking remotely sensed data to the biophysical variables and ecosystem functions associated with key ecosystem services. The book does not aim to exhaustively cover all ecosystem service categories; instead, it provides a global look into the most relevant approaches for estimating key ecosystem services from satellite data.

1.2.1 Ecosystem Services Related to the Carbon Cycle

Regional and global primary production has probably been one of the ecosystem processes most frequently studied through remote sensing techniques (for a review of estimates, see Ito 2011). Many studies have used direct estimates or surrogates of primary production in order to evaluate ecosystem services from the supply side (e.g., Malmstrom et al. 2009; Paruelo et al. 2011; Turner et al. 2011; Caride et al. 2012; Volante et al. 2012). In Chapter 2, Paruelo and Vallejos offer a review of the remote sensing theoretical bases and techniques used to estimate fluxes and stocks of the carbon cycle utilizing optical and active sensors. They present the conceptual connection between provision of ecosystem services and key processes of the carbon cycle with application to sustainable land use planning. Chapter 3 by Garbulsky, Filella, and Peñuelas and Chapter 6 by Castro et al. assess the current knowledge and the advances and challenges in remote sensing for the estimation of light use efficiency, the most difficult aspect of quantifying and modeling the service that regulates the carbon gains. In Chapter 4, Epstein covers all components of the carbon cycle that have been studied through remote sensing in the northern high latitudes—from carbon uptake and fire emissions to methane fluxes and soil microbe respiration. In Chapter 5, Irisarri et al. present a remote sensing approach to a monitoring system of forage production.

Finally, Shimabukuro et al., in Chapter 7, address the use of remote sensing data to estimate biomass burning emissions in tropical forests.

1.2.2 Ecosystem Services Related to Biodiversity

Biodiversity may play three different roles in ecosystem service assessments (Mace et al. 2012): Biodiversity can act as a regulator of underpinning ecosystem processes (e.g., primary producers or soil microbes that govern nutrient cycling), as a final ecosystem service (e.g., wild crops, livestock, or fisheries), or as a good in itself with intrinsic value (e.g., charismatic, flagship, or umbrella species). Several studies have taken advantage of remote sensing tools for ecosystem services assessment and biodiversity conservation (Naidoo et al. 2008; Krishnaswamy et al. 2009; Rocchini 2009; DeFries et al. 2010; Feld et al. 2010; Jones et al. 2010; Rocchini et al. 2010; Cabello et al. 2012; Alcaraz-Segura et al. 2013). In Chapter 8, Fernández offers a thorough review of studies that use remote sensing data to estimate the aspects that biodiversity covers in ecosystem service assessments: from surveying and modeling species distributions to their relationships with key ecosystem processes. Cabello et al. provide an original exercise on how remote sensing may help to assess the effectiveness of protected area networks in representing biodiversity and in providing services related to the carbon cycle in Chapter 9. Finally, in Chapter 10, Tormos et al. use high spatial resolution imagery and object-based image analysis to evaluate whether riparian vegetation enhances river ecosystem integrity and ecosystem services.

1.2.3 Ecosystem Services Related to the Water Cycle

Ecosystems provide many provisioning and regulating services related to the water cycle, namely freshwater provision and flood control. Many remote sensing products can be useful for estimating and modeling multiple components of the water cycle, from rainfall to snow cover, ET, runoff, or groundwater consumption. Chapter 11, by Carvalho-Santos et al., first offers a conceptualization of hydrological services and how they are linked to key elements of the water cycle. Then, these authors present a thorough review of the most relevant satellite sensors and hydrological models in order to evaluate and monitor those elements of the water cycle. Chapter 12, by Herrero et al., is centered on the important role that the assimilation of remote sensing information has for hydrological modeling and the subsequent estimation of hydrological ecosystem services. In Chapter 13, Contreras et al. propose a remote sensing approach in order to identify and quantify the reliance of ecosystems on water inputs beyond precipitation that includes both groundwater and surface water inputs. Finally, the chapters by Polo et al. (Chapter 14) and Bonet, Millares, and Herrero (Chapter 15) focus on reviewing and applying two key elements of the hydrological cycle—ET and snow cover, respectively.

1.2.4 Ecosystem Services Related to Energy Balance

Ecosystems influence climate through biogeochemical processes, by emitting/absorbing greenhouse gases and aerosols to/from the atmosphere, and through biophysical processes that rule the land surface energy balance as a function of albedo, latent heat, and sensible heat. In Chapter 16, Alcaraz-Segura et al. explain the ways in which the biophysical or energy balance component of climate regulation is estimated using ecosystem service assessments. In addition, these authors introduce an original remote sensing method in order to indirectly monitor key biophysical properties of ecosystems that are relevant for climate modeling based on ecosystem functional types. In Chapter 17, Di Bella and Beget address the estimation of land surface albedo, one of the main components of biogeophysical feedback. They also evaluate the effect that land use changes have on surface reflectance and on spectral albedo. In Chapter 18, Marchesini, Guerschman, and Sobrino discuss the latent heat component of ET, the different models used to estimate ET via remote sensing, and the energy balance equation employed to predict the partition between latent and sensible heat fluxes. In Chapter 19, Sobrino et al. address land surface temperature, a biophysical property related to the sensible heat fluxes in the energy balance. They use the temperature and emissivity separation algorithm to evaluate the urban heat-island effect and to determine the best spatial resolution and overpass time to properly monitor this phenomenon from remote sensing platforms.

1.2.5 Other Dimensions of Ecosystem Services

As noted earlier, there is a need to develop a standardized and interdisciplinary framework, or even a blueprint, for mapping the stocks and flows, and the supply and demand of ecosystem services (Crossman et al. 2012; Martínez-Harms and Balvanera 2012). In Chapter 20, Castro et al. present a review of existing methodologies for assessing ecosystem services from the supply to the demand side, considering their multidimensional nature (i.e., biophysical, sociocultural, and economic), and highlighting where remote sensing tools may help to advance in an interdisciplinary comprehension of services assessment.

Acknowledgments

We sincerely thank the many authors and reviewers that contributed to this book; their thoughtful evaluations and constructive comments substantially improved the content and presentation of the chapters of this book. We are also grateful to the numerous universities, research centers, public agencies, and taxpayers that funded our works. Finally, we acknowledge the confidence and support of Emilio Chuvieco, the series editor, and Irma Britton and Laurie Schlags for their patience and generous assistance throughout the editorial process.

References

Alcaraz-Segura, D., J. Paruelo, H. Epstein, and J. Cabello. 2013. Environmental and human controls of ecosystem functional diversity in temperate South America. *Remote Sensing* 5:127–154.

Ayanu, Y. Z., C. Conrad, T. Nauss, M. Wegmann, and T. Koellner. 2012. Quantifying and mapping ecosystem services supplies and demands: A review of remote sensing applications. *Environmental Science and Technology* 46:8529–8541.

Bian, Z., and Q. Lu. 2013. Ecological effects analysis of land use change in coal mining area based on ecosystem service valuing: A case study in Jiawang. *Environmental Earth Sciences* 68:1619–1630.

Burkhard, B., F. Kroll, S. Nedkov, and F. Müller. 2012. Mapping ecosystem service supply, demand and budgets. *Ecological Indicators* 21:17–29.

Cabello, J., N. Fernández, D. Alcaraz-Segura, et al. 2012. The ecosystem functioning dimension in conservation: Insights from remote sensing. *Biodiversity and Conservation* 21:3287–3305.

Caride, C., G. Piñeiro, and J. M. Paruelo. 2012. How does agricultural management modify ecosystem services in the Argentine Pampas? The effects on soil C dynamics. *Agriculture Ecosystems and Environment* 154:23–33.

Chuvieco, E. 2008. *Earth observation of global change: The role of satellite remote sensing in monitoring the global environment.* Berlin: Springer-Verlag.

Costanza, R., R. d' Arge, R. de Groot, et al. 1997. The value of the world's ecosystem services and natural capital. *Nature* 387:253–260.

Crossman, N. D., B. Burkhard, and S. Nedkov. 2012. Quantifying and mapping ecosystem services. *International Journal of Biodiversity Science, Ecosystem Services and Management* 8:1–4.

DeFries, R., K. K. Karanth, and S. Pareeth. 2010. Interactions between protected areas and their surroundings in human-dominated tropical landscapes. *Biological Conservation* 143:2870–2880.

De-Yong, Y., P. Yao-Zhong, W. Yan-Yan, L. Xin, L. Jing, and L. Zhong-Hua. 2005. Valuation of ecosystem services for Huzhou City, Zhejiang Province from 2001 to 2003 by remote sensing data. *Journal of Forestry Research* 16:223–227.

Du, Z., Y. Shen, J. Wang, and W. Cheng. 2009. Land-use change and its ecological responses: A pilot study of typical agro-pastoral region in the Heihe River, northwest China. *Environmental Geology* 58:1549–1556.

Duan, J., Y. Li, and J. Huang. 2013. An assessment of conservation effects in Shilin Karst of South China Karst. *Environmental Earth Sciences* 68:821–832.

Estoque, R. C., and Y. Murayama. 2012. Examining the potential impact of land use/cover changes on the ecosystem services of Baguio City, the Philippines: A scenario-based analysis. *Applied Geography* 35:316–326.

Euliss, N. H., Jr., L. M. Smith, S. Liu, et al. 2010. The need for simultaneous evaluation of ecosystem services and land use change. *Environmental Science and Technology* 44:7761–7763.

Feld, C. K., J. P. Sousa, P. M. da Silva, and T. P. Dawson. 2010. Indicators for biodiversity and ecosystem services: Towards an improved framework for ecosystems assessment. *Biodiversity and Conservation* 19:2895–2919.

Feng, X., B. Fu, X. Yang, and Y. Lü. 2010. Remote sensing of ecosystem services: An opportunity for spatially explicit assessment. *Chinese Geographical Science* 20:522–535.

Forsius, M., S. Anttila, L. Arvola, et al. 2013. Impacts and adaptation options of climate change on ecosystem services in Finland: A model based study. *Current Opinion in Environmental Sustainability* 5:26–40.

Frazier, A. E., C. S. Renschler, and S. B. Miles. 2012. Evaluating post-disaster ecosystem resilience using MODIS GPP data. *International Journal of Applied Earth Observation and Geoinformation* 21:43–52.

GEO BON ES (Group on Earth Observations. Biodiversity Observation Network. Ecosystem Services Working Group). 2013. Available from: http://www.earthobservations.org/geobon_wgs.shtml (accessed April 4, 2013).

Hao, F., X. Lai, W. Ouyang, Y. Xu, X. Wei, and K. Song. 2012. Effects of land use changes on the ecosystem service values of a reclamation farm in northeast China. *Environmental Management* 50:888–899.

Hu, H., W. Liu, and M. Cao. 2008. Impact of land use and land cover changes on ecosystem services in Menglun, Xishuangbanna, southwest China. *Environmental Monitoring and Assessment* 146:147–156.

Huang, Q., D. D. Li, and H. B. Zhang. 2009. Effects of land use and land cover change on ecosystem service values in oasis region of northwest China. *Proceedings of SPIE—The International Society for Optical Engineering* 7384. doi:10.1117/12.834897.

Huang, X., Y. Chen, J. Ma, and X. Hao. 2011. Research of the sustainable development of Tarim River based on ecosystem service function. *Procedia Environmental Sciences* 10:239–246.

IPBES (Intergovernmental Science-Policy Platform on Biodiversity and Ecosystem Services). 2013. IPBES Draft Work Programme 2014–2018. Available from: http://www.ipbes.net (accessed April 4, 2013).

Ito, A. 2011. A historical meta-analysis of global terrestrial net primary productivity: Are estimates converging? *Global Change Biology* 17:3161–3175.

Ivits, E., M. Cherlet, G. Tóth, et al. 2012. Combining satellite derived phenology with climate data for climate change impact assessment. *Global and Planetary Change* 88–89:85–97.

Jayawickreme, D. H., C. S. Santoni, J. H. Kim, E. G. Jobbágy, and R. B. Jackson. 2011. Changes in hydrology and salinity accompanying a century of agricultural conversion in Argentina. *Ecological Applications* 21:2367–2379.

Jensen, J. R. 2007. *Remote sensing of the environment: An earth resource perspective.* Boston, MA: Pearson Prentice Hall.

Jin, Y., J. F. Huang, and D. L. Peng. 2009. A new quantitative model of ecological compensation based on ecosystem capital in Zhejiang Province, China. *Journal of Zhejiang University: Science B* 10:301–305.

Jones, K. B., E. T. Slonecker, M. S. Nash, et al. 2010. Riparian habitat changes across the continental United States (1972–2003) and potential implications for sustaining ecosystem services. *Landscape Ecology* 25:1261–1275.

Konarska, K. M., P. C. Sutton, and M. Castellon. 2002. Evaluating scale dependence of ecosystem service valuation: A comparison of NOAA-AVHRR and Landsat TM datasets. *Ecological Economics* 41:491–507.

Kreuter, U. P., H. G. Harris, M. D. Matlock, and R. E. Lacey. 2001. Change in ecosystem service values in the San Antonio area, Texas. *Ecological Economics* 39:333–346.

Krishnaswamy, J., K. S. Bawa, K. N. Ganeshaiah, and M. C. Kiran. 2009. Quantifying and mapping biodiversity and ecosystem services: Utility of a multi-season NDVI based Mahalanobis distance surrogate. *Remote Sensing of Environment* 113:857–867.

Kucharik, C. J. 2003. Evaluation of a process-based agro-ecosystem model (Agro-IBIS) across the U.S. corn belt: Simulations of the interannual variability in maize yield. *Earth Interactions* 7:1–33.

Lane, C. R., and E. D'Amico. 2010. Calculating the ecosystem service of water storage in isolated wetlands using LIDAR in north central Florida, USA. *Wetlands* 30:967–977.

Li, P., L. Jiang, Z. Feng, and X. Yu. 2012. Research progress on trade-offs and synergies of ecosystem services: An overview. *Shengtai Xuebao/Acta Ecologica Sinica* 32:5219–5229.

Li, R. Q., M. Dong, J. Y. Cui, L. L. Zhang, Q. G. Cui, and W. M. He. 2007. Quantification of the impact of land-use changes on ecosystem services: A case study in Pingbian County, China. *Environmental Monitoring and Assessment* 128:503–510.

Li, W. J., S. H. Zhang, and H. M. Wang. 2011. Ecosystem services evaluation based on geographic information system and remote sensing technology: A review. *Chinese Journal of Applied Ecology* 22:3358–3364.

Liu, J., J. Gao, and Y. Nie. 2009. Measurement and dynamic changes of ecosystem services value for the Tibetan Plateau based on remote sensing techniques. *Geography and Geo-Information Science* 3:022.

Liu, Y., J. Li, and H. Zhang. 2012. An ecosystem service valuation of land use change in Taiyuan City, China. *Ecological Modelling* 225:127–132.

MA (Millennium Ecosystem Assessment). 2005. *Ecosystems and human well-being. Synthesis report.* Washington, DC: Island Press.

Mace, G. M., K. Norris, and A. H. Fitter. 2012. Biodiversity and ecosystem services: A multilayered relationship. *Trends in Ecology and Evolution* 27:19–26.

Malmstrom, C. M., H. S. Butterfield, C. Barber, et al. 2009. Using remote sensing to evaluate the influence of grassland restoration activities on ecosystem forage provisioning services. *Restoration Ecology* 17:526–538.

Martínez-Harms, M. J., and P. Balvanera. 2012. Methods for mapping ecosystem service supply: A review. *International Journal of Biodiversity Science, Ecosystem Services and Management* 8:17–25.

McNally, C. G., E. Uchida, and A. J. Gold. 2011. The effect of a protected area on the tradeoffs between short-run and long-run benefits from mangrove ecosystems. *Proceedings of the National Academy of Sciences of the United States of America* 108:13945–13950.

McPherson, E. G., J. R. Simpson, Q. Xiao, and C. Wu. 2011. Million trees Los Angeles canopy cover and benefit assessment. *Landscape and Urban Planning* 99:40–50.

Naidoo, R., A. Balmford, and R. Costanza. 2008. Global mapping of ecosystem services and conservation priorities. *Proceedings of the National Academy of Sciences of the United States of America* 105:9495–9500.

Newton, A. C., R. A. Hill, C. Echeverría, et al. 2009. Remote sensing and the future of landscape ecology. *Progress in Physical Geography* 33:528–546.

Oki, T., E. M. Blyth, E. H. Berbery, and D. Alcaraz-Segura. 2013. Land cover and land use changes and their impacts on hydroclimate, ecosystems and society. In *Climate science for serving society: Research, modeling and prediction priorities*, eds. G. R. Asrar and J. W. Hurrel, 185–203. Dordrecht, The Netherlands: Springer Science and Business Media.

Palomo, I., B. Martín-López, M. Potschin, R. Haines-Young, and C. Montes. 2012. National parks, buffer zones and surrounding landscape: Mapping ecosystem services flows. *Ecosystem Services Journal* 4:104–116.

Paruelo, J., D. Alcaraz-Segura, and J. N. Volante. 2011. El seguimiento del nivel de provisión de los servicios ecosistémicos [Monitoring ecosystem services provision]. In *Valoración de servicios ecosistémicos: Conceptos, herramientas y aplicaciones para el ordenamiento territorial* [Ecosystem services valuation: concepts, tools and applications for land use planning], eds. P. Laterra, E. G. Jobbágy, and J. M. Paruelo, 141–160. Buenos Aires: Instituto Nacional de Tecnología Agropecuaria.

Politi, E., M. E. J. Cutler, and J. S. Rowan. 2012. Using the NOAA advanced very high resolution radiometer to characterise temporal and spatial trends in water temperature of large European lakes. *Remote Sensing of Environment* 126:1–11.

Porfirio, L. L., W. Steffen, D. J. Barrett, and S. L. Berry. 2010. The net ecosystem carbon exchange of human-modified environments in the Australian capital region. *Regional Environmental Change* 10:1–12.

Rocchini, D. 2009. Commentary on Krishnaswamy et al. Quantifying and mapping biodiversity and ecosystem services: Utility of a multi-season NDVI based Mahalanobis distance surrogate. *Remote Sensing of Environment* 113:904–906.

Rocchini, D., N. Balkenhol, G. A. Carter, et al. 2010. Remotely sensed spectral heterogeneity as a proxy of species diversity: Recent advances and open challenges. *Ecological Informatics* 5:318–329.

Schneider, A., K. E. Logan, and C. J. Kucharik. 2012. Impacts of urbanization on ecosystem goods and services in the U.S. corn belt. *Ecosystems* 15:519–541.

Scullion, J., C. W. Thomas, K. A. Vogt, O. Pérez-Maqueo, and M. G. Logsdon. 2011. Evaluating the environmental impact of payments for ecosystem services in Coatepec (Mexico) using remote sensing and on-site interviews. *Environmental Conservation* 38:426–434.

Shi, Y., R. S. Wang, J. L. Huang, and W. R. Yang. 2012. An analysis of the spatial and temporal changes in Chinese terrestrial ecosystem service functions. *Chinese Science Bulletin* 57:2120–2131.

Sutton, P. C., and R. Costanza. 2002. Global estimates of market and non-market values derived from nighttime satellite imagery, land cover, and ecosystem service valuation. *Ecological Economics* 41:509–527.

Tallis, H., H. Mooney, S. Andelman, et al. 2012. A global system for monitoring ecosystem service change. *BioScience* 62:977–986.

Tenhunen, J., R. Geyer, S. Adiku, et al. 2009. Influences of changing land use and CO_2 concentration on ecosystem and landscape level carbon and water balances in mountainous terrain of the Stubai Valley, Austria. *Global and Planetary Change* 67:29–43.

Turner, D. P., W. D. Ritts, Z. Yang, et al. 2011. Decadal trends in net ecosystem production and net ecosystem carbon balance for a regional socioecological system. *Forest Ecology and Management* 262:1318–1325.

Verburg, P., S. van Asselen, E. van der Zanden, and E. Stehfest. 2013. The representation of landscapes in global scale assessments of environmental change. *Landscape Ecology* 28(6):1067–1080.

Verburg, P. H., J. van de Steeg, A. Veldkamp, and L. Willemen. 2009. From land cover change to land function dynamics: A major challenge to improve land characterization. *Journal of Environmental Management* 90:1327–1335.

Verstraete, M. M., M. Menenti, and J. Peltoniemi. 2000. *Observing land from space: Science, customers and technology*. Berlin: Springer-Verlag.

Viglizzo, E. F., F. C. Frank, and L. V. Carreño. 2011. Ecological and environmental footprint of 50 years of agricultural expansion in Argentina. *Global Change Biology* 17:959–973.

Volante, J. N., D. Alcaraz-Segura, M. J. Mosciaro, E. F. Viglizzo, and J. M. Paruelo. 2012. Ecosystem functional changes associated with land clearing in NW Argentina. *Agriculture, Ecosystems and Environment* 154:12–22.

Wang, C., P. D. Van Meer, M. Peng, W. Douven, R. Hessel, and C. Dang. 2009. Ecosystem services assessment of two watersheds of Lancang River in Yunnan, China with a decision tree approach. *Ambio* 38:47–54.

Wang, Z., B. Zhang, S. Zhang, et al. 2006. Changes of land use and of ecosystem service values in Sanjiang Plain, northeast China. *Environmental Monitoring and Assessment* 112:69–91.

Woodcock, B. A., J. Redhead, A. J. Vanbergen, et al. 2010. Impact of habitat type and landscape structure on biomass, species richness and functional diversity of ground beetles. *Agriculture, Ecosystems and Environment* 139:181–186.

Zhao, B., U. Kreuter, B. Li, Z. Ma, J. Chen, and N. Nakagoshi. 2004. An ecosystem service value assessment of land-use change on Chongming Island, China. *Land Use Policy* 21:139–148.

Zhao, X., and P. Tong. 2013. Ecosystem services valuation based on land use change in a typical waterfront town, Poyang Lake basin, China. *Applied Mechanics and Materials* 295:722–725.

Section II

Ecosystem Services Related to the Carbon Cycle

2

Ecosystem Services Related to Carbon Dynamics: Its Evaluation Using Remote Sensing Techniques

J. M. Paruelo and M. Vallejos
University of Buenos Aires, Argentina

CONTENTS

2.1   Introduction 17
2.2   The Carbon Cycle: Key Processes 18
2.3   Conceptual Frameworks to Connect Carbon Dynamics and ESs 19
    2.3.1   Production Functions to Link Intermediate to Final ESs 19
    2.3.2   Impact Functions to Link Disturbances or Stress Factors
        with ES Provision 21
2.4   Scale Issues in the Evaluation of Carbon-Related ESs 22
2.5   Which Intermediate Services Should Be Monitored? 23
    2.5.1   NPP Estimations 24
    2.5.2   AGB Estimations 27
    2.5.3   Carbon and Energy Released by Wildfires 30
2.6   Concluding Remarks 31
Acknowledgments 31
References 32

2.1 Introduction

Policies aimed at integrating social, economic, and environmental dimensions of sustainability have to explicitly consider evaluating the influence of human activities on the provision of ecosystem services (ESs). The decision-making process for land use planning requires ES inventory along with an estimation of the ES provision rates and the effects of related human activities. ESs are commonly evaluated on the basis of indicators that do not provide a proper representation of the whole territory and/or do not capture the temporal dynamics of the service supply (Carpenter and Folke 2006). In this sense, remote sensing techniques are particularly appropriated to map ESs in a fast and continuous way in time and space (Paruelo 2008).

One of the advantages of the ES framework is its direct relationship with ecosystem functioning and human well-being. Moreover, the Millennium Ecosystem Assessment (MA 2004) definition explicitly links the ES supply with ecosystem exchange of matter and energy (i.e., nutrient cycling of carbon gains). Díaz et al. (2007) identified a number of ESs and related ecosystem processes. These authors associated the variation in the level of ES provision with related functional changes, specifically with changes in plant functional diversity. McNaughton et al. (1989) proposed carbon (C) gains as an integrative aspect of ecosystem functioning because many other processes are tightly linked to this flux. C stocks (in live or dead biomass) are also integrative descriptors of the processes and disturbances that operate in the ecosystem. Changes in soil organic C reflect the influence of the disturbance regime and land use changes on inputs and outputs of C to/from the soil. The global C balance is a critical issue in the analysis of climate change due to its importance in determining atmospheric carbon dioxide (CO_2) and, consequently, the radiative forcing of the atmosphere (Canadell et al. 2004). Quantifying key fluxes and stocks of the C cycle would synthesize the condition of the ecosystem and, moreover, its ability to supply ESs (Cabello et al. 2012).

In this chapter, we discuss the opportunities to evaluate ESs related to the C dynamics using remotely sensed data. We present the key processes of the C cycle, the basis of its evaluation using spectral data, and the conceptual connection with the ES provision.

2.2 The Carbon Cycle: Key Processes

Carbon exchange dynamics between the biota and the atmosphere, which is tightly linked to energy flow and circulation of other materials, is an integrative aspect of ecosystem functioning. The balance between photosynthesis and respiration by plants, animals, and microorganisms is a major determinant of C dynamics. The final result of such balance—net ecosystem production (NEP) or net ecosystem exchange (NEE)—is a fundamental characteristic of terrestrial ecosystems because it is directly connected to C sequestration. The dynamics of C sequestration by vegetation and soil (assuming no lateral flows) can be described by two different equations:

$$\Delta C = \Delta AGB + \Delta BGB + \Delta L + \Delta S \quad \text{mass balance equation} \quad (2.1)$$

$$\Delta C = GPP - RA - RH - D \quad \text{process equation} \quad (2.2)$$

where ΔC is changes in carbon stock by vegetation and soil; ΔAGB is changes in aboveground biomass; ΔBGB is changes in belowground biomass; ΔL is changes in litter; ΔS is changes in soil carbon; GPP is gross primary production; RA is autotrophic respiration; RH is heterotrophic respiration; and D is carbon loss by disturbance. Although Equation 2.1 can be regarded as an allocation equation, where biomass is explicitly included, Equation 2.2 represents the C fluxes between the different reservoirs.

Remotely sensed data are the primary source for large-scale biomass estimations on regional to global scales (Goetz et al. 2009). However, aboveground biomass (AGB) cannot be directly measured from space by any sensor (Sun et al. 2011). Understanding terrestrial carbon processes requires integration of many types and sources of information, including ground data, ecological models, and remotely sensed data. Currently, three different remote sensing technologies are available to estimate ecosystem biomass: optical remote sensing, synthetic aperture radar (SAR), and LIDAR. These methods are highly complementary.

2.3 Conceptual Frameworks to Connect Carbon Dynamics and ESs

ESs have been defined in different ways (Fisher et al. 2009). The Millennium Ecosystem Assessment (MA 2004) definition states that ESs are the benefits that people obtain from ecosystems. The MA definition and other related definitions (Costanza et al. 1997; Daily 1997) consider subjective and cultural elements outside the ecological systems to define the benefits in the characterization of the level of ES provision. The MA classifies ESs into provisioning ESs, regulating ESs, cultural ESs, and supporting ESs (Figure 2.1). In the MA scheme, the level of ES provision, regulation, or support is not only linked to basic aspects of ecosystem functioning (e.g., ecosystem exchanges of matter and energy; Virginia and Wall 2001) but also to the societal context of values, interests, and needs.

Boyd and Banzhaf (2007) referred to ESs as the ecological components directly consumed or enjoyed to produce human well-being, without considering the subjective and cultural context. Based on this, Fisher et al. (2009) defined ESs as the aspects of ecosystems utilized (actively or passively) to produce human well-being.

2.3.1 Production Functions to Link Intermediate to Final ESs

Fisher et al. (2009) proposed an ES classification scheme where ecosystem functioning and structure are considered "intermediate" services,

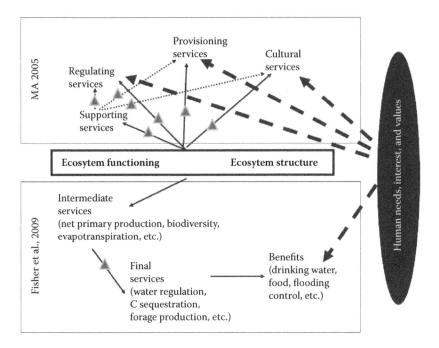

FIGURE 2.1
Main concepts related to the classification schemes of ecosystem services adopted by MA
(2005) and developed by Fisher et al. (2009). Black arrows indicate the relationship between
the different categories of ecosystem services (ESs) and the structure and functioning of ecosystems. Such a relationship is defined in terms of production functions (triangles). Dotted
lines represent the relationship between ES categories. Broken lines represent the influence of
human needs, interests, and values on the definition of ESs and on the benefits in the two classification schemes. (Redrawn from Volante, J. N., et al., *Agriculture, Ecosystems & Environment*,
154, 12–22, 2012.)

which in turn determine "final" services (Figure 2.1). Several intermediate
services (e.g., primary production or species composition) may determine
the level of provision of a final service (e.g., forage production or C sequestration; see Chapter 5). The link between ecosystem functioning and structure (intermediate services) and final services are defined by "production
functions" (Figure 2.2a). Such production functions are well defined for
final ESs with market values, such as grain production, where yields are
defined by a number of biophysical (water and nutrient availability, temperature, etc.) and management factors (sowing date, cultural practices,
etc.). The definition of production functions for final ESs (e.g., C sequestration) from intermediate ESs (e.g., net primary production, vegetation
structure, or soil characteristics) has been identified as an important step
in incorporating the ES idea into decision-making processes (Laterra
et al. 2011).

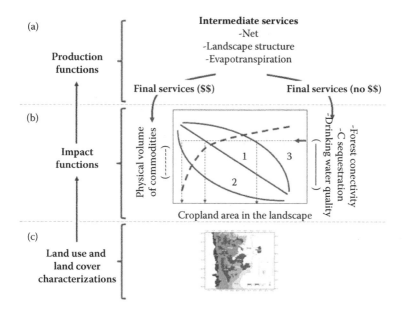

FIGURE 2.2
General scheme of the connections between (a) production functions, (b) impact functions, and (c) land use and land cover characterization. Production functions connect intermediate (ecological processes) and final services (with or without monetary value). Impact functions connect the change in the level of production of a service with the stress or disturbance factors related to human activities. The broken lines represent the change in the physical volume of commodity production, and the solid lines represent different types of change in the level of provision of other final ESs. The arrow indicates the hypothetical level of reduction in the provision of ESs that the society decides to tolerate.

2.3.2 Impact Functions to Link Disturbances or Stress Factors with ES Provision

Human activities significantly reduce the provision of some final ESs in order to increase the provision of others. Trade-offs among ESs lead to increases in the level of provision of some ESs (e.g., food production) and to reduction in others (e.g., soil protection, water regulation, C sequestration, etc.) (de Groot et al. 2010). Understanding the tradeoffs among final ESs (i.e., grain production and drinking water quality) is a critical, though difficult, task of land use planning (Viglizzo et al. 2012). Changes in the provision of final ESs are mediated by structural and functional changes, such as biodiversity losses and changes in C and water dynamics (intermediate services) (Guerschman et al. 2003; Guerschman and Paruelo 2005; Jackson et al. 2005; Nosetto et al. 2005; Fisher et al. 2009). To define the "impact functions" that account for such changes, it would be necessary to identify the main disturbance and stress factors and quantify their effects—for instance, how the level of ESs (e.g., C sequestration) changes with a particular stress or disturbance (e.g., deforested area).

Given a change in the magnitude of a stress factor or disturbance agent related to human activities (i.e., an increase in the agricultural area in a landscape or an intensification of the activity), the final ESs will change (Figure 2.2b). In general, some will increase (i.e., physical volumes of commodities, with market value) and others will decrease (i.e., water quality, biodiversity conservation, atmospheric regulation, without market value). Understanding the functional relationship between the magnitude of the stress factor and the ESs is critical to define the level of modification that the society would tolerate. The level of reduction of a given ES that society is willing to tolerate (considering society as an entity that expresses in an unified way, to simplify the analysis) is indicated by the horizontal arrow on the right axis of Figure 2.2b; the level of transformation of the landscape would differ depending on the functional relationship of ESs. If the relationship is linear (curve 1, Figure 2.2b), the amount of land converted into cropland should be different than in the case where the relationship is described by curve 2 (should be lower) or curve 3 (should be higher). Developing impact functions is a key step in incorporating the ES concept into land use planning or into other decision-making processes (Paruelo et al. 2011). Remote sensing techniques have an important role in quantifying the ecosystem processes that produce the final services (carbon gains, evapotranspiration, and albedo) (Figure 2.2a) in characterizing the magnitude of human intervention (spatial and temporal dynamics of land cover and land use changes; Figure 2.2b) and in land use and land cover characterization (Figure 2.2c).

2.4 Scale Issues in the Evaluation of Carbon-Related ESs

The definition of the spatial and temporal scale of the analysis is a critical step of an ecological study (O'Neill et al. 1986; Peterson et al. 1998). ESs have an associated spatial scale, but this scale may not be the same scale as the ecosystem processes that support these services. For example, the capacity to detoxify residues or to regulate the emissions of methane or nitrous oxide results from the activity of microorganisms. The mechanisms behind these processes involve complex metabolic pathways occurring at the subcellular level. Although the biophysical mechanisms operate at a microscopic level, the net results of the processes (in terms of regulating the concentration of atmospheric gases or detoxification) become relevant at coarser scales. In the case of the regulation of atmospheric gases, the scale is global. There also other benefits in addition to those supplied by the ecosystems that are providing the services or that are in particular configurations (e.g., downstream) (Fisher et al. 2009). Erosion control exemplifies how a spatial perspective is required in order to evaluate the ES provision: Even though the type and amount of plant cover, the slope, and the soil texture of a particular plot are key elements to characterize erosion risk, the landscape context (relative position, disturbance regime, characteristics of neighboring plots) is critical. Ecological succession, nutrient redistribution, runoff, or local extinctions are further examples of landscape context-dependent ecological processes that are directly linked to ES provision. All these examples highlight the importance of the landscape level in evaluating ESs. Although landscape dimensions may vary, often the landscapes' extension range from 10 to 10^5 ha and the limits are associated with those of a watershed. The spatial resolution of the ES observation protocol has to consider these scale issues.

2.5 Which Intermediate Services Should Be Monitored?

From an operational perspective, the ecosystem aspects to be evaluated (intermediate services) have to be reliable and simple to measure or to estimate at different scales, and should be logically connected to the final services. Some aspects of the C cycle are particularly appropriate for monitoring due to their ability to integrate the ecosystem C dynamics. Estimates of AGB and its change over time can reduce significantly any uncertainty in the mass balance equation (Le Toan et al. 2004). Nevertheless, direct estimation of carbon storage in moderate to high biomass forests remains a major challenge for remote sensing.

We highlight, in particular, two aspects for monitoring: net primary production (NPP) and the stock of biomass. These two attributes integrate several other functional and structural aspects of the ecosystems (McNaughton et al. 1989).

Breckenridge et al. (1995) enumerated several criteria to be considered in the selection of indicators:

- Generality and simplicity to be applied in different regions
- Correlation with key ecosystem processes
- Temporal and spatial variability
- Possibility to automate the record
- Relationship cost-effectiveness
- Response/sensitivity to changes
- Environmental impacts of the sampling
- Empirical and conceptual support of the protocol

These authors concluded that spectral data provided by satellite platforms are particularly well-suited to satisfy these criteria. Spectral data are able not only to characterize structural aspects of the landscapes (i.e., distribution of spatial and temporal land cover types) but also functional aspects of the ecosystems (i.e., C gain dynamics, evapotranspiration, disturbance regime) (Wessman 1992; Kerr and Ostrovsky 2003; Pettorelli et al. 2005; Paruelo 2008; Cabello et al. 2012).

In this chapter, we will review functional (NPP) and structural (biomass) ecosystem attributes that can be assimilated to intermediate services in the C cycle and that represent key terms of Equations 2.1 and 2.2. These attributes can be estimated from remotely sensed data using well-established techniques and simulation models. The existence of well-defined protocols allows the integration of these attributes into monitoring programs at the landscape level. Remote sensing techniques also allow the characterization of key aspects of the disturbance regime that modify stocks and flows of C—floods and fires (Di Bella et al. 2008). We discuss the possibilities of monitoring the release of energy (and C) through biomass combustion, particularly the evaluation of the fire radiative power (a component of the AGB change of Equation 2.1 and factor D in Equation 2.2).

2.5.1 NPP Estimations

Harvest biomass techniques are limited in the contribution they can make to the analysis of the spatial and temporal variation of NPP on large spatial scales (Singh et al. 1975; Lauenroth et al. 1986). Satellite imagery provides valuable data in order to monitor NPP in different vegetation types (Prince 1991; Running et al. 2000) with large area coverage, high temporal resolution, and moderate spatial resolution. Several optical sensors and platforms that record the reflectance in the red and near-infrared portion of the electromagnetic spectrum have been widely used (i.e., Landsat MSS, TM and ETM+, MODIS, vegetation, and AVHRR-NOAA.

Radiometric indices, particularly the Normalized Difference Vegetation Index (NDVI) and the Enhanced Vegetation Index (EVI), are closely and positively correlated with the fraction of the absorbed photosynthetically active radiation (fAPAR) by green vegetation (Sellers et al. 1992; Huete et al. 2002; Di Bella et al. 2004). Absorbed photosynthetically active radiation (APAR) may, therefore, be estimated by multiplying fAPAR by the incoming photosynthetically active radiation (PAR), available from weather stations. NPP can be estimated according to Monteith's model:

$$NPP = fAPAR \cdot PAR \cdot RUE \qquad (2.3)$$

where RUE is the radiation use efficiency, in grams of dry matter per megajoules (Monteith 1972). Remote sensing is beginning to provide estimates of RUE based on an index calculated from two bands centered at

530 and 570 nm, the photochemical reflectance index (PRI; Garbulsky et al. 2008) (see Chapter 3).

Monteith's model has been used to estimate NPP at multiple spatial resolutions, from 1 to 64 km^2 (Running et al. 2004). Piñeiro et al. (2006), Baeza et al. (2010), and Irisarri et al. (2012) provided estimates of aboveground net primary production (ANPP) over large areas in the grasslands of South America using Monteith´s model. Vasallo et al. (2013) used Monteith's model to compare C gains between native grasslands and the tree plantations that have replaced them.

Pettorelli et al. (2005) showed that the seasonal and interannual C gains, assessed using spectral indices, provide an integrative description of ecosystem functioning. Two attributes of the seasonal curve of EVI or NDVI are particularly descriptive: the annual integral and the seasonal variability (Paruelo and Lauenroth 1998). Volante et al. (2012) analyzed the impact of land clearing on these two C-related intermediate ESs in the Chaco region of South America. Although land clearing for agriculture and ranching had relatively small impacts on total annual ANPP, once deforested, parcels became significantly more seasonal than the natural vegetation that had been replaced. Such an increase in seasonality is associated with a reduction of photosynthetic activity during a portion of the year (fallow). Direct consequences of this reduction can be expected on several ESs such as erosion control and water regulation (due to greater exposure of bare soil) and biodiversity (due to the loss or decline in habitat quality and the decrease of green biomass availability for primary consumers during fallow). On a different scale, Paruelo et al. (2004) showed, again for the Chaco region, that total C gains (characterized by the annual NDVI integral) decreased as the proportion of croplands in the landscape increased (Figure 2.3). Similar patterns have been described for the Argentine Pampas (Guerschman et al. 2003) and the Great Plains in the United States (Paruelo et al. 2001a). For temperate grasslands and woodlands of South America, agricultural expansion may decrease NPP depending on the original cover and the mean annual precipitation of the landscape (Paruelo et al. 2001b). Areas with higher precipitation showed a marked decrease in NPP when compared to drier areas (Figure 2.4a). Many final ESs are directly linked to the total C gains, from C sequestration to water regulation.

The seasonality of C gains (the variation through the year) always increases with the cultivated proportion of the landscape. However, the reduction depends on the cropping system; double crops (wheat–soybean) have lower reductions than single crops (Figure 2.4b). Therefore, changes in intermediate ESs such as NPP seasonality would determine changes in final services such as erosion control (due to changes in plant cover across seasons) and climatic controls (due to changes in the leaf area index across seasons and then on the magnitude of latent heat, or of albedo), among others.

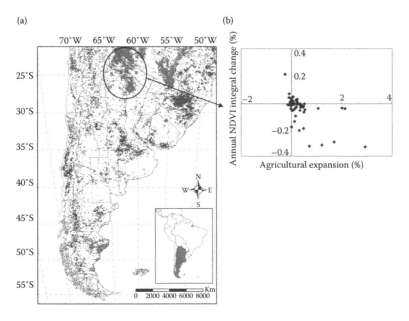

FIGURE 2.3 (See color insert.)
(a) Map of Argentina displaying the slope of the relationship between the absorbed photo-synthetically active radiation (APAR) absorbed by the vegetation and time, for the 1981–2000 period. Red and blue pixels represent negative and positive slopes, respectively. APAR was calculated from the NDVI derived from PAL series of the AVHRR/NOAA satellite. (b) Relationship between the annual change in the cropped area per county and the annual change in Normalized Difference Vegetation Index (NDVI) for counties of northwestern Argentina covered by forests (Salta, Chaco, Formosa, Jujuy y Tucuman). (Modified from Paruelo, J. M., et al., *International Journal of Remote Sensing*, 25, 2793–2806, 2004).

Grazing has been identified as a major disturbance and/or stress factor in ecosystems. The effects of grazing on the structure and functioning of grasslands, shrublands, and savannas have generated controversy and debate (McNaughton 1979; Milchunas and Lauenroth 1993; Oesterheld et al. 1999; Chase et al. 2000). NPP may have a complicated response to long-term grazing pressure depending on resource availability and long-term grazing history (Milchunas and Lauenroth 1993; Oesterheld et al. 1999). Aguiar et al. (1996) proposed an impact function for NPP in the Patagonian steppes as a function of the historical grazing pressure. Moreover, this article presents a production function of a final service (domestic herbivore biomass) from the intermediate service (grass ANPP; Figure 2.4c) using a model presented by Oesterheld et al. (1992).

Protected areas and nondegraded grasslands or shrublands showed a lower sensitivity to changes in precipitation than did heavily grazed ones (Paruelo et al. 2005; Verón et al. 2011). In this case, the intermediate ES is the buffer capacity of the ecosystem, that is, the relative variability of functional attributes with respect to environmental fluctuations, in terms of C gains

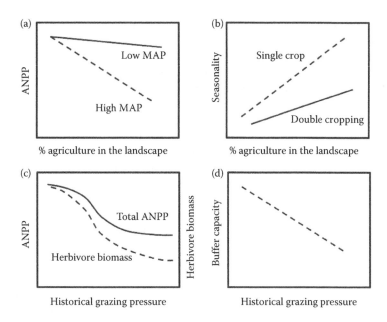

FIGURE 2.4
(a, b) Hypothetical impact functions for aboveground net primary production (ANPP) and ANPP seasonality (intermediate services) and the percent of the landscape occupied by agriculture (disturbance factor). Different lines correspond to different climatic conditions—high and low mean annual precipitation (MAP) or managements (single or double cropping). (c, d) Hypothetical impact functions for the ANPP and for the capacity of the ecosystem to buffer climatic fluctuations at the functional level (intermediate factors) as a function of the historical grazing pressure (disturbance factor) on native rangelands. (c) also shows the impact of grazing on domestic herbivore biomass (a final service). (Based on analyses presented by Aguiar, M. R., et al., *Journal of Vegetation Science*, 7, 381–390, 1996; Guerschman, J. P., et al., *International Journal of Remote Sensing*, 17, 3381–3402, 2003; Paruelo, J. M., et al., *Ecología Austral*, 21, 163–178, 2005; and Verón, S. R., et al., *Oecologia*, 165, 501–510, 2011.)

(Figure 2.4d). Land clearing also increased the magnitude of interannual differences in C gains, suggesting a lower buffer capacity against climate fluctuations of natural vegetation compared to croplands (Volante et al. 2012).

2.5.2 AGB Estimations

The difficulty in making reliable biomass estimates is well recognized, from local (Fang et al. 2006) to continental scales (Houghton et al. 2001). A lot of effort has been made to estimate biomass using field-based as well as remote sensing techniques—mainly in forests, but the development of impact functions is scarce. Estimates of biomass have been based on active sensors, meaning they generate a signal and measure the amount of energy reflected back to the sensor. The advantage of using active sensors is that they can operate day or night, and the microwaves can go across haze, smoke, and clouds. The energy transmitted can also penetrate into forest canopies and is able to measure the canopy height and vertical structure. Two main types of active sensors have been used to measure forest structure attributes and biomass at global scales: radar (SAR) and LIDAR.

SAR is an airborne or spaceborne radar system that uses its relative forward motion, between an antenna and its target region, in order to provide a high-resolution remote sensing imagery generated by recording and combining the individual signals of the sensor. Because of its penetration capability and sensitivity to water content in vegetation, SAR is sensitive to the spatial structure of forests (Le Toan et al. 2004). The backscatter is the portion of the outgoing radar signal that is redirected back to the antenna. Backscattering is influenced by surface parameters (roughness, geometric shape, and dielectric properties of the target) and radar observation parameters (frequency, polarization, and incidence angle of the electromagnetic waves emitted). SAR is known to sense the canopy volume (especially at longer wavelengths) and provide image data, with the amount of backscattered energy, which is largely dependent on the size and orientation of canopy structural elements, such as leaves, branches, and stems.

The frequency (f) of the signal defines the interaction with the forest structure and the penetration capability of the wave. The longer the wavelength, the greater the sensitivity to the vertical structure of vegetation and the greater the penetration into the forest canopy. Radar data are acquired in X, C, L, and P bands. Shorter wavelengths (X and C band, with 2.5 and 7.5 cm, respectively) are sensitive to smaller canopy elements such as leaves and small branches, and longer wavelengths (L and P band, with 23.5 and 70 cm, respectively) are sensitive to large branches and trunks. The polarization (p) is the direction of the electric field in the electromagnetic waves and is the main factor in the interaction between the signals and the reflectors. Most of the microwave sensors emit and receive signals in horizontal (H) or vertical (V) polarizations. Measuring the polarization of the transmitted and received electromagnetic waves allows for further sensitivity of AGB measurements (Goetz et al. 2009). Interferometry calculates the interference pattern caused by the differences in phase between two images acquired by a spaceborne SAR at two distinct times, and the resulting interferogram is a contour map of the change in distance between the ground and the SAR instrument (Feigl 1998). According to Kasischke et al. (1997), the best performance for biomass estimation is achieved using lower frequency (P and L band) radar systems with a cross-polarized (HV or VH) channel.

The simplest method for biomass estimation using SAR is relating the backscatter coefficient to field biomass measurement using regression analysis. This approach has been tested on different areas, and good results have been achieved in coniferous forests (Dobson et al. 1992; Le Toan 1992). Indirect methods are also used to estimate AGB, consisting of deriving forest structural estimates (e.g., tree heights or canopy heights) in order to infer forest biomass quantities through interferometry. Moreover, polarimetric and interferometric

SAR data have been used for forest biomass estimation (Dobson et al. 1992, 1995; Ranson and Sun 1994; Kasischke et al. 1995) and canopy height estimation (Treuhaft et al. 1996, 2004; Kobayashi et al. 2000; Kellndorfer et al. 2004; Walker et al. 2007). These applications require ground sampling data for training and validation purposes (Sun et al. 2011). The principal problem of using SAR to estimate biomass is the saturation level. Experimental studies with SAR over different types of forests (temperate, boreal, and tropical) indicate that saturation occurs at around 30, 50, and 150–200 tonnes ha^{-1} at C, L, and P bands, respectively (Le Toan et al. 2004). These saturation values are approximate and depend on the experimental conditions and forest characteristics. Advanced airborne SAR systems using long wavelengths or combining polarization diversity with interferometry techniques (polarimetric interferometry) have demonstrated significantly greater capabilities for estimating forest biomass (Le Toan 2002). Interferometric SAR (InSAR) is also employed to improve AGB estimations (Walker et al. 2007), where allometric equations are used to establish quantitative relations between structural patterns (e.g., tree height) and other properties of the forest (e.g., biomass).

LIDAR is a relatively new active remote sensing technology especially suitable for reproducing the three-dimensional (3D) structure of forest stands due to its ability to determine 3D measurements with high accuracy. LIDAR instrumentation uses a laser scanner that transmits pulses and records the delay time between a light pulse transmission and its reception in order to calculate elevation values. Each data point is recorded with precise horizontal position, vertical elevation, and other attribute values. The multiple returns are recorded and a classification is assigned to each point in order to identify landscape features. The intensity of the reflected energy is also captured and can be analyzed to provide additional information on terrain characteristics. LIDAR metrics are statistical measurements created from the 3D point cloud (a set of vertices in a 3D coordinate system) and are normally used when predicting forest variables from LIDAR data. Various types of LIDAR systems have been used to capture an increasingly broad range of vegetation characteristics and biomass estimations. LIDAR-based estimations of AGB can be performed by means of point cloud and rasterized data. The point collections of 3D data have to be managed and processed in a standardized binary format for storing 3D point cloud data and point attributes. The procedure of LIDAR raw point cloud-based 3D single tree modeling was first published by Wang et al. (2007). Point cloud data processing is a computationally demanding task when processing large datasets for the generation of area-wide AGB maps (Jochem et al. 2011). Using rasterized data requires the aggregation of the 3D point cloud to cells, meaning that the canopy surface is represented by a single-valued function. This procedure is accompanied by an irreversible loss of the 3D structure but makes processing less time consuming and drastically reduces the storage size. LIDAR systems are classified as either discrete return or full waveform recording and may be further divided into profiling (recording only along a narrow line directly below the sensor) or scanning systems (recording across a wide swath on either side of the sensor) (Lefsky et al. 2002; Lim et al. 2003). Within a forest, full waveform systems record the entire wave form for analysis, while discrete return systems record clouds of points representing intercepted features (Wulder et al. 2012).

LIDAR sensors have been used for extending plot-level estimates to larger spatial and ecological scales (Lefsky et al. 2002; Zhao et al. 2009). Allometric equations to estimate carbon stocks using LIDAR data are usually region-specific, involving laborious calibration methods and expensive plot inventory data (Lefsky et al. 1999; Nelson et al. 2012). Nevertheless, Asner et al. (2012) used a single universal LIDAR model to predict aboveground carbon density estimated in field inventory plots, generalizing biomass allometric equations for tropical trees. In this approach, the authors reduced the forest structural properties to mean canopy profile height (also known as MCH), which is the vertical center of the canopy volumetric profile (as opposed to a simple top-of-canopy height), and detailed a relationship with carbon density and basal area.

2.5.3 Carbon and Energy Released by Wildfires

Fires represent an important pathway of energy and C release from land ecosystems. Giglio et al. (2010) reported that, globally, between 3.3 and 4.3 million km^2 burn each year. Fires release C mainly in the form of particulate matter and greenhouse gases, including CO_2 and CH_4 (van der Werf et al. 2010) and seriously affect ESs by modifying the hydrological cycle, by triggering soil erosion, and by radiative forcing in the atmosphere (Lohmann and Feichter 1997; DeFries et al. 2002; Hoffmann et al. 2002, 2003; Mouillot and Field 2005; van der Werf et al. 2008).

The fire radiative power product (FRP) measures the radiant heat output (in megawatts) of a given fire. FRP is related to the biomass being consumed by detected fires (see also Chapter 7). It has been demonstrated (in small-scale experimental fires) that the amount of radiant heat energy liberated per unit time (FRP) is related to the rate at which fuel is being consumed (Wooster et al. 2005), and it represents a direct output of the combustion process. The integration of FRP over time provides an estimate of the fire radiative energy (FRE), which—for wildfires—should be proportional to the total biomass combusted (Verón et al. 2012).

To derive the FRP product, the process starts with the identification of fire pixels. The fire thermal anomaly (FTA) algorithm tests for elevated radiance in the mid-infrared portion of the spectrum. The algorithm includes additional tests to discriminate fires from other phenomena that may induce similar responses in this spectral band (i.e., specular reflections and cloud edges) (Roberts et al. 2005), and it works mainly on data derived from the 3.9 and 11.0 µm brightness temperatures and their differences. Thresholds for fire detection are based on contextual tests that adjust the detection from immediately neighboring nonfire background pixels. Once the fire pixel is detected, FRP is estimated from the middle-infrared (MIR, 3.9 µm) channel and the background radiance that would have been observed at the same location in the absence of fire (Giglio et al. 2003). FRP data are available from the MOD and MYD14CMG fire products (Giglio et al. 2006) generated from the MODIS (Moderate Resolution Imaging Spectro radiometer) sensor, collection 5, onboard Terra and Aqua platforms. This dataset integrates subdaily, 1 km² resolution data into monthly values for 0.5° x 0.5° grid cells.

On the basis of a global analysis of the energy generation and spatial distribution of fires, Verón et al. (2012) showed that between 2003 and 2010, global fires consumed approximately 8300 ± 592 PJ yr^{-1} of energy, equivalent to approximately 36%–44% of the global electricity consumption in 2008 and more than 100% of the national consumption in 57 countries. Forests/woodlands, cultivated areas, shrublands, and grasslands contributed 53%, 19%, 16%, and 3.5%, respectively, of the global energy released by fires.

2.6 Concluding Remarks

Remote sensing techniques provide the opportunity to estimate two critical aspects of the C balance: NPP and biomass. In terms of the definition of ESs determined by Fisher et al. (2009), these two variables represent intermediate services and capture many basic aspects of ecosystem structure and functioning. Moreover, they show a clear relationship with important final services, from forage and wood production to C sequestration. Both contribute, together with other intermediate services, to determining several other final services such as climate regulation, soil erosion control, or water provision.

Satellite-derived observations are a major step in determining indicators that cover large areas, based on the same observation protocols and estimated in almost real time. These characteristics represent a clear advantage in programs in order to monitor changes in the level of provision of ESs. Remote sensing techniques are able to estimate not only C related to intermediate ESs but also those processes related to the energy balance—such as latent heat fluxes and albedo.

Acknowledgments

This project has been funded by UBACYT, FONCYT, and CONICET. This work was carried out with the aid of a grant from the Inter-American Institute for Global Change Research (IAI) CRN-3095 which is supported by the US National Science Foundation (Grant GEO-1128040).

References

Aguiar, M. R., J. M. Paruelo, C. E. Sala, and W. K. Lauenroth. 1996. Ecosystem responses to changes in plant functional type composition: An example from the Patagonian steppe. *Journal of Vegetation Science* 7:381–390.

Asner, G. P., J. Mascaro, H. C. Muller-Landau, et al. 2012. A universal airborne LiDAR approach for tropical forest carbon mapping. *Oecologia* 168:1147–1160.

Baeza, S., F. Lezama, G. Piñeiro, A. Altesor, and J. M. Paruelo. 2010. Spatial variability of aboveground net primary production in Uruguayan grasslands: A remote sensing approach. *Applied Vegetation Science* 13:72–85.

Boyd, J., and S. Banzhaf. 2007. What are ecosystem services? The need for standardized environmental accounting units. *Ecological Economics* 63:616–626.

Breckenridge, R. P., W. G. Kepner, and D. A. Mouat. 1995. A process for selecting indicators for monitoring conditions of rangeland health. *Environmental Monitoring and Assessment* 36:45–60.

Cabello, J., D. Alcaraz-Segura, R. Ferrero, A. J. Castro, and E. Liras. 2012. The role of vegetation and lithology in the spatial and inter-annual response of EVI to climate in drylands of Southeastern Spain. *Journal of Arid Environments* 79:76–83.

Canadell, J., P. Ciais, P. Cox, and M. Heimann. 2004. Quantifying, understanding and managing the carbon cycle in the next decades. *Climatic Change* 67:147–160.

Carpenter, S. R., and C. Folke. 2006. Ecology for transformation. *TRENDS in Ecology and Evolution* 21:309–315.

Chase, J. M., M. A. Leibold, A. L. Downing, and J. B. Shurin. 2000. The effects of productivity, herbivory, and plant species turnover in grassland food webs. *Ecology* 81:2485–2497.

Costanza, R., R. d'Arge, R. de Groot, et al. 1997. The value of the world's ecosystem services and natural capital. *Nature* 387:253–260.

Daily, G. C. (ed.). 1997. *Nature's services: Societal dependence on natural ecosystems.* Washington, DC: Island Press.

DeFries, R. S., R. A. Houghton, M. C. Hansen, C. B. Field, D. Skole, and J. Townshend. 2002. Carbon emissions from tropical deforestation and regrowth based on satellite observations for the 1980s and 1990s. *Proceedings of the National Academy of Sciences* 99:14256–14261.

de Groot, R. S., R. Alkemade, L. Braat, L. Hein, and L. Willeman. 2010. Challenges in integrating the concept of ecosystem services and values in landscape planning, management and decision making. *Ecological Complexity* 7:260–272.

Díaz, S., S. Lavorel, F. S. Chapin, P. A. Tecco, D. E Gurvich, and K. Grigulis. 2007. Functional diversity at the crossroads between ecosystem functioning and environmental filters. In *Terrestrial ecosystems in a changing world*, eds. J. Canadell, L. F. Pitelka, and D. Pataki, 81–91. New York: Springer-Verlag.

Di Bella, C. M., J. M. Paruelo, J. E. Becerra, C. Bacour, and F. Baret. 2004. Experimental and simulated evidences of the effect of senescent biomass on the estimation of fPAR from NDVI measurements on grass canopies. *International Journal of Remote Sensing* 25:5415–5427.

Di Bella, C. M., G. Posse, M. E. Beget, M. A. Fischer, N. Mari, and S. Veron. 2008. La teledetección como herramienta para la prevención, seguimiento y evaluación de incendios e inundaciones [Remote sensing as a tool for the prevention, monitoring and evaluation of fires and floods]. *Ecosistemas* 17:39–52.

Dobson, M. C., F. T. Ulaby, T. Le Toan, A. Beaudoin, E. S. Kasischke, and N. Christensen. 1992. Dependence of radar backscatter on coniferous forest biomass. *IEEE Transactions on Geoscience and Remote Sensing* 30:412–415.

Dobson, M. C., F. T. Ulaby, L. E. Pierce, et al. 1995. Estimation of forest biophysical characteristics in Northern Michigan with SIR-C/X-SAR. *IEEE Transactions on Geoscience and Remote Sensing* 33:877–895.

Fang, J., S. Brown, Y. Tang, G. J. Nabuurs, X. Wang, and S. Haihua. 2006. Overestimated biomass carbon pools of the northern mid- and high latitude forests. *Climatic Change* 74:355–368.

Feigl, K. L. 1998. RADAR interferometry and its application to changes in the earth surface. *Reviews of Geophysics* 36:441–500.

Fisher, B., K. R. Turner, and P. Morling. 2009. Defining and classifying ecosystem services for decision making. *Ecological Economics* 68:643–653.

Garbulsky, M. F., J. Peñuelas, J. M. Ourcival, and I. Filella. 2008. Estimación de la eficiencia del uso de la radiación en bosques mediterráneos a partir de datos MODIS. Uso del Índice de Reflectancia Fotoquímica (PRI) [Radiation use efficiency estimation in Mediterranean forests using MODIS Photochemical Reflectance Index (PRI)]. *Ecosistemas* 17:89–97.

Giglio, L., J. Descloitres, C. O. Justice, and Y. J. Kaufman. 2003. An enhanced contextural fire detection algorithm for MODIS. *Remote Sensing of Environment* 87:273–282.

Giglio, L., J. T. Randerson, G. R. van der Werf, et al. 2010. Assessing variability and long-term trends in burned area by merging multiple satellite fire products. *Biogeosciences* 7:1171–1186.

Giglio, L., G. R. van der Werf, J. T. Randerson, G. J. Collatz, and P. Kasibhatla. 2006. Global estimation of burned area using MODIS active fire observations. *Atmospheric Chemistry and Physics* 6:957–974.

Goetz, S. J., A. Baccini, N. T. Laporte, et al. 2009. Mapping and monitoring carbon stocks with satellite observations: A comparison of methods. *Carbon Balance and Management* 4:2.

Guerschman, J. P., and J. M. Paruelo. 2005. Agricultural impacts on ecosystem functioning in temperate areas of North and South America. *Global and Planetary Change* 47:170–180.

Guerschman, J. P., J. M. Paruelo, C. M. Di Bella, M. C. Giallorenzi, and F. Pacín. 2003. Land classification in the Argentine Pampas using multitemporal landsat TM data. *International Journal of Remote Sensing* 17:3381–3402.

Hoffmann, W. A., W. Schroeder, and R. B. Jackson. 2002. Positive feedbacks of fire, climate, and vegetation change and the conversion of tropical savannas. *Geophysical Research Letters* 29:1–9.

Hoffmann, W. A., W. Schroeder, and R. B. Jackson. 2003. Regional feedbacks among climate, fire, and tropical deforestation. *Journal of Geophysical Research: Atmospheres* 108:1–11.

Houghton, R. A., K. T. Lawrence, J. L. Hackler, and S. Brown. 2001. The spatial distribution of forest biomass in the Brazilian Amazon: A comparison of estimates. *Global Change Biology* 7:731–746.

Huete, A., K. Didan, T. Miura, E. P. Rodriguez, X. Gao, and L. G. Ferreira. 2002. Overview of the radiometric and biophysical performance of the MODIS vegetation indices. *Remote Sensing of Environment* 83:195–213.

Irisarri, G., M. Oesterheld, J. M. Paruelo, and M. Texeira. 2012. Patterns and controls of above-ground net primary production in meadows of Patagonia. A remote sensing approach. *Journal of Vegetation Science* 23:114–126.

Jackson, R. B., E. G. Jobbágy, R. Avissar, et al. 2005. Trading water for carbon with biological carbon sequestration. *Science* 310:1944–1947.

Jochem, A., M. Hollaus, M. Rutzinger, K. Schadauer, and B. Maier. 2011. Estimation of aboveground biomass using airborne LiDAR data. *Sensors* 11:278–295.

Kasischke, E. S., N. L. Christensen, and L. L. Bourgeauchavez. 1995. Correlating radar backscatter with components of biomass in loblolly-pine forests. *IEEE Transactions on Geoscience and Remote Sensing* 33:643–659.

Kasischke, E. S., J. M. Melack, and M. C. Dobson. 1997. The use of imaging radars for ecological applications—A review. *Remote Sensing of Environment* 59:141–156.

Kellndorfer, J., W. Walker, L. Pierce, et al. 2004. Vegetation height estimation from Shuttle Radar Topography Mission and National Elevation Datasets. *Remote Sensing of Environment* 93:339–358.

Kerr, J. T., and M. Ostrovsky. 2003. From space to species: Ecological applications for remote sensing. *TRENDS in Ecology and Evolution* 18:299–305.

Kobayashi, Y., K. Sarabandi, L. Pierce, and M. C. Dobson. 2000. An evaluation of the JPL TOPSAR for extracting tree heights. *IEEE Transactions on Geoscience and Remote Sensing* 38:2446–2454.

Laterra, P., J. M. Paruelo, and E. G. Jobbagy (eds.). 2011. *Valoración de servicios ecosistémicos: Conceptos, herramientas y aplicaciones para el ordenamiento territorial* [Appraisal of ecosystem services: concepts, tool and applications for territorial organization]. Buenos Aires: Ediciones INTA.

Lauenroth, W. K., H. W. Hunt, D. M. Swift, and J. S. Singh. 1986. Estimating aboveground net primary productivity in grasslands: A simulation approach. *Ecological Modeling* 33:297–314.

Lefsky, M. A., W. B. Cohen, G. G. Parker, and D. J. Harding. 2002. Lidar remote sensing for ecosystem studies. *BioScience* 1:19–30.

Lefsky, M. A., D. Harding, W. B. Cohen, and G. G. Parker. 1999. Surface Lidar remote sensing of the basal area and biomass in deciduous forests of eastern Maryland, USA. *Remote Sensing of Environment* 67:83–98.

Le Toan, T. 1992. Relating forest biomass to SAR data. *IEEE Transactions on Geoscience and Remote Sensing* 30:403–411.

Le Toan, T. 2002. BIOMASCA: Biomass Monitoring Mission for Carbon Assessment. A proposal in response to the ESA Second Call for Earth Explorer Opportunity Missions. Available from: http://www.cesbio.ups-tlse.fr/data_all/pdf/biomasca.pdf (accessed July, 2013).

Le Toan, T., S. Quegan, I. A. N. Woodward, M. Lomas, N. Delbart, and G. Picard. 2004. Relating radar remote sensing of biomass to modeling of forest carbon budgets. *Climatic Change* 67:379–402.

Lim, K., P. Treitz, M. A. Wulder, B. St-Onge, and M. Flood. 2003. Lidar remote sensing of forest structure. *Progress in Physical Geography* 27:88–106.

Lohmann, U., and J. Feichter. 1997. Impact of sulfate aerosols on albedo and lifetime of clouds: A sensitivity study with the ECHAM GCM. *Journal of Geophysical Research* 102:685–700.

MA (Millennium Ecosystem Assessment). 2004. *Ecosystems and human well-being: Our human planet*. Washington, DC: Island Press.

MA (Millennium Ecosystem Assessment). 2005. *Ecosystems and human well-being: General synthesis*. Washington, DC: Island Press and World Resources Institute.

McNaughton, S. J. 1979. Grazing as an optimization process: Grass–ungulate relationships in the Serengeti. *The American Naturalist* 113:691–703.

McNaughton, S. J., M. Oesterheld, D. A. Frank, and K. J. Williams. 1989. Ecosystem-level patterns of primary productivity and herbivory in terrestrial habitats. *Nature* 341:142–144.

Milchunas, D. G., and K. W. Lauenroth. 1993. Quantitative effects of grazing on vegetation and soils over a global range of environments. *Ecological Monographs* 63:327–366.

Monteith, J. 1972. Solar radiation and productivity in tropical ecosystems. *Journal of Applied Ecology* 9:747–766.

Mouillot, F., and C. B. Field. 2005. Fire history and the global carbon budget: A 1° × 1° fire history reconstruction for the 20th century. *Global Change Biology* 11:398–420.

Nelson, R., T. Gobakken, E. Næsset, et al. 2012. Lidar sampling—Using an airborne profiler to estimate forest biomass in Hedmark County, Norway. *Remote Sensing of Environment* 123:563–578.

Nosetto, M. D., E. G. Jobbágy, and J. M. Paruelo. 2005. Land use change and water losses: The case of grassland afforestation across a soil textural gradient in central Argentina. *Global Change Biology* 11:1101–1117.

Oesterheld, M., J. Loreti, M. Semmartin, and J. M. Paruelo. 1999. Grazing, fire, and climate effects on primary productivity of grasslands and savannas. In *Ecosystems of disturbed ground*, ed. L. R. Walker, 287–306. Amsterdam: Elsevier.

Oesterheld, M., O. E. Sala, and S. J. McNaughton. 1992. Effect of animal husbandry on herbivore-carrying capacity at a regional scale. *Nature* 356:234–236.

O'Neill, R. V., D. L. de Angelis, J. B. Waide, and T. F. Allen. 1986. *A hierarchical concept of ecosystems*. Princeton, NJ: Princeton University Press.

Paruelo, J. M. 2008. La caracterización funcional de ecosistemas mediante sensores remotos. *Ecosistemas* 17:3.

Paruelo, J. M., I. C. Burke, and W. K. Lauenroth. 2001a. Land use impact on ecosystem functioning in eastern Colorado, USA-*Global Change Biology* 7:631–639.

Paruelo, J. M., M. F. Garbulsky, J. P. Guerschman, and E. G. Jobbágy. 2004. Two decades of NDVI in South America: Identifying the imprint of global changes. *International Journal of Remote Sensing* 25:2793–2806.

Paruelo, J. M., E. G. Jobbagy, and O. E. Sala. 2001b. Current distribution of ecosystem functional types in temperate South America. *Ecosystems* 4:683–698.

Paruelo, J. M., and W. K. Lauenroth. 1998. Interannual variability of NDVI and their relationship to climate for North American shrublands and grasslands. *Journal of Biogeography* 25:721–733.

Paruelo, J. M., G. Piñeiro, C. Oyonarte, D. Alcaraz, J. Cabello, and P. Escribano. 2005. Temporal and spatial patterns of ecosystem functioning in protected arid areas of southeastern Spain. *Applied Vegetation Science* 8:93–102.

Paruelo, J. M., S. R. Verón, J. N. Volante, et al. 2011. Elementos conceptuales y metodológicos para la Evaluación de Impactos Ambientales Acumulativos (EIAAc) en bosques subtropicales. El caso del Este de Salta, Argentina [Conceptual and Methodological Elements for Cumulative Environmental Effects Assessment (CEEA) in Subtropical Forests. The Case of Eastern Salta, Argentina]. *Ecología Austral* 21:163–178.

Peterson, G. D., C. R. Allen, and C. S. Holling. 1998. Ecological resilience, biodiversity and scale. *Ecosystems* 1:6–18.

Pettorelli, N., J. O. Vik, A. Mysterud, J. M. Gaillard, C. J. Tucker, and N. C. Stenseth. 2005. Using the satellite-derived Normalized Difference Vegetation Index (NDVI) to assess ecological effects of environmental change. *TRENDS in Ecology and Evolution* 20:503–510.

Piñeiro, G., M. Oesterheld, and J. M. Paruelo. 2006. Seasonal variation in aboveground production and radiation use efficiency of temperate rangelands estimated through remote sensing. *Ecosystems* 9:357–357.

Prince, S. D. 1991. A model of regional primary production for use with coarse resolution satellite data. *International Journal of Remote Sensing* 12:1313–1330.

Ranson, K. J., and G. Sun. 1994. Mapping biomass for a northern forest ecosystem using multifrequency SAR data. *IEEE Transactions on Geoscience and Remote Sensing* 32:388–396.

Roberts, G., M. J. Wooster, G. L. W. Perry, et al. 2005. Retrieval of biomass combustion rates and totals from fire radiative power observations: Application to southern Africa using geostationary SEVIRI imagery. *Journal of Geophysical Research* 110:1–19.

Running, S., R. R. Nemani, F. A. Heinsch, M. Zhao, M. Reeves, and H. Hashimoto. 2004. A continuous satellite-derived measure of global terrestrial primary production. *BioScience* 54:547–560.

Running, S. W., P. E. Thornton, R. R. Nemani, and J. M. Glassy. 2000. Global terrestrial gross and net primary productivity from the earth observing system. In *Methods in ecosystem science*, eds. O. Sala, R. Jackson, and H. Mooney, 44–57. New York: Springer-Verlag.

Sellers, P. J., J. A. Berry, G. J. Collatz, C. B. Field, and F. G. Hall. 1992. Canopy reflectance, photosynthesis, and transpiration. A reanalysis using improved leaf models and a new canopy integration scheme. *Remote Sensing of Environment* 42:187–216.

Singh, J. S., W. K. Lauenroth, and R. K. Sernhorst. 1975. Review and assessment of various techniques for estimating net aerial primary production in grasslands from harvest data. *Botanical Review* 41:181–232.

Sun, G., K. J. Ranson, Z. Guo, Z. Zhang, P. Montesano, and D. Kimes. 2011. Forest biomass mapping from Lidar and radar synergies. *Remote Sensing of Environment* 115:2906–2916.

Treuhaft, R. N., B. E. Law, and G. P. Asner. 2004. Forest attributes from radar interferometric structure and its fusion with optical remote sensing. *BioScience* 54:561–571.

Treuhaft, R. N., S. N. Madsen, M. Moghaddam, and J. J. van Zyl. 1996. Vegetation characteristics and underlying topography from interferometric radar. *Radio Science* 31:1449–1485.

van der Werf, G. R., J. T. Randerson, L. Giglio, et al. 2010. Global fire emissions and the contribution of deforestation, savanna, forest, agricultural, and peat fires (1997–2009). *Atmospheric Chemistry and Physics* 10:11707–11735.

van der Werf, G. R., J. T. Randerson, L. Giglio, N. Gobron, and A. J. Dolman. 2008. Climate controls on the variability of fires in the tropics and subtropics. *Global Biogeochemical Cycle* 22:28–36.

Vasallo, M. M., H. D. Dieguez, M. F. Garbulsky, E. G. Jobbágy, and J. M. Paruelo. 2013. Grassland afforestation impact on primary productivity: A remote sensing approach. *Applied Vegetation Science*. 16(3): 390–403.

Verón, S. R., E. G. Jobbágy, C. M. Di Bella, et al. 2012. Assessing the potential of wildfires as a sustainable bioenergy opportunity. *Global Change Biology Bioenergy* 4:634–641.

Verón, S. R., J. M. Paruelo, and M. Oesterheld. 2011. Grazing-induced losses of biodiversity affect the transpiration of an arid ecosystem. *Oecologia* 165:501–510.

Viglizzo, E. F., J. M. Paruelo, P. Laterra, and E. G. Jobbágy. 2012. Ecosystem service evaluation to support land-use policy. *Agriculture, Ecosystems & Environment* 154:78–84.

Virginia, R. A., and D. H. Wall. 2001. Principles of ecosystem function. In *Encyclopedia of biodiversity*, ed. S. A. Levin, 345–352. San Diego, CA: Academic Press.

Volante, J. N., D. Alcaraz-Segura, M. J. Mosciaro, E. F. Viglizzo, and J. M. Paruelo. 2012. Ecosystem functional changes associated with land clearing in NW Argentina. *Agriculture, Ecosystems & Environment* 154:12–22.

Walker, W. S., J. M. Kellndorfer, and L. E. Pierce. 2007. Quality assessment of SRTM C- and X-band interferometric data: Implications for the retrieval of vegetation canopy height. *Remote Sensing of Environment* 109:482–499.

Wang, Y., H. Weinacker, and B. Koch. 2007. Development of a procedure for vertical structure analysis and 3D single tree extraction within forest based on Lidar point cloud. *Proceedings of the ISPRS Workshop Laser Scanning 2007 and SilviLaser 2007* 36(3/W52):419–423.

Wessman, C. A. 1992. Spatial scales and global change: Bridging the gaps from plots to GCM grids cells. *Annual Review of Ecology and Systematics* 23:175–200.

Wooster, M. J., G. Roberts, G. L. W. Perry, and Y. J. Kaufman. 2005. Retrieval of biomass combustion rates and totals from fire radiative power observations: FRP derivation and calibration relationships between biomass consumption and fire radiative energy release. *Journal of Geophysical Research* 110:1–24.

Wulder, M. A., J. C. White, R. F. Nelson, et al. 2012. Lidar sampling for large-area forest characterization: A review. *Remote Sensing of Environment* 121:196–209.

Zhao, K., S. Popescu, and R. Nelson. 2009. Lidar remote sensing of forest biomass: A scale-invariant estimation approach using airborne lasers. *Remote Sensing of Environment* 113:182–196.

3

Recent Advances in the Estimation of Photosynthetic Stress for Terrestrial Ecosystem Services Related to Carbon Uptake

M. F. Garbulsky

University of Buenos Aires, Argentina

I. Filella and J. Peñuelas

Center for Ecological Research and Forestry Applications (CREAF), Spain

CONTENTS

3.1 Introduction 40
3.2 Alternative Ways to Remotely Estimate Photosynthetic Stress of
    Terrestrial Vegetation 42
    3.2.1 Leaf Pigments 42
        3.2.1.1 Chlorophyll Content 42
        3.2.1.2 Leaf Pigment Cycles and the Photochemical
                Reflectance Index 44
        3.2.1.3 Drawbacks, Caveats, and Cautionary Remarks 47
    3.2.2 Chlorophyll Fluorescence 49
        3.2.2.1 Basics, Origin, and Characteristics of Fluorescence 49
        3.2.2.2 Recent Advancements 51
        3.2.2.3 Future of Chlorophyll Fluorescence 52
        3.2.2.4 Final Comments on Chlorophyll Fluorescence
                from Space 54
3.3 Final Considerations 54
Acknowledgments 56
References 56

3.1 Introduction

Estimation of the carbon uptake of terrestrial vegetation is still a major challenge in evaluating ecosystem services. In general, these services are the direct and indirect contribution of ecosystems to human well-being or, in other terms, the goods and services provided by ecosystems. Because photosynthesis is the key process mediating 90% of carbon and water fluxes (Joiner et al. 2011), it is one of the main drivers of many of the services provided by the ecosystems, such as climate regulation, carbon sequestration, carbon storage, food, or livestock grassland production (Costanza et al. 1997; de Groot et al. 2002; Naidoo et al. 2008). The integrity of ecosystem services is thus fundamental to human well-being. We need to understand the links between these services and the ecosystem processes. Estimating their magnitude at the local, regional, and global scales is crucial for maintaining them over time (Haines-Young and Potschin 2010).

The distribution of photosynthetic tissues changes across spatial and temporal scales and across different land uses (e.g., Paruelo et al. 2004). Traditional remote sensing techniques allow the assessment of green plant biomass and, therefore, plant photosynthetic capacity. However, detecting how much of this capacity is actually realized is a more challenging goal that is necessary to estimate photosynthetic carbon fluxes or the gross primary productivity (GPP), the ecosystem level expression of the photosynthetic carbon uptake. The efficiency involved in the conversion of absorbed light by plant photosynthetic tissues into organic compounds (i.e., light use efficiency [LUE] of terrestrial vegetation) also varies in time and space (Runyon et al. 1994; Gamon et al. 1995; Garbulsky et al. 2010) due to the periodic environmental and physiological limitations of photosynthesis. Contrasting functional types (Gamon et al. 1997; Huemmrich et al. 2010), drought and temperature extremes (Landsberg and Waring 1997; Sims et al. 2006), and nutrient levels (Gamon et al. 1997; Ollinger et al. 2008) are the factors that contribute to this variability.

In various forms, the simple relationships between primary productivity and the product between absorbed radiation by vegetation and the efficiency that plants convert radiation into biomass (Monteith 1977), have been the basis for many evaluations of photosynthesis and primary production from the canopy scale to the global scale (Field et al. 1995; Running et al. 2004). Monteith (1977) originally proposed this relationship for the estimation of the net primary productivity (NPP). However, production efficiency models (PEMs) are based on the theory of LUE for GPP, which states that a relatively constant relationship exists between photosynthetic carbon uptake and radiation absorbed by the canopy. Different approaches for estimating carbon uptake are based on this model of LUE. Many of these approaches have assumed a constant efficiency (Myneni et al. 1995) or derived this term from constant values by biome, cited in the literature (Ruimy et al. 1994). Another approach is to downregulate the maximum efficiency by biome using meteorological variables, such as vapor pressure deficit (VPD) and temperature, as surrogates for photosynthetic stresses (Running et al. 2004). Because VPD and temperature alone are not always good surrogates of reduced efficiency (Garbulsky et al. 2010), meteorologically based methods may not always adequately explain efficiency variation. Therefore, other methods of determining photosynthetic stress or LUE for carbon uptake estimations are needed to produce better assessments of the derived ecosystem services. A direct remote estimation of LUE may thus be of great importance and may have multiple applications, such as estimation of productivity (CO_2 fixation) or detection of effects of environmental stress on vegetation carbon uptake, that may precede reduction in leaf area (Garbulsky et al. 2008b).

Since concerns have recently been voiced by the scientific community and the society at large regarding the global effects of the alteration in carbon cycle, important improvements in our capacity to remotely estimate LUE have occurred (Grace et al. 2007). The approaches to remotely estimate LUE are linked to how we can gather a signal of the photosynthetic efficiency from reflected light, heat, or chlorophyll fluorescence (Figure 3.1). Although the theory of developing remotely sensed estimations from these pathways seems clear enough and estimations are made at different scales,

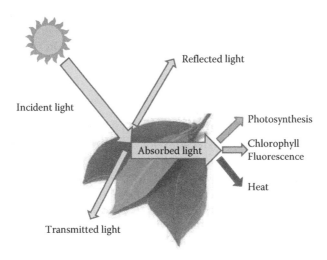

FIGURE 3.1 (See color insert.)
The incidental sunlight reaching the leaves and the complementary pathways of absorbed light after the interception by the canopy. Absorbed light by chlorophylls can be used to drive photochemistry, a process that can use up to 80% of the absorbed radiation. Alternatively, it can be lost as fluorescence (between 0.5% and 2% of absorbed radiation) or as heat (18%–98%). Photosynthesis, chlorophyll fluorescence, and heat loss are closely linked and in direct competition; thus, the increase of the rate of one process will unfailingly decrease the rate of the others.

complications still are being resolved as new platforms and sensors become available.

The main challenge related to the remote estimation of carbon uptake and primary productivity is to elucidate how to scale up the signal from leaves to the entire ecosystem and, therefore, how remote sensors can make assessments from foliar to ecosystem functional traits. In particular, the challenge is to estimate the actual photosynthetic performance or leaf photosynthesis stress and to scale up this to the ecosystem level. In this chapter, we review and synthesize recent approaches to remotely estimating LUE (i.e., the photosynthetic efficiency of carbon uptake) using current and future data. We present state-of-the-art methods of estimating LUE and recent advances in available remote sensing technologies and data.

3.2 Alternative Ways to Remotely Estimate Photosynthetic Stress of Terrestrial Vegetation

3.2.1 Leaf Pigments

Leaf pigment variation is a key tool for diagnosing a range of plant physiological properties and processes (Peñuelas and Filella 1998; Blackburn 2007). Different approaches related to contents and cycles of leaf pigments that try to estimate carbon uptake efficiency are based on the relationship with biochemical processes at the leaf level.

3.2.1.1 Chlorophyll Content

This approach is based on the remote estimation of the crop chlorophyll content (Chl) of vegetation as an estimator of GPP. In annual crops, because long- or medium-term changes in canopy Chl are related to crop phenology, canopy stresses, and photosynthetic capacity of the vegetation (e.g., Ustin et al. 1998; Zarco-Tejada et al. 2002), these can also be related to GPP. At the canopy level, Chl may appear to be the community property most relevant for the prediction of productivity (Dawson et al. 2003). Chlorophyll content is not a surrogate of LUE, but it estimates the total photosynthesis capacity. This approach does not precisely depend on LUE estimation, but it assumes that Chl is equivalent to the product of the fraction of absorbed photosynthetically active radiation (fAPAR) × LUE. In principle, it does not depend on the relationship between the widely used Normalized Difference Vegetation Index (NDVI), a spectral index derived from the red and infrared reflectances of the canopy, and fAPAR. In this context, it is proposed that the LUE model can be written as GPP = VI × PAR (Photosynthetic Active Radiation), where VI is a spectral index proxy of Chl.

The previously mentioned vegetation indices (VIs) are based on reflectance (ρ) in two spectral channels: the near-infrared (NIR) and either the green or the red edge. Several VIs, such as the MERIS Terrestrial Chlorophyll Index (MTCI = [ρNIR–ρred edge]/[ρred edge–ρred]) and chlorophyll indices (CIgreen = ρNIR/ρgreen–1; CIred edge = ρNIR/ρred edge–1) have been specifically proposed to estimate total Chl content (Gitelson et al. 2003, 2005; Dash and Curran 2004). In annual crops, such as maize, the relationships between total Chl content and VIs showed that some VIs could explain more than 87% of Chl content variation (Peng et al. 2011). The determination coefficient of the relationship between CIgreen or CIred edge and total Chl in maize and soybean exceeded 0.92 (Gitelson et al. 2005). Thus, these chlorophyll- and green Leaf Area Index (LAI)-related VIs can be used as a proxy for Chl in the model GPP = VI × PAR, specifically for herbaceous annual crops in which water or nutritional stresses will lead to a fast decrease in carbon uptake through total chlorophyll loss. At the ecosystem level, across 15 eddy covariance towers encompassing a wide variation in North American vegetation composition, the relationship between MTCI and tower GPP was stronger than that between either the Moderate Resolution Imaging Spectroradiometer (MODIS) GPP or Enhanced Vegetation Index (EVI) and tower GPP in croplands and deciduous forests, and to a lesser degree in grasslands. However, this was not the case in evergreen forests (Harris and Dash 2010) or in peatlands (Harris and Dash 2011). These analyses suggest that data from the MERIS sensor can be used as an alternative to MODIS for estimating carbon fluxes. Correlations between tower GPP and both vegetation indices (EVI and MTCI) were similar only for deciduous vegetation, indicating that physiologically driven spectral indices, such as the MTCI, may also complement existing structurally based indices in satellite-based carbon flux modeling efforts. The relationship between GPP and any chlorophyll index may thus be highly dependent in its parameters and its strength for different vegetation functional types (Figure 3.2).

The last evaluations of the model that relied on total Chl content and incident PAR showed that it can be applied to accurately estimate GPP in irrigated and rain-fed maize and soybean (Peng and Gitelson 2012). Due to differences in leaf structures and canopy architectures, the algorithms for GPP estimation are species-specific for maize and soybean, especially when using VIs with NIR and either red or green reflectance. However, it is possible to apply a unified algorithm for GPP estimation in both maize and soybean using CIred edge, MTCI, and red edge NDVI with MERIS spectral bands, which are least sensitive to different crop species. CIred edge and red edge NDVI with the red edge band around 720 nm were found to be non-species-specific for maize and soybean and to be very accurate in the estimation of GPP in maize and soybean combined.

This technique can provide accurate estimations of midday GPP in maize and soybean crops, and perhaps other annual crops, under rainfed and irrigated conditions because of a rather constant relationship between LUE

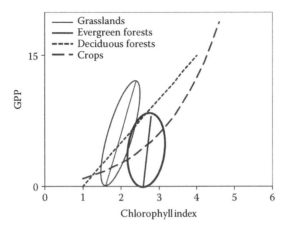

FIGURE 3.2
Schematic relationships between midday gross primary productivity (GPP) and a chlorophyll index in ecosystems with contrasting phenologies and GPP dynamics. Although the relationships for crops and deciduous forests are well supported by the literature, those for grasslands and evergreen forests are not because of the lower strength and high dispersion of the data symbolized by the ellipses.

and chlorophyll content. In contrast, there is little evidence for the utility of this approach in ecosystem types such as evergreen forest or other drought-adapted canopies. Further studies used the green CI (CIgreen)—mainly proposed for the estimation of canopy chlorophyll content (Gitelson et al. 2005)—as an estimator of midday LUE using cloud-free MODIS images (500 m) and flux measurements in maize (Wu et al. 2012). This relationship was then successfully applied for the estimation of midday LUE for coniferous forest and grassland.

3.2.1.2 Leaf Pigment Cycles and the Photochemical Reflectance Index

One proven pathway for detecting the spatial and temporal variations in LUE is the remote sensing of plant pigment cycles. The foundation of this remote sensing approach for estimating LUE is the de-epoxidation state of the xanthophyll cycle, which is linked to heat dissipation of leaves (Demmig-Adams and Adams 1996). This is a decay process of excited chlorophyll that competes with and is complementary to photosynthetic electron transport (Niyogi 1999). Hyperspectral remote sensing has been used to develop technologies and analytical methods for quantifying pigments nondestructively and repeatedly across a range of spatial scales. The recent progress in deriving predictive relationships among various characteristics and transformations of hyperspectral reflectance data related to plant pigments showed an expanding range of applications in the ecophysiological, environmental, agricultural, and forestry sciences (Blackburn 2007; Nichol et al. 2012).

During the 1990s, a series of studies at the leaf and close canopy levels using close-range remote sensing from the ground or from low platforms were able to assess this efficiency parameter (LUE) based on concurrent xanthophyll pigment changes (Gamon et al. 1990, 1992, 1997; Peñuelas et al. 1994, 1995, 1997, 1998; Filella et al. 1996; Gamon and Surfus 1999). Because reflectance at 531 nm is functionally related to the de-epoxidation state of the xanthophyll cycle (Gamon et al. 1990, 1992; Peñuelas et al. 1995), a photochemical reflectance index (PRI; typically calculated as [R531 − R570]/[R531 + R570], where R indicates reflectance and numbers indicate wavelength nanometers at the center of the bands) was developed as a method to remotely assess photosynthetic efficiency using narrow-band reflectance (Gamon et al. 1992; Peñuelas et al. 1995).

PRI measures the relative reflectance on either side of the green reflectance peak (550 nm; Figure 3.3); therefore, it also compares the reflectance in the blue (chlorophyll and carotenoids absorption) region of the spectrum with the reflectance in the red (chlorophyll absorption only) region (Peñuelas et al. 2011). Consequently, it can serve as an index of relative chlorophyll/carotenoid levels, often referred to as bulk pigment ratios. Over longer time scales (weeks–months), changes in bulk pigment content and ratios due to leaf development, aging, or chronic stress have been reported to play a significant role together with the xanthophyll pigment epoxidation in the PRI signal (Peñuelas et al. 1997; Gamon et al. 2001; Stylinski et al. 2002). Thus, PRI is also often related to carotenoid/chlorophyll ratios in

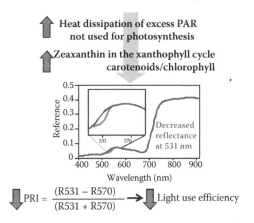

FIGURE 3.3
The pigments cycle approach to remotely sense gross light use efficiency (LUE) in terrestrial vegetation by means of the photochemical reflectance index (PRI). The remote sensing of the xantophyll cycle provides a surrogate for the estimation of dissipation of excess radiation not used for photosynthesis from leaf to primary productivity at regional scales. Dark gray for the arrows and reflectance spectrum are indicative of conditions with excess radiation and therefore low LUE while the gray spectrum around 531 nm indicates the conditions with high LUE.

leaves across a large number of species, ages, and conditions (Stylinski et al. 2002; Filella et al. 2009). Therefore, to the extent that photosynthetic activity correlates with changing chlorophyll/carotenoid ratios in response to stress, ontogeny, or senescence, PRI may provide an effective measure of relative photosynthetic rates. Seasonally varying pigment levels also strongly affect PRI. This seasonal variation may actually help explain the good performance of PRI to predict LUE because of the covarying chlorophyll/carotenoid ratios with xanthophyll pigment levels. All the relationships described here show that this approach offers great possibilities for significantly improving the monitoring of CO_2 uptake in terrestrial ecosystems globally as well as regionally.

Although the mechanics of these wavelength selections have been fully explored at the leaf scale (Gamon et al. 1993), there is less support at a canopy or greater scale, where a variety of alternate wavebands have been used that are often based on statistical correlations (Gamon et al. 1992; Inoue et al. 2008) or determined by instrument limitations (Garbulsky et al. 2008b). The lack of a clear consensus in the literature on which PRI wavelengths best estimate LUE has hindered cross-study comparisons. Consequently, it is not entirely clear if the best wavelengths for measuring this feature at the leaf scale (531 and 570 nm) are necessarily the best wavelengths at progressively larger scales, where multiple scattering and other confounding effects may alter the spectral response of the xanthophyll cycle feature, much in the way that pigment absorption peaks can vary depending upon their chemical and scattering medium. More work, therefore, may be needed to determine the ideal PRI algorithm for airborne or spaceborne platforms; these studies have been hampered by the limited availability and high costs of suitable airborne and spaceborne instruments.

Together, these responses to the de-epoxidation state of the xanthophyll cycle and to carotenoid/chlorophyll ratios ensure that PRI scales with photosynthetic efficiency vary across a wide range of conditions, species, and functional types. The available evidence shows that the PRI is a reliable estimator of ecophysiological variables closely related to the photosynthetic efficiency at the leaf and canopy levels over a wide range of species, plant functional types, and temporal scales (see cites herein; Garbulsky et al. 2011).

Since 2000, the availability of global data provided by the MODIS sensor has been the primary tool for testing the utility of PRI at the ecosystem level. The 530-nm (526–536) waveband provided by the satellite-borne MODIS sensor was used as a LUE indicator at the ecosystem scale across different vegetation types with significant success (Rahman et al. 2004; Drolet et al. 2005, 2008; Garbulsky et al. 2008a, 2008b; Goerner et al. 2009). Currently, there are few spaceborne remote sensing instruments of high spectral resolution. (Note that Hyperion and CHRIS/PROBA are exceptions, but these are demonstration instruments with limited accessibility.) But these types of data can now be collected from a range of novel helicopter and aircraft instruments (Malenovsky et al. 2009) and from the planned new satellite data. Apart from the sensors already in orbit, there are various others coming up (Table 3.1) that will provide better data for estimating LUE from space based on pigment physiology. The launching of new image spectrometers, such as the HyspIRI by NASA or the EnMAP project led by the German Aerospace Centre (DLR), will allow the calculation of PRI at high spatial resolution, thus offering great new potential for remote sensing of the vegetation pigment cycle.

3.2.1.3 Drawbacks, Caveats, and Cautionary Remarks

There are different problems that still prevent the generalization of PRI use for ecosystem and biospheric scales and its global and operational use as a LUE estimator (Grace et al. 2007). Possible multiple biochemical, ecological, and physical confounding factors operating at several levels of aggregation in the LUE–PRI relationship emerge from the literature review (Garbulsky et al. 2011). At the leaf level, biochemical processes, including photorespiration, PSI cyclic electron transport, and nitrate reduction, can compete with CO_2 fixation for reductant generated by photosynthetic electron transport (Niyogi 1999). This can cause PSII (Photosystem II) efficiency (PRI) and CO_2 assimilation to diverge. There are even other pigment cycles, such as those included in the lutein epoxide cycle of tropical trees (Matsubara et al. 2008; Esteban et al. 2009) that could also produce noise in the PRI signal. Despite these potential complications, it appears that the overall photosynthetic system is often sufficiently regulated to maintain consistent relationships between PSII processes and CO_2 fixation (Gamon et al. 1997; Stylinski et al. 2002). On the other hand, to the extent that pigment ratios are not closely related to LUE, changing pigment ratios would be a confounding variable, as mentioned earlier.

At the canopy level, the problems are related to the structural differences of the canopies, the varying background effects of the satellite data (e.g., soil color, moisture, shadows, or presence of other non-green landscape components), the different reflectance signals derived from illumination and viewing angle variations (Filella et al. 2004; Sims et al. 2006; Hilker et al. 2010), or other physical effects of canopy and stand structure (e.g., LAI changes), such as leaf movement, sun and viewing angles, soil background, and shadows that can significantly influence the PRI signal (Barton and North 2001; Gamon et al. 1995). Different studies showed that PRI reflectance could be affected by sun–target–sensor geometry and by stand structure (Asner 1998; Barton and North 2001; Drolet et al. 2005; Hall et al. 2008; Hilker et al. 2008). Recent advances (from the analysis of multiangular satellite observations from the CHRIS sensor on board the PROBA satellite at 34-m spatial resolution) showed that PRI depends on the canopy shadow fractions for low levels of LUE across evergreen or mixed forest sites. A negative logarithmic relationship was found between the slope of the relationship between

TABLE 3.1

Current and Future Satellite Tools for Estimating Photosynthetic Efficiency through Plant Pigments

Satellite	Sensor	Launching/Service Period	Reference	Main Features of the Data
Earth Observing -1 (EO-1)	Hyperion	November 2000	http://eo1.gsfc.nasa.gov	Descending polar orbit with an equatorial crossing time of 10:03. High-resolution hyperspectral imager with 220 spectral bands (from 400 to 2500 nm) at 30-m resolution with a revisit period of 16 days.
PROBA (Project for On-Board Autonomy)	CHRIS (Compact High Resolution Imaging Spectrometer)	October 22, 2001	https://earth.esa.int/web/guest/missions/esa-operational-eo-missions/proba	Sun-synchronous orbit with a revisit of seven days. Depending on the setting mode, spatial resolution is 18 or 34 m at nadir setting. Along track narrow-band spectrometric observations of PRI of up to five angles.
TERRA–AQUA	MODerate resolution Imaging Spectroradiometer (MODIS)	TERRA: January 2000 AQUA: May 2002	http://modis.gsfc.nasa.gov	Same sensor on board two satellites. Two revisits a day (morning and afternoon) for 16 spectral bands from 400 to 1000 nm at 250-m, 500-m, and 1000-m spatial resolution and 10-nm bandwidth.
ADvanced Earth Observing Satellite II (ADEOS-2)	GLobal Imager (GLI)	December 2002–October 2003	http://sharaku.eorc.jaxa.jp/ADEOS2	Six 250-m resolution channels and 30 other 1-km resolution channels and four-day revisit.
Environmental Mapping and Analysis Program (EnMAP)	HyperSpectral Imager (HSI)	Planned for 2015	http://www.enmap.org	Sun-synchronous orbit with a revisit of four days and a spatial resolution of 30 m × 30 m; 94 bands between 420 and 1000 nm.
Hyperspectral Infrared Imager (HyspIRI)	VSWIR	Currently in the study stage	http://hyspiri.jpl.nasa.gov	380 to 2500 nm in 10-nm contiguous bands with a spatial resolution of 60 m at nadir and a revisit of 19 days.

PRI and the canopy shadow fraction with the gross LUE (Hilker et al. 2011). In contrast, other recent analysis showed that MODIS, PRI, and also NDVI, in contrast to EVI, were not affected by view angle across a 10-year period in three forest sites (Sims et al. 2011), results that may be contradictory with other studies. In any case, shadow fraction could be a variable affecting spatial resolution of the data.

The consistency of the relationship between PRI, LUE, and CO_2 uptake that is increasingly found in different studies across ecosystems (Garbulsky et al. 2011; Hilker et al. 2011; Peñuelas et al. 2011) suggests a great degree of functional convergence of biochemical, physiological, and structural components affecting ecosystem C fluxes between ecosystem types (Field 1991). Emergent ecosystem properties may allow exploration of their complex photosynthetic behavior using simple spectral methods such as measurement of the plant pigment cycle through the PRI. Understanding the basis for this convergence (and finding the ecophysiological principles governing these responses) remains a primary goal of current research. Meanwhile, PRI, which is more important for the pragmatic empirical remote sensing of CO_2 uptake, especially from near-nadir satellite observations (Goerner et al. 2010) and with multiangular atmospheric correction (Lyapustin and Wang 2009; Hilker et al. 2010), can become an excellent tool for continuous global monitoring of GPP, which is essential to follow the C sequestration under changing climate.

The use of uniform protocols is needed to generate comparable data and, at the end, a possible general calibration of the PRI–LUE relationship (Peñuelas et al. 2011). Further studies are also needed to disentangle the several drivers of the PRI signal and to resolve the potentially confounding factors to improve the assessment of CO_2 fluxes in many different biomes using hyperspectral or narrow-band remote sensing.

3.2.2 Chlorophyll Fluorescence

Estimating LUE and carbon uptake from chlorophyll fluorescence viewed from space is undoubtedly a field of knowledge where great advancements have taken place in recent years. Remote sensing of terrestrial vegetation fluorescence from space is of great interest because it can provide global information on the functional status of vegetation including LUE and GPP. Global retrieval of solar-induced fluorescence emitted by terrestrial vegetation is beginning to provide an unprecedented measure for photosynthetic efficiency.

3.2.2.1 Basics, Origin, and Characteristics of Fluorescence

Improvements in fluorescence measuring techniques have made the fluorescence method an important tool for basic and applied plant physiology research (Krause and Weis 1991). The fluorescence signal originates in the core of the photosynthetic apparatus, where absorbed photosynthetically active radiation (APAR) is converted into chemical energy. Fluorescence reflects the competition among several pathways for the excitation captured in the antenna. When photochemistry occurs with maximal efficiency, excitation is passed mainly to the photoreactions. When the photochemical traps are closed, excitation is lost because of a competition between fluorescence and nonradiative dissipative pathways, the latter converting the energy to heat. Because the fluorescence yield varies inversely with the fraction of open reaction centers, it provides a useful tool for investigation of photosynthetic processes. A small part of the light absorbed by chlorophyll in an assimilating leaf can be reemitted as fluorescence by the chlorophyll molecules of the photosystem II, which adds a weak signal to reflected solar radiation. Because these processes occur in competition, by measuring chlorophyll fluorescence we can also know the efficiency of photochemistry and heat dissipation, which is linked to CO_2 assimilation (Baker 2008).

Chlorophyll fluorescence measurements are traditionally made at the leaf level using an external source of light. The fluorescence spectrum peak is of a longer wavelength than the absorbed light; therefore, fluorescence yield can be quantified by exposing a leaf to light of a defined wavelength and measuring the amount of light reemitted at longer wavelengths when the light is turned off. This technique allows the quantifying of different parameters related to the light phase of carbon fixation. Extensive experimental and theoretical studies demonstrate that chlorophyll fluorescence is a proxy to actual photosynthesis and as such directly related to LUE and CO_2 uptake (Seaton and Walker 1990); it also behaves as an indicator of plant vitality and plant stress because fluorescence emission competes with adaptation/ protection mechanisms set up by the plant. Hence, measuring fluorescence can provide access to missing information regarding photosynthesis performance variables (Maxwell and Johnson 2000).

This laboratory measurement methodology is far from the requirements necessary to estimate fluorescence of an ecosystem scale; therefore, new methods are needed to measure Chl fluorescence from space. Solar-induced Chl fluorescence (F) can provide an early and direct approach for the evaluation of the actual functional status of vegetation because of the rapid response to perturbations in the environmental conditions such as light and water stress; therefore, F can detect stress conditions before significant reductions in Chl content or LAI have occurred. The fluorescence of green vegetation consists of blue-green fluorescence (maxima at 440 and 520 nm) and of red and far-red chlorophyll fluorescence (maxima at 690 and 740 nm). To monitor vegetation photosynthesis, it is necessary to analyze the red and far-red chlorophyll fluorescence from the photosynthetically active parts of the leaf tissues. The magnitude of the two broad peaks with maxima around 685 and 740 nm can be related to photosynthetic efficiency. Because the magnitude of solar radiation reflected by vegetation and by atmosphere can be 100 to 150 times more intense than F at the top of the atmosphere, the main challenge in achieving an F estimation from remote sensing passive measurements is to decouple the F signal from the solar radiation (Meroni et al. 2009).

3.2.2.2 Recent Advancements

Even though the first attempts to quantify chlorophyll on terrestrial vegetation without an artificial excitation source date back to the 1970s, the remote sensing of chlorophyll fluorescence is still in a developmental stage, but research is advancing. The F signal can be detected passively in narrow absorption lines ($\approx$2–3 nm) of the solar and atmospheric spectrum in which irradiance is strongly reduced (i.e., the Fraunhofer lines). Three main Fraunhofer lines in the visible and NIR have been used for F estimation: Hα due to hydrogen (H) absorption in the solar atmosphere centered at 656.4 nm and two telluric oxygen (O_2) absorption bands in the Earth's atmosphere: O_2-B centered at 687.0 nm and O_2-A at 760.4 nm. A combination of Fraunhofer lines and O_2 lines would make it possible to measure all the main fluorescence bands. The two O_2 bands (A and B) and the Hα bands are considered the most useful (Meroni et al. 2009).

Radiance measurements at high spectral resolution exploit the Fraunhofer line to decouple F from the reflected flux. F retrieval in solar Fraunhofer lines is based on the evaluation of the in-filling of the Fraunhofer lines due to F and hereafter. Because the fractional depth of Fraunhofer lines is not affected by atmospheric scattering and by absorption in narrow spectral windows free from telluric absorption features, the atmospheric modeling required is much simpler than with atmospheric bands.

Although there have been many measurements of fluorescence, especially recently, from ground- and airborne-based instruments (Meroni et al. 2009), there has been scant information available from satellites. The first space-based estimation of F (Guanter et al. 2007) was performed using data from the O_2-A absorption band provided by the ENVISAT MEdium Resolution Imaging Spectrometer (MERIS). The MERIS-derived fluorescence correlated well ($R^2 = 0.85$) with data acquired by the Compact Airborne Spectrographic Imager (CASI-1500) sensor and ground-based estimates. Recent studies (Joiner et al. 2011; Frankenberg et al. 2011b) presented the first results at the global scale from the use of high spectral resolution data from the Thermal And Near-infrared Sensor for carbon Observation–Fourier Transform Spectrometer (TANSO–FTS/GOSAT). These two studies used different approaches. The method used by Joiner et al. (2011) makes use of one strong Fraunhofer line (K I at 770.1 nm). It employs real solar irradiance measurements from the TANSO–FTS in order to avoid the explicit modeling of the instrument line shape function (ILSF). In turn, the method proposed in Frankenberg et al. (2011a) extends this single-line approach to two broader spectral windows centered at 755 and 770 nm. This retrieval, making use of broader spectral windows containing several Fraunhofer lines, is expected to be less sensitive to instrumental noise than that based on one single line, whereas retrievals performed for the two separate windows provide more independent measurements to be used to enhance the signal-to-noise ratio of the final F products. At present, data from a 22-month series have shown an accurate comparison of F intensity levels and spatial patterns with physically based F retrieval approach. However, there is a need for a biome-dependent scaling from F to gross primary production (Guanter et al. 2012).

3.2.2.3 *Future of Chlorophyll Fluorescence*

Several projects are underway to develop satellite platforms and sensors to remotely determine chlorophyll fluorescence for the estimation of the carbon cycle of terrestrial vegetation (Table 3.2). Because vegetation fluorescence can be converted into an indicator of photosynthetic activity, by using fluorescence data we can achieve a better understanding of how much carbon is being taken up by plants and their role in the carbon and water cycles. In addition to the sensors in orbit at the time of this publication, there are a number of satellite projects related to the measurements of vegetation fluorescence. The availability of fluorescence data at detailed levels from regional to global scales is a great step forward and will represent a huge increase in the capability of PEMs to assess the spatial and temporal variability of carbon uptake by terrestrial vegetation. One of the major and more ambitious satellite projects, scheduled to be in orbit in 2018, is led by the European Space Agency. Called the FLuorescence EXplorer (FLEX), its objective is to observe photosynthesis for a better understanding of the carbon cycle and to provide global maps of vegetation fluorescence. The main instrument is the fluorescence imaging spectrometer (FIS) that covers the O_2-A (760 nm) and O_2-B (687 nm) absorption lines with a spectral band of 20 nm. The ground spatial resolution at nadir will be 300 m, and the revisiting period will be seven days.

Another important projected satellite is NASA's Orbiting Carbon Observatory-2 (OCO-2) that will be specifically dedicated to studying atmospheric carbon dioxide from space. Three high-resolution grating spectrometers (Day et al. 2011)—one for each spectral band located at O_2 A-Band (757–772 nm), weak CO_2 Band (1590–1621 nm), and strong CO_2 Band (2041–2082 nm)—will be combined with meteorological observations and ground-based CO_2 measurements to characterize CO_2 sources and sinks on regional scales at monthly intervals for two years at 1.29 km × 2.25 km spatial resolution.

The Geostationary Carbon Process Mapper (GCPM) has three proposed geostationary platforms. This project aims to measure key atmospheric trace gases and process tracers related to climate change and human activity at high temporal resolution. This understanding comes from contiguous maps of carbon dioxide (CO_2), methane (CH_4), carbon monoxide (CO),

TABLE 3.2

Remote Sensing Tools for Chlorophyll Fluorescence Assessment from Space

Satellite	Sensor	Launch/in Orbit	Reference	Main Features of the Data
ENVISAT	MEdium Resolution Imaging Spectrometer (MERIS)	From March 2002 to April 2012	http://wdc.dlr.de/sensors/meris	Spatial resolution: 260 m × 300 m Two channels near the O_2-A-bands centered at 753.8 and 760.6 nm; bandwidths = 7.5 and 3.75 nm, respectively.
ENVISAT	SCanning Imaging Absorption spectroMeter for Atmospheric CHartographY (SCIAMACHY)	From March 2002 to April 2012	http://www.sciamachy.org wdc.dlr.de/sensors/sciamachy	From 0.2 to 0.5 nm, from 240 nm to 1700 nm, and from 2000 to 2400 nm, Limb vertical 3 km × 132 km, Nadir horizontal 32 km × 215 km.
GOSAT (Greenhouse gases Observing SATellite)	Thermal And Near-infrared Sensor for carbon Observation–Fourier Transform Spectrometer (TANSO-FTS)	January 23, 2009	http://www.gosat.nies.go.jp	Sun-synchronous orbit with an equator crossing time 13:00, revisiting in three days.
OCO-2 (Orbiting Carbon Observatory)	Single instrument with three classical grating spectrometers	July 2014	http://oco.jpl.nasa.gov	Sun-synchronous crossing equator at noon; three measurement angles.
Earth Explorer 8	FLuorescence EXplorer (FLEX) Fluorescence Imaging Spectrometer (FIS)	2018 or later, now in development	http://esamultimedia.esa.int/docs/SP1313-4_FLEX.pdf	Descending sun-synchronous orbit with an equator crossing time 10:00.
Geostationary Carbon Process Mapper (GCPM)	Geostationary Fourier Transform Spectrometer (GeoFTS)	In project phase	Key et al. 2012	Geostationary: 10 times per day at ~4 km × 4 km.

and F collected up to 10 times per day at relatively high spatial resolution (~4 km × 4 km) from geostationary orbit (GEO). These measurements will capture the spatial and temporal variability of the carbon cycle across diurnal, synoptic, seasonal, and interannual time scales. The combination of high-resolution mapping and high measurement frequency provides quasi-continuous monitoring, effectively eliminating atmospheric transport uncertainties from source/sink inversion modeling. The CO_2/CH_4/CO/F measurements could also provide the information needed to disentangle natural and anthropogenic contributions to atmospheric carbon concentrations.

3.2.2.4 *Final Comments on Chlorophyll Fluorescence from Space*

Currently, there are a number of new, encouraging results and ongoing projects on the estimation of carbon uptake by vegetation from chlorophyll fluorescence from space. The retrieval of chlorophyll fluorescence from space is feasible through the Fraunhofer line retrieval method, which is simple, fast, and robust, and is now verified with real data on the ground. Fluorescence appears to display information that is independent of reflectance data (e.g., fAPAR). Chlorophyll fluorescence retrievals from the GOSAT and the OCO-2, in conjunction with their global atmospheric CO_2 measurements, will provide an exceptional combination of a vegetation and atmospheric perspective on the global carbon budget, constraining our model predictions for future atmospheric CO_2 abundance. Most importantly, this method is largely unaffected by atmospheric scattering and is even able to sense F through thin clouds.

Although this approach seems promising, different problems also arise from the available data. The strong correlation between F and GPP displayed is not as strong in boreal summers because of the overestimation produced in savannas and croplands and the underestimation produced in boreal needleleaf forests. Therefore, additional research is needed to disentangle the effect of disturbance factors such as illumination, canopy structure, and temporal and spatial resolution of the satellite data on the relationship between sun-induced fluorescence and photosynthesis.

3.3 Final Considerations

Although eddy covariance towers represent the current standard for ecosystem carbon flux estimation of GPP, we must learn to properly calibrate these estimations with the new remote sensing products, if our objective is to develop reliable remote sampling methods for ecosystem carbon flux.

This remains a significant challenge because flux towers sample through time, whereas remotely sensed imagery samples through space (Rahman et al. 2001). To make this calibration, we should blend these sampling domains by applying remote sensing aircraft and satellite measurements at the same temporal and spatial scales as flux tower footprint measurements, which is rarely done. Therefore, coordinated flux and optical data acquisition from different biomes are needed. In addition, standardized ground-based optical sampling programs at flux towers (Gamon et al. 2006) should be expanded. We have to properly calibrate the surrogates for LUE for different ecosystems or vegetation types, and then we will be able to apply remote sensing to extrapolate in time and space from tower sites. Actual widespread location of the towers across vegetation types (the long-term goal of this methodology) and the increasing availability of remote sensing tools at different spatial, temporal, and spectral resolutions are the main advantages and the reasons to conduct further research along this avenue.

The different approaches outlined in this chapter highlight how remote estimation of chlorophyll fluorescence will provide accurate global estimations of LUE in the near future. Although research efforts to generate estimates of chlorophyll fluorescence outweigh all other approaches, chlorophyll fluorescence estimation seems to be less ecosystem dependant than the other approaches. The temporal reaction of each of the remote sensing approaches is one issue to be considered in the estimations of LUE considering the objectives of GPP estimations. Although chlorophyll fluorescence has a very short reaction time (in the order of milliseconds) in relation to the changes in environmental conditions regulating photosynthesis, chlorophyll content change as a response to stress is a process that can take up to three days in herbaceous crops (Houborg et al. 2011). The proposed geostationary constellation of satellite platforms to capture chlorophyll fluorescence (Table 3.2) will provide, in contrast, an extraordinary daily temporal resolution database.

Even though important, specifically designed missions are providing and will provide data on LUE, it is interesting to note that three new satellite missions—Suomi National Polar-orbiting Partnership (NPP), Landsat 8, and Sentinel—will not give much information on LUE spatial and temporal variability. In the case of the Suomi NPP, as the continuity of MODIS missions, it is not designed to provide the bands to calculate PRI. In any case, there is a wide and promising avenue for the estimation of terrestrial vegetation LUE from satellite data.

In this chapter, we presented and analyzed the most recent advances related to the quantification of services related to carbon uptake by terrestrial ecosystems. The cascade effects of GPP, as the main energy input that determines many ecosystem services, warrants accurate estimation of these fluxes in time and space. Remote sensing technologies are providing new methodologies for such estimations.

Acknowledgments

This research was supported by the University of Buenos Aires project UBACyT 01/F362, the Spanish Government projects CGC2010-17172 and Consolider Ingenio Montes (CSD2008-00040), and the Catalan Government project SGR 2009-458.

References

Asner, G. P. 1998. Biophysical and biochemical sources of variability in canopy reflectance. *Remote Sensing of Environment* 64:234–253.

Baker, N. R. 2008. Chlorophyll fluorescence: A probe of photosynthesis in vivo. *Annual Review of Plant Biology* 59:89–113.

Barton, C. V. M., and P. R. J. North. 2001. Remote sensing of canopy light use efficiency using the photochemical reflectance index—Model and sensitivity analysis. *Remote Sensing of Environment* 78:264–273.

Blackburn, G. A. 2007. Hyperspectral remote sensing of plant pigments. *Journal of Experimental Botany* 58:855–867.

Costanza, R., R. d'Arge, R. de Groot, et al. 1997. The value of the world's ecosystem services and natural capital. *Nature* 387:253–260.

Dash, J., and P. J. Curran. 2004. The MERIS terrestrial chlorophyll index. *International Journal of Remote Sensing* 25:5003–5013.

Dawson, T. P., P. R. J. North, S. E. Plummer, and P. J. Curran. 2003. Forest ecosystem chlorophyll content: Implications for remotely sensed estimates of net primary productivity. *International Journal of Remote Sensing* 24:611–617.

Day, J. O., C. W. O'Dell, R. Pollock, et al. 2011. Preflight spectral calibration of the Orbiting Carbon Observatory. *IEEE Transactions on Geoscience and Remote Sensing* 49:2793–2801.

de Groot, R. S., M. A. Wilson, and R. M. J. Boumans. 2002. A typology for the classification, description and valuation of ecosystem functions, goods and services. *Ecological Economics* 41:393–408.

Demmig-Adams, B. B., and W. W. Adams. 1996. The role of xanthophyll cycle carotenoids in the protection of photosynthesis. *Trends in Plant Science* 1:21–26.

Drolet, G. G., K. F. Huemmrich, F. G. Hall, et al. 2005. A MODIS-derived photochemical reflectance index to detect inter-annual variations in the photosynthetic light-use efficiency of a boreal deciduous forest. *Remote Sensing of Environment* 98:212–224.

Drolet, G. G., E. M. Middleton, K. F. Huemmrich, et al. 2008. Regional mapping of gross light-use efficiency using MODIS spectral indices. *Remote Sensing of Environment* 112:3064–3078.

Esteban, R., J. M. Olano, J. Castresana, et al. 2009. Distribution and evolutionary trends of photoprotective isoprenoids (*xanthophylls* and *tocopherols*) within the plant kingdom. *Physiologia Plantarum* 135:379–389.

Field, C. B. 1991. Ecological scaling of carbon gain to stress and resource availability. In *Response of plants to multiple stresses*, eds. H. A. Mooney, W. E. Winner, and E. J. Pell, 35–65. San Diego, CA: Academic Press.

Field, C. B., J. T. Randerson, and C. M. Malmstrom. 1995. Global net primary production: Combining ecology and remote sensing. *Remote Sensing of Environment* 51:74–88.

Filella, I., T. Amaro, J. L. Araus, and J. Peñuelas. 1996. Relationship between photosynthetic radiation-use efficiency of barley canopies and the photochemical reflectance index (PRI). *Physiologia Plantarum* 96:211–216.

Filella, I., J. Peñuelas, L. Llorens, and M. Estiarte. 2004. Reflectance assessment of seasonal and annual changes in biomass and CO_2 uptake of a Mediterranean shrubland submitted to experimental warming and drought. *Remote Sensing of Environment* 90:308–318.

Filella, I., A. Porcar-Castell, S. Munné-Bosch, J. Bäck, M. F. Garbulsky, and J. Peñuelas. 2009. PRI assessment of long-term changes in carotenoids/chlorophyll ratio and short-term changes in de-epoxidation state of the xanthophyll cycle. *International Journal of Remote Sensing* 30:4443–4455.

Frankenberg, C., A. Butz, and G. C. Toon. 2011a. Disentangling chlorophyll fluorescence from atmospheric scattering effects in O_2 A-band spectra of reflected sunlight. *Geophysical Research Letters* 38:L03801.

Frankenberg, C., J. B. Fisher, J. Worden, et al. 2011b. New global observations of the terrestrial carbon cycle from GOSAT: Patterns of plant fluorescence with gross primary productivity. *Geophysical Research Letters* 38:L17706.

Gamon, J. A., C. B. Field, W. Bilger, O. Björkman, A. Fredeen, and J. Peñuelas. 1990. Remote sensing of the xanthophyll cycle and chlorophyll fluorescence in sunflower leaves and canopies. *Oecologia* 85:1–7.

Gamon, J. A., C. B. Field, A. L. Fredeen, and S. Thayer. 2001. Assessing photosynthetic downregulation in sunflower stands with an optically-based model. *Photosynthesis Research* 67:113–125.

Gamon, J. A., C. B. Field, M. Goulden, et al. 1995. Relationships between NDVI, canopy structure, and photosynthetic activity in three Californian vegetation types. *Ecological Applications* 5:28–41.

Gamon, J. A., I. Filella, and J. Peñuelas. 1993. The dynamic 531-nanometer delta reflectance signal: A survey of twenty angiosperm species. In *Photosynthetic responses to the environment*, eds. H. Yamamoto and C. Smith, 172–177. Rockville, MD: American Society of Plant Physiologists.

Gamon, J. A., J. Peñuelas, and C. B. Field. 1992. A narrow-waveband spectral index that tracks diurnal changes in photosynthetic efficiency. *Remote Sensing of Environment* 41:35–44.

Gamon, J. A., A. F. Rahman, J. L. Dungan, M. Schildhauer, and K. F. Huemmrich. 2006. Spectral Network (SpecNet). What is it and why do we need it? *Remote Sensing of Environment* 103:227–235.

Gamon, J. A., L. Serrano, and J. S. Surfus. 1997. The photochemical reflectance index: An optical indicator of photosynthetic radiation use efficiency across species, functional types, and nutrient levels. *Oecologia* 112:492–501.

Gamon, J. A., and J. S. Surfus. 1999. Assessing leaf pigment content and activity with a reflectometer. *New Phytologist* 143:105–117.

Garbulsky, M. F., J. Peñuelas, J. A. Gamon, Y. Inoue, and I. Filella. 2011. The photochemical reflectance index (PRI) and the remote sensing of leaf, canopy and ecosystem radiation use efficiencies: A review and meta-analysis. *Remote Sensing of Environment* 115:281–297.

Garbulsky, M. F., J. Peñuelas, J. M. Ourcival, and I. Filella. 2008a. Estimación de la eficiencia del uso de la radiación en bosques mediterráneos a partir de datos MODIS. Uso del Índice de Reflectancia Fotoquímica (PRI) [Radiation use efficiency estimation in Mediterranean forests using MODIS Photochemical Reflectance Index (PRI)]. *Ecosistemas* 17:89–97.

Garbulsky, M. F., J. Peñuelas, D. Papale, and I. Filella. 2008b. Remote estimation of carbon dioxide uptake of a Mediterranean forest. *Global Change Biology* 14:2860–2867.

Garbulsky, M. F., J. Peñuelas, D. Papale, et al. 2010. Patterns and controls of the variability of radiation use efficiency and primary productivity across terrestrial ecosystems. *Global Ecology and Biogeography* 19:253–267.

Gitelson, A. A., S. B. Verma, A. Viña, et al. 2003. Novel technique for remote estimation of CO2 flux in maize. *Geophysical Research Letter* 30:1486.

Gitelson, A. A., A. Viña, V. Ciganda, D. C. Rundquist, and T. J. Arkebauer. 2005. Remote estimation of canopy chlorophyll content in crops. *Geophysical Research Letter* 32:L08403.

Goerner, A., M. Reichstein, and S. Rambal. 2009. Tracking seasonal drought effects on ecosystem light use efficiency with satellite-based PRI in a Mediterranean forest. *Remote Sensing of Environment* 113:1101–1111.

Goerner, A., M. Reichstein, E. Tomelleri, et al. 2010. Remote sensing of ecosystem light use efficiency with MODIS-based PRI—The DOs and DON'Ts. *Biogeosciences Discusssion* 7:6935–6969.

Grace, J., C. Nichol, M. Disney, P. Lewis, T. Quaife, and P. Bowyer. 2007. Can we measure terrestrial photosynthesis from space directly, using spectral reflectance and fluorescence? *Global Change Biology* 13:1484–1497.

Guanter, L., L. Alonso, L. Gómez-Chova, J. Amorós, J. Vila, and J. Moreno. 2007. Estimation of solar-induced vegetation fluorescence from space measurements. *Geophysical Research Letters* 34:L08401.

Guanter, L., C. Frankenberg, A. Dudhia, et al. 2012. Retrieval and global assessment of terrestrial chlorophyll fluorescence from GOSAT space measurements. *Remote Sensing of Environment* 121:236–251.

Haines-Young, R., and M. Potschin. 2010. The links between biodiversity, ecosystem services and human well-being. In *Ecosystem ecology: A new synthesis*, eds. D. G. Raffaelli and C. L. J. Frid, 110–139. Cambridge, UK: Cambridge University Press.

Hall, F. G., T. Hilker, N. C. Coops, et al. 2008. Multi-angle remote sensing of forest light use efficiency by observing PRI variation with canopy shadow fraction. *Remote Sensing of Environment* 112:3201–3211.

Harris, A., and J. Dash. 2010. The potential of the MERIS Terrestrial Chlorophyll Index for carbon flux estimation. *Remote Sensing of Environment* 114:1856–1862.

Harris, A., and J. Dash. 2011. A new approach for estimating northern peatland gross primary productivity using a satellite-sensor-derived chlorophyll index. *Journal of Geophysical Research-Biogeosciences* 116:G04002.

Hilker, T., N. C. Coops, F. G. Hall, et al. 2008. Separating physiologically and directionally induced changes in PRI using BRDF models. *Remote Sensing of Environment* 112:2777–2788.

Hilker, T., N. C. Coops, F. G. Hall, et al. 2011. Inferring terrestrial photosynthetic light use efficiency of temperate ecosystems from space. *Journal of Geophysical Research* 116:G03014.

Hilker, T., F. G. Hall, N. C. Coops, et al. 2010. Remote sensing of photosynthetic light-use efficiency across two forested biomes: Spatial scaling. *Remote Sensing of Environment* 114:2863–2874.

Houborg, R., M. C. Anderson, C. S. T. Daughtry, W. P. Kustas, and M. Rodell. 2011. Using leaf chlorophyll to parameterize light-use-efficiency within a thermal-based carbon, water and energy exchange model. *Remote Sensing of Environment* 115:1694–1705.

Huemmrich, K. F., J. A. Gamon, C. E. Tweedie, et al. 2010. Remote sensing of tundra gross ecosystem productivity and light use efficiency under varying temperature and moisture conditions. *Remote Sensing of Environment* 114:481–489.

Inoue, Y., J. Peñuelas, A. Miyata, and M. Mano. 2008. Normalized difference spectral indices for estimating photosynthetic efficiency and capacity at a canopy scale derived from hyperspectral and CO_2 flux measurements in rice. *Remote Sensing of Environment* 112:156–172.

Joiner, J., Y. Yoshida, A. P. Vasilkov, Y. Yoshida, L. A. Corp, and E. M. Middleton. 2011. First observations of global and seasonal terrestrial chlorophyll fluorescence from space. *Biogeosciences* 8:637–651.

Key, R., S. Sander, A. Eldering, et al. 2012. The Geostationary Carbon Process Mapper. *IEEE Aerospace Conference Proceedings*. Big Sky, MT. 3–10 March 2012. Article no. 6187029.

Krause, G., and E. Weis. 1991. Chlorophyll fluorescence and photosynthesis: The basics. *Annual Review of Plant Biology* 42:313–349.

Landsberg, J. J., and R. H. Waring. 1997. A generalised model of forest productivity using simplified concepts of radiation-use efficiency, carbon balance and partitioning. *Forest Ecology and Management* 95:209–228.

Lyapustin, A., and Y. Wang. 2009. The time series technique for aerosol retrievals over land from MODIS. In *Satellite aerosol remote sensing over land*, eds. A. Kokhanovky and G. De Leeuw, 69–99. Heidelberg, Berlin: Springer.

Malenovsky, Z., K. B. Mishra, F. Zemek, U. Rascher, and L. Nedbal. 2009. Scientific and technical challenges in remote sensing of plant canopy reflectance and fluorescence. *Journal of Experimental Botany* 60:2987–3004.

Matsubara, S., G. H. Krause, M. Seltmann, et al. 2008. Lutein epoxide cycle, light harvesting and photoprotection in species of the tropical tree genus Inga. *Plant, Cell and Environment* 31:548–561.

Maxwell, K., and G. N. Johnson. 2000. Chlorophyll fluorescence—A practical guide. *Journal of Experimental Botany* 51:659–668.

Meroni, M., M. Rossini, L. Guanter, et al. 2009. Remote sensing of solar-induced chlorophyll fluorescence: Review of methods and applications. *Remote Sensing of Environment* 113:2037–2051.

Monteith, J. L. 1977. Climate and the efficiency of crop production in Britain. *Philosophical Transactions of the Royal Society—Biological Sciences* 281:277–294.

Myneni, R. B., S. O. Los, and G. Asrar. 1995. Potential gross primary productivity of terrestrial vegetation from 1982–1990. *Geophysical Research Letters* 22:2617–2620.

Naidoo, R., A. Balmford, R. Costanza, et al. 2008. Global mapping of ecosystem services and conservation priorities. *Proceedings of the National Academy of Sciences of the United States of America* 105:9495–9500.

Nichol, C. J., R. Pieruschka, K. Takayama, et al. 2012. Canopy conundrums: Building on the Biosphere 2 experience to scale measurements of inner and outer canopy photoprotection from the leaf to the landscape. *Functional Plant Biology* 39:1–24.

Niyogi, K. K. 1999. Photoprotection revisited: Genetic and molecular approaches. *Annual Review Plant Physiology and Plant Molecular Biology* 50:333–359.

Ollinger, S. V., A. D. Richardson, M. E. Martin, et al. 2008. Canopy nitrogen, carbon assimilation, and albedo in temperate and boreal forests: Functional relations and potential climate feedbacks. *Proceedings of the National Academy of Sciences of the United States of America* 105:19335–19340.

Paruelo, J. M., M. F. Garbulsky, J. P. Guerschman, and E. G. Jobbagy. 2004. Two decades of Normalized Difference Vegetation Index changes in South America: Identifying the imprint of global change. *International Journal of Remote Sensing* 25:2793–2806.

Peng, Y., and A. Gitelson. 2012. Remote estimation of gross primary productivity in soybean and maize based on total crop chlorophyll content. *Remote Sensing of Environment* 117:440–448.

Peng, Y., A. Gitelson, G. Keydan, D. C. Rundquist, and W. Moses. 2011. Remote estimation of gross primary production in maize and support for a new paradigm based on total crop chlorophyll content. *Remote Sensing of Environment* 115:978–989.

Peñuelas, J., and I. Filella. 1998. Visible and near-infrared reflectance techniques for diagnosing plant physiological status. *Trends in Plant Science* 3:151–156.

Peñuelas, J., I. Filella, and J. A. Gamon. 1995. Assessment of photosynthetic radiation-use efficiency with spectral reflectance. *New Phytologist* 131:291–296.

Peñuelas, J., I. Filella, J. A. Gamon, and C. Field. 1997. Assessing photosynthetic radiation-use efficiency of emergent aquatic vegetation from spectral reflectance. *Aquatic Botany* 58:307–315.

Peñuelas, J., I. Filella, J. Llusia, D. Siscart, and J. Piñol. 1998. Comparative field study of spring and summer leaf gas exchange and photobiology of the Mediterranean trees *Quercus ilex* and *Phillyrea latifolia*. *Journal of Experimental Botany* 49:229–238.

Peñuelas, J., J. A. Gamon, A. L. Fredeen, J. Merino, and C. B. Field. 1994. Reflectance indexes associated with physiological changes in nitrogen-limited and water-limited sunflower leaves. *Remote Sensing of Environment* 48:135–146.

Peñuelas, J., M. F. Garbulsky, and I. Filella. 2011. Photochemical reflectance index (PRI) and remote sensing of plant CO_2 uptake. *New Phytologist* 191:596–599.

Rahman, A. F., V. D. Cordova, J. A. Gamon, H. P. Schmid, and D. A. Sims. 2004. Potential of MODIS ocean bands for estimating CO_2 flux from terrestrial vegetation: A novel approach. *Geophysical Research Letters* 31:L10503.

Rahman, A. F., J. A. Gamon, D. A. Fuentes, D. A. Roberts, and D. Prentiss. 2001. Modeling spatially distributed ecosystem flux of boreal forest using hyperspectral indices from AVIRIS imagery. *Journal of Geophysical Research—Atmospheres* 106:33579–33591.

Ruimy, A., B. Saugier, and G. Dedieu. 1994. Methodology for the estimation of terrestrial net primary production from remotely sensed data. *Journal of Geophysical Research* 99:5263–5283.

Running, S. W., R. R. Nemani, F. A. Heinsch, M. Zhao, M. Reeves, and H. Hashimoto. 2004. A continuous satellite-derived measure of global terrestrial primary production. *BioScience* 54:547–560.

Runyon, J., R. H. Waring, S. N. Goward, and J. M. Welles. 1994. Environmental limits on net primary production and light-use efficiency across the Oregon transect. *Ecological Applications* 4:226–237.

Seaton, G. G. R., and D. A. Walker. 1990. Chlorophyll fluorescence as a measure of photosynthetic carbon assimilation. *Proceedings of the Royal Society of London* 242:29–35.

Sims, D. A., H. Luo, S. Hastings, W. C. Oechel, A. F. Rahman, and J. A. Gamon. 2006. Parallel adjustments in vegetation greenness and ecosystem CO_2 exchange in response to drought in a Southern California chaparral ecosystem. *Remote Sensing of Environment* 103:289–303.

Sims, D. A., A. F. Rahman, E. F. Vermote, and Z. Jiang. 2011. Seasonal and inter-annual variation in view angle effects on MODIS vegetation indices at three forest sites. *Remote Sensing of Environment* 115:3112–3120.

Stylinski, C. D., J. A. Gamon, and W. C. Oechel. 2002. Seasonal patterns of reflectance indices, carotenoid pigments and photosynthesis of evergreen chaparral species. *Oecologia* 131:366–374.

Ustin, S. L., D. A. Roberts, J. Pinzón, et al. 1998. Estimating canopy water content of chaparral shrubs using optical methods. *Remote Sensing of Environment* 65:280–291.

Wu, C., Z. Niu, and S. Gao. 2012. The potential of the satellite derived green chlorophyll index for estimating midday light use efficiency in maize, coniferous forest and grassland. *Ecological Indicators* 14:66–73.

Zarco-Tejada, P. J., J. R. Miller, G. H. Mohammed, T. L. Noland, and P. H. Sampson. 2002. Vegetation stress detection through chlorophyll a + b estimation and fluorescence effects on hyperspectral imagery. *Journal of Environmental Quality* 31:1433–1441.

4

Earth Observation of Carbon Cycling Pools and Processes in Northern High-Latitude Systems

H. E. Epstein

University of Virginia, Virginia

CONTENTS

4.1 Introduction to Remote Sensing of Carbon Cycling Processes at Northern High Latitudes...63
4.2 Remote Sensing of Vegetation Biomass and Primary Production in Arctic Tundra..65
4.3 Remote Sensing of Vegetation Biomass and Primary Production in Boreal Forests..68
4.4 Remote Sensing of Land–Atmosphere Exchange of Carbon in Arctic Tundra and Boreal Forests...71
4.5 Remote Sensing of Soil Carbon Processes (Soil Respiration, Decomposition, Methane Production) and Stocks in Arctic Tundra and Boreal Forests..73
4.6 Remote Sensing of Carbon Emissions from Fire....................................75
4.7 Summary and Conclusions..76
Acknowledgments...77
References..77

4.1 Introduction to Remote Sensing of Carbon Cycling Processes at Northern High Latitudes

Ecosystem services of northern terrestrial systems encompass all three main categories of service: provisioning, regulating, and cultural (Haynes-Young and Potschin 2010). The ecosystem services focused on within this chapter are largely provisioning (e.g., biomass production) and regulating (e.g., carbon sequestration and storage). One key service of the northern high latitudes has been their ability to store carbon. Arctic and subarctic ecosystems have historically acted as carbon sinks, storing large quantities of dead organic matter and protecting it from rapid decomposition. This protection of dead organic matter in northern ecosystems largely occurs due to the presence of relatively cold soils, waterlogged soils, and permafrost, all of which reduce rates of decomposition. In addition to low rates of soil organic matter breakdown and carbon release, the other component of carbon sequestration is of course uptake of CO_2 by plants through photosynthesis and the incorporation of this carbon into plant tissues. The historic carbon sink of northern ecosystems implies that the rate of CO_2 uptake by plants has been greater than the loss of CO_2 through respiratory processes, including decomposition. New estimates suggest that the amount of carbon stored in soils of high-latitude permafrost regions to 3-m depth is greater than twice the amount of carbon present in atmospheric CO_2 (Schuur et al. 2008; Tarnocai et al. 2009), and recent warming in the Arctic will most certainly alter the land–atmosphere exchange of carbon in this region (e.g., Euskirchen et al. 2009; Lee et al. 2012; Natali et al. 2012; Trucco et al. 2012).

Carbon cycling processes have been studied through the use of remote sensing at high latitudes since the mid-1990s. Since remote sensing is predominately conducive for observing properties of the land surface, it has largely been used for examining the CO_2 uptake component of the carbon budget, through vegetation indices, rather than to quantify decomposition and the rates of CO_2 efflux from northern soils—although there are examples of using remotely sensed information to estimate belowground processes, which will be discussed in Sections 4.4 and 4.5. With regard to the aboveground component, remote sensing has been used to examine vegetation at high latitudes, and investigators have used this information to infer CO_2 uptake by vegetation, as either gross primary production (GPP) or net primary production (NPP), beginning with the early years of available satellite information (see Chapter 2).

Some of the first published studies used the record from the Advanced Very High Resolution Radiometer (AVHRR) onboard U.S. National Oceanographic and Atmospheric Administration (NOAA) Earth-orbiting satellites to examine spatial patterns and temporal trends in the Normalized Difference Vegetation Index (NDVI) for northern high latitudes (Myneni et al. 1997; Tucker et al. 2001; Zhou et al. 2001; Slayback et al. 2003). The NDVI is an indicator of photosynthetic activity in vegetation, calculated from the surface reflectances of the red and near-infrared (NIR) wavelengths; vegetation is generally absorptive in the red wavelengths due to chlorophyll and reflective in the NIR from scattering by plant cell structures. The studies mentioned earlier demonstrated that the NDVI had increased between the beginning of the 1980s and the end of the 1990s at latitudes north of 35°N, implying increases in at least aboveground NPP. Zhou et al. (2001) showed that over the two-decade period, NDVI in Eurasia had increased to a greater degree than NDVI in North America (from 40°N to 70°N); however, Slayback et al. (2003) indicated that since 1992, North America exhibited greater increases in NDVI than Eurasia (from 45°N to 75°N).

Further studies began to extend the satellite record and also focused on some of the spatial variability of NDVI dynamics within the large region of the northern high latitudes. One important distinction became apparent— that the Arctic tundra and boreal forest biomes were generally moving in the opposite direction, with the tundra exhibiting nearly ubiquitous "greening" (increases in NDVI) and the boreal forest showing "browning" (decreases in NDVI) (Jia et al. 2003; Bunn et al. 2005, 2007; Goetz et al. 2005; Bunn and Goetz 2006; Verbyla 2008; Bhatt et al. 2010; Beck and Goetz 2011; Baird et al. 2012), although there is still some discussion in the literature regarding these trends (e.g., Alcaraz-Segura et al. 2010). For this reason, we separate our discussion of remote sensing of primary production at high latitudes into Arctic tundra, and boreal forest components. Therefore, the following two sections of this chapter discuss remote sensing of vegetation biomass and primary production in Arctic tundra, and remote sensing of vegetation biomass and primary production in boreal forests; these two sections address both provisioning (biomass accumulation) and regulating (carbon sequestration) ecosystem services. The next three sections address remote sensing of land–atmosphere exchange of carbon, soil carbon processes and stocks, and carbon emissions from fire; these are largely regulating ecosystem services related to atmospheric CO_2 and methane (CH_4), as well as soil carbon storage. The final section presents an overall summary of how Earth observations have been used to assess carbon-related ecosystem services for northern high-latitude environments.

4.2 Remote Sensing of Vegetation Biomass and Primary Production in Arctic Tundra

Some of the first remote sensing studies suggesting potential changes in biomass and primary production specifically for Arctic tundra were done using repeat photographs of the North Slope of Alaska (Sturm et al. 2001; Tape et al. 2006). These studies identified expansion of tall and low shrubs (alder, willow, birch) since the mid-twentieth century on hillslopes and riparian lowlands that were either previously unvegetated or occupied by low-statured tundra vegetation. Ropars and Boudreau (2012) recently used aerial photographs (from 1957) and a WorldView-1 satellite image to document birch shrub increases on sandy terraces and hilltops at the northern treeline in subarctic Québec. Tremblay et al. (2012) also used repeat aerial photographs (from 1964 and 2003) to show dramatic increases in birch shrub and eastern larch cover at a Low Arctic site in Eastern Nunavik, Canada. Concomitant with these kinds of vegetation changes, Jia et al. (2003) found that peak NDVI on the Alaskan North Slope had increased by 16.9% between 1981 and 2001, using 8 km × 8 km AVHRR data. Verbyla (2008) then extended the AVHRR record for Alaska to 2003 and confirmed the findings of increased maximum NDVI, with the greatest changes occurring on the Alaskan coastal plain.

Studies over the past decade have examined NDVI dynamics for continental North America (Alaska and Canada) using AVHRR data and have observed a consistent increase in NDVI for the tundra since 1981 (Stow et al. 2004; Bunn et al. 2005, 2007; Goetz et al. 2005; Bunn and Goetz 2006; Jia et al. 2009). Jia et al. (2009) found that tundra greening had occurred in all five Arctic tundra bioclimate subzones (Walker et al. 2005, subzones A–E ranging from north to south) within Canada over the period 1982–2006. Several of these studies also noted changes in the seasonal dynamics of NDVI, with some areas of earlier onset of vegetation growth (Goetz et al. 2005; Jia et al. 2009), greater early season NDVI (Verbyla 2008), and earlier peak NDVI (particularly in the High Arctic) (Jia et al. 2009). A few studies observed consistent increases in productivity (NDVI) in the North American Arctic tundra at higher spatial resolutions (1 km AVHRR and 30 m Landsat) (Neigh et al. 2008; Olthof et al. 2008; Pouliot et al. 2009), suggesting that the greening trend of tundra vegetation might be rather homogeneous at the landscape scale. Pouliot et al. (2009) found that 22% of the Canadian land surface had significantly positive NDVI trends from 1985 to 2006, with much of this in northern Canada. Olthof et al. (2008) noted that vascular plants contributed more to the greening signal in Canadian tundra compared to lichen-dominated landscapes. Fraser et al. (2011) found NDVI increases from 6.1% to 25.5% across four national parks in northern Canada using Landsat data ranging from 1984 to 2009.

Several studies have added information (greater temporal extent and finer spatial detail) to the earlier examinations of circumpolar NDVI at high latitudes cited earlier. Increased productivity trends have essentially continued for Arctic tundra at the global scale (Bunn and Goetz 2006; Bunn et al. 2007; Bhatt et al. 2010). Bhatt et al. (2010) most explicitly examined the spatial patterns of productivity dynamics for circumpolar Arctic tundra, dividing the Arctic into oceanic subregions *sensu* Treshnikov (1985), and analyzing both the peak NDVI and the seasonally integrated NDVI. They found a nearly ubiquitous greening of the near-coastal tundra, however, with some declines in the Bering and West Chukchi regions. Compared to the earlier studies that showed mixed results with regard to differences in NDVI trends between North America and Eurasia, these later studies concur that the tundra vegetation of North America appeared to be greening to a greater extent than that of Eurasia (Bunn et al. 2007; Bhatt et al. 2010; Goetz et al. 2011). Bhatt et al. (2010) found a 9% increase in the maximum NDVI for North American tundra from 1982 to 2008 but only a 2% increase for Eurasian tundra. Using both AVHRR and Moderate Resolution Imaging Spectroradiometer (MODIS) data from 1982 to 2008, Beck and Goetz (2011) found amplified increases in summer NDVI across the North Slope of Alaska (essentially regardless of shrub cover), since the mid-1990s. Very little has been done that has focused specifically on Eurasian tundra, with the exception of a study by Dye and Tucker (2003), which examined seasonality of NDVI, snow cover, and temperature in Eurasia from 50°N to 71°N, and a synthesis of information by Goetz et al. (2011).

Most of the aforementioned studies have focused on NDVI as the metric for aboveground vegetation, with the assumption that there is a relationship between NDVI and either aboveground primary production or aboveground biomass, or both. Several studies have developed relationships between NDVI and aboveground biomass for Arctic tundra (Hope et al. 1993; Shippert et al. 1995; Walker et al. 2003, 2012; Jia et al. 2006), and recently Raynolds et al. (2012) published a robust relationship between maximum NDVI and aboveground biomass for tundra vegetation. So, the connection between NDVI and carbon-related pools and processes can be made. Epstein et al. (2012) used the logarithmic relationships published in Raynolds et al. (2012) to essentially convert the observed NDVI changes in Arctic tundra to aboveground biomass (or carbon) changes. They found that over the period 1982–2010, the total aboveground biomass of Arctic tundra had increased by 0.40 Pg (Figure 4.1). If we assume an equal increase belowground and a carbon mass

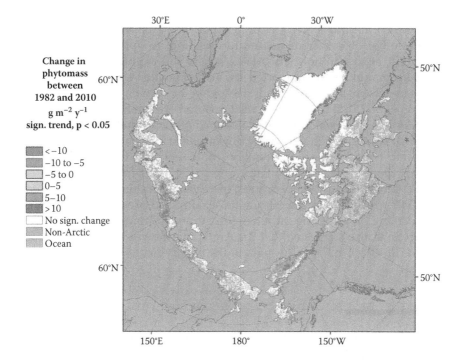

FIGURE 4.1 (See color insert.)
Changes in aboveground phytomass (g m^{-2} y^{-1}) between 1982 and 2010 for Arctic tundra, derived from relationships between AVHRR-NDVI and field-measured biomass. (From Epstein, H. E., et al., *Environmental Research Letters*, 7, 015506. With permission.)

percentage of ~50% in plant tissue, then the live plant carbon pool increased by 0.40 Pg C over this time period. At this point, it is challenging to convert this carbon difference into an actual value for additional carbon sequestered since 1982; the carbon sequestered would be a function of (1) the rate of this biomass increase over the observed time period, (2) the turnover rates of the various plant tissues, and (3) the fate of the senesced plant material. One notable result from Epstein et al. (2012) was that on average Arctic tundra aboveground biomass increased by 19.8%; however, much of this increase was seen in the more southern tundra subzones (20%–26%) compared to the more northern tundra subzones (2%–7%).

Remote sensing products have also been used in combination with simulation modeling to assess the present status and future dynamics of GPP and NPP for Arctic tundra across a variety of spatial scales (e.g., Williams et al. 2001; Turner et al. 2005; Kimball et al. 2007). Williams et al. (2001) projected the seasonal dynamics of GPP (with daily predictions) for the Kuparuk River watershed in northern Alaska, using AVHRR-derived NDVI and a soil–plant–atmosphere model. Kimball et al. (2007) used AVHRR-NDVI in conjunction with a production efficiency model (PEM), BIOME-BGC, and the terrestrial ecosystem model (TEM) to derive both GPP and NPP for Arctic tundra. With the AVHRR PEM, they estimated an increase of 7.6% (21.1 g C m^{-2}) in tundra NPP per decade between 1982 and 2000, consistent with the other observations of tundra greening. A study by Huemmrich et al. (2010), at sites in northern Alaska where soil temperatures and moisture were manipulated, supported the use of remote-sensing-driven efficiency models in assessing tundra response to climate change. Recently, Tagesson et al. (2012) developed a relationship between high-resolution satellite NDVI and the peak fraction of absorbed photosynthetically active radiation (FAPAR) for a High Arctic wet tundra site at Zackenberg, northeastern Greenland. They then used this relationship, coupled with a simple light use efficiency model, to estimate GPP from 1992 to 2008 and found that peak season GPP (along with air temperatures) increased over this time period. Sitch et al. (2007) reviewed the use of remote sensing and process-based simulation modeling for assessing carbon cycling in Arctic tundra, and a synthesis of projections also suggested that the Arctic tundra will be a small sink for carbon (~10–40 g C m^{-2} y^{-1}) throughout the twenty-first century.

4.3 Remote Sensing of Vegetation Biomass and Primary Production in Boreal Forests

There is an extensive body of literature on the use of remote sensing to estimate biomass and primary production in boreal forest ecosystems, and several reviews of this literature have been written (e.g., Gamon et al. 2004; Boyd and Danson 2005; Lutz et al. 2008), so a comprehensive discussion is far from warranted here (and also quite a challenge to construct). Therefore, focus is on studies that have been conducted at relatively coarser spatial scales and those that examine the production of biomass and the sequestration of carbon over time. One relatively early study was conducted by Goetz and Prince (1996) for a mixed evergreen-deciduous boreal forest in northeast Minnesota. They used a vegetation greenness index derived from Landsat imagery to estimate the photosynthetically active radiation intercepted annually by the canopy (IPAR), and developed relationships between IPAR and annual aboveground NPP (AANPP) data. With estimates of the light use efficiency, they derived AANPP values for both the deciduous (363 Mg C km^{-2}) and evergreen (125 Mg C km^{-2}) components of the forest, and calculated the annual net carbon uptake by aboveground vegetation (256 Mg C km^{-2}) across the 2280 km^2 study region.

A multitude of advances were made in the use of remote sensing to assess boreal vegetation biomass and production during the Boreal Ecosystem Atmosphere Study (BOREAS), summarized in Gamon et al. (2004). Hall et al. (1996) used Landsat TM reflectance data in the red and NIR bands and a radiative transfer model to estimate aboveground biomass density within the southern BOREAS study region dominated by wetland black spruce. Goetz et al. (1999) used a PEM (GLO-PEM; Prince and Goward 1995), driven by 1 km^2 of AVHRR reflectance data and PAR from the Geostationary Observational Environmental Satellite (GOES), to estimate carbon cycling variables, including aboveground biomass, GPP, and NPP, for the entire BOREAS region of central Canada. They found that average GPP for the region was 604 g C m^{-2} y^{-1}, and average NPP was 235 g C m^{-2} y^{-1}, and that these values varied substantially across land cover types. Other estimates, using a process-based model and AVHRR-derived products, yielded very similar results (average NPP of 217 g C m^{-2} y^{-1}) (Liu et al. 1999). Active optical sensors LIDAR (light detection and ranging) were also used in the BOREAS project to robustly estimate aboveground biomass in boreal black spruce stands (Lefsky et al. 2002). In addition, Leboeuf et al. (2007) used canopy shadow fraction derived from high-resolution QuickBird imagery to map the aboveground forest biomass of boreal black spruce stands in eastern Canada.

Studies using remote sensing to assess vegetation biomass and primary production in Eurasian boreal forests have been less extensive. Krankina et al. (2005) used Landsat imagery to estimate the total carbon in live forest biomass and the carbon sink rate (0.36 Mg C ha^{-1} y^{-1}) for the St. Petersburg region, and these estimates compared favorably with forest inventory data for the early 1990s. Fuchs et al. (2009) developed relationships between aboveground carbon and indices derived from both QuickBird and ASTER imagery for a forest-tundra catchment in northwestern Siberia. Expectedly, they found high spatial heterogeneity of aboveground carbon ranging from 4.3 to 28.8 tons ha^{-1}. Zheng et al. (2004) used 1 km^2 AVHRR data to estimate the NPP for boreal forests in Finland and Sweden, and they also found an order of magnitude in spatial variability across this region.

At the circumpolar scale, AVHRR-NDVI data have been used to estimate carbon in the woody biomass of northern forests (Myneni et al. 2001; Dong et al. 2003). Myneni et al. (2001) developed strong, yet saturating, relationships between growing-season integrated NDVI and forest inventory estimates of woody biomass; with these relationships, they calculated the woody carbon pool and the change in the woody carbon pool (1981–1999) for Northern Hemisphere forests. They found woody carbon gains more than 0.3 tons C ha^{-1} y^{-1} in Russian boreal forests and woody carbon losses more than 0.1 tons C ha^{-1} y^{-1} in Canadian boreal forests. In fact, the BOREAS region was found to be a source of carbon to the atmosphere in the 1990s (Gamon et al. 2004); however, this was likely due to high levels of fire disturbance during this time (Steyaert et al. 1997; Li et al. 2000; Chen et al. 2003). Combining remote sensing with process-based modeling has also been an effective methodology for estimating carbon cycling processes, such as NPP, across coarse spatial scales in boreal forests (Hall et al. 2006; Kimball et al. 2007; Zhang et al. 2007; Smith et al. 2008; Tagesson et al. 2009).

With regard to the recent temporal trends in boreal forest aboveground biomass, the AVHRR record since 1982 has been used extensively to identify these dynamics (Bunn et al. 2005, 2007; Goetz et al. 2005; Bunn and Goetz 2006; Verbyla 2008; Beck and Goetz 2011; Baird et al. 2012). Goetz et al. (2005) found that, between 1982 and 2003, the NDVI for North American boreal forest declined in areas that had not been burned over that time period. Bunn and Goetz (2006) extended this analysis spatially to the circumpolar Arctic and found "browning" largely in the more densely forested areas and generally ubiquitous declines in late season NDVI (July–August). Considering the entire growing season, the boreal forest NDVI trends were essentially mixed positive and negative. Bunn et al. (2007) calculated a predominantly greening trend for the sparse-canopied larch (deciduous) forests, in contrast to the browning of the more dense evergreen forests. A study by Verbyla (2008) confirmed a strong browning trend for boreal forests in interior Alaska, with and without considering fires, and Parent and Verbyla (2010) also found declining NDVI in boreal Alaska using higher resolution Landsat data. In contrast, Alcaraz-Segura et al. (2010), using a 1-km-resolution AVHRR dataset from the Canadian Centre for Remote Sensing, found that less than 1% of unburned forest pixels in a region of northern Canada showed a decline in NDVI between 1984 and 2006. With the most extensive record to date of AVHRR and MODIS data (1982–2008), Beck and Goetz (2011) showed continuing declines in boreal forest NDVI in North America and Eurasia, again in denser and more evergreen-dominated stands (Figure 4.2). Kimball et al. (2007) simulated a 9.1% increase in NPP from 1982 to 2000 for boreal Alaska and northwestern Canada, using a PEM with AVHRR-based inputs, however Zhang et al. (2008) simulated declines in NPP in boreal North America from the late 1990s to 2005, also using a PEM driven by AVHRR and MODIS data.

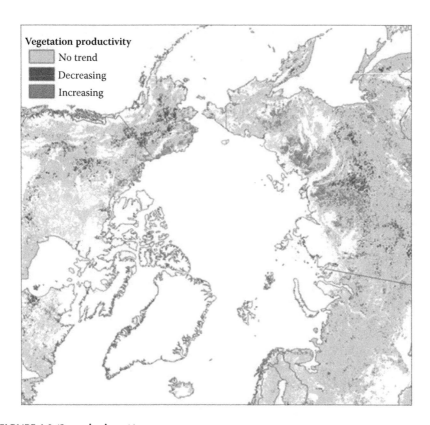

FIGURE 4.2 (See color insert.)
Trends in remotely sensed gross primary productivity between 1982 and 2008, derived from AVHRR-NDVI. Green and brown areas indicate either significantly increasing or decreasing productivity, respectively, based on a Vogelsang test with $\alpha = 0.10$. (From Beck, P. S. A., and S. J. Goetz, *Environmental Research Letters*, 7, 029501, 2012.)

4.4 Remote Sensing of Land–Atmosphere Exchange of Carbon in Arctic Tundra and Boreal Forests

In the earlier sections, we reviewed the use of remote sensing to assess the carbon content (biomass) and carbon uptake (GPP, NPP) of the aboveground component of vegetation in northern high-latitude ecosystems; however, remote sensing has also been used to examine carbon-related processes that include belowground components. Net ecosystem productivity (NEP, ecologically defined net carbon sequestration) and net ecosystem exchange (NEE, atmospherically defined net carbon sequestration) include ecosystem respiration in their calculations and, therefore, root respiration and below-ground heterotrophic respiration, making it even more challenging to assess from a remote sensing perspective. These processes are all components of the ecosystem service of regulation of atmospheric functioning, in this case, related to the concentrations of the greenhouse gases, CO_2 and CH_4.

For the Arctic tundra, a modest number of studies have attempted to use remote sensing approaches to either estimate or extrapolate values for land–atmosphere exchange of carbon dioxide. Some of the first studies were conducted on the North Slope of Alaska (Stow et al. 1998; McMichael et al. 1999). As a first step toward developing relationships between remotely sensed data and CO_2 exchange for Arctic tundra, Stow et al. (1998) related NDVI data at two different spatial resolutions to CO_2 fluxes measured by an aircraft flown along sampling transects within the Kuparuk River basin of Alaska. McMichael et al. (1999), operating at a much finer spatial resolution at two sites in northern Alaska, found relationships between field-measured (handheld spectroradiometer) NDVI and *in situ* flux measurements of both gross photosynthesis and ecosystem respiration. Soegaard et al. (2000) used peak growing season Landsat data to extrapolate field-measured CO_2 fluxes from the site to the landscape scale, across three different vegetation types in a high-Arctic ecosystem of northeast Greenland. In this system, landscape scale carbon losses in June were more than offset by a July–August carbon sink.

More recently, Kimball et al. (2009) combined inputs from MODIS (land cover, leaf area index [LAI], and NPP) and the Advanced Microwave Scanning Radiometer Earth Observing System (AMSR-E, surface wetness and temperature) in their terrestrial carbon flux (TCF) model to estimate NEE (and respiration and soil carbon stocks) for Arctic and boreal sites across North America, and they compared these estimates to eddy covariance CO_2 flux measurements. Likewise, Loranty et al. (2011) used MODIS-derived LAI as an input to a simple model (only three input variables) for simulating Arctic tundra NEE at the landscape scale. Finally, Rocha and Shaver (2011) used a MODIS Enhanced Vegetation Index (EVI) to parameterize a model of NEE and to extrapolate information from flux tower footprints across the fire scar of the Anaktuvuk River fire in northern Alaska. They found that their MODIS two-band EVI explained 86% of the variance in NEE across the observed burn severity gradient.

For the boreal forest, studies that use remote sensing to assess NEP or NEE are equally sparse, and the approaches are very similar to those used for Arctic tundra, for example, using remote sensing data to extrapolate CO_2 exchange information collected from eddy covariance flux towers, as was done by Meroni et al. (2002) with Landsat for central Siberia. Another common approach is to use remote sensing data as a simulation model input (or to derive a simulation model input), as was done by Amiro et al. (2003) with AVHRR data for western Canada. Kushida et al. (2004) used a combination of handheld spectral measurements and Landsat ETM imagery to estimate NEP in boreal forest of interior Alaska. Field-measured spectral data of forest floor and spruce needles were used to develop relationships between reflectances and LAI, and forest floor bryophytes. These relationships were combined with LAI–NPP relationships and observations of soil respiration to estimate NEP, which was extrapolated spatially using the Landsat data. As mentioned earlier, Kimball et al. (2009) used MODIS and AMSR-E to estimate NEE for a collection of both tundra and boreal forest sites with existing flux towers. Drezet and Quegan (2007) used synthetic aperture radar (SAR) data to estimate forest biomass across Britain, which was in turn used to calculate forest age structure and then NEE, providing carbon uptake values that were dramatically greater than those from prior national inventories. Finally, Richardson et al. (2010) evaluated the effects of phenology on ecosystem productivity (both NEP and gross ecosystem photosynthesis) across 21 FLUXNET sites in boreal and temperate forests, using the flux data and MODIS data as phenological indicators. They found that the productivity of evergreen needleleaf forests was less sensitive to variability in phenology than that of deciduous broadleaf forests.

4.5 Remote Sensing of Soil Carbon Processes (Soil Respiration, Decomposition, Methane Production) and Stocks in Arctic Tundra and Boreal Forests

Even more challenging to assess with remote sensing approaches are processes that occur almost exclusively belowground, such as soil respiration, organic matter decomposition, and methane production. Soil respiration is the combined CO_2 flux from root respiration and respiration from other soil flora and fauna (with a large component being heterotrophic microorganisms); soil respiration therefore includes a component (roots) that is explicitly tied to the aboveground organs of the vegetation, making remote sensing at least a reasonable option. Decomposition (in part a component of soil respiration), on the other hand is largely driven by belowground controls and indirectly by aboveground processes, such as litterfall and photosynthate exudation through roots, and therefore may be more difficult to assess with remote sensing. The few studies that have used remote sensing approaches to estimate either soil respiration or decomposition in Arctic tundra and boreal forests have already been mentioned here because these variables are often used in conjunction with assessing net ecosystem carbon fluxes.

The most common approach here is to use microwave remote sensing, such as that from AMSR-E, SAR, scanning multichannel microwave radiometer (SMMR), and Special Sensor Microwave/Imager (SSM/I), to estimate soil temperature, soil moisture, soil freeze–thaw status, and surface wetness (dominant controls on soil respiration and decomposition) in boreal and Arctic regions (e.g., Way et al. 1997; Smith et al. 2004; Jones et al. 2007; Kimball et al. 2009), which have in turn been used to model these ecosystem processes. Methane production and flux from soil to atmosphere is also largely controlled by belowground conditions, with indirect connections to aboveground processes. Many of the techniques using remote sensing to estimate methane fluxes at northern high latitudes are similar to those used for estimating soil respiration and decomposition: use remote sensing tools to assess the land surface and soil conditions, and then use these in turn to estimate or simulate the carbon-related processes. For methane fluxes from vegetated landscapes (such as wetlands), assessing surface water and soil moisture is the most common approach; however, substantive fluxes of methane in the high latitudes are generated from permanent lakes, thaw lakes, and drained lake basins, and these are often evaluated with remote sensing tools as well (McGuire et al. 2009).

Bartsch et al. (2007, 2008) used advanced synthetic aperture radar (ASAR) imagery from the European Environmental Satellite (ENVISAT) as well as data from the NASA QuickScat satellite to map the distribution of different wetland types and permanent water bodies across the Taimyr Peninsula of Russia at a spatial resolution of 150 m. Specifically for the open water bodies, a few *in situ* measurements of methane emissions from tundra lakes were used to extrapolate values for the Taimyr Peninsula. Walter et al. (2008) also used SAR data from RADARSAT-1 to estimate methane emissions from Arctic lakes but with a much higher resolution approach. They used radar backscatter to estimate the distribution of distinct types of bubble clusters in lake ice, from which the ebullition of methane can account for greater than 95% of the total methane flux (Casper et al. 2000). Combined with field measurements of bubble clusters and methane ebullition, the authors developed equations that could ultimately relate radar backscatter to whole lake methane fluxes. Finally, as was commonly done for CO_2 flux studies, Bubier et al. (2005) used remote sensing data (in this case Landsat) to extrapolate field measurements of CH_4 flux to the landscape scale using data collected during BOREAS. Numerous studies have assessed the distribution of thaw lakes and drained thaw lake basins using remote sensing tools, such as Landsat (e.g., Stow et al. 2004; Frohn et al. 2005; Hinkel et al. 2005; Plug et al. 2008; Morgenstern et al. 2011; Wang et al. 2012). However, the present literature only alludes to the use of remote sensing for quantifying methane fluxes from these land surface features.

Few studies have gotten to the point of assessing belowground carbon stocks using remote sensing approaches in high-latitude ecosystems (see the next section on fire disturbances as well). Recently, Hugelius (2012) used a Landsat-based land cover classification to spatially upscale soil organic carbon for the northern Usa River basin in European Russia. Ulrich et al. (2009) related soil properties, including soil organic carbon concentrations, to field spectrometry data for the Lena River delta in Siberia. The soils data were then scaled up to land cover units classified with Landsat ETM+ and Corona imagery. Again, we see the approach of combining simulation modeling with remote sensing inputs to estimate carbon cycling properties and process (e.g. Kimball et al. 2009), in this case for soil organic carbon.

4.6 Remote Sensing of Carbon Emissions from Fire

Fire is a major disturbance type in high-latitude systems that can have important effects on both soil carbon storage and regulation of atmospheric CO_2, and
remote sensing has been a key tool in assessing these effects (see Chapter 7).
Remote sensing has often been used for the detection of fire in boreal forests
(e.g., Li et al. 2000; Soja et al. 2004), and data from satellite remote sensing has
been a critical resource for enhancing the development of burned area maps
for much of Alaska and Canada (e.g., Epp and Lanoville 1996; Fraser et al.
2004; Soja et al. 2004; Giglio et al. 2009; Loboda et al. 2011). Extensive information on burned areas has been somewhat less reliable for much of Russia.
However, Sukhinin et al. (2004) developed the first comprehensive burned
area product for eastern Russia using AVHRR, and subsequent products from
MODIS have provided a much more extensive picture of Siberian boreal forest fires (Loboda and Csizsar 2007; Loboda et al. 2007, 2012; Giglio et al. 2010).

Area burned, in addition to several other fire-related variables garnered
from remote sensing data, can then be used to assess the carbon consumed
during fires at regional, continental, and global scales (e.g., Sukhinin et al.
2004; Kasischke et al. 2005; de Groot et al. 2007; Frolking et al. 2009; Stinson
et al. 2011; Turetsky et al. 2011). Fire severity is a key variable, and remote
sensing data have been used successfully in several studies to map fire
severity for the purpose of estimating carbon consumption during fires in
boreal forests (e.g., Michalek et al. 2000; Conrad et al. 2002; Isaev et al. 2002).
Developing remotely sensed approaches for consistently mapping fire severity across different regions and under different environmental conditions has
proved challenging (e.g., Allen and Sorbel 2008; French et al. 2008). However,
there have been concerted recent efforts to improve burn severity estimates,
such as those by Barrett et al. (2010, 2011), for spruce forests in interior Alaska.
The temporal dynamics of fire activity, based on thermal infrared remote
sensing data, are another important input for estimating carbon consumed
during fires (e.g., Loboda and Csizsar 2007; Kasischke and Hoy 2012).

Carbon lost in fires is of course not confined to the aboveground plant material, and many of these studies used remote sensing to estimate the losses of
forest floor and soil organic carbon following fires. One such study used
high-resolution airborne hyperspectral imagery to assess the loss of forest
floor material following fires in interior Alaska (Lewis et al. 2011); the results
indicated that forest floor consumption from fires was strongly related to
cover of green moss, which was estimated from the imagery. Kasischke and
Hoy (2012) used Landsat (vegetation cover and burn extent) and MODIS (date
of burn) data to estimate carbon consumed by fires also in interior Alaska;
their results included the loss of surface organic layer carbon. Other studies
have used remote sensing imagery such as Landsat to map the consumption
of the soil organic layer during fire in North American boreal forests (French
et al. 2008; Barrett et al. 2010). The use of remote sensing to assess carbon

emissions from fires at high latitudes has also not been confined to boreal forests (Palacios-Orueta et al. 2005), and some recent studies have estimated carbon losses from tundra fires using remotely sensed data (Mack et al. 2011; Rocha and Shaver 2011; Rocha et al. 2012).

4.7 Summary and Conclusions

Remote sensing has been used extensively, with a variety of different approaches, over the past few decades to assess provisioning and regulating ecosystem services related to carbon cycling in northern high-latitude ecosystems, such as Arctic tundra and boreal forests (Figure 4.3). These ecosystem services include (1) carbon accumulation by vegetation (GPP, NPP) and storage in vegetation biomass of the Arctic tundra and boreal forest (biomass provisioning, carbon sequestration/storage, and regulation of

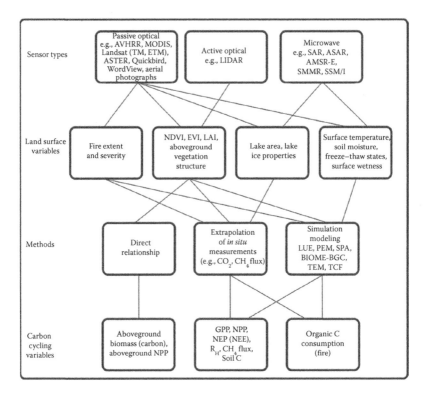

FIGURE 4.3
Conceptual diagram of the use of remote sensing techniques for evaluating carbon cycling variables in northern high-latitude ecosystems.

atmospheric CO_2); (2) the net exchange of carbon between the land and the atmosphere (NEP, NEE), the breakdown of organic matter and conversion to CO_2 (e.g., soil respiration, decomposition), and the production of methane from lakes and wetlands (carbon sequestration/storage, and regulation of atmospheric CO_2 and CH_4); and (3) the storage of carbon in soils (soil formation, carbon sequestration/storage, and regulations of atmospheric CO_2). These ecosystem services get progressively more challenging to estimate with remote sensing, as the carbon pools and processes involved move from completely aboveground-influenced (e.g., GPP, aboveground biomass) to largely belowground-influenced (e.g., methane production, soil carbon), and the literature gets more scant along this gradient as well. We therefore see numerous studies examining vegetation indices (such as the NDVI) as proxies for carbon uptake by plants across the northern high latitudes and have developed a reasonably solid understanding of the "greening" of the Arctic tundra and the concomitant "browning" of the boreal forests over the length of the satellite record. However, there are very few studies that examine the carbon cycling processes of net ecosystem exchange (or NEP), soil respiration, decomposition, and methane fluxes, using remote sensing. Often these are modeling studies that use remotely sensed inputs of vegetation properties (such as LAI) to simulate carbon cycling processes, or they are studies that use remotely sensed land surface classifications to spatially extrapolate field observations of the relevant processes. The same is true for the few studies that used remote sensing in estimating soil carbon storage. Because of the large stores of carbon in high-latitude ecosystems, the fate of this carbon is a crucial uncertainty in future land–atmosphere projections. Existing remote sensing approaches for understanding carbon cycling will hopefully continue to be refined, and new remote sensing approaches need to be developed to more directly assess carbon-related pools and processes in northern high-latitude systems.

Acknowledgments

Funding for this work was provided by the U.S. National Science Foundation grants ARC-0531166 and ARC-0902152 and NASA grants NNG6GE00A and NNX09AK56G. Jerrica Frazier contributed to the production of figures.

References

Alcaraz-Segura, D., E. Chuvieco, H. E. Epstein, E. Kasischke, and A. Trishchenko. 2010. The remotely sensed greening versus browning of the North American boreal forest. *Global Change Biology* 16:760–770.

Allen, J. L., and B. Sorbel. 2008. Assessing the differenced normalized burn ratio's ability to map burn severity in the boreal forest and tundra ecosystems of Alaska's national parks. *International Journal of Wildland Fire* 17:463–475.

Amiro, B. D., J. I. MacPherson, R. L. Desjardins, J. M. Chen, and J. Liu. 2003. Post-fire carbon dioxide fluxes in western Canadian boreal forest: Evidence from towers, aircraft and remote sensing. *Agricultural and Forest Meteorology* 115:91–107.

Baird, R. A., D. Verbyla, and T. N. Hollingsworth. 2012. Browning of the landscape of interior Alaska based on 1986–2009 Landsat sensor NDVI. *Canadian Journal of Forest Research* 42:1371–1382.

Barrett, K., E. S. Kasischke, A. D. McGuire, M. R. Turetsky, and E. S. Kane. 2010. Modeling fire severity in black spruce stands in the Alaskan boreal forest using spectral and non-spectral geospatial data. *Remote Sensing of the Environment* 114:1494–1503.

Barrett, K., A. D. McGuire, E. E. Hoy, and E. S. Kasischke. 2011. Potential shifts in dominant forest cover in interior Alaska driven by variations in fire severity. *Ecological Applications* 21:2380–2396.

Bartsch, A., R. A. Kidd, C. Pathe, K. Scipal, and W. Wagner. 2007. Satellite radar imagery for monitoring inland wetlands in boreal and sub-arctic environments. *Aquatic Conservation—Marine and Freshwater Ecosystems* 17:305–317.

Bartsch, A., C. Pathe, K. Scipal, and W. Wagner. 2008. Detection of permanent open water surfaces in central Siberia with ENVISAT ASAR wide swath data with species emphasis on the estimation of methane fluxes from tundra wetlands. *Hydrology Research* 39:89–100.

Beck, P. S. A., and S. J. Goetz. 2011. Satellite observations of high northern latitude vegetation productivity changes between 1982 and 2008: Ecological variability and regional differences. *Environmental Research Letters* 6:045501.

Beck, P. S. A., and S. J. Goetz. 2012. Corrigendum: Satellite observations of high northern latitude vegetation productivity changes between 1982 and 2008: Ecological variability and regional differences. *Environmental Research Letters* 7:029501.

Bhatt, U. S., D. A. Walker, M. K. Raynolds, et al. 2010. Circumpolar Arctic tundra vegetation change is linked to sea ice decline. *Earth Interactions* 14:8.

Boyd, D. S., and F. M. Danson. 2005. Satellite remote sensing of forest resources: Three decades of research development. *Progress in Physical Geography* 29:1–26.

Bubier, J., T. Moore, K. Savage, and P. Crill. 2005. A comparison of methane flux in a boreal landscape between a dry and a wet year. *Global Biogeochemical Cycles* 19:GB1023.

Bunn, A. G., and S. J. Goetz. 2006. Trends in satellite-observed circumpolar photosynthetic activity from 1982 to 2003: The influence of seasonality, cover type, and vegetation density. *Earth Interactions* 10:12.

Bunn, A. G., S. J. Goetz, and G. J. Fiske. 2005. Observed and predicted responses of plant growth to climate across Canada. *Geophysical Research Letters* 32:L16710.

Bunn, A. G., S. J. Goetz, J. S. Kimball, and K. Zhang. 2007. Northern high-latitude ecosystems respond to climate change. *EOS* 88:333–335.

Casper, P., S. C. Maberly, G. H. Hall, and B. J. Finlay. 2000. Fluxes of methane and carbon dioxide from a small productive lake to the atmosphere. *Biogeochemistry* 49:1–19.

Chen, J. M., W. M. Ju, J. Cihlar, et al. 2003. Spatial distribution of carbon sources and sinks in Canada's forests. *Tellus Series B—Chemical and Physical Meteorology* 55:622–641.

Conrad, S. G., A. I. Sukhinin, B. J. Stocks, D. R. Cahoon, E. P. Davidenko, and G. A. Ivanova. 2002. Determining effects of area burned and fire severity on carbon cycling and emissions in Siberia. *Climatic Change* 55:197–211.

de Groot, W. J., R. Landry, W. A. Kurz, et al. 2007. Estimating direct carbon emissions from Canadian wildland fires. *International Journal of Wildland Fire* 16:593–606.

Dong, J. R., R. K. Kaufmann, R. B. Myneni, et al. 2003. Remote sensing estimates of boreal and temperate forest woody biomass: Carbon pools, sources, and sinks. *Remote Sensing of the Environment* 84:393–410.

Drezet, P. M. L., and S. Quegan. 2007. Satellite-based radar mapping of British forest age and Net Ecosystem Exchange using ERS tandem coherence. *Forest Ecology and Management* 238:65–80.

Dye, D. G., and C. J. Tucker. 2003. Seasonality and trends of snow-cover, vegetation index, and temperature in northern Eurasia. *Geophysical Research Letters* 30:1405.

Epp, H., and R. Lanoville. 1996. Satellite data and geographic information systems for fire and resource management in the Canadian Arctic. *Geocarto International* 11:97–103.

Epstein, H. E., M. K. Raynolds, D. A. Walker, U. S. Bhatt, C. J. Tucker, and J. E. Pinzon. 2012. Dynamics of aboveground phytomass of the circumpolar Arctic tundra during the past three decades. *Environmental Research Letters* 7:015506.

Euskirchen, E. S., A. D. McGuire, F. S. Chapin, S. Yi, and C. C. Thompson. 2009. Changes in vegetation in northern Alaska under scenarios of climate change, 2003–2010: Implications for climate feedbacks. *Ecological Applications* 19:1022–1043.

Fraser, R. H., R. J. Hall, R. Landry, et al. 2004. Validation and calibration of Canada-wide coarse-resolution satellite burned-area maps. *Photogrammetric Engineering and Remote Sensing* 70:451–459.

Fraser, R. H., I. Olthof, M. Carriére, A. Deschamps, and D. Pouliot. 2011. Detecting long-term changes to vegetation in northern Canada using the Landsat satellite image archive. *Environmental Research Letters* 6:045502.

French, N. H. F., E. S. Kasischke, R. J. Hall, et al. 2008. Using Landsat data to assess fire and burn severity in the North American boreal forest region: An overview and summary of results. *International Journal of Wildland Fire* 17:443–462.

Frohn, R. C., K. M. Hinkel, and W. R. Eisner. 2005. Satellite remote sensing classification of thaw lakes and drained thaw lake basins on the North Slope of Alaska. *Remote Sensing of the Environment* 97:116–126.

Frolking, S., M. W. Palace, D. B. Clark, J. Q. Chambers, H. H. Shugart, and G. C. Hurtt. 2009. Forest disturbance and recovery: A general review in the context of space-borne remote sensing of impacts on aboveground biomass and canopy structure. *Journal of Geophysical Research—Biogeosciences* 114:G00E02.

Fuchs, H., P. Magdon, C. Kleinn, and H. Flessa. 2009. Estimating aboveground carbon in a catchment of the Siberian forest tundra: Combining satellite imagery and field inventory. *Remote Sensing of the Environment* 113:518–531.

Gamon, J. A., K. F. Huemmrich, D. R. Peddle, et al. 2004. Remote sensing in BOREAS: Lessons learned. *Remote Sensing of the Environment* 89:139–162.

Giglio, L., T. Loboda, D. P. Roy, B. Quayle, and C. O. Justice. 2009. An active-fire based burned area mapping algorithm for the MODIS sensor. *Remote Sensing of the Environment* 113:408–420.

Giglio, L., J. T. Randerson, G. van der Werf, et al. 2010. Assessing variability and long-term trends in burned area by merging multiple satellite fire products. *Biogeosciences* 7:1171–1186.

Goetz, S. J., A. G. Bunn, G. J. Fiske, and R. A. Houghton. 2005. Satellite-observed photosynthetic trends across boreal North America associated with climate and fire disturbance. *Proceedings of the National Academy of Sciences of the United States of America* 102:13521–13525.

Goetz, S. J., H. E. Epstein, U. S. Bhatt, et al. 2011. Vegetation productivity and disturbance changes across arctic northern Eurasia: Satellite observations and simulation modeling. In *Eurasian Arctic land cover and land use in a changing climate*, eds. G. Gutman, P. Groisman, and A. Reissell, 9–36. Berlin: Springer-Verlag.

Goetz, S. J., and S. D. Prince. 1996. Remote sensing of net primary production in boreal forest stands. *Agricultural and Forest Meteorology* 78:149–179.

Goetz, S. J., S. D. Prince, S. N. Goward, M. M. Thawley, J. Small, and A. Johnston. 1999. Mapping net primary production and related biophysical variables with remote sensing: Applications to the BOREAS region. *Journal of Geophysical Research—Atmospheres* 104:22719–27734.

Hall, F. G., D. R. Peddle, and E. F. Ledrew. 1996. Remotes sensing of biophysical variables in boreal forest stands of *Picea mariana*. *International Journal of Remote Sensing* 17:3077–3081.

Hall, R. J., R. S. Skakum, E. J. Arsenault, and B. S. Case. 2006. Modeling forest stand structure attributes using Landsat ETM+ data: Application to mapping of aboveground biomass and stand volume. *Forest Ecology and Management* 225:378–390.

Haynes-Young, R., and M. Potschin. 2010. The links between biodiversity, ecosystem services and human well-being. In *Ecosystem ecology: A new synthesis*, eds. D. G. Rafaelli and C. L. J. Frid, 110–139. Cambridge, UK: Cambridge University Press.

Hinkel, K. M., R. C. Frohn, F. E. Nelson, W. R. Eisner, and R. A. Beck. 2005. Morphometric and spatial analysis of thaw lakes and drained thaw lake basins in the western Arctic Coastal Plain, Alaska. *Permafrost and Periglacial Processes* 16:327–341.

Hope, A. S., J. S. Kimball, and D. A. Stow. 1993. The relationship between tussock tundra spectral reflectance properties and biomass and vegetation composition. *International Journal of Remote Sensing* 14:1861–1874.

Huemmrich, K. F., G. Kinoshita, J. A. Gamon, S. Houston, H. Kwon, and W. C. Oechel. 2010. Tundra carbon balance under varying temperature and moisture regimes. *Journal of Geophysical Research—Biogeosciences* 115:G00I02.

Hugelius, G. 2012. Spatial upscaling using thematic maps: An analysis of uncertainties in permafrost soil carbon estimate. *Global Biogeochemical Cycles* 26:GB2026.

Isaev, A. S., G. N. Korovin, S. A. Bartalev, et al. 2002. Using remote sensing for assessment of forest wildfire carbon emissions. *Climatic Change* 55:231–255.

Jia, G. S., H. E. Epstein, and D. A. Walker. 2003. Greening of Arctic Alaska, 1981–2001. *Geophysical Research Letters* 30:2067.

Jia, G. S., H. E. Epstein, and D. A. Walker. 2006. Spatial heterogeneity of tundra vegetation response to recent temperature changes. *Global Change Biology* 12:42–55.

Jia, G. S., H. E. Epstein, and D. A. Walker. 2009. Vegetation greening in the Canadian arctic related to decadal warming. *Journal of Environmental Monitoring* 11:2231–2238.

Jones, L. A., J. S. Kimball, K. C. McDonald, S. T. K. Chan, E. G. Njoku, and W. C. Oechel. 2007. Satellite microwave remote sensing of boreal and arctic soil temperatures from AMSR-E. *IEEE Transactions on Geoscience and Remote Sensing* 45:2004–2018.

Kasischke, E. S., and E. E. Hoy. 2012. Controls on carbon consumption during Alaskan wildland fires. *Global Change Biology* 18:685–699.

Kasischke, E. S., E. Hyer, P. Novelli, et al. 2005. Influences of boreal fire emissions on Northern Hemisphere atmospheric carbon and carbon monoxide. *Global Biogeochemical Cycles* 19:GB1012.

Kimball, J. S., L. A. Jones, K. Zhang, F. A. Heinsch, K. C. McDonald, and W. C. Oechel. 2009. A satellite approach to estimate land-atmosphere CO_2 exchange for boreal and arctic biomes using MODIS and AMSR-E. *IEEE Transactions on Geoscience and Remote Sensing* 47:569–587.

Kimball, J. S., M. Zhao, A. D. McGuire, et al. 2007. Recent climate-drive increases in vegetation productivity for the western Arctic: Evidence of an acceleration of the northern terrestrial carbon cycle. *Earth Interactions* 11:4.

Krankina, O. N., R. A. Houghton, M. E. Harmon, et al. 2005. Effects of climate, disturbance, and species on forest biomass across Russia. *Canadian Journal of Forest Research* 35:2281–2293.

Kushida, K., Y. Kim, N. Tanaka, and M. Fukida. 2004. Remote sensing of net ecosystem productivity based on component spectrum and soil respiration observation in a boreal forest, interior Alaska. *Journal of Geophysical Research—Atmospheres* 109:D06108.

Leboeuf, A., A. Beaudoin, R. A. Fournier, L. Guindon, J. E. Luther, and M. C. Lambert. 2007. A shadow fraction method for mapping biomass of northern boreal black spruce forests using QuickBird imagery. *Remote Sensing of the Environment* 110:488–500.

Lee, H., E. A. G. Schuur, K. S. Inglett, M. Lavoie, and J. P. Chanton. 2012. The rate of permafrost carbon release under aerobic and anaerobic conditions and its potential effects on climate. *Global Change Biology* 18:515–527.

Lefsky, M. A., W. B. Cohen, D. J. Harding, G. G. Parker, S. A. Acker, and S. T. Gower. 2002. Lidar remote sensing of above-ground biomass in three biomes. *Global Ecology and Biogeography* 11:393–399.

Lewis, S. A., A. T. Hudak, R. D. Ottmar, et al. 2011. Using hyperspectral imagery to estimate forest floor consumption from wildfire in boreal forests of Alaksa, USA. *International Journal of Wildland Fire* 2:255–271.

Li, Z., S. Nadon, and J. Cihlar. 2000. Satellite-based detection of Canadian boreal forest fires: Development and application of the algorithm. *International Journal of Remote Sensing* 21:3057–3069.

Liu, J., J. M. Chen, J. Cihlar, and W. Chen. 1999. Net primary productivity distribution in the BOREAS region from a process model using satellite and surface data. *Journal of Geophysical Research—Atmospheres* 104:27735–27754.

Loboda, T., K. J. O'Neal, and I. Csiszar. 2007. Regionally adaptable dNBR-based algorithm for burned area mapping from MODIS data. *Remote Sensing of the Environment* 109:429–442.

Loboda, T. V., and I. A. Csiszar. 2007. Reconstruction of fire spread within wildland fire events in Northern Eurasia from the MODIS active fire product. *Global and Planetary Change* 56:258–273.

Loboda, T. V., E. E. Hoy, L. Giglio, and E. S. Kasischke. 2011. Mapping burned area in Alaska using MODIS data: A data limitations-driven modification to the regional burned area algorithm. *International Journal of Wildland Fire* 20:487–496.

Loboda, T. V., Z. Zhang, K. J. O'Neal, et al. 2012. Reconstructing disturbance history using satellite-based assessment of the distribution of land cover in the Russian Far East. *Remote Sensing of the Environment* 118:241–248.

Loranty, M. M., S. J. Goetz, E. B. Rastetter, et al. 2011. Scaling an instantaneous model of tundra NEE to the Arctic landscape. *Ecosystems* 14:76–93.

Lutz, D. A., R. A. Washington-Allen, and H. H. Shugart. 2008. Remote sensing of boreal forest biophysical and inventory parameters: A review. *Canadian Journal of Remote Sensing* 34:S286–S313.

Mack, M. C., M. S. Bret-Harte, T. N. Hollingsworth, et al. 2011. Carbon loss from an unprecedented Arctic tundra wildfire. *Nature* 475:489–492.

McGuire, A. D., L. G. Anderson, T. R. Christenson, et al. 2009. Sensitivity of the carbon cycling in the Arctic to climate change. *Ecological Monographs* 79:523–555.

McMichael, C. E., A. S. Hope, D. A. Stow, J. B. Fleming, G. Vourlitis, and W. Oechel. 1999. Estimating CO_2 exchange at two sites in Arctic tundra ecosystems during the growing season using a spectral vegetation index. *International Journal of Remote Sensing* 20:683–698.

Meroni, M., D. Mollicone, L. Belelli, et al. 2002. Carbon and water exchanges of regenerating forests in central Siberia. *Forest Ecology and Management* 169:115–122.

Michalek, J. L., N. H. F. French, E. S. Kasischke, R. D. Johnson, and J. E. Colwell. 2000. Using Landsat TM data to estimate carbon release from burned biomass in an Alaskan spruce complex. *International Journal of Remote Sensing* 21:323–338.

Morgenstern, A., G. Grosse, F. Gunther, I. Federova, and L. Schirmeister. 2011. Spatial analyses of thermokarst lakes and basins in Yedoma landscapes of the Lena Delta. *Cryosphere* 5:849–867.

Myneni, R. B., J. Dong, C. J. Tucker, et al. 2001. A large carbon sink in the woody biomass of Northern forests. *Proceedings of the National Academy of Sciences of the United States of America* 98:14784–14789.

Myneni, R. B., C. D. Keeling, C. J. Tucker, G. Asrar, and R. R. Nemani. 1997. Increased plant growth in the northern high latitudes from 1981 to 1991. *Nature* 386:698–702.

Natali, S. M., E. A. G. Schuur, and R. L. Rubin. 2012. Increased plan productivity in Alaska as a result of experimental warming of soil and permafrost. *Journal of Ecology* 100:488–498.

Neigh, C. S. R., C. J. Tucker, and J. R. G. Townshend. 2008. North American vegetation dynamics observed with multi-resolution satellite data. *Remote Sensing of the Environment* 112:1749–1772.

Olthof, I., D. Pouliot, R. Latifovic, and W. Chen. 2008. Recent (1986–2006) vegetation-specific NDVI trends in northern Canada from satellite data. *Arctic* 61:381–394.

Palacios-Orueta, A., E. Chuvieco, A. Parra, and C. Carmona-Moreno. 2005. Biomass burning emissions: A review of models using remote-sensing data. *Environmental Monitoring and Assessment* 104:189–209.

Parent, M. B., and D. Verbyla. 2010. The browning of Alaska's boreal forest. *Remote Sensing* 2:2729–2747.

Plug, L. J., C. Walls, and B. M. Scott. 2008. Tundra lake changes from 1978 to 2001 on the Tuktoyaktuk Peninsula, western Canadian Arctic. *Geophysical Research Letters* 35:L03502.

Pouliot, D., R. Latifovic, and I. Olthof. 2009. Trends in vegetation NDVI from 1 km AVHRR data over Canada for the period 1985–2006. *International Journal of Remote Sensing* 30:149–168.

Prince, S. D., and S. N. Goward. 1995. Global primary production: A remote sensing approach. *Journal of Biogeography* 22:815–835.

Raynolds, M. K., D. A. Walker, H. E. Epstein, J. E. Pinzon, and C. J. Tucker. 2012. A new estimate of tundra-biome phytomass from trans-Arctic field data and AVHRR NDVI. *Remote Sensing Letters* 5:403–411.

Richardson, A. D., T. A. Black, P. Ciais, et al. 2010. Influence of spring and autumn phenological transitions on forest ecosystem productivity. *Philosophical Transactions of the Royal Society B—Biological Sciences* 365:3227–3246.

Rocha, A. V., M. M. Loranty, P. E. Higuera, et al. 2012. The footprint of Alaskan tundra fires during the past half-century: Implications for surface properties and radiative forcing. *Environmental Research Letters* 7:044039.

Rocha, A. V., and G. S. Shaver. 2011. Burn severity influences postfire CO_2 exchange in arctic tundra. *Ecological Applications* 21:477–489.

Ropars, P., and S. Boudreau. 2012. Shrub expansion at the forest-tundra ecotone: Spatial heterogeneity linked to local topography. *Environmental Research Letters* 7:015502.

Schuur, E. A. G., J. Bockheim, J. G. Canadell, et al. 2008. Vulnerability of permafrost carbon to climate change: Implications for the global carbon cycle. *BioScience* 58:701–714.

Shippert, M. M., D. A. Walker, N. A. Auerbach, and B. E. Lewis. 1995. Biomass and leaf-area index maps derived from SPOT images for Toolik Lake and Imnavait Creek areas, Alaska. *Polar Record* 31:147–154.

Sitch, S., A. D. McGuire, J. Kimball, et al. 2007. Assessing the carbon balance of circumpolar Arctic tundra using remote sensing and process modeling. *Ecological Applications* 17:213–234.

Slayback, D. A., J. E. Pinzon, S. O. Los, and C. J. Tucker. 2003. Northern hemisphere photosynthetic trends 1982–99. *Global Change Biology* 9:1–15.

Smith, B., W. Knorr, J. L. Widlowski, B. Pinty, and N. Gobron. 2008. Combining remote sensing data with process modeling to monitor boreal conifer forest carbon balances. *Forest Ecology and Management* 255:3985–3994.

Smith, N. V., S. S. Saatchi, and J. T. Randerson. 2004. Trends in high northern latitude soil freeze and thaw cycles from 1988 to 2002. *Journal of Geophysical Research—Atmospheres* 109:D12101.

Soegaard, H., C. Nordstroem, T. Friborg, B. U. Hansen, T. R. Christensen, and C. Bay. 2000. Trace gas exchange in a high-Arctic valley 3. Integrating and scaling CO_2 fluxes from canopy to landscape using flux data, footprint modeling, and remote sensing. *Global Biogeochemical Cycles* 14:725–744.

Soja, A. J., A. I. Sukhinin, D. R. Cahoon, H. H. Shugart, and P. W. Stackhouse. 2004. AVHRR-derived fire frequency, distribution and area burned in Siberia. *International Journal of Remote Sensing* 25:1939–1960.

Steyaert, L. T., F. G. Hall, and T. R. Loveland. 1997. Land cover mapping, fire regeneration, and scaling studies in the Canadian boreal forest with 1 km AVHRR and Landsat TM data. *Journal of Geophysical Research—Atmospheres* 102:29581–29598.

Stinson, G., W. A. Kurz, C. E. Smyth, et al. 2011. An inventory-based analysis of Canada's managed forest carbon dynamics, 1990 to 2008. *Global Change Biology* 17:2227–2244.

Stow, D., A. Hope, W. Boynton, S. Phinn, D. Walker, and N. Auerbach. 1998. Satellite-derived vegetation index and cover type maps for estimating carbon dioxide flux for arctic tundra regions. *Geomorphology* 21:313–327.

Stow, D. A., A. Hope, D. McGuire, et al. 2004. Remote sensing of vegetation and land-cover change in Arctic Tundra Ecosystems. *Remote Sensing of the Environment* 89:281–308.

Sturm, M., C. Racine, and K. Tape. 2001. Climate change—Increasing shrub abundance in the Arctic. *Nature* 411:546–547.

Sukhinin, A. I., N. H. F. French, E. S. Kasischke, et al. 2004. AVHRR-based mapping of fires in Russia: New products for fire management and carbon cycle studies. *Remote Sensing of the Environment* 93:546–564.

Tagesson, T., M. Mastepanov, M. P. Tamstorf, et al. 2012. High-resolution satellite data reveal an increase in peak growing season gross primary production in a high-Arctic wet tundra ecosystem 1992–2008. *International Journal of Applied Earth Observation and Geoinformation* 18:407–416.

Tagesson, T., B. Smith, A. Lofgren, A. Rammig, L. Eklundh, and A. Lindroth. 2009. Estimating net primary production of Swedish forest landscapes by combining mechanistic modeling and remote sensing. *Ambio* 38:316–324.

Tape, K., M. Sturm, and C. Racine. 2006. The evidence for shrub expansion in Northern Alaska and the Pan-Arctic. *Global Change Biology* 12:686–702.

Tarnocai, C., J. G. Canadell, E. A. G. Schuur, P. Kuhry, G. Mazhitova, and S. Zimov. 2009. Soil organic carbon pools in the northern circumpolar permafrost region. *Global Biogeochemical Cycles* 23:GB2023.

Tremblay, B., E. Lévesque, and S. Boudreau. 2012. Recent expansion of erect shrubs in Low Arctic: Evidence from Eastern Nunavik. *Environmental Research Letters* 7:035501.

Treshnikov, A. F. 1985. *Atlas of the Arctic* (in Russian). Moscow: Administrator of Geodesy and Cartography of the Soviet Ministry.

Trucco, C., E. A. G. Schuur, S. M. Natali, E. F. Belshe, R. Bracho, and J. Vogel. 2012. Seven-year trends of CO_2 exchange in a tundra ecosystem affected by long-term permafrost thaw. *Journal of Geophysical Research—Biogeosciences* 117:G02031.

Tucker, W. B., J. W. Weatherly, D. T. Eppler, L. D. Farmer, and D. L. Bently. 2001. Evidence for rapid thinning of sea ice in the western Arctic Ocean at the end of the 1980s. *Geophysical Research Letters* 28:2851–2854.

Turetsky, M. R., E. S. Kane, J. W. Harden, et al. 2011. Recent acceleration of biomass burning and carbon losses in Alaskan forests and peatlands. *Nature Geosciences* 4:27–31.

Turner, D. P., W. D. Ritts, W. B. Cohen, et al. 2005. Site-level evaluation of satellite-based global terrestrial gross primary production and net primary production monitoring. *Global Change Biology* 11:666–684.

Ulrich, M., G. Grosse, S. Chabrillat, and L. Schirmeister. 2009. Spectral characterization of periglacial surfaces and geomorphological units in the Arctic Lena Delta using field spectrometry and remote sensing. *Remote Sensing of the Environment* 113:1220–1235.

Verbyla, D. 2008. The greening and browning of Alaska based on 1982–2003 satellite data. *Global Ecology and Biogeography* 17:547–555.

Walker, D. A., H. E. Epstein, G. J. Jia, et al. 2003. Phytomass, LAI, and NDVI in northern Alaska: Relationships to summer warmth, soil pH, plant functional types, and extrapolation to the circumpolar Arctic. *Journal of Geophysical Research—Atmospheres* 108:8169.

Walker, D. A., H. E. Epstein, M. K. Raynolds, et al. 2012. Environment, vegetation and greenness (NDVI) along the North America and Eurasia Arctic transects. *Environmental Research Letters* 7:015504.

Walker, D. A., M. K. Raynolds, F. J. A. Daniels, et al. 2005. The circumpolar Arctic vegetation map. *Journal of Vegetation Science* 16:267–282.

Walter, K. M., M. Engram, C. R. Duguay, M. O. Jeffries, and F. S. Chapin. 2008. The potential use of synthetic aperture radar for estimating methane ebullition from Arctic lake. *Journal of the American Water Resources Association* 44:305–315.

Wang, J. D., Y. W. Sheng, K. M. Hinkel, and E. A. Lyons. 2012. Drained thaw lake basin recovery on the western Arctic Coastal Plain of Alaska using high-resolution digital elevation models and remote sensing imagery. *Remote Sensing of the Environment* 119:325–336.

Way, J., R. Zimmerman, E. Rignot, K. McDonald, and R. Oren. 1997. Winter and spring thaw as observed with imaging radar at BOREAS. *Journal of Geophysical Research—Atmospheres* 102:29673–29684.

Williams, M., E. B. Rastetter, G. R. Shaver, J. E. Hobbie, E. Carpino, and B. L. Kwiatkowski. 2001. Primary production of an arctic watershed: An uncertainty analysis. *Ecological Applications* 11:1800–1816.

Zhang, K., J. S. Kimball, E. H. Hogg, et al. 2008. Satellite-based detection of recent climate-driven changes in northern high-latitude vegetation productivity. *Journal of Geophysical Research—Biogeosciences* 113:G03033.

Zhang, K., J. S. Kimball, M. S. Zhao, W. C. Oechel, J. Cassano, and S. W. Running. 2007. Sensitivity of pan-Arctic terrestrial net primary productivity simulations to daily surface meteorology from NCEP-NCAR and ERA-40 reanalyses. *Journal of Geophysical Research—Biogeosciences* 112:G01011.

Zheng, D., S. Prince, and T. Hame. 2004. Estimating net primary production of boreal forests in Finland and Sweden from field data and remote sensing. *Journal of Vegetation Science* 15:161–170.

Zhou, L. M., C. J. Tucker, R. K. Kaufmann, D. Slayback, N. V. Shabanov, and R. B. Myneni. 2001. Variations in northern vegetation activity inferred from satellite data of vegetation index during 1981 to 1999. *Journal of Geophysical Research—Atmospheres* 106:20069–20083.

5

Monitoring the Ecosystem Service of Forage Production

J. G. N. Irisarri and M. Oesterheld
University of Buenos Aires, Argentina

M. Oyarzabal and J. M. Paruelo
University of Buenos Aires, Argentina

M. Durante
National Institute of Agricultural Technology (INTA), Argentina

CONTENTS

5.1 Introduction..87
5.2 ANPP Estimation through Successive Biomass Harvests....................89
5.3 ANPP Estimation through Remote Sensing..92
5.4 RUE Estimation through ANPP and APAR..93
    5.4.1 ANPP Estimation..94
    5.4.2 APAR Estimation 94
    5.4.3 RUE Estimation...95
5.5 Forage Monitoring System Based on Remote Sensing..........................98
5.6 Other Uses of Remote Sensing for Livestock Systems 99
5.7 Conclusions...99
Acknowledgments 100
References..100

5.1 Introduction

Worldwide, 80% of the energy required by cattle to reach market weight is derived from rangelands and pastures (Wheeler et al. 1981; Oltjen and Beckett 1996). Managing these forage resources requires knowing their production. Forage production, also known as forage growth rate, is a portion of the aboveground net primary production (ANPP), a key ecosystem variable. At the individual level, ANPP represents the difference between photosynthesis and respiration (Chapin et al. 2002), whereas at the landscape level, ANPP represents the rate of biomass production per unit of area and time. At the ecosystem level, ANPP represents the rate of production of available energy for primary consumers.

ANPP is an integrator of ecosystem functioning and determines the input level of many ecosystem services (Costanza et al. 1997). For example, the availability of information on ANPP for pastures and rangelands is critical to establish adequate stocking rates (Oesterheld et al. 1998) and to manage excesses or deficits of forage. Considering the high relevance of ANPP in rangeland management, few data on its spatial and temporal variation are available. This lack of information is due to the difficulty in estimating ANPP in the field and in extrapolating the few available data to other spatial or temporal situations. ANPP models derived from remote sensing data (Paruelo et al. 1997, 2000; Running et al. 2000; Piñeiro et al. 2006) can tackle this lack of information because they cover large areas and long periods.

ANPP varies at different scales in space and time. Across wide resource gradients, ANPP consistently increases with precipitation (Lauenroth 1979; Sala et al. 1988). Within a landscape, ANPP widely varies with regard to topographic gradients (Knapp et al. 1993; Buono et al. 2010; Irisarri et al. 2012), edaphic differences (Aragón and Oesterheld 2008), or disturbance frequency (Knapp et al. 1993; Oesterheld et al. 1999). The interannual variation of ANPP of steppes, grasslands, or savannas is large and generates the greatest uncertainties for range management (Lauenroth and Sala 1992; Knapp and Smith 2001; Bai et al. 2008). The seasonal variation of ANPP is also large and restricts management. For example, in temperate areas of the Río de la Plata grasslands, spring ANPP is 10 times larger than winter's (Sala et al. 1981; Oesterheld and León 1987; Altesor et al. 2005; Semmartin et al. 2007).

The objectives of this chapter are (1) to introduce two approaches for ANPP estimations—the classic way through biomass harvests and a more recent practice based on remote sensing data, and (2) to describe a forage production monitoring system that routinely observes ranches of Argentina and Uruguay based on the second approach. In this chapter, we first describe the logic and difficulties involved in estimating ANPP through sequential biomass harvests. Second, we describe the logic of an approach based on remote sensing. Third, we discuss in some detail the estimation of radiation use efficiency (RUE). Finally, we present a monitoring system of forage production currently functioning in Argentina and Uruguay.

5.2 ANPP Estimation through Successive Biomass Harvests

ANPP results from the transformation of incoming photosynthetic active radiation (PAR) into biomass. The generated biomass may either be consumed by herbivores or may follow the successive steps of the detritus flow: standing dead biomass, litter, and organic matter, with the potential of being decomposed in any of those steps (Figure 5.1).

Clipping biomass in order to estimate ANPP is the most frequent, and at the same time, the most disputed method (Singh et al. 1975; McNaughton et al. 1996; Sala and Austin 2000; Scurlock et al. 2002; Knapp et al. 2007). The objective is to capture the accumulated biomass, a consequence of vegetation growth, in a certain period of time. Using this approach involves at least two difficulties. One difficulty is that the same data may result in different

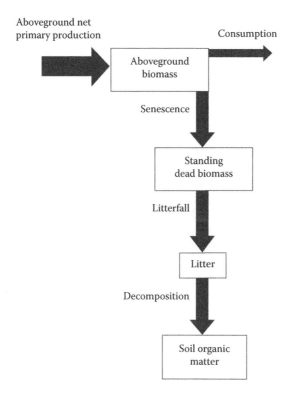

FIGURE 5.1
Diagram of the flows and compartments related to biomass generation and transformation in pastures or grasslands. The size of the arrows and boxes does not represent the size of flows and state variables. Losses of biomass through respiration or other process such as leaching or photodegradation (Austin and Vivanco 2006) are not considered in this model.

estimations of ANPP depending on the initial assumptions on the magnitude of some of the flows (Scurlock et al. 2002). For example, it is possible to consider ANPP as a single live biomass peak, or it can include other flows such as senescence or decomposition (Sala and Austin 2000). Depending on these assumptions, it is possible to generate widely different ANPP estimations. For example, based on the same data, the ANPP of grasslands varies between 200 to 400 g m^{-2} y^{-1} and 350 to 1000 g m^{-2} y^{-1} depending on the initial assumptions that are made (Scurlock et al. 2002).

The second difficulty is that many rangelands and pastures do not have a short growing season compatible with the assumption that single-peak biomass equates ANPP (Semmartin et al. 2007). As a result, ANPP must be quantified on the basis of at least two successive harvests, one at the beginning of the period under study, and the other at the end of it (Sala and Austin 2000). This situation necessarily involves quantifying the different boxes in Figure 5.1. The changes in the sizes of the boxes, between two consecutive dates, will represent an estimation of the size of the arrows (fluxes).

Under certain circumstances, it may be difficult to capture ANPP using the successive biomass harvest approach. When consumption and senescence equal zero, the biomass differences between two successive dates clearly represent ANPP. This would happen if consumption is suppressed after a very intense defoliation and the period between harvests is short enough to keep senescence at minimum levels. As a result, when trying to estimate ANPP through successive biomass harvests, grazing must be excluded (Sala and Austin 2000). As mentioned earlier, it is also necessary to discriminate among live biomass per species, standing dead biomass, and litter (Oesterheld and León 1987; Sala and Austin 2000). These compartments are the basis for estimating some of the flows described in Figure 5.1. However, quantifying separate compartments of biomass demands a bigger sampling effort, and there are always some flows that cannot be totally estimated (senescence or decomposition).

As more partial fluxes need quantification, the sampling challenge grows. ANPP represents the combination of other variables, at least biomass values of two consecutive dates. Thus, ANPP variance is the sum of biomass variances from two consecutive dates, assuming no covariance. If standing dead or litter is included, their variances must also be included. Furthermore, biomass variance is generally high in grazed grasslands at the spatial resolution of most sampling procedures (<1 m^2). Therefore, biomass estimations usually have low precision (Knapp et al. 2007).

The logic and difficulties that the successive harvest method represents can be explained by an analogy with the estimation of incoming water flow into a tank, from which a cow can drink water freely (Figure 5.2a). In this analogy, the incoming flow represents ANPP, and the water volume at a certain moment represents biomass. If the water tank does not have any outgoing flow, the incoming flow may be estimated through volume differences between two consecutive periods. However, estimation of incoming flow faces problems from three different sources. First, the tank loses water

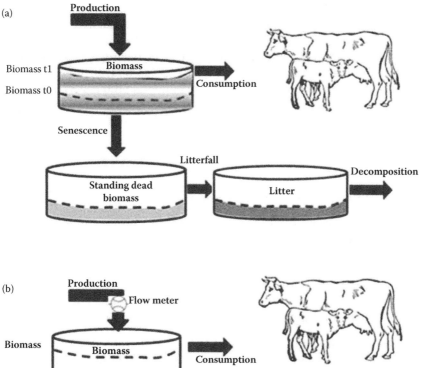

FIGURE 5.2
Diagram illustrating the type of problems encountered when estimating ANPP using an analogy of estimating the incoming water flow into a tank, where cows can drink freely. (a) For successive harvests, incoming water is measured as the difference in volume for two moments and (b) using remote sensing, incoming water is estimated by attaching a flow meter to the incoming pipe. Water level represents biomass at a given time. The water inflow in the first tank represents ANPP. Each tank represents a certain state of biomass: green biomass (where the principal incoming flow is ANPP), standing dead (where the incoming flow is senescence), and litter (where the incoming flow is litterfall). The difficulties related with the successive harvests approach are that water leaks through the "holes" of consumption, senescence, and decomposition. Thus, the cow must be excluded—with a subsequent loss of realism. Seasonal and spatial variations are analogous to water movements and increase the difficulty of mea-suring the water volume. The difficulties related with the remote sensing approach are the accuracy of the flow meter (the temporal and spatial resolutions of images) and the need to calibrate the flow meter readings into water flow rate (knowing radiation use efficiency).

through the "holes" of senescence and decomposition. Second, the cow must be excluded to eliminate water loss due to drinking. Inevitably, this leads to a loss of realism (Figure 5.2a). Third, it is very difficult to measure the volume with high precision. This sampling problem may be regarded as if water were in continuous agitation and there were only a few seconds to make measurements. The major consequence of the aforementioned problems is that ANPP estimations for rangeland areas are rare in space and time, and these estimations are frequently inaccurate (Singh et al. 1975, Scurlock et al. 2002).

5.3 ANPP Estimation through Remote Sensing

To understand the approach based on remote sensing, it is necessary to understand the Monteith model (Monteith 1972). This model indicates that ANPP is the product of three components, the photosynthetically active radiation (PAR), the proportion of the incoming radiation that is absorbed by active photosynthetic tissues (fPAR), and the RUE:

$$\text{ANPP} \ (g \cdot m^{-2} \cdot y^{-1}) = \text{PAR} \ (MJ \cdot m^{-2} \cdot y^{-1}) \times \text{fPAR} \times \text{RUE} \ (g \cdot MJ^{-1}) \quad (5.1)$$

The model indicates that new biomass is generated (ANPP) as photosynthetic radiation is absorbed and converted into new tissues. In the special case of ANPP, it is necessary to take into account that RUE is the conversion of absorbed energy into new aboveground biomass, which includes the resource partition between belowground and aboveground biomass.

ANPP estimation through this approach requires solving three problems. First, it is necessary to know the photosynthetic incoming radiation. This is relatively simple because average daily values are available from several sources and for many locations. In addition, the interannual variation of the photosynthetic incoming radiation is low. Therefore, the value for a certain month of the year can be replaced by its average. If more precision is needed, there are models that estimate incoming radiation based on meteorological data, temperature, and precipitation (Thornton and Running 1999).

Second, it is necessary to know the proportion of the incoming radiation that is absorbed by photosynthetically active tissues. Fortunately, remote sensing provides this type of information (see details in Piñeiro et al. 2006). Sensors capable of estimating the reflectance from the red and the near infrared portions of the spectrum allow estimation of the proportion of the incoming photosynthetic radiation absorbed by green vegetation (Sellers et al. 1992).

Third, it is necessary to know the RUE. This is the hardest and the least known factor. RUE variations are associated with photosynthetic pathways (C_3–C_4), the ratio between photosynthesis and respiration associated with life form type, and abiotic factors, such as water, nutrient availability, or temperature. It may be estimated from the literature through calibrations or remote sensing (for details, see Garbulsky et al. 2010). Fortunately, this component is the least variable, both in space and time, at the scales usually relevant for range management (Chapin et al. 2002).

The logic and difficulties of estimating ANPP through remote sensing can be seen using the same water tank analogy (Figure 5.2). In this case, it is not necessary to know the water levels (biomass). The use of remote sensing is analogous to setting a flow meter on the incoming water pipe (Figure 5.2b). In this case, the cow can drink freely because the flow was measured beforehand. Consumption has an effect on ANPP, but this is exactly what remote sensing is estimating through the estimation of fPAR. What are the difficulties for ANPP estimation through remote sensing? The first difficulty is that the flow meter has a maximum resolution related to the pixel size used to estimate fPAR. The maximum spatial resolution is set by the pixel size, so it is impossible to detect differences within a pixel. This could be inconvenient in situations where the managed unit is smaller than the pixel. For example, the pixel area of some Moderate Resolution Imaging Spectroradiometer (MODIS) products (5.3 ha) may be too large for dairy system paddocks yet adequate for extensive rangelands devoted to cow-calf operations.

The second difficulty is estimating RUE. There are two basic strategies for dealing with this limitation. The first strategy is not to utilize RUE at all and only to use the information on absorbed radiation. There are many instances in which the evaluated rangeland units or moments can be compared because it can be assumed that they have similar RUE. For example, pastures with similar species composition and similar landscape position should have similar RUE. Probable differences should be smaller than other sources of variation. In these situations, the first two components of Monteith model, PAR and fPAR, represent the absorbed photosynthetically active radiation, APAR ($MJ\,ha^{-1}\,d^{-1}$), which in many cases could be enough for certain comparisons. The second strategy is to estimate the RUE as noted in the following section.

5.4 RUE Estimation through ANPP and APAR

RUE varies across vegetation types and environmental conditions. It shows major variations among biomes, from $0.27\,g\cdot MJ^{-1}$ for deserts to $0.71\,g\cdot MJ^{-1}$ for tropical forests (Ruimy and Saugier 1994; Field et al. 1995). RUE of semi-arid grassland varies yearly from 0.27 to $0.35\,g\cdot MJ^{-1}$ (Nouvellon et al. 2000). Among seasons, variations are even higher; for example, in humid grasslands, RUE varies between 0.2 and $1.2\,g\cdot MJ^{-1}$ (Piñeiro et al. 2006). RUE also varies across pastures with different species composition and topographic position. For example, cultivated pastures set on uplands dominated by lucerne had higher RUE than those cultivated on lowlands dominated by wheatgrass (Grigera et al. 2007). In grassland areas, changes in species composition through intercropping techniques also affect RUE (Baeza et al. 2011). Grazing modifies RUE, as reported by numerous cases of compensatory growth. All these sources of variation should be considered when trying to estimate ANPP through APAR, as explained earlier.

RUE may be estimated through three approaches (see also Chapter 6). One, derived from Monteith's (1972) equation (see Equation 5.1), requires independent estimates of ANPP and APAR (Turner et al. 2003; Bradford et al. 2005; Piñeiro et al. 2006; Grigera et al. 2007; Irisarri et al. 2012). The other two approaches are based on either meteorological data or spectral indices, such as the photochemical radiation index (Garbulsky et al. 2011) (see Chapter 3). In this chapter, we will focus on the first approach.

The following example represents an estimation of RUE utilizing the first approach. Three independent sources of information were used: successive biomass harvests for ANPP, remote sensing to estimate fPAR, and a meteorological station for PAR. Once the RUE was estimated, and the model calibrated (Equation 5.1), it was used to independently estimate the forage production of several paddocks through remote sensing (Piñeiro et al. 2006; Grigera et al. 2007). The example corresponds to upland cultivated pastures from the southwest end of the humid Pampas region of Argentina (Grigera et al. 2007), composed of a mixture of cool season grasses (*Festuca arundinacea*, *Dactylis glomerata*, and *Lolium multiflorum*) and legumes (*Medicago sativa*, *Trifolium pratense*, and *Trifolium repens*). The data used in this example belongs to a previous study (Grigera et al. 2007). In contrast with that study, here we will develop a step-by-step explanation of the relation between ANPP and each component of the Monteith model.

5.4.1 ANPP Estimation

Biomass harvests were performed from October 2000 to September 2003, across four different paddocks. The harvest design tried to copy the rotational grazing system performed in the area. It involved biomass harvests after a two-month rest period. Within each of the four paddocks, eight 1-m² cages were used in order to restrict grazing. At the beginning of each rest period, aboveground biomass was clipped at a 4-cm height within each cage. At the end of the rest period, the accumulated biomass was harvested to the same height. ANPP was estimated as the biomass harvested during each rest period.

5.4.2 APAR Estimation

APAR was estimated for each paddock and rest period through a combination of MODIS and meteorological data. For each paddock, MODIS pixels (5.3 ha) entirely located on an upland area were selected. Thus, pixels influenced by forestation, construction, or roads were eliminated. The Normalized Difference Vegetation Index (NDVI) was used for fPAR estimation, particularly the

MODIS product: NDVI; Collection 5 of MOD13Q1 (https://lpdaac.usgs.gov/products/modis_products_table/mod13q1, vegetation indices product). This MODIS product consists of a 16-day composite and includes a quality layer used to discard NDVI values. The discarded values were interpolated with consecutive values. Only 1.5% of the NDVI values were interpolated. NDVI was calculated as a weighted average for each rest period and paddock. For example, in order to estimate NDVI for the rest period from 11 January to 14 March, five NDVI values were used. Each NDVI value was weighted considering its temporal overlap for the described period. In turn, NDVI values for the four paddocks were averaged.

The proportion of the incoming photosynthetic radiation absorbed by green vegetation (fPAR, Equation 5.1) was estimated for each rest period as a nonlinear function of NDVI. Potter et al. (1993) proposed an empiric model to translate NDVI values into fPAR. Grigera et al. (2007) calibrated this model for the study area. They assigned 0 fPAR to NDVI values from bare soil areas dedicated to agriculture. The maximum fPAR, 95%, was assigned to NDVI values from highly covered cultivated pastures and wheat fields (LAI > 3). Finally, fPAR at the pixel level was averaged for each paddock. photosynthetically active radiation (PAR) was measured by a weather station, located 5 km to the nearest paddock and 50 km to the farthest (37° 24′S, 61° 26′W, 200 m a.s.l.). PAR of each rest period was estimated as the sum of the total incoming radiation multiplied by 0.48 (McCree 1972). The absorbed photosynthetically active radiation (APAR) was estimated for each paddock as the product between fPAR and PAR (Equation 5.1).

A certain degree of synchronicity was observed between NDVI and ANPP (Figure 5.3a). NDVI explained 55% of ANPP variability ($R^2 = 0.55$, $p < 0.001$, $n = 18$). When compared to NDVI, fPAR increased the explained proportion of ANPP variability ($R^2 = 0.58$, $p < 0.001$, $n = 18$). But certain differences still remained, for example, between February and March 2001 (Figure 5.3b). The seasonal dynamics of PAR and ANPP were slightly similar (Figure 5.3c), indicating that ANPP variations were only partially associated with PAR ($R^2 = 0.25$; $p < 0.03$; $n = 18$). The seasonal dynamics of ANPP and APAR were very similar (Figure 5.3d) (Grigera et al. 2007). This indicates that APAR better described ANPP seasonal variations than NDVI, fPAR, or PAR individually (Figure 5.3). A major consequence of the observed pattern was that APAR explained most of ANPP seasonal variations ($R^2 = 0.87$; $p < 0.0001$; $n = 18$). In contrast, working in a different region, Piñeiro et al. (2006) observed that NDVI described ANPP seasonal variations as closely as APAR. This was related to a strong and positive correlation between NDVI and PAR ($r = 0.56$; $n = 21$) that was absent in the pastures of our example ($r = 0$; $n = 18$).

5.4.3 RUE Estimation

RUE was estimated in two ways. The first way considered the ratio between ANPP and APAR for each date. This approach resulted in a value of RUE

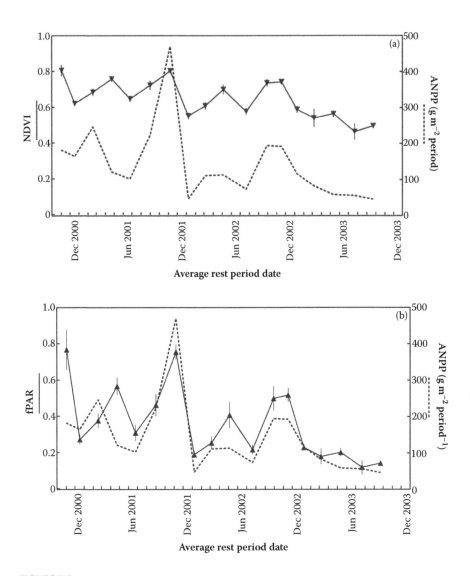

FIGURE 5.3
Seasonal dynamics of aboveground net primary production (ANPP) and some of its compo-
nents for upland pastures from the southwestern Buenos Aires province from October 2000 to
September 2003. (a) Normalized Difference Vegetation Index (NDVI) from MODIS sensor and
ANPP; (b) proportion of the incoming photosynthetic radiation absorbed by green vegetation
(fPAR) and seasonal ANPP; fPAR was estimated from NDVI values through an empiric model.

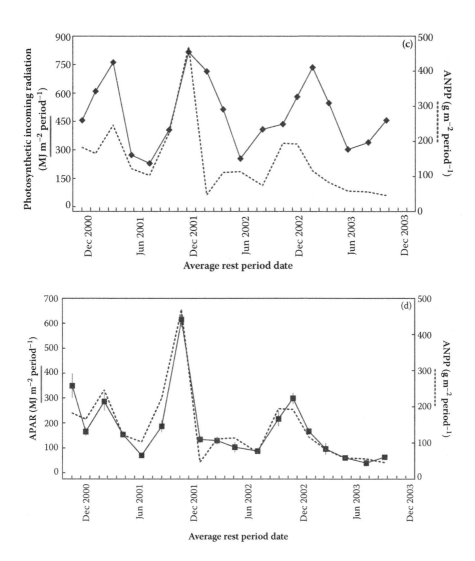

FIGURE 5.3 (*Continued*)
Seasonal dynamics of aboveground net primary production (ANPP) and some of its components for upland pastures from the southwestern Buenos Aires province from October 2000 to September 2003. (c) Photosynthetically active radiation (PAR) was measured by a nearby weather station. Each value represents the accumulated PAR for each rest period and (d) absorbed photosynthetic radiation for upland pastures (adapted from Grigera, G., et al., *Agricultural Systems*, 94, 637–648, 2007). The bars indicate the standard error; when absent, the errors were smaller than the symbols.

for each period that had ANPP and APAR values (Figure 5.3d). The RUE estimated in this way varied between 0.3 g·MJ^{-1} (in summer) and 1.3 g·MJ^{-1} (in winter), with an average value of 0.82 g·MJ^{-1}. The advantage of this first approach is that the seasonal and interannual variations of RUE were captured.

The second way to estimate the RUE is through a regression model, in which ANPP is the predicted variable ("y" axes) and APAR is the observed variable ("x" axes). If the Y-intercept of the model is not different from 0, then the RUE is equal to the slope of the regression model, and the slope should be similar to the average ratio between ANPP and APAR. But, if the Y-intercept is larger than 0, the slope will be an underestimation of the RUE because the regression model predicts a positive ANPP value when APAR is equal to 0. If the Y-intercept is smaller than 0, the slope will be an overestimation of the RUE because the regression model predicts ANPP equals 0 when APAR is positive. In both cases, the estimation error of the RUE is higher at low APAR values than at high values (Verón et al. 2005). As expected, in the case of the pastures, ANPP was positively related to APAR. It explained 87% of ANPP variations ($R^2 = 0.87$; n = 18). The slope of the model was equal to 0.7 g·MJ^{-1} (p < 0.0001) and the Y-intercept was equal to 22.3 g DM. m^{-2}·period^{-1} and was not different from 0 (p = 0.15). In this case, the slope of the regression model represents the RUE (Grigera et al. 2007).

5.5 Forage Monitoring System Based on Remote Sensing

Remote sensing monitoring of fPAR at the ranch scale has been limited both in space (pixel size too big) and time (fPAR values sparse within the growing season). The MODIS sensor onboard NASA Terra and Aqua satellites combines a good spatial resolution for paddock monitoring (~5 ha) with daily data. This type of remote sensing information provided the opportunity to develop an ANPP monitoring system for rangeland areas, in Argentina and Uruguay, based on the aforementioned Monteith model (Paruelo et al. 2000; Piñeiro et al. 2006; Grigera et al. 2007).

This monitoring system is based on three sources of information. The first source is a geographic information system (GIS) that includes ranches and their paddocks. The second source is a database with information from each paddock on its land use, yearly base from 2000, on its area, and on a variable number of pixels that represent it. The third source is a database on PAR, NDVI values, and the RUE calibration models for different types of pasture or rangeland areas and geographic regions (Grigera et al. 2007; Irisarri et al. 2012). These three sources of information are combined to estimate ANPP or APAR (for pastures and rangeland areas where there are no calibrated models) for each paddock. Specific software developed for this purpose combines and stores the three data sources and generates output.

ANPP estimations are generated by the fifteenth day of each month, once meteorological and satellite information of the previous month is available. Results are sent by e-mail to the users and also hosted in a website (http://larfile.agro.uba.ar/lab-sw/sw/gui/Inicial.page). In this site, each specific user has a keyword to access data from his/her ranch. But there is also public data on the average ANPP value for various pastures and rangelands within different regions.

5.6 Other Uses of Remote Sensing for Livestock Systems

There are at least two other variables related to ANPP that represent future challenges for its estimation through remote sensing techniques: the proportion of ANPP that corresponds to forage species and the nutritive value of those forage species. As was mentioned before, at the ranch scale, seasonal ANPP estimation through remote sensing was achieved because limitations associated with fPAR estimation were removed.

Several examples demonstrated that forage species identification (Schmidt and Skidmore 2001, 2003; Mutanga et al. 2003) and nutrient value quantification (Curran 1989; Asner 1998; Curran et al. 2001; Hansen and Schjoerring 2003; Mutanga et al. 2004; Ferwerda et al. 2005; Beeri et al. 2007; Knyazikhin et al. 2013) are possible through remote sensing considering a particular season. But only two studies addressed the remote sensing capability to either differentiate forage species (Irisarri et al. 2009) or to quantify its nutrient value in relation to seasonal changes (Pullanagari et al. 2013). During the growing season, the spectral ranges sensitive to discriminate grass species or to quantify the nutrient value of pastures changed, suggesting that *ad hoc* calibrations are needed. The spectral resolution required to capture these changes is finer than the resolution of most ongoing satellite missions.

5.7 Conclusions

The lack of information on ANPP spatial and temporal variation is due to the difficulties in estimating the variation by successive biomass harvests. ANPP estimations derived from remote sensing necessarily involve estimating each Monteith's model component. Two of them, PAR and fPAR, are easy to estimate because meteorological stations provide PAR data, and remote sensing information estimates fPAR. The estimation of the third component,

RUE, is the hardest because it requires calibrations based on ANPP field estimations through successive biomass harvests. In this chapter, we showed a case in which the product of PAR and fPAR, APAR, together with *ad hoc* calibrations, provided estimations of the temporal and spatial variation of ANPP to be used by decision makers.

Acknowledgments

This project was funded by ANPCyT, University of Buenos Aires, CREA, and IPCVA. Gonzalo Irisarri received funding from CONICET through a postdoctoral scholarship.

References

Altesor, A., M. Oesterheld, E. Leoni, F. Lezama, and C. Rodríguez. 2005. Effect of grazing on community structure and productivity of a Uruguayan grassland. *Plant Ecology* 179:83–91.

Aragón, R., and M. Oesterheld. 2008. Linking vegetation heterogeneity and functional attributes of temperate grasslands through remote sensing. *Applied Vegetation Science* 11:115–128.

Asner, G. P. 1998. Biophysical and biochemical sources of variability in canopy reflectance. *Remote Sensing of Environment* 64:234–253.

Austin, A. T., and L. Vivanco. 2006. Plant litter decomposition in a semi-arid ecosystem controlled by photodegradation. *Nature* 442:555–558.

Baeza, S., J. M. Paruelo, and W. Ayala. 2011. Eficiencia en el uso de la radiación y productividad primaria en recursos forrajeros del este del Uruguay [Radiation use efficiency and primary production of forage resources from Eastern Uruguay]. *Agrociencia* 15:48–59.

Bai, Y., J. Wu, Q. Xing, et al. 2008. Primary production and rain use efficiency across a precipitation gradient on the Mongolia Plateau. *Ecology* 89:2140–2153.

Beeri, O., R. Phillips, J. Hendrickson, A. B. Frank, and S. Kronberg. 2007. Estimating forage quantity and quality using aerial hyperspectral imagery for northern mixed-grass prairie. *Remote Sensing of Environment* 110:216–225.

Bradford, J. B., J. A. Hicke, and W. K. Lauenroth. 2005. The relative importance of light-use efficiency modifications from environmental conditions and cultivation for estimation of large-scale net primary productivity. *Remote Sensing of Environment* 96:246–255.

Buono, G., M. Oesterheld, V. Nakamatsu, and J. M. Paruelo. 2010. Spatial and temporal variation of primary production of Patagonian wet meadows. *Journal of Arid Environments* 74:1257–1261.

Chapin, F. S., P. A. Matson, and H. A. Mooney. 2002. *Principles of terrestrial ecosystem ecology*. New York: Springer.

Costanza, R., R. d'Arge, R. de Groot, et al. 1997. The value of the world's ecosystem services and natural capital. *Nature* 387:253–260.

Curran, P. J. 1989. Remote sensing of foliar chemistry. *Remote Sensing of Environment* 30:271–278.

Curran, P. J., J. L. Dungan, and D. L. Peterson. 2001. Estimating the foliar biochemical concentration of leaves with reflectance spectrometry: Testing the Kokaly and Clark methodologies. *Remote Sensing of Environment* 76:349–359.

Ferwerda, J. G., A. K. Skidmore, and O. Mutanga. 2005. Nitrogen detection with hyperspectral normalized ratio indices across multiple plant species. *International Journal of Remote Sensing* 26:4083–4095.

Field, C. B., J. T. Randerson, and C. M. Malmström. 1995. Global net primary production: Combining ecology and remote sensing. *Remote Sensing of Environment* 51:74–88.

Garbulsky, M. F., J. Peñuelas, J. Gamon, Y. Inoue, and I. Filella. 2011. The photochemical reflectance index (PRI) and the remote sensing of leaf, canopy and ecosystem radiation use efficiencies: A review and meta-analysis. *Remote Sensing of Environment* 115:281–297.

Garbulsky, M. F., J. Peñuelas, D. Papale, et al. 2010. Patterns and controls of the variability of radiation use efficiency and primary productivity across terrestrial ecosystems. *Global Ecology and Biogeography* 19:253–267.

Grigera, G., M. Oesterheld, and F. Pacín. 2007. Monitoring forage production for farmers' decision making. *Agricultural Systems* 94:637–648.

Hansen, P. M., and J. K. Schjoerring. 2003. Reflectance measurement of canopy biomass and nitrogen status in wheat crops using normalized difference vegetation indices and partial least squares regression. *Remote Sensing of Environment* 86:542–553.

Irisarri, J. G. N., M. Oesterheld, J. M. Paruelo, and M. A. Texeira. 2012. Patterns and controls of above-ground net primary production in meadows of Patagonia. A remote sensing approach. *Journal of Vegetation Science* 23:114–126.

Irisarri, J. G. N., M. Oesterheld, S. R. Verón, and J. M. Paruelo. 2009. Grass species differentiation through canopy hyperspectral reflectance. *International Journal of Remote Sensing* 30:5959–5975.

Knapp, A. K., J. M. Briggs, D. L. Childers, and O. E. Sala. 2007. Estimating aboveground net primary production in grassland and herbaceous dominated ecosystems. In *Principles and standards for measuring net primary production*, eds. T. J. Fahey and A. K. Knapp, 27–48. New York: Oxford University Press.

Knapp, A. K., J. T. Fahenstock, S. P. Hamburg, L. B. Statland, T. R. Seastedt, and D. S. Schimel. 1993. Landscape patterns in soil–plant water relations and primary production in tallgrass prairie. *Ecology* 74:549–560.

Knapp, A. K., and M. D. Smith. 2001. Variation among biomes in temporal dynamics of aboveground primary production. *Science* 291:481–484.

Knyazikhin, Y., M. A. Schull, P. Stenberg, et al. 2013. Hyperspectral remote sensing of foliar nitrogen content. *Proceedings of the National Academy of Sciences of the United States of America* 110:E185–E192.

Lauenroth, W. 1979. Grassland primary production: North American grasslands in perspective. In *Perspectives in grassland ecology*, ed. N. French, 3–24. New York: Springer-Verlag.

Lauenroth, W. K., and O. E. Sala. 1992. Long-term forage production of North American shortgrass steppe. *Ecological Applications* 2:397–403.

McCree, K. J. 1972. Test of current definitions of photosynthetically active radiation against leaf photosynthesis data. *Agricultural Meteorology* 10:442–453.

McNaughton, S. J., D. G. Milchunas, and D. A. Frank. 1996. How can net primary productivity be measured in grazing ecosystems? *Ecology* 77:974–977.

Monteith, J. L. 1972. Solar radiation and productivity in tropical ecosystems. *Journal of Applied Ecology* 9:747–766.

Mutanga, O., A. K. Skidmore, and H. H. T. Prins. 2004. Predicting in situ pasture quality in the Kruger National Park, South Africa, using continuum-removed absorption features. *Remote Sensing of Environment* 89:393–408.

Mutanga, O., A. K. Skidmore, and S. van Wieren. 2003. Discriminating tropical grass (*Cenchrus ciliaris*) canopies grown under different treatments using spectroradiometry. *ISPRS Journal of Photogrammetry and Remote Sensing* 57:263–272.

Nouvellon, Y., D. L. Seen, S. Rambal, et al. 2000. Time course of radiation use efficiency in a shortgrass ecosystem: Consequences for remotely sensed estimation of primary production. *Remote Sensing of Environment* 71:43–55.

Oesterheld, M., C. Di Bella, and K. Herdiles. 1998. Relation between NOAA-AVHRR satellite data and stocking rate of rangelands. *Ecological Applications* 8:207–212.

Oesterheld, M., and R. J. León. 1987. El envejecimiento de las pasturas implantadas: Su efecto sobre la Productividad Primaria [Sown pastures aging: its effect on primary production]. *Turrialba* 37:29–35.

Oesterheld, M., J. Loreti, M. Semmartin, and J. Paruelo. 1999. Grazing, fire, and climate effects on primary productivity of grasslands and savannas. In *Ecosystems of disturbed ground*, ed. L. Walker, 287–306. Amsterdam: Elsevier.

Oltjen, J. W., and J. L. Beckett. 1996. Role of ruminant livestock in sustainable agricultural systems. *Journal of Animal Science* 74:1406–1409.

Paruelo, J. M., H. E. Epstein, W. K. Lauenroth, and I. C. Burke. 1997. ANPP estimates from NDVI for the central grassland region of the United States. *Ecology* 78:953–958.

Paruelo, J. M., M. Oesterheld, C. M. Di Bella, et al. 2000. Estimation of primary production of subhumid rangelands from remote sensing data. *Applied Vegetations Science* 3:189–195.

Piñeiro, G., M. Oesterheld, and J. M. Paruelo. 2006. Seasonal variation in aboveground production and radiation-use efficiency of temperate rangelands estimated through remote sensing. *Ecosystems* 9:357–373.

Potter, C. S., J. T. Randerson, C. B. Field, et al. 1993. Terrestrial ecosystem production: A process model based on global satellite and surface data. *Global Biogeochemical Cycles* 7:811–841.

Pullanagari, R. R., I. J. Yule, M. P. Tuohy, M. J. Hedley, R. A. Dynes, and W. M. King. 2013. Proximal sensing of the seasonal variability of pasture nutritive value using multispectral radiometry. *Grass and Forage Science* 68:110–119.

Ruimy, A., and B. Saugier. 1994. Methodology for the estimation of terrestrial net primary production from remotely sensed data. *Journal of Geophysical Research* 99:5263–5283.

Running, S. W., P. E. Thornton, R. R. Nemani, and J. Glassy. 2000. Global terrestrial gross and net primary productivity from the Earth observing system. In *Methods in ecosystem science*, eds. O. E. Sala, R. B. Jackson, H. A. Mooney, and R. W. Howarth, 44–57. New York: Springer-Verlag.

Sala, O. E., and A. T. Austin. 2000. Methods of estimating aboveground primary production. In *Methods in ecosystem science*, eds. O. E. Sala, R. B. Jackson, H. A. Mooney, and R. W. Howarth, 31–43. New York: Springer-Verlag.

Sala, O. E., V. A. Deregibus, T. Schlichter, and H. Alippe. 1981. Productivity dynamics of a native temperate grassland in Argentina. *Journal of Range Management* 34:48–51.

Sala, O. E., W. J. Parton, L. A. Joyce, and W. K. Lauenroth. 1988. Primary production of the central grassland region of the United States. *Ecology* 69:40–45.

Schmidt, K. S., and A. K. Skidmore. 2001. Exploring spectral discrimination of grass species in African rangelands. *International Journal of Remote Sensing* 22:3421–3434.

Schmidt, K. S., and A. K. Skidmore. 2003. Spectral discrimination of vegetation types in a coastal wetland. *Remote Sensing of Environment* 85:92–108.

Scurlock, J. M. O., K. Johnson, and R. J. Olson. 2002. Estimating net primary productivity from grasslands biomass dynamics measurements. *Global Change Biology* 8:736–753.

Sellers, P., J. A. Berry, G. J. Collatz, C. B. Field, and F. G. Hall. 1992. Canopy reflectance, photosynthesis, and transpiration. III. A reanalysis using improved leaf models and a new canopy integration scheme. *Remote Sensing of Environment* 42:187–216.

Semmartin, M. A., M. Oyarzabal, J. Loreti, and M. N. Oesterheld. 2007. Controls of primary productivity and nutrient cycling in a temperate grassland with year-round production. *Austral Ecology* 32:416–428.

Singh, J. S., W. K. Lauenroth, and R. K. Steinhorst. 1975. Review and assessment of various techniques for estimating net aerial primary production in grasslands from harvest data. *The Botanical Review* 41:231–237.

Thornton, P. E., and S. W. Running. 1999. An improved algorithm for estimating incident daily solar radiation from measurements of temperature, humidity, and precipitation. *Agricultural and Forest Meteorology* 93:211–228.

Turner, D. P., S. Urbanski, D. Bremer, et al. 2003. A cross-biomes comparison of daily light use efficiency for gross primary production. *Global Change Biology* 9:383–395.

Verón, S. R., M. Oesterheld, and J. M. Paruelo. 2005. Production as a function of resource availability: Slopes and efficiencies are different. *Journal of Vegetation Science* 16:351–354.

Wheeler, R. D., G. L. Kramer, K. B. Young, and E. Ospina. 1981. *The world livestock product, feedstuff, and food grain system*. Morrilton, AR: Winrock International Livestock Research and Training Center.

6

Missing Gaps in the Estimation of the Carbon Gains Service from Light Use Efficiency Models

A. J. Castro Martínez
University of Oklahoma, Oklahoma; University of Almería, Spain

J. M. Paruelo
University of Buenos Aires, Argentina

D. Alcaraz-Segura
University of Granada, Spain; University of Almería, Spain

J. Cabello
University of Almería, Spain

M. Oyarzabal
University of Buenos Aires, Argentina

E. López-Carrique
University of Almería, Spain

CONTENTS

6.1   Introduction 106
6.2   Material and Methods 107
6.3   Results 108
  6.3.1   Estimation Methods and LUE Units 108
  6.3.2   LUE Estimates across Organizational Levels and
          Land Cover Types 110
  6.3.3   Time Interval of LUE Estimates 112
6.4   Conclusions 115

Acknowledgments 115
Appendix 6.1 (Articles Reviewed from 1972 to 2007) 115
References 121

6.1 Introduction

The scientific community is being urged to invest more time and economic resources to improve current estimates of global and regional carbon budgets (Scurlock et al. 1999). Carbon gains are considered either as an intermediate service (Fisher et al. 2009) or as supports of provision and regulating services (MA 2005). In addition, net primary production (NPP), an estimate of ecosystem carbon gains, is often considered the most integrative descriptor of ecosystem function (McNaughton et al. 1989). NPP estimates are derived from biomass harvesting, flux tower measurements, remote sensing, and model simulation (Ruimy et al. 1995; Sala et al. 2000; Still et al. 2004). Biomass harvesting is expensive and not exempt from errors and methodological problems. These methods are limited in their spatial and temporal coverage. Given the linear relationship between the fraction of solar radiation absorbed by vegetation and spectral vegetation indices (Sellers et al. 1992), Monteith's model (Monteith 1972) offers the possibility of estimating seasonal variation in carbon gains from remote sensing data (Potter 1993). Monteith's model states that carbon gains (Equation 6.1) of vegetation cover are a function of the quantity of incoming photosynthetically active radiation (PAR), the fraction of this radiation intercepted by vegetation (fPAR), and the light use efficiency (LUE; Still et al. 2004). The flux estimated using the Monteith's model included net and gross primary production and net ecosystem exchange (NEE) (Ruimy et al. 1999; see Equations 6.2 and 6.3).

$$NPP = PAR*fPAR*LUE \qquad (6.1)$$

$$GPP = PAR*fPAR*LUE \qquad (6.2)$$

$$NEE = PAR*fPAR*LUE \qquad (6.3)$$

PAR can be directly measured using radiometers; fPAR can be estimated from spectral indices such as the Normalized Difference Vegetation Index (NDVI; Asrar et al. 1984) or the Enhanced Vegetation Index (EVI). The relationship of fPAR-spectral indices may vary between land cover types, but several authors have proposed different empirical relationships: (a) linear (Choudhury 1987); (b) nonlinear (Potter 1993; Sellers et al. 1994); and

(c) a combination of both (Los et al. 2000). The LUE term has a maximum value comparable to the photosynthetic efficiency or quantum yield at leaf level under optimum conditions (Gower et al. 1999). However, low temperatures and water and nutritional stress reduce LUE value (Field et al. 1995; Gamon et al. 1995). Field et al. (1995) reported LUE differences from 0.27 g C/MJ APAR for deserts to 0.70 g C/MJ APAR for tropical forests.

LUE was first defined at the species level and mainly for crop species (Andrade et al. 1993; Kiniry et al. 1998). The use of Monteith's model as the conceptual framework for remotely sensed estimates of aboveground net primary productivity (ANPP) (see Chapter 5) requires a definition of LUE at the ecosystem level (Ruimy et al. 1999; Sala et al. 2000; Fensholt et al. 2006). Often a single fixed value of approximately 1 g C/MJ APAR is used for a wide range of spatiotemporal situations (Maselli et al. 2009). Several authors showed that LUE varied in space (Field et al. 1995; Paruelo et al. 2004; Tong et al. 2008; Garbulsky et al. 2010) and time (Nouvellon et al. 2000; Piñeiro et al. 2006) and that the use of a single value may lead to substantial errors in regional (Hilker et al. 2008) and global (Turner et al. 2002, 2003, 2005; Tong et al. 2008) estimates of carbon gains.

Many factors affect the spatiotemporal patterns of LUE variation. Species composition, plant structure and physiology, including leaf form and RUBISCO (Ribulose-1,5-bisphosphate carboxylase oxygenase) content (Zhao et al. 2007), and environmental factors (i.e., water stress, CO_2 concentration, temperature) modify LUE at the ecosystem level. Measuring LUE is not a simple task. LUE can be estimated at different levels of organization (e.g., from individuals to ecosystems), using leaf-level estimates for single individuals or eddy covariance towers to derive LUE values at the ecosystem level (Garbulsky et al. 2010).

LUE is the more uncertain parameter of the Monteith's model since it is not possible to measure it directly, and it depends on estimates of GPP/NPP/NEE and absorbed radiation (Gower et al. 1999; Ruimy et al. 1999). This chapter reviews the reported estimates of LUE, and the effect of the time interval in LUE estimation at different organizational levels, from individuals to ecosystems. We sought to answer the following questions: (1) How was LUE estimated? (2) How did LUE differ across land cover types and levels of organization? (3) How variable are LUE estimates according to the time interval of estimation?

6.2 Material and Methods

We reviewed 125 articles from 1972 to 2007 containing the terms "light use efficiency" and "radiation use efficiency," but only 101 provided quantitative LUE data (Appendix 6.1). The review included 65 different journals primarily in the field of ecology (72% of total studies) and remote sensing (22% of total studies). From the published studies, we developed a database that included the organizational level, the flux estimated, LUE estimates, and the geographical coordinates of the study site (Table 6.1). We characterized LUE values at three organizational levels: (1) "Individual" referred to local-scale studies that estimated LUE based on individuals of a single species; (2) "Single-species-dominated ecosystems" referred to when the study focused on plots with one dominant species (e.g., NEE of agroecosystem in eddy covariance flux tower with a footprint of 100 m²); and (3) "Multispecies-dominated ecosystems." Assuming that 50% of the dry biomass corresponds to carbon, LUE values were transformed to the most common unit system: grams of carbon fixed per megajoules of absorbed PAR (g C/MJ APAR). To analyze the variability in LUE data, we assigned each data to one category of Archibold's (1995) classification of terrestrial land cover types. We calculated the mean, maximum, and minimum LUE values, as well as the deviation for each organizational level and land cover type. Kruskal–Wallis tests were applied to detect significant differences in LUE estimates (n = 185) across organizational levels and land cover types.

6.3 Results

6.3.1 Estimation Methods and LUE Units

LUE values were estimated using two main approaches. In the first approach (82% of total studies), LUE values were calculated at a local scale using the Monteith's equation and based on previous field estimates of a carbon flux (i.e., NPP or GPP). Here, the 20% of carbon flux estimates were derived from CO_2 flux between atmosphere and vegetation observations using eddyco-variance techniques (Ruimy et al. 1995; Zhao et al. 2007); fPAR data were calculated as a lineal function of satellite-derived NDVI in 44% of the total data. The remaining studies used fPAR data reported from other studies or compiled from direct measurements of the canopy. In most of the studies, PAR was calculated by radiometers.

In the second approach (18% of total studies), LUE was estimated based on correlative models with other variables such as leaf area index (LAI) or the photochemical reflectance index (PRI; Gu et al. 2002; Filella et al. 2004; Grace et al. 2007) (see Chapter 3). Here, LUE was also derived as a ratio between the harvested biomass at the plot scale and the incoming APAR throughout an entire year or a growing season.

Most of the articles reviewed (77% of total studies) offered quantitative estimates of LUE. From these studies, we obtained 185 LUE values that

TABLE 6.1

Summary Sample of Light Use Efficiency Data Reviewed for 1972–2007

Land Cover Types	Number of Studies	Locations	LUE Units	Carbon Flux Model (expressed as percentage of total studies within biome)
PHM	1	Alaska	g C/MJ APAR	(100%) Other*
CF	49	Durham, Canada, Wisconsin, Sweden	g C/MJ APAR Mol C/mol photons Moles CO_2/ mol PAR	(8%) GPP = APAR*LUE (16%) NEE = APAR*LUE (32%) NPP = APAR*LUE (44%) Other*
TW	12	Canada, Europe, EEUU	g C/MJ APAR Mol C/mol photons Moles CO_2/ mol PAR	(8%) ANPP = APAR*LUE (29%) NPP = APAR*LUE (29%) Other*
Cr	38	Ireland, EEUU, China, Italy, Australia, United Kingdom, South Africa, India	g C/MJ Kg (CO_2/ha·h)/ (J/m²·sg)	(30%) NPP = APAR*LUE (70%) Other*
TFE	65	Europe, EEUU, Japan, China, New Zealand	g C/MJ APAR Mol CO_2/ mol APAR mmol CO_2/ mmol photons	(8%) GPP = APAR*LUE (32%) NEE = APAR*LUE (27%) NPP = APAR*LUE (33%) Other*
ME	12	Spain, Italy, India, EEUU	g C/MJ APAR Mol C/mol APAR	(20%) NEE = APAR*LUE (70%) NPP = APAR*LUE (10%) Other*
TG	16	Canada, EEUU, Argentina	g C/MJ APAR g DM/MJ	(13%) NEE = APAR*LUE (40%) NPP = APAR*LUE (47%) Other*
TpF	4	Panama, Colombia, EEUU	g C/MJ APAR Kg(CO_2/ha·h)/ (J/m²·sg)	(67%) GPP = APAR*LUE (33%) NEE = APAR*LUE
TpS	9	Senegal, Argentina	g C/MJ APAR Mol C/mol APAR	(100%) Other*
AR	8	Sahara, Southern Australia, Mali, Mexico	g C/MJ APAR gr DM/MJ	(17%) GPP = APAR*LUE (50%) NPP = APAR*LUE (33%) Other*

Note: Other* expresses (a) when the study did not specify the carbon flux model for LUE estimation, (b) modifications of Monteith's model such as NASA-CASA model (i.e., Carnegie-Ames-Stanford Approach) simulates net primary productivity and the soil heterotrophic respiration at regional to global scales or the TURC model for the estimation of the continental gross primary productivity and net primary productivity, or derived models by Monteith's approach based on the inclusion of other physiological parameters, and (c) a constant LUE value. Archibold's land cover type classification: PHM = polar and high mountain tundra, CF = coniferous forests, TW = terrestrial wetlands, Cr = crops, TFE = temperate forest ecosystems, ME = Mediterranean ecosystems, TG = temperate grasslands, TpF = tropical forests, TpS = tropical savannas, and AR = arid regions.

were originally expressed in four different units, including g C/MJ APAR (65%), mol CO_2/mol APAR (14%), g of dry matter/MJ APAR (9%), and mol C/mol absorbed photons per minute (3%). After converting to g C/MJ APAR, the average LUE value was 0.99 g C/MJ APAR (SD = 1.09), with an absolute maximum of 8.2 g C/MJ APAR and an absolute minimum of 0.05 g C/MJ APAR.

6.3.2 LUE Estimates across Organizational Levels and Land Cover Types

The number of studies varied among organizational levels and land cover types (Figure 6.1). The multispecies-dominated ecosystems level was the most commonly studied (62%) followed by single-species-dominated ecosystems (19%) and individuals (19%). Temperate forests, coniferous forests, and croplands were the most highly represented land cover types in the literature reviewed, and the least represented were polar and high mountain tundra, tropical forests, and arid regions (Figure 6.1). Croplands and temperate forest ecosystems were the unique land cover types studied at all levels of organization.

We found that the average LUE values at the individual level (1.7 g C/MJ APAR; SD = 1.6) were significantly higher than at the multispecies-dominated (0.8 g C/MJ APAR; SD = 0.9) and single-species-dominated

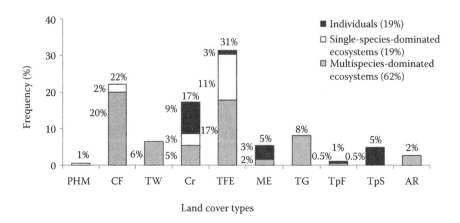

FIGURE 6.1
Frequency of articles for land cover types at individual, single-species-dominated ecosystems, and multispecies-dominated ecosystems levels. Total LUE values = 185. Archibold's land cover type classification: PHM = polar and high mountain tundra, CF = coniferous forests, TW = terrestrial wetlands, Cr = crops, TFE = temperate forest ecosystems, ME = Mediterranean ecosystems, TG = temperate grasslands, TpF = tropical forests, TpS = tropical savannas, and AR = arid regions.

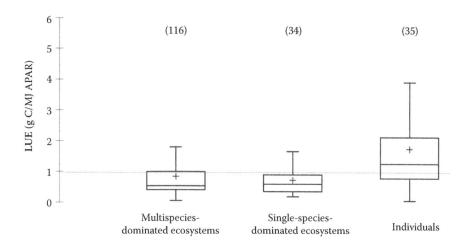

FIGURE 6.2

Box plot of light use efficiency (LUE) values at multispecies-dominated ecosystems, single-species-dominated ecosystems, and individual levels. The graphic explains the minimum, first quartile (25%), median, mean, and third quartile (75%) of LUE values. The mean is displayed with a +, and a black line corresponds to the median. The maximum LUE value at the individual level was 8.2 g C/MJ APAR; at the single-species-dominated ecosystems level, it was 2 g C/MJ APAR; and at multispecies-dominated ecosystem level, it was 5.7 g C/MJ APAR. The horizontal dotted line represents the total average of LUE values. The total LUE values per box plot appear in brackets.

(0.7 g C/MJ APAR; SD = 0.4) ecosystems levels (Figure 6.2). Average LUE values by land cover type exhibited significant differences (Figure 6.3). The average LUE values varied between 2.20 g C/MJ APAR (SD = 1.67) in croplands and 0.55 g C/MJ APAR (SD = 0.23) in terrestrial wetlands. The maximum LUE was found in crops (8.20 g C/MJ APAR) and temperate grasslands (5.20 g C/MJ APAR), whereas the minimum LUE value was observed in tropical savannas (0.05 g C/MJ APAR) and temperate grasslands (0.06 g C/MJ APAR) (Figure 6.3).

At the individual level, significant differences in LUE were observed between tropical savannas, tropical forests, and crops (Figure 6.4). At the single-species-dominated ecosystems level, coniferous forests and crops were significantly different in LUE values (Figure 6.4). At the multispecies-dominated ecosystems level, crops were significantly different from coniferous forests, terrestrial wetlands, tropical forests, and Mediterranean ecosystems. Temperate grasslands and tropical forest ecosystems were significantly different from coniferous forest, terrestrial wetlands, and temperate forest ecosystems (Figure 6.4).

Although LUE estimates were not available in the literature for all organizational levels, we observed significant differences within land

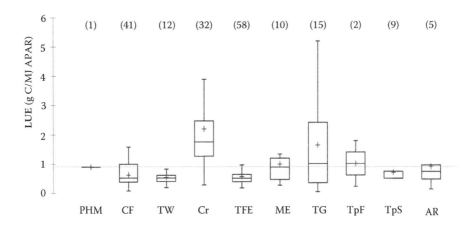

FIGURE 6.3
Box plot of light use efficiency (LUE) values by land cover types. The graphic explains the minimum, first quartile (25%), median, mean, and third quartile (75%) of LUE values. The mean is displayed with a +, and a black line corresponds to the median. Total LUE values = 185. Archibold's land cover type classification: PHM = polar and high mountain tundra, CF = coniferous forests, TW = terrestrial wetlands, Cr = crops, TFE = temperate forest ecosystems, ME = Mediterranean ecosystems, TG = temperate grasslands, TpF = tropical forests, TpS = tropical savannas, and AR = arid regions.

cover types. Coniferous forests showed significant differences between single-species-dominated and multispecies-dominated ecosystems levels. In temperate forest ecosystems, we observed significant differences between the individual and single-species-dominated ecosystems and multispecies-dominated ecosystems levels (Figure 6.4). In Mediterranean ecosystems, differences were observed between the individual and multispecies-dominated ecosystems levels. Crops, represented at all levels of organization, were the only land cover type showing no significant differences. Tropical forest ecosystems with values for individual and multispecies-dominated ecosystems levels did not exhibit significant differences (Figure 6.4).

6.3.3 Time Interval of LUE Estimates

LUE values significantly differed according to the time interval of the estimation (i.e., sampled within a day, season, or year). Annual and seasonal estimates were obtained for all organizational levels. Daily LUE estimates were only found in the literature at the multispecies-dominated ecosystems level. Significant differences between measurement time intervals were detected at the single-species-dominated ecosystems level between annual and seasonal LUE estimates. Kruskal–Wallis tests did not reveal significant differences in annual and seasonal estimates at the individual level and for annual, seasonal, and daily estimates at

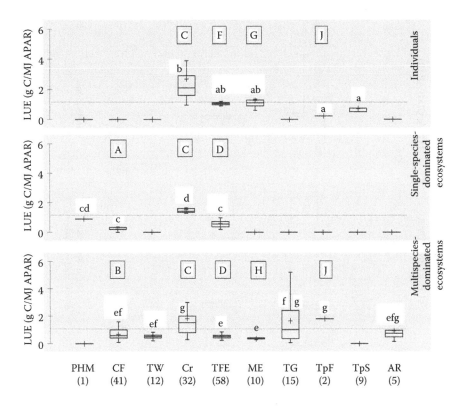

FIGURE 6.4

Box plot of light use efficiency (LUE) values comparing values between organizational levels and land cover types. The graphic explains the minimum, maximum, first quartile (25%), median, mean, and third quartile (75%) of LUE values. The mean is displayed with a +, and a black line corresponds to the median. Letters indicate significantly different groups (Kruskal–Wallis test, $p < 0.05$). Lowercase letters indicate significantly different groups within each organizational level and between land cover types. Uppercase letters indicate significantly different groups between organizational levels and per each land cover type. Total LUE values = 185. Archibold's land cover type classification: PHM = polar and high mountain tundra, CF = coniferous forests, TW = terrestrial wetlands, Cr = crops, TFE = temperate forest ecosystems, ME = Mediterranean ecosystems, TG = temperate grasslands, TpF = tropical forests, TpS = tropical savannas; and AR = arid regions.

the multispecies-dominated ecosystems level. However, we observed significant differences between annual and seasonal estimates at the single-species-dominated ecosystems level. A comparison between the time interval of the estimation and the organizational levels (Figure 6.5) revealed significant differences between LUE obtained in annual and seasonal observations. Annual and seasonal estimates were not significantly different at the multispecies-dominated and single-species-dominated ecosystems levels, but we did detect significant differences with the results obtained at the individual level.

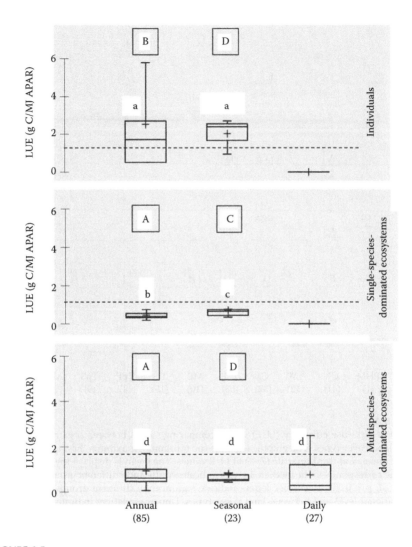

FIGURE 6.5
Box plot of temporal variation of average light use efficiency (LUE) values in each organizational level and the estimate periods. The graphic explains the minimum, maximum, first quartile (25%), median, mean, and third quartile (75%) of LUE values. The mean is displayed with a +, and a black line corresponds to the median. Different letters indicate significant differences (Kruskal–Wallis test, $p < 0.05$). Lower case letters indicate significantly different groups between time interval of estimation and within each organizational level. Capital letters indicate significantly different groups between organizational levels and the time interval of estimation. Total LUE values = 185. Archibold's land cover type classification: PHM = polar and high mountain tundra, CF = coniferous forests, TW = terrestrial wetlands, Cr = crops, TFE = temperate forest ecosystems, ME = Mediterranean ecosystems, TG = temperate grasslands, TpF = tropical forests, TpS = tropical savannas, and AR = arid regions.

6.4 Conclusions

Monteith's model, based on LUE and remotely sensed estimates of fPAR, constitutes the most widely used approach for mapping the terrestrial carbon cycle (Jenkins et al. 2007; Pereira et al. 2007), though it is not free of uncertainties. The inherent spatiotemporal variability found among different methodologies may explain the variability of LUE estimates found in this study. The time interval of the estimates and the level of organization are two clear sources of such variation. In such a way, annual estimates of NPP at regional scale should not be used for LUE estimation at individual level and derive for short-term (e.g., days) measurements. The variability of LUE estimates related to environmental and physiological factors (such as leaf form, ribulose diphosphate carboxylase content, temperature, and/or moisture) (Ito and Oikawa 2007; Tong et al. 2008) can result in large errors if these values are extrapolated to global or regional scales (Nouvellon et al 2000; Piñeiro et al. 2006). Our results indicated that high temporal variation in LUE estimates at the individual and multispecies-dominated ecosystems levels across land cover types (see also Grace et al. 2007; Cook et al. 2008; Hilker et al. 2008) do not account for regional and global NPP estimates, which typically apply a constant LUE value (Drolet et al. 2008; Maselli et al. 2009).

Acknowledgments

The authors gratefully acknowledge the staff of the Andalusian Government's Department of the Environment for providing the facilities required to obtain the necessary assistance. They also thank Gervasio Piñeiro, Dolores Arocena, Carlos Di Bella, Piedad Cristiano, and two anonymous reviewers for their useful comments. Financial support was provided by the ERDF (FEDER), Andalusian Regional Government (Junta de Andalucía GLOCHARID & SEGALERT Projects, P09–RNM-5048), and the Ministry of Science and Innovation (Project CGL2010-22314). Support for A. J. C. was also provided by the Centro Andaluz para la Evaluación y Seguimiento del Cambio Global (CAESCG) and the Oklahoma Biological Survey (OBS) at the University of Oklahoma.

Appendix 6.1 (Articles Reviewed from 1972 to 2007)

Aalto, T., P. Ciais, A. Chevillard, and C. Moulin. 2004. Optimal determination of the parameters controlling biospheric CO_2 fluxes over Europe using eddy covariance fluxes and satellite NDVI measurements. *Tellus B* 56:93–104.

Ahl, D. E., S. T. Gower, D. S. Mackay, S. N. Burrows, J. M. Norman, and G. R. Diak. 2004. Heterogeneity of light use efficiency in a northern Wisconsin forest: Implications for modeling net primary production with remote sensing. *Remote Sensing of Environment* 93:168–178.

Ahl, D. E., S. T. Gower, D. S. Mackay, S. N. Burrows, J. M. Norman, and G. R. Diak. 2005. The effects of aggregated land cover data on estimating NPP in northern Wisconsin. *Remote Sensing of Environment* 97:1–14.

Anderson, M. C., W. P. Kustas, and J. M. Norman. 2007. Upscaling flux observations from local to continental scales using thermal remote sensing. *Agronomy Journal* 99:240–254.

Asner, G. P., K. M. Carlson, and R. E. Martin. 2005. Substrate age and precipitation effects on Hawaiian forest canopies from spaceborne imaging spectroscopy. *Remote Sensing of Environment* 98:457–467.

Asrar, G., M. Fuchs, E. T. Kanemasu, and J. L. Hatfield. 1984. Estimating absorbed photosynthetic radiation and leaf-area index from spectral reflectance in wheat. *Agronomy Journal* 76:300–306.

Baldocchi, D. D. 2003. Assessing the eddy covariance technique for evaluating carbon dioxide exchange rates of ecosystems: Past, present and future. *Global Change Biology* 9:479–492.

Black, T. A., D. Gaumont-Guay, R. S. Jassal, et al. 2005. Measurement of carbon dioxide exchange between the boreal forest and the atmosphere. In *Carbon balance of forest biomes*, eds. H. Griffiths and P. G. Jarvis, 151–185. Oxfordshire, UK: BIOS Scientific Publishers.

Boschetti, L., P. A. Brivio, H. D. Eva, J. Gallego, A. Baraldi, and J. M. Gregoire. 2006. A sampling method for the retrospective validation of global burned area products. *IEEE—Transactions on Geoscience and Remote Sensing* 44:1765–1773.

Bradford, J. B., J. A. Hickec, and W. K. Lauenroth. The relative importance of light-use efficiency modifications from environmental conditions and cultivation for estimation of large-scale net primary productivity. *Remote Sensing of Environment* 96:246–255.

Cannell, M. G. R., R. Milne, L. J. Sheppard, and M. H. Unsworth. 1987. Radiation interception and productivity of willow. *Journal of Applied Ecology* 24:261–278.

Christensen, S., and J. Goudriaan. 1993. Deriving light interception and biomass from spectral reflectance ratio. *Remote Sensing of the Environment* 43:87–95.

D'Antuono, L. F., and F. Rossini. 2006. Yield potential and ecophysiological traits of the Altamurano linseed (*Linum usitatissimum L.*), a landrace of southern Italy. *Genetic Resources and Crop Evolution* 53:65–75.

Drolet, G. G., K. F. Huemmrich, F. G. Hall, et al. 2005. A MODIS-derived photochemical reflectance index to detect inter-annual variations in the photosynthetic light-use efficiency of a boreal deciduous forest. *Remote Sensing of Environment* 98:212–224.

Dungan, R. J., and D. Whitehead. 2006. Modelling environmental limits to light use efficiency for a canopy of two broad-leaved tree species with contrasting leaf habit. *New Zealand Journal of Ecology* 30:251–259.

Fang, S., X. Xizeng, X. Xiang, and L. Zhengcai. 2005. Poplar in wetland agroforestry: A case study of ecological benefits, site productivity, and economics. *Wetlands Ecology and Management* 13:93–104.

Fensholt, R., I. Sandholt, and M. S. Rasmussen. 2004. Evaluation of MODIS LAI, fAPAR and the relation between fAPAR and NDVI in a semi-arid environment using in situ measurements. *Remote Sensing of Environment* 91:490–507.

Fensholt, R., I. Sandholt, M. S. Rasmussen, S. Stisen, and A. Diouf. 2006. Evaluation of satellite based primary production modeling in the semi-arid Sahel. *Remote Sensing of Environment* 105:173–188.

Fernández, M. E., J. E. Gyenge, and T. M. Schlichter. 2006. Growth of the grass *Festuca pallescens* in silvopastoral systems in a semi-arid environment, Part 1: Positive balance between competition and facilitation. *Agroforestry Systems* 66:259–269.

Field, C. B., J. T. Randerson, and C. M. Malmstrom. 1995. Global net primary production: Combining ecology and remote sensing. *Remote Sensing of Environment* 51:74–88.

Fleisher, D. H., D. J. Timlin, and V. R. Reddy. 2006. Temperature influence on potato leaf and branch distribution and on canopy photosynthetic rate. *Agronomy Journal* 98:1442–1452.

Fuentes, D., J. A. Gamon, Y. Cheng, et al. 2006. Mapping carbon and water flux in a chaparral ecosystem using vegetation indices derived from AVIRIS. *Remote Sensing of Environment* 103:312–323.

Gamon, J. A., K. Kitajima, S. S. Mulkey, L. Serrano, and S. J. Wright. 2005. Diverse optical and photosynthetic properties in a neotropical dry forest during the dry season: Implications for remote estimation of photosynthesis. *Biotropica* 37:547–560.

Gebremichael, M., and A. P. Barros. 2006. Evaluation of MODIS gross primary productivity (GPP) in tropical monsoon regions. *Remote Sensing of Environment* 100:150–166.

Goetz, S. J., and S. D. Prince. 1999. Modelling terrestrial carbon exchange and storage: Evidence and implications of functional convergence in light-use efficiency. *Advances in Ecological Research* 28:57–92.

Goetz, S. J., S. D. Prince, S. N. Goward, M. M. Thawle, J. Small, and A. Johnston. 1999. Mapping net primary production and related biophysical variables with remote sensing: Application to the Boreas region. *Journal of Geophysical Research— Atmospheres* 104:27719–27734.

Goulden, M. L., J. W. Munge, S. M. Fan, B. C. Daube, and S. C. Wofsy. 1996. Exchange of carbon dioxide by a deciduous forest: Response to interanual climate variability. *Science* 271:1576–1578.

Goward, S. N., and D. G. Dye. 1987. Evaluating North American net primary productivity with satellite observations. *Advances in Space Research* 7:165–174.

Goward, S. N., and K. F. Huemmrich. 1992. Vegetation canopy PAR absorbance and the normalized difference vegetation index: An assessment using the SAIL model. *Remote Sensing of the Environment* 39:119–140.

Guo, J., and C. M. Trotter. 2004. Estimating photosynthetic light-use efficiency using the photochemical reflectance index: Variations among species. *Functional Plant Biology* 31:255–265.

Hill, M., A. A. Held, R. Leuning, et al. 2006. MODIS spectral signals at a flux tower site: Relationships with high-resolution data, and CO_2 flux and light use efficiency measurements. *Remote Sensing of Environment* 103:351–368.

Inoue, Y., and J. Peñuelas. 2006. Relationships between light use efficiency and photochemical reflectance index in soybean leaves as affected by soil water content. *International Journal of Remote Sensing* 27:5109–5114.

Ito, A., and T. Oikawa. 2004. Global mapping of terrestrial primary productivity and light-use efficiency with a process-based model. In *Global environmental change in the ocean and on land,* eds. M. Shiyomi, H. Kawahata, H. Koizumi, A. Tsuda, and Y. Awaya, 343–358. Tokyo: Terrapub.

Kato, T., Y. Tang, S. Gu, et al. 2006. Temperature and biomass influences on interannual changes in CO_2 exchange in an alpine meadow on the Qinghai-Tibetan Plateau. *Global Change Biology* 12:1285–1298.

Kinirya, J. R., C. E. Simpsonb, A. M. Schubertc, and J. D. Reed. Peanut leaf area index, light interception, radiation use efficiency, and harvest index at three sites in Texas. *Field Crops Research* 91:297–306.

Krishnan, P., T. A. Black, N. J. Grant, et al. 2006. Carbon dioxide and water vapour exchange in a boreal aspen forest during and following severe drought. *Agricultural and Forest Meteorology* 139:208–223.

Lagergren, F. 2005. Net primary production and light use efficiency in a mixed coniferous forest in Sweden. *Plant, Cell & Environment* 28:412–423.

Leuning, R., H. A. Cleugh, S. J. Zegelin, and D. Hughes. 2005. Carbon and water fluxes over a temperate *Eucalyptus* forest and a tropical wet/dry savanna in Australia: Measurements and comparison with MODIS remote sensing estimates. *Agricultural and Forest Meteorology* 129:151–173.

Li, S. G., J. Asanuma, W. Eugster, et al. 2005. Net ecosystem carbon dioxide exchange over grazed steppe in central Mongolia. *Global Change Biology* 11:1941–1955.

Li, S. G., W. Eugster, J. Asanuma, et al. 2006. Energy partitioning and its biophysical controls above a grazing steppe in central Mongolia. *Agricultural and Forest Meteorology* 137:89–106.

Monteith, J. L. 1972. Solar-radiation and productivity in tropical ecosystems. *Journal of Applied Ecology* 9:747–766.

Monteith, J. L. 1977. Climate and the efficiency of crop production in Britain. *Philosophical Transactions of the Royal Society of London B* 281:277–297.

Monteith, J. L. 1994. Validity of the correlation between intercepted radiation and biomass. *Agricultural and Forest Meteorology* 68:213–220.

Myneni, R. B., S. Hoffman, Y. Knyazikhin, et al. 2002. Global products of vegetation leaf area and fraction absorbed PAR from year one of MODIS data. *Remote Sensing of Environment* 83:214–231.

Myneni, R. B., R. R. Nemani, and S. W. Running. 1997. Estimation of global leaf área index and absorbed PAR using radiative transfer models. *IEEE Transactions on Geoscience and Remote Sensing* 35:1380–1393.

Myneni, R. B., and D. L. Williams. 1994. On the relationship between fAPAR and NDVI. *Remote Sensing of Environment* 49:200–11.

Nakaji, T., H. Oguma, and Y. Fujinuma. 2006. Seasonal changes in the relationship between photochemical reflectance index and photosynthetic light use efficiency of Japanese larch needles. *International Journal of Remote Sensing* 27:493–509.

Nakaji, T., T. Takeda, Y. Fujinuma, and H. Oguma. 2005. Effect of autumn senescence on the relationship between the PRI and LUE of young Japanese larch trees. *Phyton* 45:535–542.

Nemani, R. R., and S. W. Running. 1989. Estimation of regional surface-resistance to evapotranspiration from NDVI and thermal-IR AVHRR data. *Journal of Applied Meteorology* 28:276–284.

Nichol, C. J., K. F. Huemmrich, T. A. Black, et al. 2002. Sensing of photosynthetic-light-use efficiency of boreal forest. *Agricultural and Forest Meteorology* 101:131–142.

Nichol, C. J., J. Lloyd, O. Shibistova, et al. 2002. Remote sensing of photosynthetic-light-use efficiency of a Siberian boreal forest. *Tellus B* 54:677–687.

Niinemets, U., A. Cescatti, and R. Christian. 2004. Constraints on light interception efficiency due to shoot architecture in broad-leaved Nothofagus species. *Tree Physiology* 24:617–630.

Niinemets, U., and L. Sack. 2006. Structural determinants of leaf light-harvesting capacity and photosynthetic potentials. *Progress in Botany* 67:385–419.

Norby, R. J., J. Ledford, C. D. Reilly, et al. 2004. Fine-root production dominates response of a deciduous forest to atmospheric CO_2 enrichment. *Proceedings of the National Academy of Sciences of the United States of America* 101:9689–9693.

Nouvellon, Y., D. Lo Seen, S. Rambal, et al. 2000. Time course of radiation use efficiency in a shortgrass ecosystem: Consequences for remotely sensed estimation of primary production. *Remote Sensing of Environment* 71:43–55.

Paruelo, J. M., H. E. Epstein, W. K. Lauenroth, and I. C. Burke. 1997. ANPP estimates from NDVI for the central grassland region of the US. *Ecology* 78:953–958.

Paruelo, J. M., M. Oesterheld, C. M. Di Bella, et al. 2000. Estimation of primary production of subhumid rangelands from remote sensing data. *Applied Vegetation Science* 3:189–195.

Patel, N. R. 2006. Investigating relations between satellite derived land surface parameters and meteorological variables *Geocarto International* 21:47–53.

Pereira, J. S., J. A. Mateus, L. M. Aires, et al. 2007. Net ecosystem carbon exchange in three contrasting Mediterranean ecosystems: The effect of drought. *Biogeosciences* 4:791–802.

Piñéiro, G., M. Oesterheld, and J. M. Paruelo. 2006. Seasonal variation in aboveground production and radiation use efficiency of temperate rangelands estimated through remote sensing. *Ecosystems* 9:357–373.

Pitman, J. I. 2000. Absorption of photosynthetically active radiation, radiation use efficiency and spectral reflectance of bracken [Pteridium aquilinum (L.) Kuhnl] canopies. *Annals of Botany* 85:101–111.

Potter, C., S. Klooster, A. Huete, and V. Genovese. 2007. Terrestrial carbon sinks for the United States predicted from MODIS satellite data and ecosystem modeling. *Earth Interactions* 11:1–21.

Potter, C. S., J. T. Randerson, C. B. Field, et al. 1993. Terrestrial ecosystem production— a process model-based on global satellite and surface data. *Global Biogeochemical Cycles* 7:811–841.

Prince, S. D., S. J., Goetz, and S. N. Goward. 1995. Monitoring primary production from Earth observing satellites. *Water, Air and Soil Pollution* 82:509–522.

Ruimy, A., B. Saugier, and G. Dedieu. 1994. Methodology for the estimation of terrestrial net primary production from remotely sensed data. *Journal of Geophysical Research* 99:5263–5283.

Running, S. W., D. D. Baldocchi, D. P. Turner, S. T. Gower, P. S. Bakwin, and K. A. Hibbard. 1999. A global terrestrial monitoring network integrating tower fluxes, flask sampling, ecosystem modeling and EOS satellite data. *Remote Sensing of Environment* 70:108–127.

Running, S. W., and R. R. Nemani. 1988. Relating seasonal patterns of the AVHRR vegetation index to simulated photosynthesis and transpiration of forests in different climates. *Remote Sensing of Environment* 24:347–367.

Running, S. W., R. R. Nemani, F. A. Heinsch, M. S. Zhao, M. Reeves, and H. Hashimoto. 2004. A continuous satellite-derived measure of global terrestrial primary production. *BioScience* 54:547–560.

Running, S. W., L. Queen, and M. Thornton. 2000. The Earth observing system and forest management. *Journal of Forestry* 98:29–31.

Russell, G., P. G. Jarvis, and J. L. Monteith. 1989. Absorption of radiation by canopies and stand growth. In *Plant canopies: Their growth, form and function,* eds. G. Russell, B. Marshall, and P. G. Jarvis, 21–39. Cambridge, UK: Cambridge University Press.

Sakai, T. 2005. Microsite variation in light availability and photosynthesis in a cooltemperate deciduous broadleaf forest in central Japan. *Ecological Research* 20:537–545.

Salazar, M. R., B. Chaves, J. W. Jones, and A. Cooman. 2006. A simple potential production model of Cape gooseberry (*Physalis peruviana* L.). *Acta Hortic* 718:105–112.

Schwalm, C. R., T. A. Black, B. D. Amiro, et al. 2006. Photosynthetic light use efficiency of three biomes across an east–west continental-scale transect in Canada. *Agricultural and Forest Meteorology* 140:269–286.

Seaquist, J., L. Olsson, and J. Ardö. 2003. A remote sensing-based primary production model for grassland biomes. *Ecological Modelling* 169:131–155.

Seaquist, J. W., L. Olsson, J. Ardo, and L. Eklundh. 2006. Broad-scale increase in NPP quantified for the African Sahel, 1982–1999. *International Journal of Remote Sensing* 27:5115–5122.

Sims, D., and J. Gamon. 2002. Relationships between leaf pigment content and spectral reflectance across a wide range of species, leaf structures and developmental stages. *Remote Sensing of Environment* 81:337–354.

Sims, D., H. Luo, S. Hastings, W. Oechel, A. Rahman, and J. Gamon. 2006. Parallel adjustments in vegetation greenness and ecosystem CO_2 exchange in response to drought in a southern California chaparral ecosystem. *Remote Sensing of Environment* 103:289–303.

Sims, D. A., A. F. Rahman, V. D. Cordova, et al. 2005. Midday values of gross CO_2 flux and light use efficiency during satellite overpasses can be used to directly estimate eight-day mean flux. *Agricultural and Forest Meteorology* 131:1–12.

Still, C. J., J. Randerson, and I. Fung. 2004. Large-scale plant light-use efficiency inferred from the seasonal cycle of atmospheric CO_2. *Global Change Biology* 10:1240–1252.

Storkey, J. 2006. A functional group approach to the management of UK arable weeds to support biological diversity. *Weed Research* 46:513–522.

Tracol, Y., E. Mougin, P. Hiernaux, and L. Jarlan. 2006. Testing a Sahelian grassland functioning model against herbage mass measurements. *Ecological Modelling* 193:437–446.

Tsubo, M., S. Walker, and H. O. Ogindo. 2005. A simulation model of cereal–legume intercropping systems for semi-arid regions I. Model development. *Field Crops Research* 93:10–22.

Tucker, C. J., I. Y. Fung, C. D. Keeling, and R. H. Gammon. 1986. Relationship between atmospheric CO_2 variations and a satellite-derived vegetation index. *Nature* 319:195–199.

Tucker, C. J., W. H. Jones, W. A. Kley, and G. J. Sundstrom. A 3-band hand-held radiometer for field use. *Science* 211:281–283.

Tucker, C. J., C. O. Justice, and S. D. Prince. 1986. Monitoring the grasslands of the Sahel 1984–1985. *International Journal of Remote Sensing* 7:1571–1581.

Tucker, C. J., and P. J. Sellers. 1986. Satellite remote-sensing of primary production. *International Journal of Remote Sensing* 7:1395–1416.

Turner, D. P., W. D. Ritts, W. B. Cohen, et al. 2003. Scaling gross primary production (GPP) over boreal and deciduous forest landscapes in support of MODIS GPP product validation. *Remote Sensing of Environment* 88:256–270.

Turner, D. P., W. D. Ritts, W. B. Cohen, T. K. Maeirsperger, S. T. Gower, and A. Kirschbaum. 2005. Site-level evaluation of satellite-based global terrestrial gross primary production and net primary production monitoring. *Global Change Biology* 11:666–684.

Turner, D. P., W. D. Ritts, W. B. Cohen, et al. 2006. Evaluation of MODIS NPP and GPP products across multiple biomes. *Remote Sensing of Environment* 102:282–292.

Turner, D. P., S. Urbanski, D. Bremer, et al. 2003. Cross-biome comparison of daily light use efficiency for gross primary production. *Global Change Biology* 9:383–395.

Ueyama, M., Y. Harazono, E. Ohtaki, and A. Miyata. 2006. Controlling factors on the inter-annual CO_2 budget at a sub-arctic black spruce forest in interior Alaska. *Tellus B* 58:491–450.

Veroustraete, F., H. Sabbe, and H. Eerens. 2002. Estimation of carbon mass fluxes over Europe using the C-Fix model and Euroflux data. *Remote Sensing of Environment* 83:376–399.

Walcroft, A. S., K. J. Brown, W. S. F. Schuster, et al. 2005. Radiative transfer and carbon assimilation in relation to canopy architecture, foliage area distribution and clumping in a mature temperate rainforest canopy in New Zealand. *Agricultural and Forest Meteorology* 135:326–339.

Xiao, X. M. 2006. Light absorption by leaf chlorophyll and maximum light use efficiency. *IEEE Transactions on Geoscience and Remote Sensing* 44:1933–1935.

Xiao, X. M., Q. Y. Zhang, B. Braswell, et al. 2004. Modeling gross primary production of temperate deciduous broadleaf forest using satellite images and climate data. *Remote Sensing of Environment* 91:256–270.

Zhang, Q. Y., X. M. Xiao, B. Braswell, E. Linder, F. Baret, and B. Moore. 2005. Estimating light absorption by chlorophyll, leaf and canopy in a deciduous broadleaf forest using MODIS data and a radiative transfer model. *Remote Sensing of Environment* 99:357–371.

Zhao, M., F. A. Heinsch, R. R. Nemani, and S. W. Running. 2005. Improvements of the MODIS terrestrial gross and net primary production global data set. *Remote Sensing of Environment* 95:164–175.

References

Andrade, F. H., S. A. Uhart, and A. Cirilo. 1993. Temperature affects radiation use efficiency in maize. *Field Crops Research* 32:17–25.

Archibold, O. W. 1995. *Ecology of world vegetation.* London: Chapman & Hall.

Asrar, G., M. Fuchs, E. T. Kanemasu, and J. L. Hatfield. 1984. Estimating absorbed photosynthetic radiation and leaf area index from spectral reflectance in wheat. *Agronomy Journal* 76:300–306.

Choudhury, B. J. 1987. Relationships between vegetation indices, radiation absorption, and net photosynthesis evaluated by a sensitivity analysis. *Remote Sensing of Environment* 22:209–233.

Cook, B. D., P. V. Bolstad, J. G. Martin, et al. 2008. Using light-use and production efficiency models to predict photosynthesis and net carbon exchange during forest canopy disturbance. *Ecosystems* 11:26–44.

Drolet, G. G., E. M. Middleton, K. F. Huemmrich, F. G. Hall, and H. A. Margolis. 2008. Regional mapping of gross light-use efficiency using MODIS spectral indices. *Remote Sensing of Environment* 112:3064–3078.

Fensholt, R., I. Sandholt, M. S. Rasmussen, S. Stisen, and A. Diouf. 2006. Evaluation of satellite based primary production modelling in the semi-arid Sahel. *Remote Sensing of Environment* 105:173–188.

Field, C. B., R. B. Jackson, and H. A. Mooney. 1995. Stomatal responses to increased CO_2—Implications from the plant to the global scale. *Plant Cell and Environment* 18:1214–1225.

Filella, I., J. Peñuelas, L. Llorens, and M. Estiarte. 2004. Reflectance assessment of seasonal and annual changes in biomass and CO_2 uptake of a Mediterranean shrubland submitted to experimental warming and drought. *Remote Sensing of Environment* 90:308–318.

Fisher, B., R. K. Turner, and P. Morling. 2009. Defining and classifying ecosystem services for decision making. *Ecological Economics* 3:643–653.

Gamon, J. A., C. B. Field, M. L. Goulden, et al.1995. Relationships between NDVI, canopy structure, and photosynthesis in 3 Californian vegetation types. *Ecological Applications* 5:28–41.

Garbulsky, M. F., J. Peñuelas, D. Papale, et al. 2010. Patterns and controls of the variability of radiation use efficiency and primary productivity across terrestrial ecosystems. *Global Ecology and Biogeography* 19:253–267.

Gower, S. T., C. J. Kucharik, and J. M. Norman. 1999. Direct and indirect estimation of leaf area index, $f(APAR)$, and net primary production of terrestrial ecosystems. *Remote Sensing of Environment* 70:29–51.

Grace, J., C. Nichol, M. Disney, L. T. Quaife, and P. Bowyer. 2007. Can we measure terrestrial photosynthesis from space directly, using spectral reflectance and fluorescence? *Global Change Biology* 13:1484–1497.

Gu, L., D. Baldocchi, S. B. Verma, et al. 2002. Advantages of diffuse radiation for terrestrial ecosystem productivity. *Journal of Geophysical Research: Atmospheres* 107:2–23.

Hilker, T., N. C. Coops, M. A. Wulder, T. A. Black, and R. G. Guy. 2008. The use of remote sensing in light use efficiency based models of gross primary production: A review of current status and future requirements. *Science of the Total Environment* 404:411–423.

Ito, A., and T. Oikawa. 2007. Absorption of photosynthetically active radiation, dry-matter production, and light-use efficiency of terrestrial vegetation: A global model simulation. *Elsevier Oceanography Series* 73:335–359; 503–505.

Jenkins, J. P., A. D. Richardson, B. H. Braswell, S. V. Ollinger, D. Y. Hollinger, and M. L. Smith. 2007. Refining light-use efficiency calculations for a deciduous forest canopy using simultaneous tower-based carbon flux and radiometric measurements. *Agricultural and Forest Meteorology* 143:64–79.

Kiniry, J. R., J. A. Landivar, M. Witt, T. J. Gerik, J. Cavero, and L. J. Wade. 1998. Radiation-use efficiency response to vapor pressure deficit for maize and sorghum. *Field Crops Research* 56:265–270.

Los, S. O., G. J. Collatz, P. J. Sellers, et al. 2000. A global 9-yr biophysical land surface dataset from NOAA AVHRR data. *Journal of Hydrometeorology* 1:183–199.

MA (Millennium Ecosystem Assessment). 2005. *Ecosystems and human well-being: The assessment series (four volumes and summary).* Washington, DC: Island Press.

Maselli, F., M. Chiesi, M. Moriondo, L. Fibbi, M. Bindi, and S. W. Running. 2009. Modelling the forest carbon budget of a Mediterranean region through the integration of ground and satellite data. *Ecological Modelling* 220:330–342.

McNaughton, S. J., M. Oesterheld, D. A. Frank, and K. J. Williams. 1989. Ecosystem-level patterns of primary productivity and herbivory in terrestrial habitats. *Nature* 341:142–144.

Monteith, J. L. 1972. Solar radiation and productivity in tropical ecosystems. *Journal of Applied Ecology* 9:747–766.

Nouvellon, Y., D. L. Seen, S. Rambal, et al. 2000. Time course of radiation use efficiency in a shortgrass ecosystem: Consequences for remotely sensed estimation of primary production. *Remote Sensing of Environment* 71:43–55.

Paruelo, J. M., M. F. Garbulsky, J. P. Guerschman, and E. G. Jobbágy. 2004. Two decades of Normalized Difference Vegetation Index changes in South America: Identifying the imprint of global change. *International Journal of Remote Sensing* 25:2793–2806.

Pereira, J. S., J. A. Mateus, L. Aires, et al. 2007. Net ecosystem carbon exchange in three contrasting Mediterranean ecosystems—The effect of drought. *Biogeosciences* 4:791–802.

Piñeiro, G., M. Oesterheld, and J. M. Paruelo. 2006. Seasonal variation in aboveground production and radiation-use efficiency of temperate rangelands estimated through remote sensing. *Ecosystems* 9:357–373.

Potter, C. S. 1993. Terrestrial ecosystem production: A process model based on global satellite and surface data. *Global Biogeochemical Cycles* 7:811–841.

Ruimy, A., P. Jarvis, D. D. Baldocchi, and B. Saugier. 1995. CO_2 fluxes over plant canopies and solar radiation: A review. *Advances in Ecological Research* 26:1–68.

Ruimy, A., L. Kergoat, A. Bondeau, et al. 1999. Comparing global NPP models of terrestrial net primary productivity (NPP): Analysis of differences in light absorption and light-use efficiency. *Global Change Biology* 5:56–64.

Sala, O. E., R. B. Jackson, H. A. Mooney, and R. W. Howarth. 2000. Methods in ecosystem science: Progress, tradeoffs, and limitations. In *Methods in ecosystem science,* eds. O. E. Sala, R. B. Jackson, H. A. Mooney, and R. W. Howarth, 1–3. New York: Springer-Verlag.

Scurlock, J. M. O., W. Cramer, R. J. Olson, W. J. Parton, and S. D. Prince. 1999. Terrestrial NPP: Toward a consistent data set for global model evaluation. *Ecological Applications* 9:913–919.

Sellers, P. J., J. A. Berry, G. J. Collatz, C. B. Field, and E. G. Hall. 1992. Canopy reflectance, photosynthesis, and transpiration. III. A reanalysis using improved leaf models and a new canopy integration scheme. *Remote Sensing of Environment* 42:187–216.

Sellers, P. J., C. J. Tucker, G. J. Collatz, et al. 1994. A global 1-degrees by 1 degrees NDVI data set for climate studies. The generation of global fields of terrestrial biophysical parameters from the NDVI. *International Journal of Remote Sensing* 15:3519–3545.

Still, C. J., J. T. Randerson, and I. Y. Fung. 2004. Large-scale plant light-use efficiency inferred from the seasonal cycle of atmospheric CO_2. *Global Change Biology* 10:1240–1252.

Tong, X. J., J. Li, and L. Wang. 2008. A review on radiation use efficiency of the cropland. *Chinese Journal of Ecology* 27:1021–1028.

Turner, D. P., S. T. Gower, W. B. Cohen, M. Gregory, and T. K. Maiersperger. 2002. Effects of spatial variability in light use efficiency on satellite-based NPP monitoring. *Remote Sensing of Environment* 80:397–405.

Turner, D. P., W. D. Ritts, W. B. Cohen, et al. 2005. Site-level evaluation of satellite-based global terrestrial gross primary production and net primary production monitoring. *Global Change Biology* 11:666–684.

Turner, D. P., S. Urbanski, D. Bremer, et al. 2003. A cross-biome comparison of daily light use efficiency for gross primary production. *Global Change Biology* 9:383–395.

Zhao, Y. M., S. K. Niu, J. B. Wang, H. T. Li, and G. C. Li. 2007. Light use efficiency of vegetation: A review. *Chinese Journal of Ecology* 26:1471–1477.

7

Biomass Burning Emission Estimation in Amazon Tropical Forest

Y. E. Shimabukuro

National Institute for Space Research (INPE), Brazil

G. Pereira

National Institute for Space Research (INPE), Brazil; Federal University of São João del-Rei (UFSJ), Brazil

F. S. Cardozo, R. Stockler, S. R. Freitas, and S. M. C. Coura

National Institute for Space Research (INPE), Brazil

CONTENTS

7.1 Introduction 126
7.2 Methods for Estimation of Biomass Burning Emission 128
7.3 Biomass Distribution in Amazon Tropical Forest 129
7.4 Materials and Methods 129
     7.4.1 Thermal Anomalies Detections 129
     7.4.2 FRP Integration 130
     7.4.3 CCATT-BRAMS Model and Biomass Burned Estimation 134
     7.4.4 Field Data and Inventory Comparison 135
7.5 Results and Discussion 137
     7.5.1 FRE Distribution 137
     7.5.2 Aerosols and Trace Gases Emission Estimation for
            Amazon Tropical Forest 139
     7.5.3 Emission Model Assessment 142
7.6 Conclusions 143
Acknowledgments 144
References 144

7.1 Introduction

Ecosystem services are processes by which the environment produces resources that are useful for humans. These services are extensive and diverse and determine the quality of our land, water, food, and health in general. There are a wide range of benefits obtained from the environment (CIFOR 2009; Haines-Young and Potschin 2013). In the "ecosystem services cascade" (Figure 7.1), services can have social as well as economic values. Most ecosystems, whether they are artificial, seminatural, or wholly natural, are multifunctional and capable of delivering market and nonmarket benefits. Figure 7.1 depicts a reasonable perspective where the ecosystem services are interrelated with the things that people directly use and value. These services can represent inputs to the economy (provisioning services, such as timber) or services to the economy such as the assimilation and processing of waste (regulating services). Social values can include cultural significance as well as moral and aesthetic worth for people. For instance, the carbon sequestration and water regulating services of plantation woodlands would not seem to be regarded as flows from the environment in the System of Economic and Environmental Accounts (SEEA 2003) model if we apply the term "natural flows" in a strict way, whereas they would be under the more conventional ecosystem service paradigm of Haines-Young and Potschin (2013).

Regarding forested ecosystems, Fearnside (1985) first proposed plans for managing forest timber and nontimber products. The same author categorizes the environmental services into three categories: biodiversity, water,

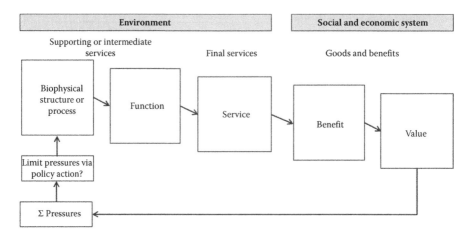

FIGURE 7.1
Ecosystem services cascade. (Adapted from Haines-Young, R. H., and M. Potschin. 2013. Common International Classification of Ecosystem Services (CICES): Consultation on Version 4, August–December 2012.)

and avoiding global warming (Fearnside 1997, 2000). Discussion of ecosystem services has since evolved to focus on "payments for ecosystem services" programs where the government pays landholders a stipend for such services as watershed maintenance. The concept of ecosystem services potentially helps to describe some of the ways that humans are linked to, and depend on, nature. It is also challenging because the connection between people and nature is complex, and different experts look at it in different ways. The ecosystem services, based on a new and hierarchical structure, are designed to eliminate overlapping and redundancies in categories levels. As a result, the Common International Classification of Ecosystem Services (CICES) methodology can be regarded as a classification sensu stricto (Haines-Young and Potschin 2010, 2013).

In this context, land use and land cover changes are considered as the major modifiers of the environment. Biomass burning is continuously used to clear large areas of vegetated regions, consuming large amounts of biomass and releasing a high amount of trace gases and aerosols into the atmosphere. Biomass burning plays a trade-off role in the ecosystem services arena. On the one side, it is used for opening spaces for agriculture or livestock, for controlling pests, or for recycling nutrients. On the other side, fires affect biodiversity patterns, alter the atmospheric composition and climate regulation services, modify the carbon-cycling, affect the energy balance between the atmosphere and the land surface as well as the biogeochemical and hydrological cycles, and may damage houses, infrastructures, and human lives.

Biomass burning modifies the physical–chemical and biological characteristics of Earth's surface, the atmosphere energy budget, and the climatic system (Andreae and Merlet 2001; Ichoku and Kaufman 2005; Cardoso et al. 2009). Furthermore, as the biomass burning emissions can be transported to distant regions, this phenomenon extrapolates from the local scale and modifies the energy balance in regional and global scales (Kaufman et al. 1995; Andreae et al. 2004). In addition, some gases emitted during the combustion process (CO, CO_2, CH_4, non-methane hydrocarbons, nitric acid, among others) are chemically active and interact with the hydroxyl concentration (OH), altering the efficiency of oxidation and modifying the tropospheric ozone, one of the greenhouse gases.

As noted, the biomass burning triggers depletory effects in several ways in three sections summarized by CICES. The consequences of the burning depend on the size and durability of the fire, causing damage to the biodiversity even under the soil, in micro and macro fauna, and in nutrient cycling, and affecting air quality and atmospheric composition. Moreover, there are some studies linking respiratory diseases and other health problem outbreaks in regions with annual high-intensity burning activities (Ignotti et al. 2010).

In the Amazon tropical forest, biomass burning is frequently associated with the process of agricultural expansion and production, such as deforestation, pasture renewal, and pest controls (Sampaio et al. 2007;

Cardoso et al. 2008; Marengo et al. 2010). The biomass burning period occurs during the dry season, between June and October (Crutzen and Andreae 1990); this is mainly due to a lack of significant precipitation and to the presence of low moisture in vegetation, which increases the vulnerability to fires (Nobre et al. 1991; Moraes et al. 2004). Moreover, initiation and maintenance of fire are influenced by factors such as the type of biomass, air temperature, humidity, and wind velocity (Freitas et al. 2005; Werf et al. 2006; Fearnside et al. 2009).

The general goal of this chapter is to show how remote sensing tools can provide valuable information in the assessment of the direct and indirect effects of biomass burning on ecosystem services. In particular, this work analyzes two different methodologies to estimate and assimilate biomass burning emissions into numerical models of air quality on Earth. In this work, Moderate Resolution Imaging Spectroradiometer (MODIS) sensors onboard Terra and Aqua platforms and Geostationary Operational Environmental Satellite Imager sensors are used to estimate the amount of aerosols and trace gases released into atmosphere from Amazon tropical forest biomass burning.

7.2 Methods for Estimation of Biomass Burning Emission

Traditional biomass emission methods for estimating aerosols and trace gases generally utilize emission factors associated with fuel load characteristics and burned biomass of dry mass (Andreae and Merlet 2001). Furthermore, while emission factors for different species are accurately known, burn efficiency depends on fuel load moisture content (Chuvieco et al. 2004) and burned area, which is usually available a long time after the fire is over (Roy et al. 2002; Silva et al. 2005). Recently, new methods have been developed for deriving the burned biomass and fire emissions by estimating fire radiative power (FRP) from environmental satellites (Wooster et al. 2003, 2005; Ichoku and Kaufman 2005). FRP can be defined as that part of the chemical energy emitted as radiation in the biomass burning process. The temporal integration of FRP gives the fire radiative energy (FRE).

In theory, radiative intensity released by fires is linearly correlated with the burned biomass and might be independent of vegetation type (Wooster et al. 2005; Freeborn et al. 2008). Also, measurements by satellites of FRE released rates could be associated with aerosol optical depth (AOD) and biomass burned coefficients to provide regional smoke emission (Wooster et al. 2003; Ichoku and Kaufman 2005). These methods allow near-real-time estimation of the concentration of aerosols and trace gases emitted into the atmosphere using chemistry transport models (Chatfield et al. 2002; Horowitz et al. 2003; Freitas et al. 2009).

7.3 Biomass Distribution in Amazon Tropical Forest

The total of aboveground biomass in the Amazon basin and in other ecosystems exhibits significant variations depending on the methodology adopted. Factors such as the amount of carbon in vegetation and the carbon sequestration in a burned area are difficult to calculate. Saatchi et al. (2007) estimated the total biomass in the Amazon basin as 86 petagrams (Pg), approximately 300 to 400 tons per hectare (ha). Furthermore, the total amount of aboveground biomass for the Amazon basin was estimated at 4 kgC·m^{-3} to 15 kgC·m^{-3} by Olson et al. (2002). Many studies such as Houghton et al. (1999, 2001) and Fernandes et al. (2007) analyzed the amount of biomass consumed by fires or land use and land cover changes in South America; however, the agreement in biomass estimation is not so evident. Houghton et al. (1999) and Tian et al. (1998) estimated the annual net flux of carbon to the atmosphere from changes in land use as 0.2 and 0.3 PgC·y^{-1}, respectively. Moreover, Pereira et al. (2011) and Kaiser et al. (2012), using FRE, estimated 0.28 PgC·y^{-1} (for 2002) and 0.33 PgC·y^{-1} (annual average) for the total consumed by fires in South America biomass burning.

In South America, several weather systems can change the transport of burning biomass emissions, modifying the chemical composition of air, the radiation budgets, and the clouds' microphysical properties (Freitas et al. 2005). In the Amazon basin, fire flame temperature can exceed 1600 K and usually ranges between 830 and 1440 K (Riggan et al. 2004). With the high energy intensity released in the biomass burning activity, the plume emitted into the atmosphere can reach more than 6000 m above ground level. As a consequence, to analyze the effect of biomass burning at different scales, it is essential to estimate (spatially and temporally) the emission of trace gases and aerosols. In this context, this work aims to analyze two different methodologies to estimate and assimilate biomass burning emissions into numerical models of air quality on Earth.

7.4 Materials and Methods

7.4.1 Thermal Anomalies Detections

The MODIS sensors onboard Terra and Aqua platforms have polar orbit, an imaging angle of ±55° and an altitude of 705 km. The time overpasses vary over a given point of the surface according to the platforms: Although the Terra platform (whose products receive the denomination MOD) crosses the equator on its descending orbit at 10:30 am and 10:30 pm, the Aqua platform (whose products are denominated MYD), in its ascending orbit, crosses the equator at 01:30 pm and 01:30 am (Giglio 2005). The products that contain information about fires and FRP are called MOD14 and MYD14. The Terra and Aqua MODIS instruments acquire data twice daily according to the platforms' overpasses. These four daily MODIS fire observations serve to advance global monitoring of the fire process and its effects on ecosystems, the atmosphere, and on climate. The science datasets in these products include fire-mask, algorithm quality, radiative power, and numerous layers describing fire pixel attributes and utilizing the method proposed by Kaufman et al. (1996). The MODIS fire products present a spatial resolution of 1 km² at nadir, reaching 3 km at high view zenith angles, and, commonly, provide two to four observations of determined land surface.

The Geostationary Operational Environmental Satellite (GOES) image sensor onboard the GOES constellation acquires information in five spectral bands located in the range of visible to thermal infrared of the electromagnetic spectrum. The algorithm for thermal anomalies detection uses the spectral bands centered at 3.9 and 10.7 μm (Prins and Menzel 1992) and is based on the method proposed by Roberts et al. (2005). The Wildfire Automated Biomass Burning Algorithm (WFABBA) (available at http://wfabba.ssec.wisc.edu/wfabba.html) is a product for fire detection based on the GOES satellite and utilizes the method proposed by Wooster et al. (2005) for FRP estimation, described in Xu et al. (2010). Currently, the GOES program maintains three satellites in operation that acquire information from South, Central, and North America at regular intervals (15–30 minutes); the METEOSAT satellite can be used for Africa and Europe. The estimation of thermal anomalies has been performed since 1997, covering a 16-year historic series. One advantage of using geostationary satellites with low spatial resolution (4 km at NADIR) is the high frequency of observations, which allows the life cycle characterization of particular fires (Figure 7.2). However, one disadvantage is that about 5% – 10% of Amazon fires detected by the sensor saturate the detectors centered at 3.9 μm, making it impossible to estimate large fire emissions (Pereira et al. 2009). Saturated pixels occur when the size and temperature of the fire exceed the sensitivity of the sensor to capture electromagnetic radiation, making it impossible to derive the FRP actual by the WFABBA method. Moreover, MODIS and GOES FRP estimations present some uncertainties such as simultaneous observations of fire and smoke (approximate error of ±11%), cloud cover (±11%), and the accuracy and consistency of FRP (±16%), among other factors (Vermote et al. 2009).

7.4.2 FRP Integration

Figure 7.3 shows the flowchart of FRP integration methodology. In the first step, FRP observations for a specific day are grouped according to time acquisition and product of origin. In this step, low-confidence fire pixels (values below 50% for MOD14 and MYD14 products and flags 4 and 5 for the WFABBA/GOES product) were eliminated from FRP integration.

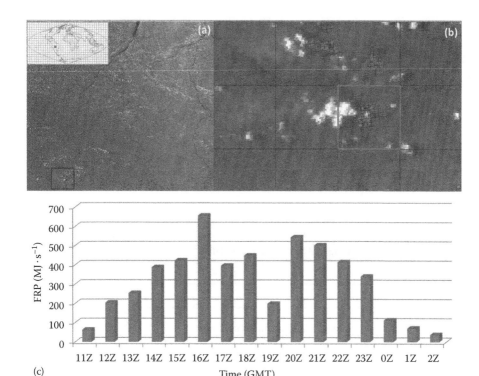

FIGURE 7.2 (See color insert.)
(a) MODIS/Aqua sensor image, 1B2G7R composition, characterizing the fires in the arc of deforestation on September 29, 2007, (in black); (b) detected and available fires by MYD14 product and squares used by convolution mask to cluster all fires; and (c) characterization of the life cycle related to fires found in red square and used for fire radiative power (FRP) integration ($MJ \cdot s^{-1}$) throughout the temporal series.

For MODIS FRP values, Equation 7.1 is applied to minimize the bow-tie effects, as described in Freeborn et al. (2011):

$$MODIS_{FRP} = FRP \ cos^2(\theta) \qquad (7.1)$$

where θ represents the MODIS view zenith angle for a particular fire pixel.

Pereira et al. (2009) analyzed 9 years of WFABBA data and found that 6.6% of detected fires saturate. Therefore, saturation percentage could vary due to variations in seasonal weather characteristics and biomass burning activities, such as atypical rainy periods or strong droughts in the Amazon basin. However, instead of removing GOES-saturated pixels without FRP estimation and ignoring important episodes of biomass burning, an alternative was used to estimate FRP based on the method of middle-infrared (MIR) radiance (Equation 7.2). This methodology is based on the premise that emitted

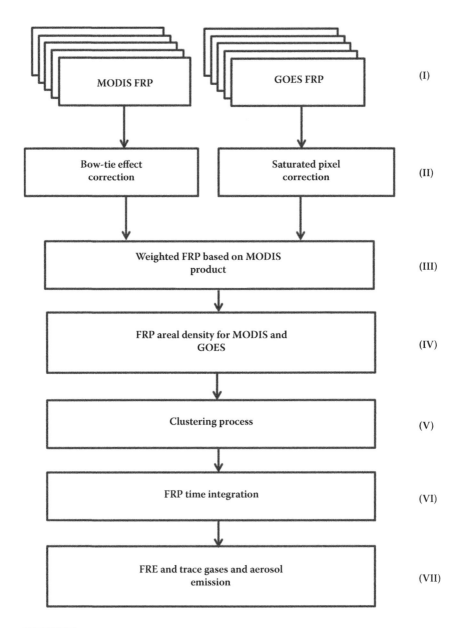

FIGURE 7.3
Flowchart of the methodology divided into seven steps: (I) Data acquisition; (II) correction of
bow-tie effect and pixel saturation; (III) FRP adjustment; (IV) FRP areal density estimation;
(V) clustering process; (VI) FRP integration; and (VII) FRE and trace gases and aerosol emission estimation.

spectral radiance (M_λ) in a spectral band centered at 3.9 μm is proportional to FRP (Wooster et al. 2003, 2005):

$$FRP_{MR} = \frac{Ag}{a}\sigma \int_{3.76}^{4.03} M(\lambda, T)_d \lambda - M_b \qquad (7.2)$$

$$M(\lambda, T) = \frac{c_1}{\lambda^5 \left(\exp\left(\frac{c_2}{\lambda T}\right) - 1 \right)}$$

where Ag represents the area of GOES pixel; a is a constant fit based in GOES MIR spectral band; M_λ is the emitted spectral radiance; c_1 and c_2 are constants (3.74×10^8 W·m^{-2} and 1.44×10^4 μm·K, respectively); λ is the wavelength (μm); T represents the temperature (Kelvin), and M_b is the radiance emitted by the background (110 MW).

After correcting bow-tie effects for MODIS fire products and assigning values to GOES-saturated pixels without FRP estimation (second step), the third step consists of correcting GOES FRP values according to their range values. Xu et al. (2010) compared the FRP coincident values of MODIS and GOES with time differences less than 10 minutes and a viewing zenith angle less than 30° and found that approximately 90% of GOES FRP values differ by less than 50% from the coincident MODIS value. In this study, FRP values estimated by GOES satellites below 1000 MW are corrected by 17%, and FRP values higher than 1000 MW are corrected by 41% (Xu et al. 2010). This procedure is also applied in the Spinning Enhanced Visible and InfraRed Imager (SEVIRI) onboard Meteosat Second Generation (MSG), but due to spatial coverage we decided not to use these data in Amazon biomass burning emission estimation.

For GOES and MODIS FRP integration, the method proposed by Kaiser et al. (2012) for MODIS was used for both satellites. In the fourth step, MODIS and GOES FRP areal density is calculated by weighting the FRP values by pixel area. In this step, a water bodies map derived from a Brazilian official land use and land cover map (MMA) is utilized to correct eventual errors in areal density. In this context, the fifth step of the algorithm consists of determining the grid configuration necessary for the clustering process. In this study, we decided to estimate the FRE on regular grids with a spatial resolution of 0.17°, which is the current operational Coupled Chemistry-Aerosol-Tracer Transport Model coupled to Brazilian Regional Atmospheric Modeling System (CCATT-BRAMS) model resolution.

The clustering process performs a combination of all detected fires from different sensors. In this step, the size of a matrix that merges FRP data could be defined according to CCATT-BRAMS spatial resolution and grid configuration. Consequently, the convolution mask $\eta(\gamma, \kappa)$, of size M × N (rows × columns), running over the grid with FRP areal density values estimated by different satellites $\xi(\text{lon,lat})$ will result in the grid $(\text{FRP}_{\text{grid}})$ containing all clustered fires for a given time step

$$\text{FRP}_{\text{grid(lon,lat,t)}} = \sum_{\gamma=-\alpha}^{\alpha} \sum_{\kappa=-\beta}^{\beta} \eta(\gamma,\kappa)\xi(\text{lon}+\gamma, \text{lat}+\kappa, t) \qquad (7.3)$$

where the clustered grid is defined to all points where the mask of $M \times N$ size overlaps the image completely (lon ε $[\alpha, M -\alpha]$, lat ε $[\beta, N -\beta]$). In this example, the algorithm developed in this study runs through the matrix in a mask size of 400 km² (20 km × 20 km). Accordingly, clustered FRP temporal evolution values for all fires detected by Aqua, Terra, GOES-10, GOES-11, GOES-12, GOES-13, and GOES-15 satellites are stored. Based on the FRP values and their respective times of occurrence for each grid point, the FRE was calculated by the following equation (sixth step):

$$\text{FRE}_{\text{grid(lon,lat)}} = \frac{1}{2}\sum_{i=1}^{n}(\text{FRP}_n + \text{FRP}_{n+1})\cdot(T_{n+1} - T_n) \qquad (7.4)$$

where $\text{FRE}_{\text{lon,lat}}$ represents the geographic location (longitude and latitude) of a particular point of the regular grid; T is the interval between the observations; and n represents the observation. For this estimation, it is assumed that the spatial distribution observed in part of the regular grid is representative for its totality. However, if the interval between two acquisitions is greater than four hours ($\Delta T > 14400$ s), it is assumed the hypothesis of two or more independent fires, then the algorithm initiate a new integration of FRP values (T = 0). After FRE estimation, these data are assimilated into a CCATT-BRAMS model and the injection of trace gases and aerosols is calculated (the seventh step).

7.4.3 CCATT-BRAMS Model and Biomass Burned Estimation

The CCATT-BRAMS, developed to simulate the atmospheric circulation at various scales, is based on BRAMS numerical model (Freitas et al. 2009). In this model, the transport of trace gases and aerosols is made simultaneously with the evolution of the atmospheric state, using the same time step and the same physical and dynamics parameterizations of the atmosphere. The mass conservation equation for CO and for PM_{25} is calculated by the following tendency equation (Freitas et al. 2009):

$$\frac{\partial \bar{s}}{\partial t} = \underbrace{\left(\frac{\partial \bar{s}}{\partial t}\right)_{\text{adv}}}_{\text{I}} + \underbrace{\left(\frac{\partial \bar{s}}{\partial t}\right)_{\text{PBL diff}}}_{\text{II}} + \underbrace{\left(\frac{\partial \bar{s}}{\partial t}\right)_{\text{deep conv}}}_{\text{III}} + \underbrace{\left(\frac{\partial \bar{s}}{\partial t}\right)_{\text{shallow conv}}}_{\text{IV}} + \underbrace{\left(\frac{\partial \bar{s}}{\partial t}\right)_{\text{chem CO}}}_{\text{V}}$$

$$+ \underbrace{W_{PM2.5\mu m}}_{\text{VI}} + \underbrace{R}_{\text{VII}} + \underbrace{Q_{\text{pr}}}_{\text{VIII}} \qquad (7.5)$$

where s is the grid box mean tracer mixing ratio, term (I) is the 3D transport (advection by mean wind), term (II) represents the subgrid-scale diffusion in the planetary boundary layer (PBL), and terms (III) and (IV) are the subgrid transport by deep and shallow convection, respectively. Term (V) is applied to CO, which is treated as a passive tracer with a lifetime of 30 days (Seinfeld and Pandis 1998), term (VI) is the wet removal applied to $PM_{2.5\ \mu m}$; term (VII) refers to the dry deposition applied to gases and aerosol particles; and term (VIII) is the source term that includes the plume rise mechanism associated with the vegetation fires (Freitas et al. 2009; Longo et al. 2010).

In CCATT-BRAMS, the module that performs the burned biomass estimation is denominated Brazilian Biomass Burning Emission Model (3BEM); see Longo et al. (2010) for more details. In this model, the burned biomass is estimated from the carbon present in living vegetation (Olson et al. 2000; Houghton et al. 2001) data. For a given pixel in the CCATT-BRAMS model, the total emission source for a given gas or aerosol and its variation during the days is estimated by 3BEM using FRE (Equation 7.6) and using the method that utilizes combustion factors and the burned biomass method (Equation 7.7):

$$Q^{[\epsilon]}_{plumerise}(t) = \frac{gf(t)}{p_0 \Delta V} \cdot (EF^{[\epsilon]} \cdot \vartheta \cdot FRE_{grid(lon,lat)}) \tag{7.6}$$

$$Q^{[\epsilon]}_{plumerise}(t) = \frac{gf(t)}{p_0 \Delta V} \cdot (EF^{[\epsilon]} \cdot BBurned) \tag{7.7}$$

where t is the time (s); gf(t) represents a Gaussian function, centered at the period of maximum emission and used to simulate the diurnal cycle; p_0 is related to the weather conditions; ΔV is the volume of the pixel; EF represents the emission factor for a given species (ϵ); ϑ is the coefficient related to the rate of biomass consumption (1.37 $kg \cdot MJ^{-1}$; Kaiser et al. 2012); and BBurned is the burned biomass estimated by the 3BEM model.

7.4.4 Field Data and Inventory Comparison

The Acre State, located in western Brazilian Amazonia (marked as 1 in Figure 7.4), has three, well-preserved phytoecological regions: (1) the area of campinarana, shrub, and forested physiognomies with no palm trees in the northwest of the state; (2) the rainforest region, presenting the alluvial formation with emergent canopy, located in two small areas, in the northwest and in a small strip in the central area, close to the border of Amazonas State (marked as 2 in Figure 7.4); and (3) the open rainforest region, which is the main phytoecological region of Acre, located throughout the state, presenting the alluvial formation with palm trees, the submontane formation with emergent canopy, the lowlands formation with uniform and emergent canopy, and the alluvial

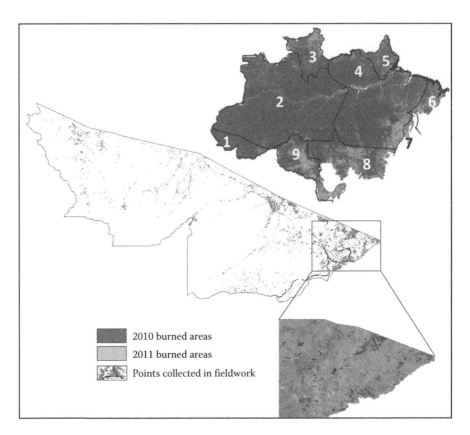

FIGURE 7.4 (See color insert.)
Brazilian State boundaries: (1) Acre; (2) Amazonas; (3) Roraima; (4) Pará; (5) Amapá; (6) Maranhão; (7) Tocantins; (8) Mato Grosso; and (9) Rondônia; and 2010 (red) and 2011 (yellow) burned areas map of Acre State with points collected in the fieldwork.

formation with the presence of palms trees and bamboos. Moreover, the Acre State has anthropic areas, especially pasturelands for livestock use (which is the main activity of the state), and has insignificant secondary vegetation located in areas associated with grasslands (IBGE 2005).

The burned areas maps for Acre State were derived from thematic mapper (TM)/Landsat 5 images for the time periods 2010 and 2011. The estimate of burned areas was performed by selecting all available images that presented burned scars in the study area for these two years, which accounted for 15 path/row and 56 images. Initially, the TM images were loaded in the SPRING 4.3.1 software developed by INPE (Câmara et al. 1996). In this software, all images were geometrically corrected using a polynomial model and were interpolated by the nearest neighbor algorithm.

The burned areas for Acre State were classified using image segmentation, a process that creates polygons with homogeneous spectral characteristics through a similarity threshold (minimum image gray value for which two regions are considered similar and grouped into a single polygon) and an area threshold (minimum area to be individualized defined by number of image pixels). In this study, the similarity and area thresholds were defined as 12 and 8, respectively, after several tests for achieving the best classification results of burned scars.

Thus, the burned scars were defined by polygons and a manual image editing was performed to minimize commission and omission errors resulting from the digital classification process (Almeida-Filho and Shimabukuro 2002). To validate the burned area maps, two field studies were performed (Figure 7.4). In these field studies, 33 ground points were visited in order to validate the burned area maps derived from TM/Landsat 5 images. Furthermore, due to the inaccessibility of some ground points regions, the collected samples were located near the main roads. The comparison of collected field data and mapped area presented an overall accuracy of 93%, with 6% omission and 1% commission errors. In this case, omission errors could be explained by regrowth of vegetation smoothing the burn scars in the satellite images.

The statistical analysis of 2010 polygons indicated that mapped burn scars showed a minimal size of 0.01 ha and a maximum size of 3660 ha, with an average value of 15 ha. Individually, the polygons with a size between 2 and 6 ha represented 54% of total mapped areas, while polygons with a size between 7 and 20 ha represented 29% of mapped areas. Also, the smaller burn scar areas, covering 0.01–1 ha, represent only 2% of the total number of polygons and larger burn scar areas (>20 ha) comprise 13% of the total mapped polygons, indicating that for the year 2010, the highest number of occurrences of burned areas was observed for small areas.

For the year 2011, the average value of burned areas was 14 ha, with a minimum size of 0.01 ha and a maximum size of 1383 ha, a smaller area when compared to 2010 burn scars. Also, the burned areas between 1 and 8 ha represented 58% of all occurrences, followed by burned polygons between 9 and 20 ha, with 27% of the total; moreover, larger burn scars (>20 ha) comprised 14% of total mapped polygons, maintaining the same pattern as 2010 but with a lower occurrence of burned areas in the study area.

7.5 Results and Discussion

7.5.1 FRE Distribution

Figure 7.5 shows the spatial distribution of FRE for the Amazon biome in the 2000 to 2011 time periods, estimated from MODIS and GOES satellites data. Commonly, the highest values of FRE are located at the border of the Amazon rainforest, known as the arc of deforestation; it is in this region that the main

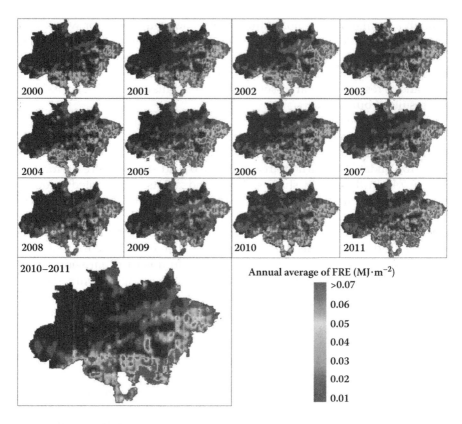

FIGURE 7.5 (See color insert.)
Spatial distribution of fire radiative energy (FRE) (MJ·m⁻²) for Amazon biome for the period
between 2000 and 2011, estimated from daily observations.

forest logging and subsequent agricultural expansion occur. In many areas,
the annual average of FRE reaches values higher than 0.07 MJ·m^{-2} in the
Brazilian Amazon. Thus, considering the grid area (400 km²), the energy
released in some grid points is equivalent to approximately 0.2 PJ·y^{-1}.
Integrating these values, we can estimate the energy released per year for
Amazon biome as approximately 400 PJ. Kaiser et al. (2012), using MODIS
fire products, estimated the monthly FRE for South America at 100–450 PJ;
however, these values change according to burning season and rainfall
regime.

In Figure 7.5, it is possible to verify whether the FRE values present a tem-
poral and spatial variability resulting from economic processes (Ewers et al.
2008) and climatic factors (Barlow and Peres 2004; Good et al. 2008). In the
Amazon biome, burning events occur mainly in the dry season, between
July and October, with some incidences of fires in the dry-to-wet transition.
Analyzing the spatial variability, the years with higher incidences of fires

and biomass consumption were 2002, 2005, and 2007. In these years, the spatial distribution of FRE is located mainly in the Brazilian states of Mato Grosso, Pará, and Rondônia, which comprise the frontier of agriculture and livestock expansion in Brazil. In the temporal distribution of FRE, it is possible to verify anomalous episodes of burning; for example (marked with the symbol ◻): (1) the episode occurred in the Acre State in 2005, exceeding considerably the annual average for the region; (2) the intense FRE values in 2003 and 2007 in Roraima State. Moreover, it was observed that in the rainiest years, the energy released was reduced significantly in all regions (such as observed in 2008 and 2009).

7.5.2 Aerosols and Trace Gases Emission Estimation for Amazon Tropical Forest

To calculate the trace gases and aerosols released into atmosphere, the CCATT-BRAMS model, in its original form, uses the 3BEM method developed by Longo et al. (2010). Thus, for a given pixel detected as burned by WFABBA or MOD14/MYD14 products, after a filter is applied to remove repeated fires in a radius of 1 km and the type of land use and land cover is determined according to Belward (1996) and updated with MODIS products, the emission estimate is obtained by the following equation:

$$M^{[\epsilon]} = \alpha_{veg} \cdot \beta_{veg} \cdot EF^{[\epsilon]} \cdot a_{fogo} \qquad (7.8)$$

where $M^{[\epsilon]}$ represents the species emission, α_{veg} e β_{veg} characterizes the aboveground biomass fraction and the burning efficiency of vegetation described in Table 7.1, $EF^{[\epsilon]}_{veg}$ is the emission factor for each species, and a_{fogo} is

TABLE 7.1

Combustion Factor and Aboveground Biomass Utilized by 3BEM

IGBP LULC Legend	Combustion Factor	Aboveground Biomass $(kg \cdot m^{-2})$
Evergreen broadleaf forest	0.5	29.24
Deciduous broadleaf forest	0.43	12.14
Mixed forest	0.43	12.14
Closed shrublands	0.87	7.40
Open shrublands	0.72	0.86
Woody savannas	0.45	10.00
Savannas	0.52	6.20
Grasslands	1.00	0.71
Permanent wetlands	0.40	3.80
Croplands	0.40	3.80
Cropland/natural vegetation mosaic	0.40	3.80
Barren or sparsely vegetated	0.84	1.00

the burned area. Correspondingly, in FRP processing and integration, the total biomass consumed was obtained for each grid of 20 km. Thus, for each detection, the land use and land cover were estimated, and these were applied to the emission factors for each species.

Figure 7.6 shows the temporal variation (January 2000 to December 2011) of CO estimation (10^{-6} kg · m^{-2}) for all of the Amazon biome, estimated by 3BEM and FRE methods. The comparison of these two methods indicates that, in general patterns, both generally agree (Figure 7.6c), with a trend of greater values of trace gases and aerosols estimated by the 3BEM method (*t*-student, p > 0.05, n = 144). Figure 7.6a and b shows the values of the angular coefficient and correlation, respectively, estimated from the bootstrap technique originally developed by Efron (1982). In this technique, a population of 1.0×10^4 reconstructs the original curve and provides the parameters to create the confidence interval and error analysis for the model estimation, providing measures of accuracy for sample estimates.

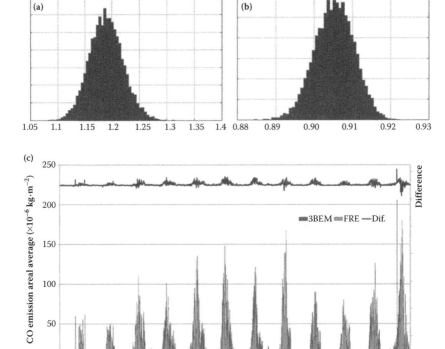

FIGURE 7.6
(a) Angular coefficient distribution and (b) correlation for CO data estimated by 3BEM (x axis) and by FRP (y axis) from the bootstrap technique; (c) estimation of the mean values of CO from 3BEM and FRE for the period between 2000 and 2011.

Looking at Figure 7.6c, we can see that, in low burning periods, the two models presented very similar values, and the years 2000 and 2001 can be cited as examples in which the estimated values of CO from FRE are 12% below those estimated by 3BEM for the months from October to December, as seen in the difference graph (3BEM minus FRE) at the top of Figure 7.5c. In general, a difference of 20% is detected for CO estimation with both methods, except for the year 2011, in which the CO values estimated from FRE in some days are larger than the values estimated by 3BEM, especially in the months of September and October. The factors that may cause this variation may be the greater number of observations at shorter intervals (new data from GOES-13) and the change in the detection algorithm and in FRP estimation (Xu et al. 2010), among others.

Figure 7.7 shows the CO average ($kg \cdot m^{-2}$) estimated for the period between January 2000 and December 2011 based on FRE (Figure 7.6a) and 3BEM (Figure 7.6b). Among the main differences between the two methods are the increase in biomass burned emissions in the eastern portion of the image (the Tocantins and Maranhão States, marked with ¤) that represents a transition between the Amazon forest and the Brazilian cerrado; and the increase in emissions in the southern Mato Grosso State and central Bolivia. In South America, the fires are mainly linked to agricultural activities, which presents a high variability. Thus, the highest incidence of fires occurs in the arc of deforestation located primarily in the Mato Grosso, Pará, and Rondônia States. In these areas, the density of fires can reach 125 observations per km^2. Over the past 16 years, Mato Grosso State has been the region with the largest number of fires, exceeding 1.3×10^6 observations, followed by Pará State with 0.9×10^6 observations (INPE 2012).

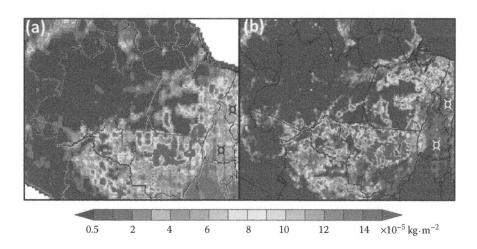

FIGURE 7.7 (See color insert.)
Average distribution of CO ($kg \cdot m^{-2}$) for the Amazon biome between 2000 and 2011 estimated from the (a) FRE and (b) 3BEM methods.

Although we can notice some differences in spatial distribution of CO between the two methods, in general, there is an agreement between the analyzed models. Among the factors that could cause such differences are the use of parameters such as the fraction of aboveground biomass, burning efficiency of vegetation, and areas burned by traditional methods. The characterization and estimation of these variables are very complex, depending on weather and vegetation moisture. Also, the fraction of aboveground biomass is unlikely to be homogeneous for large areas, and near-real-time burned area is one of the major uncertainties in remote sensing data estimation (Chuvieco et al. 2004; Yebra et al. 2008).

The trace gases and aerosol estimations derived from FRP directly relate to biomass burning. In addition, factors that affect the burning efficiency directly influence the energy released by fires, such as the decrease of soil and vegetation moisture. However, in this method, the sources of error are associated with characterization of the diurnal cycle of FRP and different sensitivities of orbital sensors. In addition, some factors could generate some uncertainties, such as the simultaneous observations of biomass burning and the smoke plume (approximate error ±11%), the accuracy and consistency of FRP (±16%), and cloud cover (±11%), among other factors, as noted by Vermote et al. (2009).

7.5.3 Emission Model Assessment

Figure 7.8 shows the mean concentration of CO (ppb) for July 15, 2010, to November 11, 2010 (Figure 7.8a), and July 15, 2011, to November 15, 2011 (Figure 7.8b), and the location of fires detected by MODIS and GOES for the same period (Figure 7.8c and d, respectively). The occurrence of fires was higher in 2010 than 2011. According to Figure 7.8c, the highest incidence of burned areas occurred in the eastern region of the Acre State, coinciding with the area that presents the highest rates of deforestation, as calculated by the Program for Deforestation Assessment in the Brazilian Legal Amazonia (PRODES). For the year 2010, 2000 km^2 of burned areas were mapped, representing 1.3% of the total Acre State area. Similarly, it was noted that the occurrence of fires decreased considerably in the following year (2011), but the spatial distribution of fires remained similar to 2010, with the eastern region presenting the highest incidence of fires. The burned area decreased to 643 km^2, a reduction of 68%. Among the factors that contributed to this decreased incidence of fires is the difference in annual rainfall; 2010 was drier than 2011, according the National Institute of Meteorology (INMET).

Furthermore, it is important to emphasize that another factor that could be primordial in the reduction of burned areas is the implementation of public policies, such as the increase of surveillance, the certification of properties, and alternative techniques to the use of fire. Recently, Acre State has created a series of programs called the Valuation Policy of Acre Environmental Forest Asset (PVAAFA), the result of a partnership with government institutions

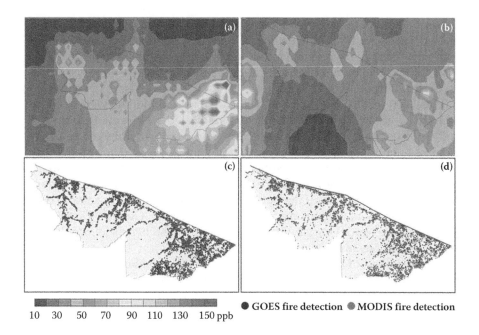

FIGURE 7.8 (See color insert.)
CO concentration modeled by CCATT-BRAMS to (a) 2010 and (b) 2011 using FRE approach; location of fires detected by MODIS and GOES for (c) 2010 and (d) 2011.

and civil society organizations. From the inventory of burned areas for 2010 and 2011, the total biomass burned was estimated from the spatial distribution of live aboveground biomass (Saatchi et al. 2007). Consequently, it was estimated that for 2010 and 2011, only in Acre State, 28.3 and 8.7 Tg of biomass were exposed to the combustion process, respectively. In general terms, comparing these values with those estimated by 3BEM and FRE, the models can reproduce 82% and 76% of all biomass consumed for 2010 and 85% and 78% for 2011, showing an agreement between observed and modeled data.

7.6 Conclusions

Many human activities disrupt, impair, or reengineer ecosystems every day. Biomass burning modifies many ecological processes that are vital to maintaining ecosystem services and the biological sustainability of Earth. In this study, we demonstrate the large number of direct and indirect effects of biomass burning on the ecosystem services at distinct spatial and temporal scales. In this context, remote sensing products can provide valuable information for estimating fire detection and its emission into atmosphere. Remote sensing high spatial temporal repetitiveness allows characterization of the life cycle of biomass burning and its assimilation into numerical models of air quality.

From 2000 to 2011, the Amazon biome displayed a high incidence of fires that varied according to climatic factors associated with rainfall anomalies and socioeconomic processes. In the Amazon biome, the highest values of biomass burning emissions are concentrated in the arc of deforestation and linked with forest logging and subsequent agricultural expansion. In many regions, daily FRE values could be higher than $0.07\ MJ \cdot m^{-2}$ or approximately $0.2\ PJ \cdot y^{-1}$ and occur essentially between July and October (dry season) with some incidences of fires in the dry-to-wet transition (October–December). Analyzing the spatial variability, the years with a higher incidence of fires and biomass consumption were 2002, 2005, and 2007.

The evaluation of the emission method that uses combustion factors and biomass and the FRE-based method indicates that, in general patterns, both methods exhibit similarities, with a tendency toward greater values of trace gases and aerosols when estimated by the 3BEM method. Inventory data acquired from Acre State indicate that 3BEM and FRE methods can reproduce 82% and 76% of all biomass consumed for 2010 and 85% and 78% for 2011. Thus, new methods that utilize FRP represent a substantial improvement in biomass burning emission estimation, especially for real-time applications. However, the FRE-based method shows promise for accuracy development and new parameterizations, minimizing errors in several approximations.

Acknowledgments

The authors acknowledge the financial support from CAPES, CNPQ (479626-2011-1), and FAPESP (2010/07083-0, 2010/17437-4).

References

Almeida-Filho, R., and Y. E. Shimabukuro. 2002. Digital processing of a Landsat-TM time series for mapping and monitoring degraded areas caused by independent gold miners, Roraima State, Brazilian Amazon. *Remote Sensing of Environment* 79:42–50.

Andreae, M., D. Rosenfeld, P. Artaxo, et al. 2004. Smoking rain clouds over the Amazon. *Science* 303:1342–1345.

Andreae, M. O., and P. Merlet. 2001. Emission of trace gases and aerosols from biomass burning. *Global Biogeochemical Cycles* 15:955–966.

Barlow, J. and C. A. Peres. 2004. Ecological responses to El Niño-induced surface fires in central Brazilian Amazonia: management implications for flammable tropical forests. *Philosophical Transactions of the Royal Society* 359:367–380.

Belward, A. 1996. The IGBP-DIS global 1 km land cover data set (DISCover) proposal and implementation plans. IGBP-DIS Working Paper 13.

Câmara, G., R. C. M. Souza, U. M. Freitas et al. 1996. Spring: Integrating Remote Sensing and GIS with Object-Oriented Data Modelling. *Computers and Graphics* 15(6):13–22.

Cardoso, M., C. A. Nobre, D. Lapola et al. 2008. Long-term potential for fires in estimates of the occurrence of savannas in the tropics. *Global Ecology and Biogeography* 17:222–235.

Cardoso, M., C. Nobre, G. Sampaio et al. 2009. Long-term potential for tropical-forest degradation due to deforestation and fires in the Brazilian Amazon. *Biologia (Bratislava)* 64:433–437.

Chatfield, R., Z. Guo, G. Sachse et al. 2002. The subtropical global plume in the Pacific Exploratory Mission-Tropics A (PEM-Tropics A), PEM-Tropics B, and the Global Atmospheric Sampling Program (GASP): How tropical emissions affect the remote Pacific. *Journal of Geophysical Research* 107:42–78.

Chuvieco, E., D. Cocero, D. Riano et al. 2004. Combining NDVI and surface temperature for the estimation of live fuel moisture content in forest fire danger rating. *Remote Sensing of Environment* 92(3):322–331.

CICES (Common International Classification of Ecosystem Services): Consultation on Version 4, August–December 2012. Available from: http://cices.eu/wp-content/uploads/2012/07/CICES-V43_Revised-Final_Report_29012013.pdf (accessed March 15, 2013).

CIFOR, 2009. Center for International Forestry Research: Realising REDD+: National strategy and policy options, ed. A. Angelsen, 363 p. Bogor, Indonesia.

Crutzen, P. J., and M. O. Andreae. 1990. Biomass burning in the tropics: Impact on atmospheric chemistry and biogeochemical cycles. *Science* 250:1669–1678.

Efron, B. 1982. *The jackknife, the bootstrap, and other resampling plans*. Society of Industrial and Applied Mathematics. CBMS-NSF Monographs, 38, California: Stanford University.

Ewers, R. M., W. F. Laurance, and C. M. Souza Jr. 2008. Temporal fluctuations in Amazonian deforestation rates. *Environmental Conservation* 35:303–310.

Fearnside, P. M. 1985. Uma estrutura para Avaliação de opções de Desenvolvimento Florestal na Amazônia [A Framework for Evaluation Forestry Development options in Amazonia]. Presentation at the 1° Seminário Internacional de Manejo em Floresta Tropical – SEMA/WWF, Serra do Carajás, Pará.

Fearnside, P. M. 1997. Greenhouse gas from deforestation in Brazilian Amazonia: Net committed emissions. *Climate Change* 35:321–360.

Fearnside, P. M. 2000. Global warming and tropical land use change: Greenhouse gas emissions from biomass burning, decomposition and soils in forest conversion, shifting cultivation and secondary vegetation. *Climate Change* 46:115–158.

Fearnside, P. M., C. A. Righi, P. M. L. A. Graça, et al. 2009. Biomass and greenhouse gas emissions from land-use change in Brazil's Amazonian "arc of deforestation": The states of Mato Grosso and Rondônia. *Forest Ecology and Management* 258:1968–1978.

Fernandes, S. D., N. M. Trautmann, D. G. Streets, et al. 2007. Global biofuel use, 1850–2000. *Global Biogeochemical Cycles* 21:GB2019.

Freeborn, P. H., M. J. Wooster, W. M. Hao, et al. 2008. Relationships between energy release, fuel mass loss, and trace gas and aerosol emissions during laboratory biomass fires. *Journal of Geophysical Research Atmospheres* 113:D01301.

Freeborn, P. H., M. J. Wooster, and G. Roberts. 2011. Addressing the spatiotemporal sampling design of MODIS to provide estimates of the fire radiative energy emitted from Africa. *Remote Sensing of Environment* 115:475–498.

Freitas, S. R., K. M. Longo, M. A. F. Silva Dias, et al. 2005. Monitoring the transport of biomass burning emissions in South America. *Environmental Fluid Mechanics* 5:135–167.

Freitas, S. R., K. M. Longo, M. A. F. Silva Dias, et al. 2009. The Coupled Aerosol and Tracer Transport model to the Brazilian developments on the Regional Atmospheric Modeling System (CATT-BRAMS). Part 1: Model description and evaluation. *Atmospheric Chemistry and Physics* 9:2843–2861.

Giglio, L. 2005. MODIS collection 4 active fire product user's guide. Version 2.2. 2005. Available from: http://maps.geog.umd.edu/products/MODIS_Fire_Users_Guide_2.2.pdf (accessed August 3, 2010).

Good, P., J. Lowe, M. Collins, et al. 2008. An objective tropical Atlantic SST gradient index for studies of South Amazon dry season climate variability and change. *Philosophical Transactions of the Royal Society* 363:1761–1766.

Haines-Young, R. H., and M. Potschin. 2010. *Proposal for a common international classification of ecosystem goods and services (CICES) for integrated environmental and economic accounting*. European Environment Agency, New York: Department of Economic and Social Affairs.

Haines-Young, R. H., and M. Potschin. 2013. Common International Classification of Ecosystem Services (CICES): Consultation on Version 4, August–December 2012.

Horowitz, L., S. Walters, D. Mauzerall, et al. 2003. A global simulation of tropospheric ozone and related tracers: Description and evaluation of MOZART, version 2. *Journal of Geophysical Research* 108:47–84.

Houghton, R. A., K. T. Lawrence, J. L. Hackler, et al. 2001. The spatial distribution of forest biomass in the Brazilian Amazon: A comparison of estimates. *Global Change Biology* 7:731–746.

Houghton, R. A., D. L. Skole, C. A. Nobre, et al. 1999. Annual fluxes of carbon from deforestation and regrowth in the Brazilian Amazon. *Nature* 403:301–304.

IBGE (Instituto Brasileiro de Geografia e Estatística). 2005. Vegetação do Estado do Acre. Escala 1:1.000.000. Avaliable from: ftp://geoftp.ibge.gov.br/.../manual_tecnico_vegetacao_brasileira.pdf (accessed 16 March 2013).

Ichoku, C., and Y. J. Kaufman. 2005. A method to derive smoke emission rates from MODIS fire radiative energy measurements. *IEEE Transactions on Geoscience and Remote Sensing* 43:2636–2649.

Ignotti, E., S. S. Hacon, W. L. Junger, et al. 2010. Air pollution and hospital admissions for respiratory diseases in the subequatorial Amazon: A time series approach. *Caderno de Saúde Pública* 26:747–761.

INPE (Instituto Nacional de Pesquisas Espaciais). 2012. Portal do Monitoramento de Queimadas e Incêndios. Available from: http://www.inpe.br/queimadas (accessed November 8, 2012).

Kaiser, J. W., A. Heil, M. O. Andreae, et al. 2012. Biomass burning emissions estimated with a global fire assimilation system based on observed fire radiative power. *Biogeosciences* 9:527–554.

Kaufman, J. B., D. L. Cummings, D. E. Ward, et al. 1995. Fire in the Brazilian Amazon: Biomass, nutrient pools, and losses in slashed primary forests. *Oecologia* 104:397–408.

Kaufman, Y. J., L. Remer, R. Ottmar, et al. 1996. Relationship between remotely sensed fire intensity and rate of emission of smoke: SCAR-C experiment. In *Global biomass burning*, ed. J. Levine, 685–696. Cambridge, MA: MIT Press.

Longo, K. M., S. R. Freitas, M. O. Andreae, et al. 2010. The Coupled Aerosol and Tracer Transport model to the Brazilian developments on the Regional Atmospheric Modeling System (CATT-BRAMS). Part 2: Model sensitivity to the biomass burning inventories. *Atmospheric Chemistry and Physics* 10:5785–5795.

Marengo, J. A., C. A. Nobre, and L. F. Salazar. 2010. Regional climate change scenarios in South America in the late XXI century: Projections and expected impacts. *Nova Acta Leopoldina* 112:251–265.

Moraes, E. C., S. H. Franchito, and V. Brahmananda Rao. 2004. Effects of biomass burning in Amazonia on climate: A numerical experiment with a statistical-dynamical model. *Journal of Geophysical Research* 109:1–12.

Nobre, C. A., P. J. Sellers, and J. Shukla. 1991. Amazonian deforestation and regional climate change. *Journal of Climate* 4:957–988.

Olson, J. S., J. A. Watts, and L. J. Allison. 2000. Major world ecosystem complexes ranked by carbon in live vegetation: A database. Available from: http://cdiac.esd.ornl.gov/ndps/ndp017.html (accessed August 7, 2010).

Olson, J. S., J. A. Watts, and L. J. Allison. 2002. Major world ecosystem complexes ranked by carbon in live vegetation: A database (Revised November 2000). Available from: http://cdiac.esd.ornl.gov/ndps/ndp017.html (accessed November 7, 2012).

Pereira, G., N. J. Ferreira, F. S. Cardozo, et al. 2011. Aerosol and trace gas retrievals from remote sensing fire products. In *Fire detection,* ed. R. P. Bennett, 103–118. New York: Nova Science Publishers.

Pereira, G., S. R. Freitas, E. C. Moraes, et al. 2009. Estimating trace gas and aerosol emissions over South America: Relationship between fire radiative energy released and aerosol optical depth observations. *Atmospheric Environment* 43:6388–6397.

Prins, E. M., and W. P. Menzel. 1992. Geostationary satellite detection of biomass burning in South America. *International Journal of Remote Sensing* 13:2783–2799.

Riggan, P., R. Tissell, R. Lockwood, et al. 2004. Remote measurement of energy and carbon flux from wildfires in Brazil. *Ecological Applications* 14:855–872.

Roberts, G., M. J. Wooster, G. L. W. Perry, et al. 2005. Retrieval of biomass combustion rates and totals from fire radiative power observations: Application to southern Africa using geostationary SEVIRI imagery. *Journal of Geophysical Research* 110:D21111.

Roy, D. P., P. E. Lewis, and C. O. Justice. 2002. Burned area mapping using multi-temporal moderate spatial resolution data—A bi-directional reflectance model-based expectation approach. *Remote Sensing of Environment* 83:263–286.

Saatchi, S., R. Houghton, R. Avala, et al. 2007. Spatial distribution of live aboveground biomass in Amazon Basin. *Global Change Biology* 13:816–837.

Sampaio, G., C. Nobre, M. H. Costa, et al. 2007. Regional climate change over eastern Amazonia caused by pasture and soybean cropland expansion. *Geophysical Research Letters* 34:1–7.

SEEA. 2003. *Handbook of national accounting system of integrated environmental and economic accounting.* United Nations European Commission, International Monetary Fund, Organisation for Economic Co-operation and Development, World Bank, 572 p., New York: United Nations.

Seinfeld, J., and S. Pandis. 1998. *Atmospheric chemistry and physics.* New York: John Wiley & Sons.

Silva, J. M. N., A. C. L. Sa, and J. M. C. Pereira. 2005. Comparison of burned area estimates derived from SPOT-VEGETATION and Landsat ETM+ data in Africa: Influence of spatial pattern and vegetation type. *Remote Sensing of Environment* 96:188–201.

Tian, H., J. M. Melillo, D. W. Kicklighter, et al. 1998. Effect of interannual climate variability on carbon storage in Amazonian ecosystems. *Nature* 396:664–667.

Vermote, E., E. Ellicott, O. Dubovik, et al. 2009. An approach to estimate global biomass burning emissions of organic and black carbon from MODIS fire radiative power. *Journal of Geophysical Research* 114:205–227.

Werf, G. R., J. T. Randerson, L. Giglio, et al. 2006. Interannual variability in global biomass burning emissions from 1997 to 2004. *Atmospheric Chemistry and Physics* 6:3423–3441.

Wooster, M. J., G. Roberts, G. Perry, et al. 2005. Retrieval of biomass combustion rates and totals from fire radiative power observations: Calibration relationships between biomass consumption and fire radiative energy release. *Journal of Geophysical Research* 110:D24311.

Wooster, M. J., B. Zhukov, and D. Oertel. 2003. Fire radiative energy for quantitative study of biomass burning: Derivation from the BIRD experimental satellite and comparison to MODIS fire products. *Remote Sensing of Environment* 86:83–107.

Xu, W., M. Wooster, G. Roberts, et al. 2010. New GOES imager algorithms for cloud and active fire detection and fire radiative power assessment across North, South and Central America. *Remote Sensing of Environment* 114:1876–1895.

Yebra, M., E. Chuvieco, and D. Riano. 2008. Estimation of live fuel moisture content from MODIS images for fire risk assessment. *Agricultural and Forest Meteorology* 148:523–536.

Section III

Ecosystem Services Related to Biodiversity

8

Earth Observation for Species Diversity Assessment and Monitoring

N. Fernández

Doñana Biological Station, Spanish National Research Council EBD-CSIC, Spain

CONTENTS

8.1 Introduction 151
8.2 Finding Species from Space 153
8.3 EO of Land Cover and Species Niches 157
    8.3.1 Species Distribution Modeling 159
    8.3.2 Global Species Assessment 161
8.4 Measuring Ecosystem Functioning in Animal Ecology and
    Conservation 165
    8.4.1 Relationships between Ecosystem Functioning and
          Species Richness 165
    8.4.2 Ecology and Conservation of Wildlife Populations 168
8.5 Concluding Remarks 171
Acknowledgments 172
References 172

8.1 Introduction

Species loss and the decline of populations are among the most important threats to the preservation of ecosystem processes and their services to humans (Chapin et al. 2000). These threats will increase in the near future: even the most conservative estimates indicate that current extinction rates have no precedent since the Cretaceous event (Barnosky et al. 2011), thus supporting the idea that anthropogenic activity is triggering the sixth mass extinction on Earth. As an example, recent assessments of the status of animal species showed that, among all species included in the catalog of the International Union for Conservation of Nature (IUCN), 19% of vertebrates and 26% of invertebrates were threatened, and a large number of these move to a higher risk category every year (Hoffmann et al. 2010; Collen et al. 2012). Overall, diversity is suffering an additional reduction through the loss of populations. The "Living Planet Index," a trend indicator of population abundance based on data from more than 1400 vertebrate species, documents around 30% global decline since 1970, although observed declines double the global rate in some regions of the world like the tropics (Collen et al. 2009). Other taxa have also suffered widespread declines (e.g., as documented for pollinator insects), affecting ecological processes with paramount roles in the maintenance of wild plant communities and crop production (Potts et al. 2010; Cameron et al. 2011). Scientific consensus indicates that these processes are widespread among the different biological taxa and that they affect ecosystems throughout the world; still the vast majority of species, populations, and ecosystems have never been assessed (Dobson 2005). Halting these declines is paramount for the key role of species diversity as a regulator of ecosystem processes and as an ecosystem service and a good in itself (Mace et al. 2012). Species number and population abundance determine, among other things, provisioning services such as food and genetic resources and regulating services such as disease regulation and pollination, whereas the critical importance of species diversity in delivering supporting services, such as primary production and nutrient cycling, is increasingly recognized (Chapin et al. 2000).

Despite great progress during the past few decades, our ability to quantify and forecast changes in biodiversity is very limited at all scales, from local to global (Balmford and Bond 2005). In particular, more work is needed on understanding how species, their diversity, and their populations may respond to alterations in the environment, including climate and land use changes, harvesting, and biological invasions. Earth observation (EO) based on satellite imagery has much to contribute in this direction. Remote sensing represents one of the greatest methodological advances in environmental research of the past three decades, although its use in species–environment studies has only become more widespread since the beginning of the twenty-first century. Building on earlier reviews by Kerr and Ostrovsky (2003) and Turner et al. (2003), the application of satellite technologies to biodiversity science and conservation can be summarized in three main areas: (1) the direct detection of organisms, species, and community assemblages; (2) the classification of the variety, type, and extent of species habitats; and (3) measurements of ecosystem functions such as primary production as an indicator of environmental heterogeneity and biological diversity. Each of these areas has experienced great progress during the past decade (e.g., Pettorelli et al. 2011; Pfeifer et al. 2012), although several important challenges in the study of biodiversity through remote sensing still remain. Data acquisition costs have dropped dramatically and some satellite products are available for free, but high costs still discourage the more extensive use of some useful products such as hyperspectral imagery. Processing capacity of computers and specific software have both evolved immensely; however, applying modern analysis techniques requires highly specialized personnel. Finally, ecologists and conservation biologists have gradually incorporated satellite data into their studies, although these scientists still do not receive sufficient practical or theoretical training in remote sensing to exploit the full potential of rapidly increasing techniques and products for their research (Cabello et al. 2012).

This chapter provides a general overview of the application of optical satellite-based remote sensing to detect, analyze, and monitor patterns in terrestrial biodiversity. A broad (although nonexhaustive) array of uses are explored with the aim of bringing EO from space closer to the actors and practitioners of biodiversity management and conservation. For the purposes of biodiversity assessment, focus is on the species level and species assemblages, which have particularly important implications for the delivery of ecosystem services (Pereira and Cooper 2006). To be reviewed are aspects such as how EO systems can help to detect species remotely, to model their distributions, and to understand how changes in the biophysical environment may affect populations and species assemblages.

8.2 Finding Species from Space

The direct detection from space of species and their diversity continues to be one of the greatest challenges of EO systems. Survey protocols based on remote sensing have the potential to offer a comprehensive, systematic, and repeatable method for mapping species distributions and quantifying diversity. They can provide information throughout extensions difficult to cover using traditional field surveys, at different moments—even retrospectively—and from remote areas. Two important drawbacks are the small spatial resolution of most sensors in relation to the size of many individual species organisms, their colonies, and their groups (Figure 8.1) and technical limitations for the spectral separation of optically similar species. Remote sensing of species has evolved to a great extent in the framework of species invasion assessment (Huang and Asner 2009). Detecting and mapping invasive species in a timely manner are critical in the fight against their spread. For this purpose, changes in the reflective properties of the land surface following the invasion are particularly useful as an indicator of the presence of the invading organism.

Some invasive species may show highly conspicuous signals that can be detected at a spatial resoultion much coarser than the size of the individual organism. For example, data from the Moderate Resolution Imaging Spectroradiometer (MODIS) of NASA, with a nominal resolution of 250 m, allowed detection of abnormal increases in vegetation greenness produced by the activity of *Eragrostis lehmanniana*, an alien grass species in the Arizona desert (Huang and Geiger 2008). Mapping the affected area was

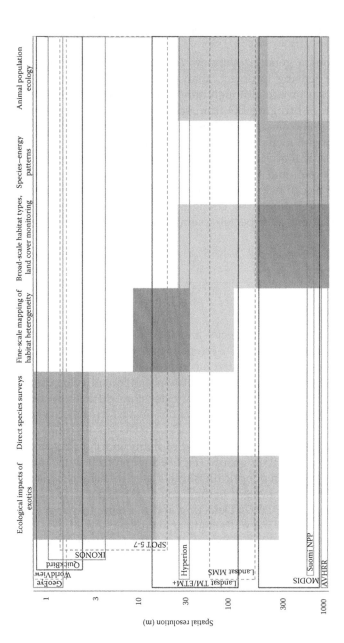

FIGURE 8.1 (See color insert.)

Overview of the spatial resolution requirements for biodiversity studies compared with selected satellite sensors. Empty boxes encompass the resolution range at NADIR for each satellite series, with the highest resolution corresponding to the panchromatic channel in GeoEye, WorldView, QuickBird, IKONOS, SPOT, and Landsat ETM+. In columns, colored areas indicate the required resolution for each application, with darker colors corresponding to better suited resolutions. In some cases, such as population studies in animals, the lower suitability of low-resolution data is not due to the spatial scale but due to a lower frequency of data acquisition, which hinders a detailed analysis of ecosystem phenology. A higher resolution does not always imply benefits, such as in the case of broad-scale species–energy relationships, where species distribution data are usually recorded at lower resolution than most sensors provide; a lot of processing can be saved using lower resolution products.

possible because the invader produced new tissues during the cold season in response to rare rainfall events, while native grasses remained senescent or dormant. The correlation between field measurements of the biomass and the increase in vegetation greenness further contributed to predict species abundance. Similarly, the invasion dynamics of cheatgrass (*Bromus tectorum*) in the Great Basin of North America has been characterized retrospectively using Landsat imagery through the analysis of amplified vegetation responses to precipitation: Cheatgrass greens up earlier than other grasses and shrubs. This characteristic, as observed from Landsat images between 1973 and 2001, was used to map the species expansion and to create predictive models of the future risk of spread based on the relationship between past invasions and characteristics of the landscape (Bradley and Mustard 2006). In summary, unique phenological characteristics of plants can greatly help to track invasions in space and time using relatively simple techniques and widely available satellite products.

Other approaches have taken advantage of the spectral distinctiveness of species to detect their presence. The saltcedar (*Tamarix ramosissima*) has expanded throughout extensive areas of the southwestern United States and Mexico and, as a consequence, has threatened native plant communities, reduced the abundance and diversity of animal populations, and altered the hydrology of riparian ecosystems. Controlling and eradicating saltcedar are considered a high-priority conservation goal, and efforts toward the remote monitoring of infected areas are highly encouraged (Morisette et al. 2006). The lower canopy reflectance of this species during the leafless period has served as a means of detecting it on the basis of spectral classifications of 30-m resolution Landsat TM5 images. Saltcedar stems are darker than any other riparian species making it distinguishable from native vegetation in the near-infrared absorption wavelength. A simple threshold approach applied to bands 4 and 5 of Landsat TM5 showed very high accuracy (≥90%) in the identification of large infested patches (Groeneveld and Watson 2008). More sophisticated methodologies have focused on spectral unmixing techniques to quantify more precisely saltcedar fractional cover throughout entire landscapes (Silvan-Cardenas and Wang 2010). Quantification of subpixel species coverage from Landsat ETM+ scenes made it possible to detect low-density stands with a high accuracy (~70%). Detecting these stands is critical for managing biological invasions during early stages when control and eradication measures are most effective.

EO products with spatial resolutions near or below 1 m, sometimes referred to as "hyperspatial" imagery, hold even a greater potential to detect species directly (Figure 8.1). It seems obvious that matching data to the size of the focal organism will improve their identification from space, reducing the problem of mixed signatures in pixels encompassing individuals from different species. For example, it has been shown that QuickBird data with a 2.5-m resolution outperformed coarser-grain Landsat TM and Hyperion data for estimating saltcedar distributions in the Colorado River in the United States, yielding results only comparable to those obtained from airborne sensors (Carter et al. 2009). Similarly, hyperspatial imagery has provided the opportunity to develop species-level monitoring of invasive plants in wetlands where coarser-resolution information often perform poorly (Laba et al. 2008). However, dealing with increased spatial resolution also has some disadvantages that may compromise the precision of surveys. For example, higher resolution may increase signature variability due to within-object reflectance heterogeneity (reviewed in Nagendra and Rocchini 2008). Signature variability may arise, for example, as a result of differences in shading in the plant canopy or dissimilarities in the optical properties of different plant structures such as bark versus leaves or vigorous versus senescent tissues. Therefore, despite the fact that shrinking grain size at or below the size of the focal organism holds great potential in species detection, there is also a trade-off with classification uncertainty that may counteract the benefits of high-resolution data.

Hyperspectral remote sensing (Underwood et al. 2003) allows the acquisition of information from many narrow, contiguous spectral bands throughout a relatively broad portion of the electromagnetic spectrum, typically the visible near-infrared and mid-infrared portions. This feature provides an unparalleled potential to discriminate organisms on the basis of differences in biochemical and structural constituents including pigment composition, water, and dry matter content and nitrogen concentrations. In this way, the structural and chemical characteristics of canopies can be used to resolve subtle differences and to construct highly precise spectral signatures (Ustin et al. 2004). Hyperspectral remote sensing holds great promise for mapping those species occurring at low densities or scattered in heterogeneous communities (reviewed by He et al. 2011); however, translating the biochemical and structural properties of vegetation into species segregation remains challenging. A study on Hawaiian subtropical forests using airborne imagery found that a group of alien tree species displayed spectral signatures different from those of native species, and it was also possible to distinguish between functional groups of nitrogen-fixing and nonfixing trees (Asner et al. 2008b). Mapping tree species distributions based on these insights was not straightforward, though: Canopy senescence, gaps, shadows, and terrain introduced great uncertainty in the classification. These difficulties can potentially be overcome by using a complex combination of hyperspectral and three-dimensional data on forest canopies acquired through the use of active light detection and ranging (LIDAR). Thus, tree species of the Hawaiian forests were mapped based on their spectral and structural properties, and the distinctive ways in which plants modify the three-dimensional structure of the forest (Asner et al. 2008a). This result has relevant implications for detecting long-lasting impacts of invasive plants on the structure, biological diversity, and functioning of ecosystems.

Hyperspectral applications are emerging beyond the field of plant invasions as a more general means of mapping species diversity. For example, three mangrove species were investigated using these techniques in a small area of Queensland, Australia, yielding distribution maps with an overall accuracy of more than 75% (Kamal and Phinn 2011). However, the accuracy of the produced maps highly depends on the classification algorithm and better methods are still needed to improve classifications in order to exploit the full potential of hyperspectral imagery (Heumann 2011).

Currently, most hyperspectral research is based on airborne sensors (He et al. 2011). Acquisition of this data is expensive and logistically complex, and spatial coverage is reduced. The Earth Observing-1 (EO-1) satellite Hyperion is at present the only source of hyperspectral data from space, although at a moderate resolution of 30 m. Nonetheless, several species surveys have benefitted from these data. A study conducted in an Amazonian forest of Peru found that taxonomic separation of large-crowned trees was possible using Hyperion imagery (Papes et al. 2010). By accounting for seasonal spectral differences between dry and wet seasons, discriminant analyses accurately separated three selected species and two additional genera of trees. However, locating species in a context of higher tree diversity is still challenging due to the spatial resolution and mixed spectra (Carter et al. 2009). In addition, analysis of Hyperion data is problematic due to a low signal-to-noise ratio resulting from signals lost to atmospheric absorption (Pengra et al. 2007).

The remote detection of animal species is obviously more challenging than detection of plants because their locations are dynamic are often unpredictable. Nevertheless, remote sensing has allowed the detection of particular populations. For example, the distinctive reflective properties of fecal stains produced on sea ice helped to survey emperor penguin colonies throughout the Antarctic coastline using Landsat ETM+ images (Fretwell and Trathan 2009). Through visual inspections of the images and simply subtracting the blue band from the visible red band, it was possible to detect colonies that had previously remained unrecorded. A deeper examination including supervised classifications of QuickBird images allowed for the direct individualization of penguin groups and the estimation of their sizes after separating signals from snow, shadow, and guano (Fretwell et al. 2012).

8.3 EO of Land Cover and Species Niches

Land use change is expected to be the main cause of biodiversity loss in terrestrial ecosystems during the next century due to its devastating effects on species habitats (Sala 2001). Against this backdrop, EO systems should play a pivotal role in the assessment and mitigation of species loss by providing accurate and spatially consistent land cover data for predicting the fate of species in a changing environment.

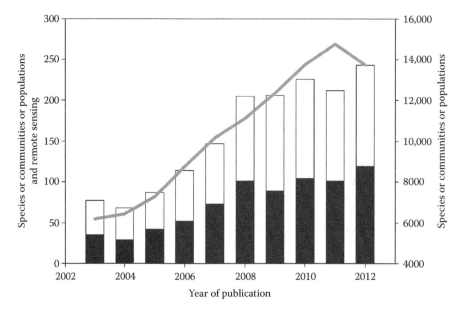

FIGURE 8.2
Number of publication records in a search in Scopus relative to the year of publication. The solid line shows publication counts, including terms in the title, keywords, and abstract, frequently used in species, communities, and populations. Bars refer to the number of these records that included remote sensing terms. The black stick refers to the number of studies with terms referring to land cover and habitat structure, and the white stick refers to the rest. Note the different scale of the axes. A great increase in biodiversity publications can be observed for the 10-year period. Of these, studies relating to the use of remote sensing have also increased significantly; however, the proportion with respect to the total has increased at a smaller rate (from 1.25% in 2003 to 1.77% in 2012).

The mapping of land cover is actually the most popular application of remote sensing in general and in the area of biodiversity conservation in particular (Figure 8.2). The launch of the first Landsat and NOAA satellites in the 1970s and 1980s fuelled the development of land cover mapping (Defries and Townshend 1999). More recently, long-term data acquisition, new moderate and high-resolution satellites, and progress in image classification methods have promoted land cover mapping as a core subject area of EO systems (Giri 2012). In general, classification methods are based on the categorization of the land surface into a discrete number of land cover types according to differences in the spectral and textural properties of satellite images, either using a single date or multiple dates to account for temporal variability. The resulting maps have been gradually incorporated as baseline data for many different ecological applications, including biodiversity studies of terrestrial vascular plants (Nagendra 2001) and terrestrial animals (Leyequien et al. 2007). They have contributed to the analysis of a broad array of topics, from the conservation planning of endangered species to the assessment of the status of biodiversity. Moreover, the need for regional and global land cover monitoring has been emphasized in the context of biodiversity monitoring aimed at guiding and assessing transnational biodiversity agreements such as those included in the Convention on Biological Diversity (Pereira and Cooper 2006).

8.3.1 Species Distribution Modeling

Products derived from remote sensing, such as land cover maps, are a valuable source of information for the analysis of species–habitat relationships. Great progress has been made in the development of timely and accurate land cover datasets at local, regional, and global scales (Giri 2012), and the advent of more capable satellites and techniques will certainly enhance land cover mapping in the near future. Land cover data have proven to be instrumental in studying aspects such as species–habitat selection and habitat use, environmental constrains of species distributions, predicting their occurrence in remote areas, and monitoring the consequences of land use and land use change. However, current land cover data do not always match the requirements of species–environment studies depending on factors such as map characteristics (e.g., spatial and temporal resolution, accuracy, and thematic legend), the ecological characteristics of the species, and the specific purposes of species–habitat assessments.

As a subdiscipline of conservation biology, *conservation biogeography* has contributed notably to the recognition of the importance of investigating the spatial distribution of species, their changes, and their drivers for adopting management decisions concerning biodiversity (Whittaker et al. 2005). It has also provided a theoretical and analytical framework for studying species distributions from global to regional scales (Richardson and Whittaker 2010). Predictive species distribution models (SDMs) have gained popularity as a method for inferring how species and their diversity may be affected by changes in the global environment. Typically, SDMs aim to uncover the associations between the occurrence of a given species or species group and the characteristics of the environment, with the goal of predicting the suitability of habitats to support species and describing realized ecological niches (in the Hutchinsonian concept; Franklin 2010). These models are based on the analysis of presence–absence data or, in some cases, only presences, in relation to variables describing climate, vegetation, human activity, and other biotic and abiotic factors relevant to the ecological requirements of the species (Elith and Leathwick 2009). Furthermore, SDMs can be used to predict the potential impact of environmental change on species and to support conservation strategies, including, for example, the design of reserve networks, developing species conservation plans, and combating invasive species.

Generic products such as the European CORINE (Coordination of Information on the Environment) land cover inventory have been used in

biogeographic models of species distributions with the aim of complement-
ing more traditional bioclimatic approaches that were exclusively based on
climate data (Luoto et al. 2007). CORINE is a general-purpose dataset coordinated by the European Environment Agency of the European Union that
classifies European land cover into a minimum of 44 categories. In Finland,
a study on 88 bird species distributions showed that adding CORINE data
to bioclimatic models significantly enhanced the accuracy of spatial distribution predictions for nearly 90% of species (Luoto et al. 2007). Similar
approaches have been successful in the modeling of anuran and turtle distributions in New Scotland (Tingley and Herman 2009), confirming the importance of land cover for identifying areas with suitable climate but unsuitable
habitat. However, broader scale studies have questioned the importance of
land cover for predicting species distributions; a European-wide analysis of
all terrestrial vertebrates and about 20% of plants showed that climate variables were the main driver of both land cover and species distributions at
a 50-km resolution, whereas land cover variables per se generally did not
improve model performance (Thuiller et al. 2004). It has been argued that
the determinants of spatial distributions are hierarchically structured, with
climate dominating distributions at broad spatial scales and land cover at
finer scales (Pearson et al. 2004). However, the relative importance of climate
versus land cover has seldom been evaluated. Clearly, more assessments are
needed with a higher variety of species groups to gather better understanding of the impact of environmental change on species diversity and distributions (Heikkinen et al. 2006).

Regional land cover data are still scarce for assessing the vast majority
of species in most regions of the world and a significant number of studies
require specific land cover and vegetation information according to the species of interest. The production of this data can be beneficial because it provides a closer connection between the ecological requirements of species and
the environmental predictors, for example, through a better delineation of
the relevant vegetation types (Fernández et al. 2003). In a review of 112 publications on bird distributions, Gottschalk et al. (2005) found that the majority of studies used specific satellite image classifications of vegetation based
on predefined schemes according to previous knowledge about the focus
species. Landsat images, which were the first satellite datasets available to a
wide research community, were preferred in nearly 80% of studies. In contrast, data from the Advanced Very High Resolution Radiometer (AVHRR)
were used only in 20% of studies probably due to the lower spatial resolution and number of bands (Figure 8.1). The precision of classification results
also varied greatly with the number of land cover classes ranging from
only 2 classes to more than 70. By capturing more detail, higher spatial and
thematic resolutions have the potential to reveal more species–habitat relationships, whereas coarser maps with fewer classes are less likely to detect
important habitat components such as small patches, linear structures, and
heterogeneous landscapes (Gottschalk et al. 2005).

Selection of ecologically significant and proximal predictors contributes to the statistical robustness of models, enhances the accuracy of model extrapolations, and reduces the likelihood of including irrelevant variables in the modeling process (Elith and Leathwick 2009). However, improving detail above a certain limit may not always entail significant advantages: An analysis of 54 bird distributions in Spain comparing CORINE data and specific vegetation classifications from Landsat showed that the supplementary classes did not significantly contribute to predictive accuracy, except for a few species closely linked to very particular habitats such as riparian (Seoane et al. 2004). The same study found that increasing the spatial resolution of CORINE did not improve the accuracy of bird SDMs. Thus, considering the high costs involved and the specialization required for producing land cover datasets, generic EO products might represent a cost-effective alternative for many applications. This solution will depend on the particular ecology of the study species, the spatial scale of interest, and map characteristics such as accuracy, thematic legend, and spatial resolution. Surprisingly, the overall suitability of environmental data sources in the context of SDMs has been evaluated only in a small number of studies, which contrasts with the strong effort put forth for other issues such as the impact of accuracy and resolution of species data and the choice of statistical modeling methods (Franklin 2010).

8.3.2 Global Species Assessment

Several EO operational programs have been developed in the past few years aiming to provide global, spatially consistent information on land use and land cover (Table 8.1). Although based on different sources of remote sensing data and with different methodologies, they all focus on characterizing vegetation types globally with the fundamental purpose of serving a large number of user needs. They are all are freely available on the Internet.

1. The International Geosphere–Biosphere Programme dataset (IGBP-DISCover) was the first effort to map the land surface at a 1-km resolution. It was based on the IGBP land cover classification scheme covering 17 categories. Monthly global AVHRR Normalized Difference Vegetation Index (NDVI) composites for the period between April 1992 and March 1993 were used to produce the map (Loveland et al. 2000). A related global product was produced from the same dataset by the University of Maryland (UMD) using a different classification scheme with 14 categories (Hansen et al. 2000).

2. The Global Land Cover 2000 (GLC2000) was the result of an international research partnership coordinated by the European Commission's Joint Research Center aimed at producing one single global map at a 1-km resolution for 2000. The map combines 19

TABLE 8.1

Global Land Cover Products

Product	Sensor	Satellite Data Period	Classification Type	Nominal Spatial Resolution (km)	Number of Classes	Overall Area-Weighted Accuracy[a] (%)
IGBP-DISCover	NOAA-AVHRR	1992–1993	Unsupervised	1.1	17	66.9
UMD Global Land Cover	NOAA-AVHRR	1992–1993	Supervised	1.1	14	65
GLC2000	SPOT-VEGETATION	1999–2000	Unsupervised	1.5	22[b]	68.6
MCD12Q1	EOS-MODIS	Annual, from 2001 to the present	Supervised	0.5	17/14/11/9[c]	75
GlobCover	Envisat-MERIS	2005–2006 and 2009	Unsupervised	0.3	22[b]	67.1

[a] Accuracy assessments used different methods and are not directly comparable among datasets. IGBP DISCover: Verification from expert image interpreters using Landsat and SPOT images (Scepan et al. 1999). GLC2000: Quantitative assessment based on a stratified random sampling of reference data using Landsat images, aerial photographs, and thematic maps (Mayaux et al. 2006). MCD12Q1: As estimated for the version 5 of the product from cross-validation analysis of the supervised classification (Friedl et al. 2010). GlobCover 2005: Bicheron et al. (2006).

[b] Refers to the number of classes of the global product.

[c] Number of classes of the type 1, 2, 3, and 4 classification schemes, respectively.

different regional products into one single dataset based on SPOT 4-VEGETATION images. It comprises two sets of products: a global land cover dataset that harmonizes regional products into 22 classes following the United Nations Land Cover Classification System (LCCS), and regional maps with optimized land cover legends for each region (Bartholome and Belward 2005). The GLC2000 database was utilized as a core dataset of the Millennium Ecosystem Assessment to define boundaries between ecosystems.

3. The MODIS Land Cover Product Collection 5 (MCD12Q1) contains a suite of datasets produced at a 500-m pixel resolution for each calendar year since 2001. The classification is based on supervised algorithms using a set of nearly 2000 land training sites distributed across the Earth (Friedl et al. 2010). Five different land cover classifications are available for each year including, among others, classification schemes of the IGBP (Land Cover Type 1 with 17 classes) and the UMD (Type 2, 14 classes). These schemes allow comparisons with products derived from AVHRR.

4. GlobCover is a joint initiative from different international institutions to produce a map of the Earth comprising 22 land cover types at a 300-m resolution, making it the most detailed global product (Bicheron et al. 2006). MERIS satellite data were used for its production. The thematic legend is compatible with the GLC2000. In addition, 11 regional datasets are also available at the same resolution containing a second-level legend with up to 51 classes. Two GlobCover maps have been developed, one for the period 2005–2006 and one for 2009.

All these products can greatly help in the global assessment of species diversity and distributions. One example is a recent study on mammals that analyzes, for the first time, habitat suitability for nearly all known terrestrial species (Rondinini et al. 2011). Based on habitat preferences as described in the IUCN Red List, the GlobCover dataset was used to develop a model that distinguished between high, medium, and low suitability for each species according to land cover data and information on elevation and hydrology. The resulting maps depicted potential species distributions and were compared with the IUCN dataset of species geographic ranges on which many global biodiversity assessments are based. Potential distribution maps predicted known mammal occurrences better than IUCN geographic ranges showing that IUCN maps frequently overestimated the amount of area occupied by species. This result stresses the importance of taking into account the environmental heterogeneity within geographic ranges in the conservation assessment of biodiversity. For example, the estimated amount of area required for species conservation in Africa increased by a factor of 2 when the habitat preferences and the distribution of land cover types were considered in addition to IUCN geographic extents (Rondinini et al. 2005).

Gobal land cover monitoring is also critical to forecast the fate of species in a changing environment. A global assessment of bird exposure to environmental change identified that 10%–20% of all 8750 land species will experience >50% range reductions by the year 2100 according to climate and land use projections of the Millennium Ecosystem Assessment (Jetz et al. 2007). Land cover change was the most serious threat in tropical countries, corresponding to the area where the vast majority of species are found. Similarly, projections for amphibian diversity revealed that changes in land cover were a major threat for species in tropical Central and South America and Africa and will aggravate the effects of climate change in other areas (Hof et al. 2011).

However, current general-purpose land cover products have some drawbacks that require special consideration in the context of biodiversity research:

1. Uncertainties in the land cover data may have a serious impact on the results of species distribution analyses. Although evaluation studies on the different global products have all reported relatively high overall accuracies (Table 8.1), significant disagreement arises in individual classes and in the overall distribution of land cover when comparing datasets with each other. For example, the two newest and highest resolution global maps, namely GlobCover and MODIS MCD12Q1, showed greater than 35% disagreement in the mapped distribution of croplands, whereas the disagreement in forests was only around 10% (Fritz et al. 2011). Land cover maps are often used without concern for their accuracy; however, error propagation from baseline data obviously compromises the reliability of SDM predictions. Thus, using any of these products for any specific application will require a cautious examination of the sensitivity to misclassifications.

2. Low thematic resolution may hamper the detection of some relevant habitats and limit the separation between potentially suitable and unsuitable cover classes. For example, the probabilistic distribution of the Eurasian brown bear can be depicted from land cover information because the distribution and degree of fragmentation of forests determine the suitability of the species habitats (Fernández et al. 2012). However, standard land cover products do not distinguish between native forests and logging plantations unsuitable for the species.

3. Spatial resolution may limit the identification of habitat heterogeneity at the influential scale, emphasizing the need to match the objectives and spatial scale of the modeling exercise to the information available. For instance, species may be affected by habitat changes at finer spatial scales than those registered by land cover maps.

4. Different land cover maps are difficult to compare due to the different classification systems, resolution, and class definitions, thus affecting our capacity to analyze temporal changes in species

habitats. Methods have been developed to compare classifications with different legends and to find areas of high thematic uncertainty (Fritz and See 2008), but, as described earlier, caution must be exercised to avoid confounding land cover changes with classification discrepancies.

8.4 Measuring Ecosystem Functioning in Animal Ecology and Conservation

Spectral information provided by satellite sensors is used regularly to model ecosystem processes and properties related to the energy and matter balance in the land surface, including the dynamics of net primary production (NPP), nutrient dynamics, and heat partition into latent and sensible fluxes. These processes ultimately describe the energetics of life activities in ecosystems and, at the same time, set the amount of resources available to heterotrophic organisms. A growing number of studies have recently incorporated remotely sensed descriptors of ecosystem functioning with the aim of understanding the ecological processes that maintain species populations and taxonomic diversity (Pettorelli et al. 2011; Cabello et al. 2012).

8.4.1 Relationships between Ecosystem Functioning and Species Richness

The species–energy hypothesis predicts that species richness is directly related to the available energy in an area (Wright 1983). It relies on the idea that (1) the density of individual organisms is a function of the available energy, and (2) species richness is set by the total number of individuals that can be supported in that area. Available energy is defined in this context as the rate at which resources available to the species are produced. This rate should be estimated by considering the unique energy requirements and constraints of each particular taxonomic group. For example, evapotranspiration has been used as a relative measure of available energy for plants since it provides information concerning the total amount of incident solar energy available for photosynthesis and water limitations as determined by the amount of precipitation, evaporation, transpiration, and soil water storage (Wright 1983) (see Chapter 18). For animals, species richness has been related to net primary productivity either directly or indirectly through climate covariates such as temperature, precipitation, potential evapotranspiration, and actual evapotranspiration (Currie et al. 2004). Supporting evidence for the species–energy hypothesis is still under discussion, and several studies have recently contributed to this debate through the estimation of available energy using remote sensing.

By analyzing the seasonal relationship between primary productivity and bird species richness in North America, and assuming that NDVI was positively correlated with food resources available to birds, Hurlbert and Haskell (2003) found that (1) the number of resident species was positively correlated with the minimum monthly NDVI, an estimator of the minimum amount of food resources available at the site; (2) the number of migrant species was positively correlated with the difference between the NDVI for June and the minimum monthly NDVI, a measure of the seasonal pulse of production during the breeding season; and (3) the percentage of migrants in the community was positively correlated with the ratio of the seasonal production pulse to NDVI for June, an index of breeding season production relative to the lowest production availability. All these results supported the hypothesis that birds distribute themselves according to fluctuating resources, with seasonal environmental production determining the number of species that can coexist in an area at a given time period. However, Phillips et al. (2010) found that the bird species–energy relationship in North America was not linear but unimodal: The slope of the relationship was positive in less productive regions, then flattened and turned negative in high-energy environments. A possible explanation for this pattern is that increasing primary production in low-productive environments allows for more complex habitats and increases food production (e.g., seeds and invertebrates). At a certain level, productivity does not represent a critical limitation to diversity, whereas high environmental productivity may involve reduced structural complexity of the habitats and, as a result, a higher competitive dominance of certain species reducing overall richness.

Understanding species–energy relationships across productivity gradients is highly relevant to biodiversity conservation. Ecosystems differing in the sign and strength of the relationship require disparate management strategies (Phillips et al. 2010), yet these aspects have been rarely addressed. Remarkably, remotely sensed estimates of primary production can greatly contribute to the monitoring and prediction of the impact of climate change on taxonomic richness, by offering an indicator of the status, variability, and changes in the energy balance of ecosystems that pure climatic variables cannot provide (Ivits et al. 2013). Further research is also needed regarding the relationship between ecosystem energy and other conservation values different from species richness, such as the number of rare species and species evenness. For example, rare bird species exhibited a weaker relationship with NDVI than common species in North America, which contradicts the prediction that increased energy availability decreases extinction risk in these species (Evans et al. 2006).

The use of vegetation indices such as NDVI as an indicator of available energy is based on their direct relationship with the fraction of photosynthetically active radiation absorbed by vegetation (f_{PAR}). The latter is a measure of the maximum potential assimilation of CO_2 per area unit (Goetz and Prince 1998), which has been used as a proxy for NPP in different ecosystems (e.g., Paruelo et al. 1998). However, the actual productivity rate is also limited by an efficiency term regulated by temperature, water availability, and the availability of other resources (Sellers et al. 1995). The effects of these factors are summarized in the Monteith light use efficiency model for primary production (Monteith 1972) (see also Chapter 6):

$$GPP = \varepsilon \times F_{PAR} \times PAR \tag{8.1}$$

where GPP = gross primary productivity, PAR = incoming photosynthetically active radiation ($\mu mol \cdot m^{-2} s^{-1}$), F_{PAR} = fraction of PAR absorbed, and ε = efficiency term describing the rate at which the absorbed PAR is converted into biomass ($gC \cdot MJ^{-1}$).

According to this equation, estimates of energy availability using only NDVI assume that differences in the conversion of absorbed radiation into biomass (ε) are negligible across ecosystems and time. This assumption introduces a significant bias when estimating energy availability in many ecosystems where the potential plant photosynthetic capacity is not fully reached (e.g., due to environmental and physiological limitations) (see Chapter 3). Yet, the importance of violating this assumption in the context of species–energy relationships remains largely unexplored.

To overcome these problems, MODIS now offers global NPP and GPP (MOD17). MOD17 is based on a light use efficiency algorithm that estimates the efficiency term from meteorological data and vegetation types in an attempt to provide a more direct estimator of productivity. The first study on species diversity comparing NDVI versus MODIS GPP and NPP found that the latter product correlated better with bird richness patterns in sparse and dense vegetation areas of North America, whereas the advantages were not clear when all vegetation types were analyzed together (Phillips et al. 2008). The inferior performance of NDVI was also observed for sparsely vegetated areas, regions where background soil reflectance introduces a significant noise, and in densely vegetated areas where NDVI saturates (Huete et al. 2002). These findings call for a more general use of direct productivity estimates in species–energy models. However, MOD17 is no panacea, and several uncertainties should be carefully considered before using this and similar products uncritically. MODIS algorithms used to estimate GPP and NPP are based on meteorological and vegetation parameters calculated from coarser-resolution data, which are known to introduce additional observational errors. In addition, they are based on assumptions concerning the effect of climate and vegetation type on the efficiency term. A study comparing MOD17 GPP against eddy covariance flux data in nine North American ecosystems (Sims et al. 2006) found no significant improvement over the enhanced vegetation index (EVI), an index that reduces the problem of background signal and saturation by adding the blue band to gain sensitivity to vegetation structural properties (Huete et al. 2011). These results suggest that EVI can provide estimates of productivity at least as good as the more complex MOD17 while relying exclusively on remote sensing data.

Recently, several additional approaches have been developed to model productivity exclusively on the basis of remote sensing data. Environmental stresses depressing ε can be estimated using remote sensing rather than meteorological reanalysis data (Hilker et al. 2008). A GPP model based on EVI and land surface temperature (LST) data from MODIS fits reasonably well with carbon flux data at 11 eddy covariance tower sites of North America (Sims et al. 2008). Other attempts have focused on estimating radiation use efficiency directly from satellite data based on the photochemical reflectance index (PRI), which responds to the chemical state of carotenoid pigments indicative of the photosynthetic efficiency of vegetation (Gamon et al. 1997) (see Chapter 3). An analysis of the published literature found that the PRI from MODIS accounted for 53%–67% of the variability in radiation use efficiency at the ecosystem level as calculated from eddy covariance data (Garbulsky et al. 2011). A version of the Monteith equation has been proposed accordingly, in which the efficiency term can be substituted by PRI:

$$CO_2 \text{ uptake} = f(PRI \times F_{PAR}) \times PAR \qquad (8.2)$$

Although these approaches are still under examination, they hold great potential to improve productivity estimates from remote sensing data, and they are readily available for implementation in biodiversity research.

8.4.2 Ecology and Conservation of Wildlife Populations

Ecosystem functioning descriptors such as rates of primary production provide critical information on the spatial and temporal variation of processes directly linked to population dynamics (Figure 8.3), although population ecologists have only recently incorporated remote sensing for measuring these key population drivers. Recent reviews have highlighted the advantages of NDVI (Pettorelli et al. 2011) and more generally of satellite-derived ecosystem functioning estimates (Cabello et al. 2012) for gathering ecological information useful for conservation planning at the population and species levels. Both reviews emphasized the particular importance of ecosystem functioning descriptors for a better understanding and prediction of species habitats and population demography. These aspects are central to conservation biology, which aims to preserve and restore the ecological conditions required to sustain populations at risk. For example, vegetation indices have been used to model the habitat characteristics of endangered species such as the European brown bear in northern Spain (Wiegand et al. 2008). High-quality habitats where bears breed more frequently corresponded to highly marginal NDVI seasonal patterns

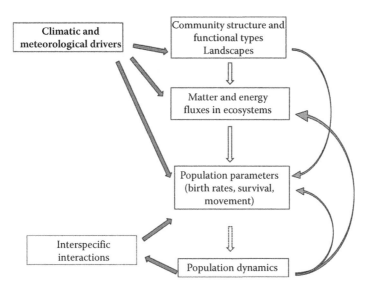

FIGURE 8.3
A schematic representation of the environmental controls on the species population dynamics. Remote sensing indices of ecosystem functioning can be used to characterize matter and energy fluxes influencing population demographic parameters. In turn, these fluxes are determined by top-down abiotic controls (e.g., climate, landscape physiognomy) and by feedbacks with biodiversity.

in a manner consistent with the species' ecological adaptations to track pulses in available energy. In contrast, areas of decreasing quality, that is, those with less reproduction, without reproduction but frequent species presence, or with sporadic presence, showed gradually less specialized and less marginal NDVI characteristics. For the endangered giant panda (Viña et al. 2008), NDVI phenological variables also contributed to descriptions of the species' habitat because they were associated with the availability of food resources and with habitat changes caused by human disturbances. These findings are important for focusing conservation efforts at the most valuable habitats and for developing more sensitive habitat monitoring schemes that take advantage of remote sensing.

Remote sensing indices of ecosystem functioning can also contribute to addressing population dynamic responses to environmental variability on the basis of the resource limitation hypothesis (Dunning and Brown 1982), which predicts a positive correlation between ecosystem productivity and population abundance and demography. Conception rates in Savanna elephants in northern Kenya were predicted with a high confidence in a modeling study that used NDVI time series data as a measure of "ecological quality" (Rasmussen et al. 2006). This index was also useful to predict elephant reproductive phenology and population recruitment (Wittemyer et al. 2007).

The relationship between productivity and reproductive indices often presents temporal lags given that the effect of energy availability on reproduction may operate prior to conception (Rasmussen et al. 2006). Analysis of these lags provides a unique opportunity to anticipate the demographic responses to previous ecosystem states and allows the opportunity to make timely conservation decisions.

Survival has also been correlated with vegetation greenness in the endangered Egyptian vulture (Grande et al. 2009) and in the white stork (Schaub et al. 2005), with comparable results for both species; survival of all age classes was positively correlated with seasonal NDVI in the wintering areas. This pattern appears to be related to food shortages during years of low productivity probably coinciding with low abundance of small wild prey and insects available to storks, and of wild and domestic herbivores for vulture scavenging. Interestingly, the survival-NDVI correlation in wintering stakeholders could explain the demographic synchrony observed in stork breeding areas.

Mueller et al. (2011) found that the population-level dynamics of ungulate movements covaried in different species with broad-scale spatiotemporal variability of vegetation productivity, and also that unpredictability in the dynamics of primary production was associated with a lack of movement coordination among individuals of the same species. Their results suggested that species in less predictable environments may tend to adopt nomadic behavior, whereas migration behavior is better adapted to highly dynamic but predictable productivity patterns. Understanding the different responses of animal movement to environmental productivity has important conservation implications for adopting strategies focused on (1) the maintenance of both critical corridors and overwintering and calving areas for migrating animals, and (2) large unfragmented habitat patches for nomadic animals with more erratic movements (Mueller et al. 2011).

In most of the studies outlined here, a high frequency of data acquisition is needed to track changes in primary production because population responses to environmental fluctuations may occur at very narrow time scales. In this regard, daily data captured by AVHRR and MODIS sensors at 1-km and 250-m resolutions, respectively, are well suited, although composites from 8 to 16 daily images are commonly used to increase the likelihood of capturing relevant vegetation signals. This resolution has proven suitable for broad-scale research concerning the dynamics of populations, although it may limit the utility of remote sensing indices in more detailed studies of resource availability (e.g., for species associated with small habitat patches or living in highly heterogeneous environments).

In addition to vegetation indices, remote sensing of ecosystem functioning offers much wider possibilities for the characterization of environmental variability, an aspect that has been largely ignored in conservation studies (Cabello et al. 2012). Besides the aforementioned NPP and GPP estimates, descriptors based on the energy balance in the land surface can provide very useful information, such as on the functional heterogeneity (Fernández et al. 2010), losses in ecosystem functionality due to degradation (Garcia et al. 2008), and fire disturbances in ecosystems (Mildrexler et al. 2009) with the potential to contribute to species–habitat assessment and monitoring.

8.5 Concluding Remarks

Species diversity is a central ecosystem service in itself, and it sustains many other services through the regulation of ecosystem processes, including primary production, nutrient and water cycling, climatic regulation, and much more. Accordingly, the assessment and conservation of species diversity should constitute an integral part of the management of ecosystems and their services. Herein outlined are the many opportunities provided by remote sensing for studying species through their direct detection, modeling their distributions and assemblages, and analyzing their responses to environmental variability in space and time. However, the potential of EO systems is certainly much greater than that which has been exploited thus far.

Currently available data and techniques are still too limited to address some key questions in species diversity assessment and conservation. Spatial resolution and coverage of sensors, spectral coverage, acquisition costs of some data, and the lack of standardized procedures are important drawbacks for a wider implementation of species and population surveys. Most global and regional land cover datasets cannot (and are not designed to) deliver more precise data on specific land cover and vegetation classes that are needed to assess species–environment relationships. The study of species responses to matter and energy fluxes in ecosystems relies on remote sensing indicators of ecosystem processes that still require further refinements.

New satellite missions are currently being scheduled with the aim of addressing some of these shortcomings. The HyspIRI mission (http://hyspiri. jpl.nasa.gov) intends to capture new hyperspectral data from space to support studies concerning the characteristics and functioning of vegetation at a 60-m resolution. Landsat 8–LDCM, launched in February 2013, collects data through two on-board sensors providing spectral information in nine bands at a spatial resolution of 15 m for the panchromatic channel, 30 m for eight additional shortwave bands, and at 100 m for two thermal infrared bands (http://landsat.gsfc.nasa.gov/about/ldcm.html). This mission provides continuity with previous Landsat data, thus supporting the continuation of a wide number of applications to biodiversity science, particularly the mapping of land cover and other species–habitat characteristics. These datasets are highly valuable for monitoring habitat changes and predicting the fate of species in a changing environment. However, research on habitat changes using long-term EO data series remains scarce.

Finally, studies on species, populations, and communities incorporating remote sensing data and techniques, have experienced a significant increase during the past decade, but they are not yet systematically used for biodiversity assessment, monitoring, and conservation planning. Therefore, increasing the diffusion of remote sensing systems in these fields can serve as a greater impetus to new avenues in biodiversity–EO applications.

Acknowledgments

The author thanks M. Delibes, N. Horning, M. García, and two anonymous reviewers for their constructive comments on earlier versions of the manuscript. Junta de Andalucía provided financial support through the Excellence Research Project RNM-6685.

References

Asner, G. P., R. F. Hughes, P. M. Vitousek, et al. 2008a. Invasive plants transform the three-dimensional structure of rain forests. *Proceedings of the National Academy of Sciences of the United States of America* 105:4519–4523.

Asner, G. P., M. O. Jones, R. E. Martin, D. E. Knapp, and R. F. Hughes. 2008b. Remote sensing of native and invasive species in Hawaiian forests. *Remote Sensing of Environment* 112:1912–1926.

Balmford, A., and W. Bond. 2005. Trends in the state of nature and their implications for human well-being. *Ecology Letters* 8:1218–1234.

Barnosky, A. D., N. Matzke, S. Tomiya, et al. 2011. Has the Earth's sixth mass extinction already arrived? *Nature* 471:51–57.

Bartholome, E., and A. S. Belward. 2005. GLC2000: A new approach to global land cover mapping from Earth observation data. *International Journal of Remote Sensing* 26:1959–1977.

Bicheron, P., M. Leroy, C. Brockmann, et al. 2006. GLOBCOVER: A 300 m global land cover product for 2005 using ENVISAT MERIS time series. *Proceedings of the Recent Advances in Quantitative Remote Sensing Symposium.* Valencia, Spain: Universidad de Valencia, 538–542.

Bradley, B. A., and J. F. Mustard. 2006. Characterizing the landscape dynamics of an invasive plant and risk of invasion using remote sensing. *Ecological Applications* 16:1132–1147.

Cabello, J., N. Fernández, D. Alcaraz-Segura, et al. 2012. The ecosystem functioning dimension in conservation: Insights from remote sensing. *Biodiversity and Conservation* 21:3287–3305.

Cameron, S. A., J. D. Lozier, J. P. Strange, et al. 2011. Patterns of widespread decline in North American bumble bees. *Proceedings of the National Academy of Sciences of the United States of America* 108:662–667.

Carter, G. A., K. L. Lucas, G. A. Blossom, et al. 2009. Remote sensing and mapping of tamarisk along the Colorado River, USA: A comparative use of summer-acquired hyperion, Thematic Mapper and QuickBird Data. *Remote Sensing* 1:318–329.

Chapin, F. S., E. S. Zavaleta, V. T. Eviner, et al. 2000. Consequences of changing biodiversity. *Nature* 405:234–242.

Collen, B., M. Böhm, R. Kemp, and J. E. M. Baillie. 2012. *Spineless: Status and trends of the world's invertebrates*. London: Zoological Society of London.

Collen, B., J. Loh, S. Whitmee, L. Mcrae, R. Amin, and J. E. M. Baillie. 2009. Monitoring change in vertebrate abundance: The living planet index. *Conservation Biology* 23:317–327.

Currie, D. J., G. G. Mittelbach, H. V. Cornell, et al. 2004. Predictions and tests of climate-based hypotheses of broad-scale variation in taxonomic richness. *Ecology Letters* 7:1121–1134.

Defries, R. S., and J. R. G. Townshend. 1999. Global land cover characterization from satellite data: From research to operational implementation? *Global Ecology and Biogeography* 8:367–379.

Dobson, A. 2005. Monitoring global rates of biodiversity change: Challenges that arise in meeting the Convention on Biological Diversity (CBD) 2010 goals. *Philosophical Transactions of the Royal Society—Biological Sciences* 360:229–241.

Dunning, J. B., and J. H. Brown. 1982. Summer rainfall and winter sparrow densities—A test of the food limitation hypothesis. *The Auk* 99:123–129.

Elith, J., and J. R. Leathwick. 2009. Species distribution models: Ecological explanation and prediction across space and time. *Annual Review of Ecology Evolution and Systematics* 40:677–697.

Evans, K. L., N. A. James, and K. J. Gaston. 2006. Abundance, species richness and energy availability in the North American avifauna. *Global Ecology and Biogeography* 15:372–385.

Fernández, N., M. Delibes, F. Palomares, and D. J. Mladenoff. 2003. Identifying breeding habitat for the Iberian lynx: Inferences from a fine-scale spatial analysis. *Ecological Applications* 13:1310–1324.

Fernández, N., J. M. Paruelo, and M. Delibes. 2010. Ecosystem functioning of protected and altered Mediterranean environments: A remote sensing classification in Doñana, Spain. *Remote Sensing of Environment* 114(1):211–220.

Fernández, N., N. Selva, C. Yuste, H. Okarma, and Z. Jakubiec. 2012. Brown bears at the edge: Modeling habitat constrains at the periphery of the Carpathian population. *Biological Conservation* 153:134–142.

Franklin, J. 2010. *Mapping species distributions. Spatial inference and prediction*. Cambridge, UK: Cambridge University Press.

Fretwell, P. T., M. A. LaRue, P. Morin, et al. 2012. An emperor penguin population estimate: The first global, synoptic survey of a species from space. *Plos One* 7:e33751.

Fretwell, P. T., and P. N. Trathan. 2009. Penguins from space: Faecal stains reveal the location of emperor penguin colonies. *Global Ecology and Biogeography* 18:543–552.

Friedl, M. A., D. Sulla-Menashe, B. Tan, et al. 2010. MODIS Collection 5 global land cover: Algorithm refinements and characterization of new datasets. *Remote Sensing of Environment* 114:168–182.

Fritz, S., and L. See. 2008. Identifying and quantifying uncertainty and spatial dis-agreement in the comparison of Global Land Cover for different applications. *Global Change Biology* 14:1057–1075.

Fritz, S., L. See, I. McCallum, et al. 2011. Highlighting continued uncertainty in global land cover maps for the user community. *Environmental Research Letters* 6:044005.

Gamon, J. A., L. Serrano, and J. S. Surfus. 1997. The photochemical reflectance index: An optical indicator of photosynthetic radiation use efficiency across species, functional types, and nutrient levels. *Oecologia* 112:492–501.

Garbulsky, M. F., J. Penuelas, J. Gamon, Y. Inoue, and I. Filella. 2011. The photochemi-cal reflectance index (PRI) and the remote sensing of leaf, canopy and ecosystem radiation use efficiencies. A review and meta-analysis. *Remote Sensing of Environment* 115:281–297.

Garcia, M., C. Oyonarte, L. Villagarcia, S. Contreras, F. Domingo, and J. Puigdefabregas. 2008. Monitoring land degradation risk using ASTER data: The non-evaporative fraction as an indicator of ecosystem function. *Remote Sensing of Environment* 112:3720–3736.

Giri, C. (ed.). 2012. *Remote sensing of land use and land cover: Principles and applications.* Boca Raton, FL: CRC Press.

Goetz, S. J., and S. D. Prince. 1998. Modeling terrestrial carbon exchange and storage: Evidence and implications of functional convergence in light use efficiency. *Advances in Ecological Research* 28:57–92.

Gottschalk, T. K., F. Huettmann, and M. Ehlers. 2005. Thirty years of analysing and modelling avian habitat relationships using satellite imagery data: A review. *International Journal of Remote Sensing* 26:2631–2656.

Grande, J. M., D. Serrano, G. Tavecchia, et al. 2009. Survival in a long-lived territorial migrant: Effects of life-history traits and ecological conditions in wintering and breeding areas. *Oikos* 118:580–590.

Groeneveld, D. P., and R. P. Watson. 2008. Near-infrared discrimination of leafless saltce-dar in wintertime Landsat TM. *International Journal of Remote Sensing* 29:3577–3588.

Hansen, M. C., R. S. Defries, J. R. G. Townshend, and R. Sohlberg. 2000. Global land cover classification at 1km spatial resolution using a classification tree approach. *International Journal of Remote Sensing* 21:1331–1364.

He, K. S., D. Rocchini, M. Neteler, and H. Nagendra. 2011. Benefits of hyperspectral remote sensing for tracking plant invasions. *Diversity and Distributions* 17:381–392.

Heikkinen, R. K., M. Luoto, M. B. Araujo, R. Virkkala, W. Thuiller, and M. T. Sykes. 2006. Methods and uncertainties in bioclimatic envelope modelling under climate change. *Progress in Physical Geography* 30:751–777.

Heumann, B. W. 2011. Satellite remote sensing of mangrove forests: Recent advances and future opportunities. *Progress in Physical Geography* 35:87–108.

Hilker, T., N. C. Coops, M. A. Wulder, T. A. Black, and R. D. Guy. 2008. The use of remote sensing in light use efficiency based models of gross primary produc-tion: A review of current status and future requirements. *Science of the Total Environment* 404:411–423.

Hof, C., M. B. Araujo, W. Jetz, and C. Rahbek. 2011. Additive threats from pathogens, climate and land-use change for global amphibian diversity. *Nature* 480:516–U137.

Hoffmann, M., C. Hilton-Taylor, A. Angulo, et al. 2010. The impact of conservation on the status of the world's vertebrates. *Science* 330:1503–1509.

Huang, C. Y., and G. P. Asner. 2009. Applications of remote sensing to alien invasive plant studies. *Sensors* 9:4869–4889.

Huang, C. Y., and E. L. Geiger. 2008. Climate anomalies provide opportunities for large-scale mapping of non-native plant abundance in desert grasslands. *Diversity and Distributions* 14:875–884.

Huete, A., K. Didan, T. Miura, E. P. Rodriguez, X. Gao, and L. G. Ferreira. 2002. Overview of the radiometric and biophysical performance of the MODIS vegetation indices. *Remote Sensing of Environment* 83:195–213.

Huete, A., K. Didan, W. van Leeuwen, T. Miura, and E. Glenn. 2011. MODIS vegetation indices. In *Land remote sensing and global environmental change*, eds. B. Ramachandran, C. O. Justice, and M. J. Abrams, 579–602. New York: Springer-Verlag.

Hurlbert, A. H., and J. P. Haskell. 2003. The effect of energy and seasonality on avian species richness and community composition. *American Naturalist* 161:83–97.

Ivits, E., M. Cherlet, W. Mehl, and S. Sommer. 2013. Ecosystem functional units characterized by satellite observed phenology and productivity gradients: A case study for Europe. *Ecological Indicators* 27:17–28.

Jetz, W., D. S. Wilcove, and A. P. Dobson. 2007. Projected impacts of climate and land-use change on the global diversity of birds. *Plos Biology* 5:1211–1219.

Kamal, M., and S. Phinn. 2011. Hyperspectral data for mangrove species mapping: A comparison of pixel-based and object-based approach. *Remote Sensing* 3:2222–2242.

Kerr, J. T., and M. Ostrovsky. 2003. From space to species: Ecological applications for remote sensing. *Trends in Ecology and Evolution* 18:299–305.

Laba, M., R. Downs, S. Smith, et al. 2008. Mapping invasive wetland plants in the Hudson River National Estuarine Research Reserve using quickbird satellite imagery. *Remote Sensing of Environment* 112:286–300.

Leyequien, E., J. Verrelst, M. Slot, G. Schaepman-Strub, I. M. A. Heitkonig, and A. Skidmore. 2007. Capturing the fugitive: Applying remote sensing to terrestrial animal distribution and diversity. *International Journal of Applied Earth Observation and Geoinformation* 9:1–20.

Loveland, T. R., B. C. Reed, J. F. Brown, et al. 2000. Development of a global land cover characteristics database and IGBP DISCover from 1 km AVHRR data. *International Journal of Remote Sensing* 21:1303–1330.

Luoto, M., R. Virkkala, and R. K. Heikkinen. 2007. The role of land cover in bioclimatic models depends on spatial resolution. *Global Ecology and Biogeography* 16:34–42.

Mace, G. M., K. Norris, and A. H. Fitter. 2012. Biodiversity and ecosystem services: A multilayered relationship. *Trends in Ecology and Evolution* 27:19–26.

Mayaux, P., H. Eva, J. Gallego, et al. 2006. Validation of the global land cover 2000 map. *IEEE Transactions on Geoscience and Remote Sensing* 44:1728–1739.

Mildrexler, D. J., M. S. Zhao, and S. W. Running. 2009. Testing a MODIS Global Disturbance Index across North America. *Remote Sensing of Environment* 113:2103–2117.

Monteith, J. L. 1972. Solar-radiation and productivity in tropical ecosystems. *Journal of Applied Ecology* 9:747–766.

Morisette, J. T., C. S. Jarnevich, A. Ullah, et al. 2006. A tamarisk habitat suitability map for the continental United States. *Frontiers in Ecology and the Environment* 4:11–17.

Mueller, T., K. A. Olson, G. Dressler, et al. 2011. How landscape dynamics link individual-to population-level movement patterns: A multispecies comparison of ungulate relocation data. *Global Ecology and Biogeography* 20:683–694.

Nagendra, H. 2001. Using remote sensing to assess biodiversity. *International Journal of Remote Sensing* 22:2377–2400.

Nagendra, H., and D. Rocchini. 2008. High resolution satellite imagery for tropical biodiversity studies: The devil is in the detail. *Biodiversity and Conservation* 17:3431–3442.

Papes, M., R. Tupayachi, P. Martinez, A. T. Peterson, and G. V. N. Powell. 2010. Using hyperspectral satellite imagery for regional inventories: A test with tropical emergent trees in the Amazon Basin. *Journal of Vegetation Science* 21:342–354.

Paruelo, J. M., E. G. Jobbágy, O. E. Sala, W. K. Lauenroth, and I. Burke. 1998. Functional and structural convergence of temperate grassland and shrubland ecosystems. *Ecological Applications* 8:194–206.

Pearson, R. G., T. P. Dawson, and C. Liu. 2004. Modelling species distributions in Britain: A hierarchical integration of climate and land-cover data. *Ecography* 27:285–298.

Pengra, B. W., C. A. Johnston, and T. R. Loveland. 2007. Mapping an invasive plant, *Phragmites australis*, in coastal wetlands using the EO-1 Hyperion hyperspectral sensor. *Remote Sensing of Environment* 108:74–81.

Pereira, H. M., and H. D. Cooper. 2006. Towards the global monitoring of biodiversity change. *Trends in Ecology and Evolution* 21:123–129.

Pettorelli, N., S. Ryan, T. Mueller, et al. 2011. The Normalized Difference Vegetation Index (NDVI): Unforeseen successes in animal ecology. *Climate Research* 46:15–27.

Pfeifer, M., M. Disney, T. Quaife, and R. Marchant. 2012. Terrestrial ecosystems from space: A review of Earth observation products for macroecology applications. *Global Ecology and Biogeography* 21:603–624.

Phillips, L. B., A. J. Hansen, and C. H. Flather. 2008. Evaluating the species energy relationship with the newest measures of ecosystem energy: NDVI versus MODIS primary production. *Remote Sensing of Environment* 112:3538, 4381–4392.

Phillips, L. B., A. J. Hansen, C. H. Flather, and J. Robison-Cox. 2010. Applying species-energy theory to conservation: A case study for North American birds. *Ecological Applications* 20:2007–2023.

Potts, S. G., J. C. Biesmeijer, C. Kremen, P. Neumann, O. Schweiger, and W. E. Kunin. 2010. Global pollinator declines: Trends, impacts and drivers. *Trends in Ecology and Evolution* 25:345–353.

Rasmussen, H. B., G. Wittemyer, and I. Douglas-Hamilton. 2006. Predicting time-specific changes in demographic processes using remote-sensing data. *Journal of Applied Ecology* 43:366–376.

Richardson, D. M. and R. J. Whittaker. 2010. Conservation biogeography—Foundations, concepts and challenges. *Diversity and Distributions* 16:313–320.

Rondinini, C., M. Di Marco, F. Chiozza, et al. 2011. Global habitat suitability models of terrestrial mammals. *Philosophical Transactions of the Royal Society—Biological Sciences* 366:2633–2641.

Rondinini, C., S. Stuart, and L. Boitani. 2005. Habitat suitability models and the shortfall in conservation planning for African vertebrates. *Conservation Biology* 19:1488–1497.

Sala, O. E. 2001. Price put on biodiversity. *Nature* 412:34–36.

Scepan, J., G. Menz, and M. C. Hansen. 1999. The DISCover validation image interpretation process. *Photogrammetric Engineering and Remote Sensing* 65:1075–1081.

Schaub, M., W. Kania, and U. Koppen. 2005. Variation of primary production during winter induces synchrony in survival rates in migratory white storks *Ciconia ciconia*. *Journal of Animal Ecology* 74:656–666.

Sellers, P. J., B. W. Meeson, F. G. Hall, et al. 1995. Remote sensing of the land surface for studies of global change models, algorithms, experiments. *Remote Sensing of Environment* 51:3–26.

Seoane, J., J. Bustamante, and R. Diaz-Delgado. 2004. Are existing vegetation maps adequate to predict bird distributions? *Ecological Modelling* 175:137–149.

Silvan-Cardenas, J. L., and L. Wang. 2010. Retrieval of subpixel *Tamarix* canopy cover from Landsat data along the Forgotten River using linear and nonlinear spectral mixture models. *Remote Sensing of Environment* 114:1777–1790.

Sims, D. A., A. F. Rahman, V. D. Cordova, et al. 2006. On the use of MODIS EVI to assess gross primary productivity of North American ecosystems. *Journal of Geophysical Research-Biogeosciences* 111:G04015.

Sims, D. A., A. F. Rahman, V. D. Cordova, et al. 2008. A new model of gross primary productivity for North American ecosystems based solely on the enhanced vegetation index and land surface temperature from MODIS. *Remote Sensing of Environment* 112:1633–1646.

Thuiller, W., M. B. Araujo, and S. Lavorel. 2004. Do we need land-cover data to model species distributions in Europe? *Journal of Biogeography* 31:353–361.

Tingley, R., and T. B. Herman. 2009. Land-cover data improve bioclimatic models for anurans and turtles at a regional scale. *Journal of Biogeography* 36:1656–1672.

Turner, W., S. Spector, N. Gardiner, M. Fladeland, E. Sterling, and M. Steininger. 2003. Remote sensing for biodiversity science and conservation. *Trends in Ecology and Evolution* 18:306–314.

Underwood, E., S. Ustin, and D. DiPietro. 2003. Mapping nonnative plants using hyperspectral imagery. *Remote Sensing of Environment* 86:150–161.

Ustin, S. L., D. A. Roberts, J. A. Gamon, G. P. Asner, and R. O. Green. 2004. Using imaging spectroscopy to study ecosystem processes and properties. *Bioscience* 54:523–534.

Viña, A., S. Bearer, H. M. Zhang, Z. Y. Ouyang, and J. G. Liu. 2008. Evaluating MODIS data for mapping wildlife habitat distribution. *Remote Sensing of Environment* 112:2160–2169.

Whittaker, R. J., M. B. Araujo, J. Paul, R. J. Ladle, J. E. M. Watson, and K. J. Willis. 2005. Conservation biogeography: Assessment and prospect. *Diversity and Distributions* 11:3–23.

Wiegand, T., J. Naves, M. F. Garbulsky, and N. Fernandez. 2008. Animal habitat quality and ecosystem functioning: Exploring seasonal patterns using NDVI. *Ecological Monographs* 78:87–103.

Wittemyer, G., H. B. Rasmussen, and I. Douglas-Hamilton. 2007. Breeding phenology in relation to NDVI variability in free-ranging African elephant. *Ecography* 30:42–50.

Wright, D. H. 1983. Species-energy theory—An extension of species-area theory. *Oikos* 41:496–506.

9

Ecosystem Services Assessment of National Parks Networks for Functional Diversity and Carbon Conservation Strategies Using Remote Sensing

J. Cabello, P. Lourenço, and A. Reyes
University of Almería, Spain

D. Alcaraz-Segura
University of Granada, Spain; University of Almería, Spain

CONTENTS

9.1   Introduction 180
9.2   Methodology 182
    9.2.1   Identification of EFTs and Quantification of Carbon Gains
        from a Satellite-Derived Enhanced Vegetation Index 182
    9.2.2   Functional Diversity Assessment of National Park
        Networks 185
    9.2.3   Congruence between Functional Diversity and Carbon
        Gains in the Networks 185
9.3   Results 186
    9.3.1   Spatial Patterns of EFTs in Natural Areas of Portugal,
        Spain, and Morocco 186
    9.3.2   Representativeness and Rarity of the National Park
        Networks 189
    9.3.3   Carbon Gains in the National Park Networks 192
    9.3.4   Spatial Congruence between Diversity and Carbon Gains
        in National Park Networks 192
9.4   Discussion 194
9.5   Conclusions 196
Acknowledgments 196
References 196

9.1 Introduction

Conservation practice faces the challenge of protecting biodiversity while promoting human welfare. So far, these two targets have remained differentiated, but now, conservation biologists are pushing a paradigm shift viewing conservation biodiversity in a broader perspective (Mace et al. 2012). Conservation strategies based solely on biodiversity conservation are now being perceived as unrealistic, and their goals unattainable, as shown, for example, by the failure of the European and worldwide 2010 Biodiversity Target. This strategy for halting the decline of biodiversity by the end of 2010 has given way to new, wider targets for 2020 and 2050 (CBD COP 10, 2010), which also include curbing degradation of ecosystem services (ESs) and their adequate evaluation as essential contributions to human welfare. This new conservation goal perspective offers the possibility of contributing to sustainability in the current scenario of global change, while not ignoring human–ecosystem interaction.

The ES concept refers to the benefits that mankind derives from ecosystems (MA 2005), and it represents the keystone of methodological steps combining biodiversity conservation with human well-being. Although there have been considerable advances in the science of ESs during the past decade, their incorporation into conservation practices still faces important challenges. First, the complexity of the concepts of biodiversity and ESs has led to a confusing representation of biodiversity in ecosystem assessments. As reported by Mace et al. (2012), the approaches developed so far have alternated between equating biodiversity with ESs (the ecosystem service perspective) and considering biodiversity itself as an ES (the conservation perspective). Although the ecosystem service perspective (biodiversity and ESs are the same thing) implies that the management of one automatically enhances the other (e.g., The Economics of Ecosystems and Biodiversity, or TEEB; http://www.teebweb.org), the conservation perspective (biodiversity as an ES) reflects the intrinsic value of biodiversity. Second, the areas that supply ESs must be mapped to quantify the conservation benefits they provide and to explore their congruence with regions that maximize biodiversity. The conservation perspective emphasizes the need for ecosystem management and policy decisions focusing on combining biodiversity and ES priorities for this (Chan et al. 2006; Turner et al. 2007; Naidoo et al. 2008; Egoh et al. 2009). Nevertheless, several studies have shown that conservation priorities exclusively designed to conserve biodiversity may not maintain optimal ES levels and vice versa (Naidoo et al. 2008). Therefore, the trade-offs and synergies between biodiversity and ESs represent key information for management and policy decisions. Whereas trade-offs usually imply strong constraints for biodiversity conservation, synergies can offer important "win–win" opportunities for both nature conservation and human welfare (Cowling et al. 2008; Egoh et al. 2009; Strassburg et al. 2009; de Groot et al. 2010; Reyers et al. 2012).

Scientific progress in the field of ESs has led to considering net primary production (NPP) as the basic ecological process supporting ES delivery (Richmond et al. 2007; Egoh et al. 2008; Fisher et al. 2009). In particular, widespread growing concern for global climate change has focused on the role of NPP. The difference between CO_2 fixed by photosynthesis and CO_2 lost to autotrophic respiration is one of the most important components of the carbon sequestration service. Thus, in the context of the biodiversity and climate change crisis, the analysis of the congruence between biodiversity and carbon conservation efforts emerges as one of the most important priorities for managers (Harvey et al. 2009; Thomas et al. 2013). Common protocols easily implemented worldwide are currently required to develop connected biodiversity and carbon conservation strategies (Harvey et al. 2009).

Satellite imagery is a valuable tool for quantifying the regional patterns of ecosystem-scale proxies of both biodiversity and carbon sequestration (Cabello et al. 2012). Remote sensing tools have been successfully used for both purposes by means of spectral vegetation indices, such as the Normalized Difference Vegetation Index (NDVI) or the Enhanced Vegetation Index (EVI). These indices are linear estimators of the fraction of photosynthetically active radiation (fPAR) intercepted by vegetation (Wang et al. 2004), the main control of carbon gains (Monteith 1981) (see Chapter 6). These indices not only enable estimation of ecosystem carbon gains (Huete et al. 2002) but can also derive functional attributes strongly related to key ecological processes, such as primary production (Ruimy et al. 1994), seasonality (Potter and Brooks 1998), and phenology (Reed et al. 1994). These functional attributes been used successfully to identify ecosystem functional types (EFTs) (Paruelo et al. 2001; Alcaraz-Segura et al. 2006, 2010). EFTs are groups of ecosystems that share functional diversity characteristics related to the dynamics of matter and energy exchanges between biota and the atmosphere (e.g., primary production, trophic transfer from plants to animals, nutrient cycling, water dynamics, and heat transfer). EFTs offer specific coordinated response to environmental factors and have been used as landscape-scale biological entities to quantify regional diversity (Alcaraz-Segura et al. 2013).

The existing protected-area networks could acquire added value for the public by incorporating ESs other than biodiversity (such as carbon gains) into their design and management (Eigenbrod et al. 2009; Sutherland et al. 2009). In practice, although frequent efforts have been made to assess the effectiveness of protected-area networks in maximizing biodiversity (e.g., Rodrigues et al. 2004; Langhammer et al. 2007), few cases actually incorporate other ESs in such assessments. This study analyzes how current national park networks preserve ecosystem functional diversity and carbon gains at the country level. We chose national parks as a study case because they aim to represent the best examples of a country's major natural systems. We first mapped the spatial heterogeneity of EFTs in each country. Then, we estimated park contributions to maximizing protected diversity in terms of the representativeness (and, conversely, the conservation gaps) and rarity of their EFTs, and carbon gains. Finally, we

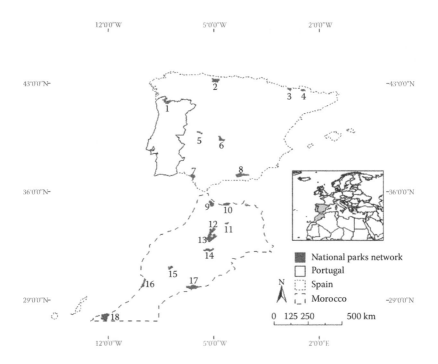

FIGURE 9.1
Study area. The analysis was done for the national park networks in three countries: Portugal: Peneda Gerês (1); Spain: Picos de Europa (2), Ordesa (3), Aigüestortes (4), Monfragüe (5), Cabañeros (6), Doñana (7), and Sierra Nevada (8); and Morocco: Talassemtane (9), Al Hoceima (10), Tazekka (11), Ifrane (12), Khenifra (13), Haut Atlas Oriental (14), Toubkal (15), Sous Massa (16), Iriqui (17), and Khenifiss (18).

analyzed the spatial congruence between ecosystem functional diversity and carbon gains across the networks to find any trade-offs or synergies between diversity and carbon conservation strategies in Portugal, Spain, and Morocco (Figure 9.1 and Table 9.1). These countries share some biodiversity and environmental traits but have different conservation policies. In addition, although Portugal and Spain have a long history of nature conservation and respond to European conservation policies, in Morocco, prior knowledge of biodiversity and ESs is still scarce.

9.2 Methodology

9.2.1 Identification of EFTs and Quantification of Carbon Gains from a Satellite-Derived Enhanced Vegetation Index

The identification of EFTs and the quantification of carbon gains were based on the EVI derived from Moderate Resolution Imaging Spectroradiometer

TABLE 9.1

National Park Networks of Portugal, Spain, and Morocco

Country	Code	National Park	Year	Environmental Characteristics	IUCN[a]
Portugal	1	Peneda-Gerês	1971	Mediterranean and Eurosiberian transition region (mainly oak forests, heathlands)	II
Spain	2	Picos de Europa	1918/1995	Atlantic forests (temperate deciduous forests, Eurosiberian grassland, montane and subalpine scrubs)	II
	3	Ordesa	1918/1982	Alpine coniferous forests and Eurosiberian grasslands	II
	4	Aigüestortes	1955/1996	Alpine coniferous forests and Eurosiberian grasslands	II
	5	Monfragüe	2007	Mediterranean oak forests and scrublands	V
	6	Cabañeros	1995	Extensive lowlands with Mediterranean forests and scrublands	II
	7	Doñana	1969/2004	Salt marshes, coastal dunes, maquis, and garrigue	II
	8	Sierra Nevada	1999	Mediterranean oak forests, scrublands, and grasslands linked to medium and high Mediterranean mountains	V
Morocco	9	Talassemtane	2004	Moroccan fir forests (*Abies maroccana*), and Atlas cedar and black pine forests	NR
	10	Al Hoceima	2004	Maquis and garrigue, high coastal cliffs	NR
	11	Tazekka	1950	Oak forests and Atlas cedar forests	NR
	12	Ifrane	2004	Atlas cedar forests and mixed forests	V
	13	Khenifra	2008	Atlas cedar forests	NR
	14	Haut Atlas Oriental	2004	Atlas cedar, oak, and *Juniper* forests	NR
	15	Toubkal	1942	Strong altitudinal gradient from infra-Mediterranean to the highest peaks of north Africa. Oak and cedar forests, scrublands, and grasslands	V
	16	Sous Massa	1991	Coastal dunes and wetlands	NA

(Continued)

TABLE 9.1 (*Continued*)

National Park Networks of Portugal, Spain, and Morocco

Country	Code	National Park	Year	Environmental Characteristics	IUCN[a]
	17	Iriqui	1994	Desert vegetation (*Acacia radiana* steppes and savanna), temporary wetlands, and dunes	NR
	18	Khenifiss	2006	Wetlands, salt flats, coastal dunes, and deserts	NA

Source: World Database on Protected Areas. http://www.wdpa.org

Note: NA = not applicable; NR = not reported.

[a] International Union for the Conservation of Nature (IUCN) in accordance with World Database on Protected Areas.

(MODIS) images. We used the MOD13C1 product for 2001–2010, which consists of 16-day maximum composite global images at a spatial resolution of 0.05° × 0.05° (~25 km² at the equator). The temporal dynamics of spectral vegetation indices have been widely used to characterize ecosystem primary production, seasonality, and phenology (Pettorelli et al. 2005). We used these three attributes to identify EFTs following the approach explained in Alcaraz-Segura et al. (2013). First, mean annual EVI (EVI_Mean) was used as a linear estimator of annual primary production, one of the most integrative descriptors of ecosystem functioning (Virginia and Wall 2001). EVI_Mean was also used as our proxy for carbon gains quantification. Second, the EVI seasonal coefficient of variation (EVI_sCV) was used as a descriptor of seasonality or the difference between carbon gains in the growing and nongrowing seasons. Third, the date of maximum EVI (DMAX) was used as a phenological indicator of the growing season (Paruelo et al. 2001; Pettorelli et al. 2005; Alcaraz-Segura et al. 2006). These three EVI metrics are widely known to capture most of the variability in the EVI time series (Alcaraz-Segura et al. 2006, 2011).

The ranges of the three EVI metrics were divided into four intervals using the approach proposed by Alcaraz-Segura et al. (2013), potentially leading to 4 × 4 × 4 = 64 EFTs. For DMAX, the four intervals agreed with the four seasons of the year in temperate ecosystems. For EVI_Mean and EVI_sCV, we calculated the first, second, and third quartiles of the histograms for each year. Then the limits between intervals were set as the 10-year median of each quartile for each metric. Three-character codes were assigned to each EFT following the terminology suggested by Paruelo et al. (2001) based on two letters and a number. The first letter (capital) of the EVI_Mean code ranged from "A" to "D" for low to high (increasing) productivity. The second letter (small) showed the EVI_sCV, ranging from "a" to "d" for high to low (decreasing) seasonality. The numbers referred to the season of maximum EVI (1–4: spring, summer, autumn, and winter). The definition and coding of EFTs were based only on descriptors of ecosystem functioning and allowed for an ecological interpretation. Once we had set the limits among the four intervals of each metric, we applied them to each year to produce ten EFTs maps (2001–2010). Because the EFTs maps showed minor changes over the years, in the end we used the interannual medians to summarize ecosystem functional heterogeneity for 2001–2010.

9.2.2 Functional Diversity Assessment of National Park Networks

In each country, our study was restricted to EFTs in natural areas (wild lands). This restriction was based on the human influence index (HIi) (dataset at 1 km × 1 km spatial resolution), which estimates the direct human influence, on ecosystems by integrating eight indicators of human presence (Sanderson et al. 2002). The HIi ranges from 0 to 64. Zero represents no human influence, and 64 is the strongest influence. Our analysis therefore only used MODIS pixels with a mean HIi below 20. The threshold for natural ecosystems was HIi < 20 because this was the maximum HIi found in national parks. In addition, we double-checked the naturalness of the pixel selections by using orthophotos and Google Earth. In each national park, we only included those MODIS pixels with more than 75% of their area within the park limits.

The functional analysis was based on richness, representativeness, and rarity. These criteria are widely used in conservation biology to quantify the ability of national protected areas to maximize a country's biodiversity. The effectiveness of each network in representing EFTs' diversity in each country (representativeness) was evaluated as the simple ratio of the number of EFTs included in the national park network to the total number of EFTs present in all of the country's natural areas (Cabello et al. 2008). As a result, we also identified the gaps in EFT's conservation at the country level. The rarity (singularity or distinctiveness) of EFTs composition in each national park was calculated as the sum of the relative rarities of the EFTs present in the park. First, the absolute rarity of each EFT was estimated according to the following equation:

$$\text{Rarity EFT}_i = (\text{AEFT}_{max} - \text{AEFT}_i)/\text{AEFT}_{max} \tag{9.1}$$

where AEFT_{max} is the area occupied by the most abundant EFT and AEFT_i is the area of the EFT being evaluated. This index grades rarity as low to high in a range of 0 to 1. Then the sum of absolute rarities in all EFTs in each country was found, and the relative rarities of each EFT were calculated as a percentage of the whole country (Rarity EFT$_i$ [%] = Rarity EFT$_i$*100/Σ Rarity EFTs in the country). In addition, to visualize the parks as rare or common, all the natural pixels in each country were plotted in the functional space defined by EVI_Mean and EVI_sCV.

9.2.3 Congruence between Functional Diversity and Carbon Gains in the Networks

To explore the spatial congruence between functional diversity and carbon gains across the networks, we plotted the pairwise associations between

EFTs richness (our estimator for ecosystem functional diversity) and EVI_ Mean (our estimator for carbon gains) for the national parks in each country. With this analysis, we were able to identify the parks that were most representative of "win–win" opportunities for biodiversity and conservation strategies and those parks that maximize only one of the two strategies.

9.3 Results

9.3.1 Spatial Patterns of EFTs in Natural Areas of Portugal, Spain, and Morocco

The EFT maps in Figure 9.2 present a synthetic characterization of the spatial patterns of ecosystem functioning in natural areas (wild lands) in Portugal, Spain, and Morocco in 2001–2010. In Portugal, 58 out of

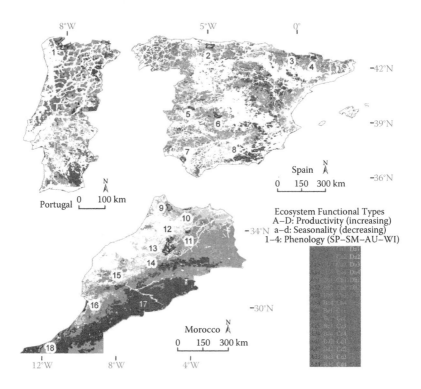

FIGURE 9.2 (See color insert.)
Spatial patterns of ecosystem functional types (EFTs) in Portugal, Spain, and Morocco. EFTs were classified by country. Thus, the range of values used to establish the productivity, seasonality, and maximum categories varied for each country.

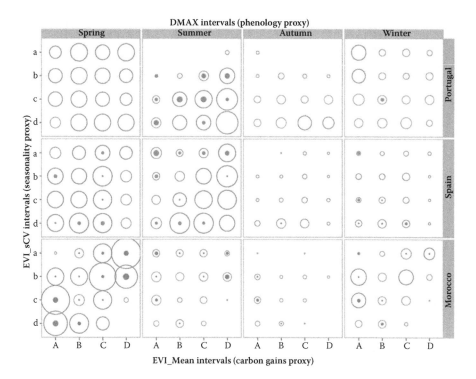

FIGURE 9.3 (See color insert.)
Gap analysis of the national park networks in Portugal, Spain, and Morocco based on the classification of ecosystem functional types (EFTs). EFTs were classified by country. Thus, the range of values used to establish the carbon gains (increasing from "A" to "D"), seasonality (decreasing from "a" to "d"), and phenology intervals varied depending on the country. Outer circles indicate the percentage of the country occupied by natural areas (i.e., human influence index less than 20) corresponding to each EFT. Inner solid circles indicate the percentage of the national park network occupied by each EFT.

the 64 possible combinations of EVI_ Mean, EVI_sCV, and DMAX were found (Figure 9.3 and Table 9.2), and most EFTs were equally abundant. The most abundant EFTs presented spring and summer maxima. Highly productive ecosystems tended to have low seasonality with spring and summer maxima and were found in the northwest quadrant of the country in the transition between the Mediterranean and Eurosiberian biogeographical regions (northward from the Serra da Lousã and Serra da Estrela mountains). EFTs with low productivity were located in the northeast (Duero River basin) and southeast quadrants (Guadiana River basin) in the Mediterranean region. Seasonality in these regions was low to medium with autumn maxima, or high with winter maxima. Almost all possible combinations of productivity and seasonality were found in the ecosystems with winter and spring maxima.

TABLE 9.2

National Park Network Assessment by Conservation Strategies Based on Ecosystem Functional Diversity and Carbon Gains

Country	National Parks	Representativeness (Number of EFTs)	Rarity Mean	Rarity Cumulative (%)	Carbon Gains (EVI Mean)	Win–Win Parks[a]
Portugal	Network	10/58[b] (17.2%)	0.789	18.4	1/1[c] (100%)	—
	Peneda-Gerês	10	0.789 ± 0.201	18.4	0.264 ± 0.034	
Spain	Network	24/63 (38.1%)	0.596 ± 0.254	44.8	6/7 (86%)	4/7
	Aigüestortes	2	0.833 ± 0.094	3.6	0.162 ± 0.028	
	Cabañeros	6	0.361 ± 0.302	4.7	0.270 ± 0.015	
	Doñana	8	0.599 ± 0.266	9.0	0.166 ± 0.042	
	Monfragüe	2	0.178 ± 0.042	0.8	0.245 ± 0.013	
	Ordesa	3	0.841 ± 0.075	5.5	0.170 ± 0.091	
	Picos de Europa	6	0.562 ± 0.351	7.3	0.288 ± 0.066	
	Sierra Nevada	8	0.798 ± 0.297	13.9	0.1668 ± 0.062	
Morocco	Network	31/57 (54.4%)	0.613 ± 0.202	66.3	5/10 (50%)	1*/10
	Al Hoceima	4	0.559 ± 0.434	4.6	0.159	
	Haut Atlas Oriental	6	0.687 ± 0.314	8.4	0.098	
	Ifrane	3	0.456 ± 0.467	2.9	0.219	
	Iriqui	3	0.422 ± 0.346	2.6	0.060	
	Khenifiss	15	0.784 ± 0.256	24.0	0.073	
	Khenifra	3	0.783 ± 0.303	4.8	0.224	
	Sous Massa	4	0.822 ± 0.193	6.7	0.064	
	Talassemtane	4	0.587 ± 0.463	4.8	0.277	
	Tazekka	2	0.217 ± 0.307	0.9	0.194	
	Toubkal	4	0.810 ± 0.171	6.6	0.074	

[a] Number of parks that show both high functional diversity (EFTs) and carbon gains.

[b] Number of EFTs included in the national park networks over the total EFTs of the natural areas of the country.

[c] Number of parks that show an EVI mean over the country median.

Spain had 63 out of the 64 possible EFTs combinations (Figure 9.3 and Table 9.2), though EFTs with spring and summer maxima were more abundant. A clear contrast was observed in the pattern between the Eurosiberian (in the north and northwest) and Mediterranean (in central, south, and east) biogeographical regions (Alcaraz-Segura et al. 2006). In general, the productivity was high for the Eurosiberian region and Mediterranean mountains. Productivity was low in the watersheds associated with major rivers, such as Ebro in the east and Duero in the north, and in the La Mancha inland plain. The semiarid zones in the southeast showed the lowest productivity. Seasonality, which was highly variable, was high in the peaks of the Eurosiberian ranges (Pyrenees and Picos de Europa) and was highest in the Mediterranean region, river basins, wetlands (Doñana, Ebro River Delta, and Albufera de Valencia), and the semiarid southeastern portion of Iberia. The majority of the country showed EVI maxima in spring and summer. The Eurosiberian ecosystems were characterized by a clear summer maximum. In the Mediterranean region, high mountains, wetlands, and water areas (riverine areas) also showed summer maxima. The rest of Spain had two main phenological patterns. Mediterranean mountains peaked in autumn to early winter, while semiarid zones, river basins, and continental plains had a spring maxima.

Morocco had 57 out of the 64 possible EFT combinations (Figure 9.3 and Table 9.2). Moroccan productivity was lower than Portugal's or Spain's. A clear contrast was observed between the more productive north and northwest regions, which have a Mediterranean climate and Atlantic influence, and the less productive south and southeast regions, which are influenced by the Sahara Desert climate. The highest EVI_Mean was reached in some localities near the Atlantic coast (Mamora Forest) on the north face of the High Atlas Mountains and in the Talassemtane National Park with summer maxima (all three of them in the westernmost Mediterranean woodlands ecoregion). The lowest means were south of the High Atlas Mountains in the provinces of Ouarzazate, Tata, and Tan Tan (North Saharan steppe ecoregion). The northern and northwestern regions were characterized by high seasonality and spring maxima, and the southern and southeastern regions were characterized by mean to low seasonality and winter and spring maxima. EFTs with EVI maxima in winter and spring were the most abundant.

9.3.2 Representativeness and Rarity of the National Park Networks

The representativeness and rarity of the national park networks varied among the three countries (Table 9.2). The Portuguese network, with only one national park, offered few opportunities for comparing the Portuguese conservation strategy to the other countries. Nevertheless, this was the network that showed the lowest representativeness (17.2%) and accumulated rarity (18.4%) of national ecosystem functional diversity (Table 9.2). Thus, even though Peneda-Gerês, a park that includes oak forest and heathlands in the Mediterranean and Eurosiberian region, was large enough to capture an important portion of the country's EFT heterogeneity and showed a relatively high mean rarity (0.789 ± 0.201), many gaps were found in the country's ecosystem protection (Figure 9.3). The park comprised almost 20% of the country's rarity; however, most of the pixels for this park showed functional combinations of EVI_Mean and EVI_sCV common in Portugal (Figure 9.4). Only six EFTs with summer maxima and one EFT with winter maxima were included in the park, whereas EFTs corresponding to seasonality or those with spring or autumn maxima would have to be represented in additional protected areas to meet more biodiversity priorities (Figure 9.3).

The Spanish network covered 24 out of 63 EFTs (Table 9.2 and Figure 9.3) and was more biased toward embracing rare ecosystems (44.8% of accumulated rarity) than representing the national functional heterogeneity (38.1% of representativeness). Sierra Nevada Park, which protects a wide range of Mediterranean ecosystems from forest to grasslands along an altitudinal gradient, and Doñana, with salt marshes, coastal, and nonforest ecosystems, each including eight EFTs, were the two parks that most contributed to the representativeness of the network. Picos de Europa and Cabañeros Parks, Atlantic and Mediterranean forests, respectively, had six EFTs, showing slightly less functional diversity. Ordesa, with 3 EFTs, and especially, Aigüestortes and Monfragüe, typically forested areas with only 2 EFTs, were the poorest parks in terms of EFTs richness. The network's mean rarity was 0.596, although for the purposes of this conservation goal, variations among parks are considered important (±0.254). This network held 44.8% of the functional rarity of national natural ecosystems. Rarity was associated with low productivity (EVI_Mean) ecosystems and also with medium seasonality (EVI_sCV) (Figure 9.4). Although Aigüestortes (0.833 ± 0.094) and Ordesa (0.841 ± 0.075) were the parks with the highest mean rarity, the parks with the most accumulated rarity were Sierra Nevada (13.9%) and Doñana (9%). In this network, the gaps in EFTs conservation referred to the ecosystems with an autumn EVI maximum (Figure 9.3).

The Moroccan network covered most EFTs (31 out of 57) (Table 9.2 and Figure 9.3). The two criteria related to maximizing functional diversity were the highest in this network, with 54.4% of representativeness and 66.3% of cumulative rarity. Khenifiss, which includes a mosaic of wetlands, salt flats, coastal dunes, and deserts, was the richest park with 15 EFTs. The others ranged from six EFTs in Haut Atlas Oriental to two EFTs in Tazekka. Although the rarity of this network (0.613 ± 0.202) was similar to Spain, its cumulative rarity was the highest of the three countries (66.3%). Cumulative rarity was highest in Khenifiss Park (24%) and lowest in Tazzekka. Overall, rarity was associated with ecosystems with very low productivity and medium to high seasonality (Figure 9.4). Nevertheless, in Khenifiss, high rarity was related to phenology because its pixels were associated with common combinations of productivity and seasonality (Figure 9.4), but the four phenology

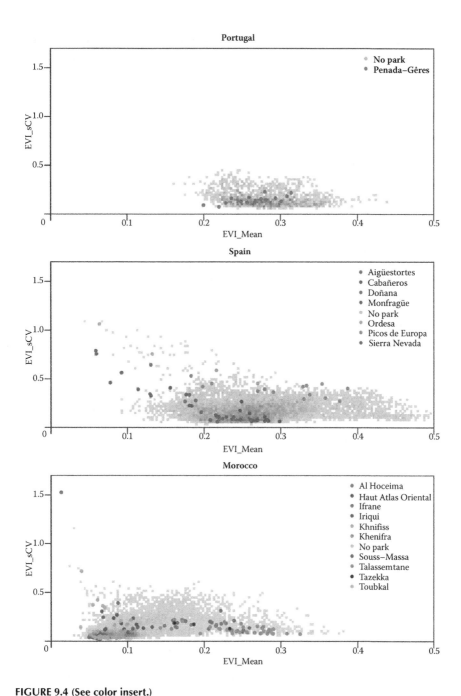

FIGURE 9.4 (See color insert.)
Functional characterization by mean annual Enhanced Vegetation Index (EVI) as an estimator of carbon gains (EVI_Mean) and EVI Spatial Variation Coefficient, a descriptor of seasonality (EVI_sCV) of natural vegetation in Portugal, Spain, and Morocco and in their national parks.

classes showed differences. In this network, the gaps were mainly found in ecosystems with maximum EVI in summer and autumn (Figure 9.3).

9.3.3 Carbon Gains in the National Park Networks

Peneda-Gerês, the only national park in Portugal, was very productive, with a mean carbon gain above the country's median (Table 9.2 and Figure 9.5). In Spain, 86% of the parks showed carbon gains above the median for the country (Table 9.2 and Figure 9.5). The national parks with the most productive ecosystems (with EVI_Mean above the country's median) were Picos de Europa, Cabañeros, and Monfragüe (which have extensive forests) and Doñana (which has a mosaic of scrublands and vegetation linked to dunes and wetlands). Ordesa and Sierra Nevada were at the median of the country's carbon gains, while Aigüestortes (with mosaics of forests, scrublands, and mountain wetlands) showed the lowest EVI means. On the other hand, 50% of the parks in the Moroccan network had carbon gains above the country median (Table 9.2 and Figure 9.5). The most productive parks (with the EVI_Mean above the country median) were Talassemtane, Khenifra, Ifrane, and Tazekka, which are typical Mediterranean coniferous forests (fir, cedar, and pine forests), and Al Hoceima, which is mostly Mediterranean scrublands. The least productive national parks were Iriqui, Sous Massa, Khenifiss, Toubkal, and Haut Atlas Oriental, all of them located in the Sahara Desert biogeographical region and mainly scrublands, wetlands, and desert vegetation. On the whole, the Moroccan national park network was less biased to productive ecosystems than the Portuguese and Spanish networks.

9.3.4 Spatial Congruence between Diversity and Carbon Gains in National Park Networks

The congruence between EFT richness and carbon gains varied among the three countries. In Portugal, Peneda-Gerês had 10 EFTs, a high number considering the results for the other two countries, and also showed high carbon gains. In the Spanish network, the parks with the highest EFT richness (eight EFTs)—Sierra Nevada and Doñana—are areas with medium to low carbon gains, although they were always over the country median (Table 9.2 and Figure 9.5). Nevertheless, Picos de Europa and Cabañeros, which also had relatively high EFT richness (six EFTs), had the highest carbon gains. This was not the case in Monfragüe, which had relatively high carbon gains but only two EFTs. Two parks, Aigüestortes and Ordesa, with low EFT richness (two and three, respectively) also had low carbon gains. Finally, in Morocco, Haut Atlas, especially Khenifiss Park—with the highest EFTs richness (15 and 6, respectively)—had mainly carbon gains below the median of the country. The remaining parks, with fewer than four EFTs, had highly variable carbon gains.

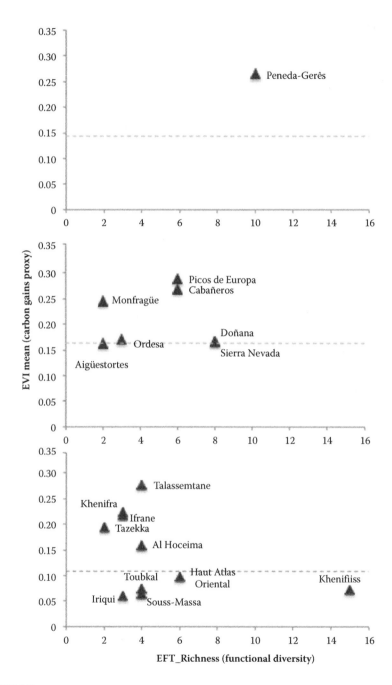

FIGURE 9.5
Analysis of spatial congruence between functional diversity and carbon gains in the national park networks. Dotted line indicates the median of the Enhanced Vegetation Index mean for each country.

9.4 Discussion

Global change and the links between ecological processes and human welfare emphasize the need for incorporating ESs as conservation targets. For this purpose, finding the spatial congruence between conservation efforts and ESs emerges as a new challenge. We now need additional knowledge about how biodiversity and ESs correlate with existing protected-area networks. Any such evaluation must estimate the level of biodiversity and ESs in the protected areas. However, this may be slower and more complicated than desirable because of the huge amount of biodiversity and environmental information available and the errors associated with ES mapping. As shown earlier, these limitations can be overcome by converting spectral information derived from satellite imagery into ecosystem functional variables. First (by identifying EFTs), landscape diversity measurements can be estimated and used as diversity proxies because they respond to the environmental controls the same way as species diversity (Alcaraz-Segura et al. 2013) (see Chapter 8). Second, satellite-derived information minimizes the errors associated with the quantification of spatial variability of ESs in large regions (Cabello et al. 2012). At this point, the level of carbon gains was found for each protected area in the three different countries, as an example of an essential ES to incorporate in the new conservation paradigm. We calculated its spatial variability directly from the spectral features of each portion of land area, avoiding generalizations made from different land covers (Eigenbrod et al. 2010).

From the regional EFT patterns in Portugal, Spain, and Morocco, we identified the gaps, biases, and spatial congruence of the three conservation objectives related to biodiversity and ES priorities—representativeness, rarity, and carbon gains—in the national park networks of these countries. The national parks studied matched these objectives to different degrees, and in any case, were not clear options of win–win parks in terms of ecosystem functional diversity and carbon gain service. In Portugal, Peneda-Gerês, the only park in the country, was very rich and responded albeit feebly to conservation strategies based on representativeness (17.2%) and rarity (18.4%). This park could be considered nearly win–win, once it was also matched to a carbon conservation strategy (Table 9.2 and Figure 9.5). However, the network should include other protected areas to fill the numerous gaps identified in functional diversity (Figure 9.3). From the carbon gains perspective, high productivity ecosystems are well represented in this country (Figure 9.2). Nevertheless, a conservation strategy based exclusively on carbon gains may not include important biodiversity features (Naidoo et al. 2008), such as ecosystems with medium to low productivity, which were distributed over more than half of the country's area (Figure 9.2).

The Spanish national park network responded to the three conservation strategies on a different level, and it was the network that best matched a carbon conservation strategy, probably because it includes mainly forest ecosystems; 80% of the parks had carbon gains above the country's median. The network was also more biased to rarity than to EFT representativeness (Table 9.2 and Figure 9.4). Sierra Nevada Park, with eight EFTs (Table 9.2) and areas with uncommon ecosystem functioning in terms of productivity (EVI_Mean), seasonality (EVI_sCV) (Figure 9.4), and phenology, contributed the most to the representation of the country's ecosystem functional diversity in terms of richness, representativeness, and rarity. The Spanish system was the network with the most clear win–win candidate parks in terms of ecosystem functional diversity and carbon gains (Figure 9.5). Picos de Europa and Cabañeros parks had the highest carbon gain, and Doñana and Sierra Nevada had the highest ecosystem functional diversity. Nevertheless, a trade-off seems to emerge here because the increase in functional diversity or biodiversity (implicit), which corresponds to a mosaic of different types of vegetation in Doñana and Sierra Nevada, could result in decreasing carbon gains compared to the forests of Picos de Europa and Cabañeros.

In Morocco, the national park network had the highest ecosystem functional diversity (53% of the representativeness and 49.5% of the cumulative rarity) (Table 9.2) and carbon gains (50% of the parks) criteria. Khenifiss Park, in particular, includes a mosaic of wetlands, salt flats, coastal dunes and deserts, and, associated with it, highly diversified fauna (http://ma.chm-cbd. net). This park, which had the highest EFT richness and the lowest carbon gains (Figure 9.5), is an example of the lack of spatial congruence between biodiversity and carbon gains in temperate and arid regions. Outside tropical regions, high ecosystem functional diversity (and implicit biodiversity) is associated with distinctive environmental characteristics or ecotone environments (e.g., Alcaraz-Segura et al. 2013), minimizing the importance of forested areas, which typically have high carbon gains, for biodiversity. Moreover, Talassemtane, which includes the most representative Moroccan coniferous forests (*Abies*, *Cedrus*, and *Pinus*), with four EFTs, had the national median for EFT richness and could be considered the park closest to the win–win option.

This evaluation of national park networks adds value to current conservation efforts. It also provides information on how current climate change mitigation strategies, such as carbon sequestration, could affect protected areas (Thomas et al. 2012). Nevertheless, a broader analysis encompassing multiple ESs would be needed (Bennett et al. 2009) to design win–win networks. In nontropical countries, as we have shown, high biodiversity may be concentrated in nonforested ecosystems, or there may simply not be any win–win areas. Biodiversity could therefore be at risk under carbon conservation-only strategies (CBD–UNEP 2011). Moreover, protected areas, such as national parks, could shelter other ESs (in addition to biodiversity and carbon sequestration) for managing global change and human welfare. For example, in Morocco, some mountain parks were created to protect water resources, while others provide areas for scientific study, and in practice, generate ecotourism (http://ma.chm-cbd.net). Considering the multiple dimension of ESs assessments (see Chapter 20), conservation policies should promote protected areas that contribute to human well-being, from the viewpoint of biodiversity, carbon, or other services.

9.5 Conclusions

As we have shown, remote sensing tools can be useful for the analysis of current national park networks under the new conservation paradigm that considers ES conservation targets. The translation of spectral information into vegetation indices enabled us to quantify national biodiversity proxies and key ESs such as carbon gains. Based on the use of ecosystem functional traits derived from spectral vegetation indices, we have identified the biases, synergies, and trade-offs between representativeness, rarity, and carbon gains in the national park networks of Portugal, Spain, and Morocco. In all cases, the networks showed gaps in conservation, although this was particularly important for Portugal. Moreover, the lack of clear synergies between ecosystem functional diversity and carbon conservation strategies pointed out the need to consider network-wide win–win options instead of just for the individual parks. This strategy could be particularly important outside the tropical regions where high biodiversity and other ESs may be associated with nonforested areas.

Acknowledgments

Comments from anonymous reviewers improved the final manuscript. Funding was received from the ERDF (FEDER), Programa de Cooperación Transfroteriza España-Fronteras Exteriores (POXTEFEX-Transhabitat), Andalusian Regional Government (Junta de Andalucía GLOCHARID and SEGALERT Projects, P09–RNM-5048), and the Ministry of Science and Innovation (Project CGL2010-22314). We thank Deborah Fuldauer for the English revision.

References

Alcaraz-Segura, D., E. H. Berbery, S. J. Lee, and J. M. Paruelo. 2011. Use of ecosystem functional types to represent the interannual variability of vegetation biophysical properties in regional models. *CLIVAR Exchanges* 17:23–27.

Alcaraz-Segura, D., E. Liras, S. Tabik, J. Paruelo, and J. Cabello. 2010. Evaluating the consistency of the 1982–1999 NDVI trends in the Iberian Peninsula across four time-series derived from the AVHRR Sensor: LTDR, GIMMS, FASIR, and PAL-II. *Sensors* 10:1291–1314.

Alcaraz-Segura, D., J. M. Paruelo, and J. Cabello. 2006. Identification of current ecosystem functional types in the Iberian Peninsula. *Global Ecology and Biogeography* 15:200–212.

Alcaraz-Segura, D., J. M. Paruelo, H. E. Epstein, and J. Cabello. 2013. Environmental and human controls of ecosystem functional diversity in temperate South America. *Remote Sensing* 5:127–154.

Bennett, E. M., G. D. Peterson, and L. J. Gordon. 2009. Understanding relationships among multiple ecosystem services. *Ecology Letters* 12:1394–1404.

Cabello, J., D. Alcaraz-Segura, A. Altesor, M. Delibes, and E. Liras. 2008. Funcionamiento ecosistémico y evaluación de prioridades geográficas en conservación [Ecosystem functioning and assessment of conservation geographical priorities]. *Ecosistemas* 17:53–63.

Cabello, J., N. Fernández, D. Alcaraz, et al. 2012. The ecosystem functioning dimension in conservation biology: Insights from remote sensing. *Biodiversity and Conservation* 21:3287–3305.

CBD–UNEP (Convention on Biological Diversity–United Nations Environment Programme). 2011. *REDD-plus and biodiversity.* CBD Technical Series No. 59. Montreal, Canada: CBD–UNEP.

Chan, K. M. A., M. R. Shaw, D. R. Cameron, E. C. Underwood, and G. C. Daily. 2006. Conservation planning for ecosystem services. *PLoS Biology* 4:14.

Convention on Biological Diversity, Conference of the Parties, 2010. The strategic Plan for Biodiversity 2011–2020 and the Aichi Biodiversity Targets.

Cowling, R. M., B. Egoh, A. T. Knight, et al. 2008. An operational model for mainstreaming ecosystem services for implementation. *Proceedings of the National Academy of Sciences of the United States of America* 105:9483–9488.

de Groot, R. S., R. Alkemade, L. Braat, L. Hein, and L. Willemen. 2010. Challenges in integrating the concept of ecosystem services and values in landscape planning, management and decision making. *Ecological Complexity* 7:260–272.

Egoh, B., B. Reyers, M. Rouget, M. Bode, and D. M. Richardson. 2009. Spatial congruence between biodiversity and ecosystem services in South Africa. *Biological Conservation* 142:553–562.

Egoh, B., B. Reyers, M. Rouget, D. M. Richardson, D. C. Le Maitre, and A. S. van Jaarsveld. 2008. Mapping ecosystem services for planning and management. *Agriculture, Ecosystems and Environment* 127:135–140.

Eigenbrod, F., B. J. Anderson, P. R. Armsworth, et al. 2009. Ecosystem service benefits of contrasting conservation strategies in a human-dominated region. *Proceedings of the Royal Society—Biological Sciences* 276:8.

Eigenbrod, F., P. R. Armsworth, B. J. Anderson, et al. 2010. The impact of proxy-based methods on mapping the distribution of ecosystem services. *Journal of Applied Ecology* 47:8.

Fisher, B., R. K. Turner, and P. Morling. 2009. Defining and classifying ecosystem services for decision making. *Ecological Economics* 68:643–653.

Harvey, C. A., B. Dickson, and C. Kormos. 2009. Opportunities for achieving biodiversity conservation through REDD. *Conservation Letters* 3:53–61.

Huete, A., K. Didan, T. Miura, E. P. Rodriguez, X. Gao, and L. G. Ferreira. 2002. Overview of the radiometric and biophysical performance of the MODIS vegetation indices. *Remote Sensing of Environment* 83:195–213.

Langhammer, P. F., M. I. Bakarr, L. A. Bennun, et al. 2007. *Identification and gap analysis of key biodiversity areas: Targets for comprehensive protected area systems.* Gland, Switzerland: IUCN.

MA (Millennium Ecosystem Assessment). 2005. Ecosystems and human well-being: Desertification Synthesis. Washington, DC: Island Press.

Mace, G. M., K. Norris, and A. H. Fitter. 2012. Biodiversity and ecosystem services: A multilayered relationship. *Trends in Ecology and Evolution* 27:19–26.

Monteith, J. L. 1981. Climatic variation and the growth of crops. *Quarterly Journal of the Royal Meteorological Society* 107:749–774.

Naidoo, R., A. Balmford, R. Costanza, et al. 2008. Global mapping of ecosystem services and conservation priorities. *Proceedings of the National Academy of Sciences of the United States of America* 105:5.

Paruelo, J. M., E. G. Jobbagy, and O. E. Sala. 2001. Current distribution of ecosystem functional types in temperate South America. *Ecosystems* 4:683–698.

Pettorelli, N., J. O. Vik, A. Mysterud, J. M. Gaillard, C. J. Tucker, and N. C. Stenseth. 2005. Using the satellite-derived NDVI to assess ecological responses to environmental change. *Trends in Ecology and Evolution* 20:503–510.

Potter, C. S., and V. Brooks. 1998. Global analysis of empirical relations between annual climate and seasonality of NDVI. *International Journal of Remote Sensing* 19:2921–2948.

Reed, B. C., J. F. Brown, D. Vanderzee, T. R. Loveland, J. W. Merchant, and D. O. Ohlen. 1994. Measuring phenological variability from satellite imagery. *Journal of Vegetation Science* 5:703–714.

Reyers, B., S. Polasky, H. Tallis, H. A. Mooney, and A. Larigauderie. 2012. Finding common ground for biodiversity and ecosystem services. *BioScience* 62:503–507.

Richmond, A., R. K. Kaufmann, and R. B. Myneni. 2007. Valuing ecosystem services: A shadow price for net primary production. *Ecological Economics* 64:454–462.

Rodrigues, A. S. L., H. R. Akçakaya, S. J. Andelman, et al. 2004. Global gap analysis: Priority regions for expanding the global protected-area network. *BioScience* 54:1092–1100.

Ruimy, A., B. Saugier, and G. Dedieu. 1994. Methodology for the estimation of terrestrial net primary production from remotely sensed data. *Journal of Geophysical Research* 99:5263–5283.

Sanderson, E. W., M. Jaiteh, M. A. Levy, K. H. Redford, A. V. Wannebo, and G. Woolmer. 2002. The human footprint and the last of the wild. *BioScience* 52:891–904.

Strassburg, B. B. N., A. Kelly, A. Balmford, et al. 2009. Global congruence of carbon storage and biodiversity in terrestrial ecosystems. *Conservation Letters* 3:98–105.

Sutherland, W., W. Adams, R. Aronson, et al. 2009. One hundred questions of importance to the conservation of global biological diversity. *Conservation Biology* 23:557–567.

Thomas, C. D., B. J. Anderson, A. Moilanen, et al. 2013. Reconciling biodiversity and carbon conservation. *Ecology Letters* 16:39–47. doi:10.1111/ele.12054.

Turner, W. R., K. Brandon, T. M. Brooks, R. Costanza, G. A. B. Da Fonseca, and R. Portela. 2007. Global conservation of biodiversity and ecosystem services. *BioScience* 57:868–873.

Virginia, R. A., and D. H. Wall. 2001. Principles of ecosystem function. In *Encyclopedia of biodiversity*, ed. S. A. Levin, 345–352. San Diego, CA: Academic Press.

Wang, Q., J. Tenhunen, N. Q. Dinh, M. Reichstein, T. Vesala, and P. Keronen. 2004. Similarities in ground- and satellite-based NDVI time series and their relationship to physiological activity of a Scots pine forest in Finland. *Remote Sensing of Environment* 93:225–237.

10

Catchment Scale Analysis of the Influence of Riparian Vegetation on River Ecological Integrity Using Earth Observation Data

T. Tormos

*National Research Institute of Science and Technology
for Environment and Agriculture (IRSTEA), France*

K. Van Looy

*National Research Institute of Science and Technology for
Environment and Agriculture (IRSTEA), France*

P. Kosuth

*National Research Institute of Science and Technology for
Environment and Agriculture (IRSTEA), France*

B. Villeneuve and Y. Souchon

*National Research Institute of Science and Technology for
Environment and Agriculture (IRSTEA), France*

CONTENTS

10.1 Introduction 202
10.2 Study Area 204
10.3 Datasets and Method 205
    10.3.1 Datasets 205
        10.3.1.1 Remotely Sensed and Ancillary Data 205
        10.3.1.2 Biological Data 208
    10.3.2 Classification of RALC 208
        10.3.2.1 Classification Procedure Overview 208
        10.3.2.2 OBIA Designed for RALC 209
        10.3.2.3 Classification Automation and Accuracy
                Assessment 211

        10.3.3 Modeling Large-Scale Relationships between Land Cover
                and Stream Ecological Integrity 212
                10.3.3.1 Model Overview 212
                10.3.3.2 River Ecological Integrity Assessment 213
                10.3.3.3 Spatial Indicators 213
                10.3.3.4 Statistical Techniques 215
10.4 Results 216
        10.4.1 RALC Map 216
        10.4.2 Land Cover Effect 217
10.5 Discussion 218
        10.5.1 Mapping RALC over Broad Territories 218
        10.5.2 Large-Scale Riparian Vegetation Influence on River
                Ecological Integrity 219
10.6 Conclusion 220
Acknowledgments 221
References 221

10.1 Introduction

During the past decade, river ecosystems have been identified as delivering
ecosystem services that are fundamental to human well-being (Postel and
Carpenter 1997; Aylward et al. 2005). River ecosystems provide cultural (e.g.,
recreation, tourism, existence values), regulating (e.g., maintenance of water
quality, buffering of flood flows, erosion control), and supporting (e.g., role
in nutrient cycling, predator–prey relationships, ecosystem resilience, and
maintenance of biodiversity) services that contribute greatly, directly and
indirectly, to human well-being (Aylward et al. 2005). Although the links
between biodiversity, ecological functions, and the provision of ecosystem
services are often poorly understood (Mertz et al. 2007), it has become evident that maintaining the river ecosystem integrity can support the protection of river ecosystem services.

    The riparian zone, as an interface between the terrestrial and aquatic
ecosystems, encompasses the strip of land containing distinctive vegetation along the margin of a stream. The vegetation may include trees,
woody shrubs, herbs/forbs, grasses, and sedges. The ecological importance
of riparian zones to ecosystem functions has been well recognized (e.g.,
Naiman and Décamps 1997; Naiman et al. 2005; Shearer and Xiang 2007).
On the one hand, riparian vegetation contributes to the regulation of trophic status and food chains (organic debris), temperature (providing shade
and cover for the aquatic communities), and habitat (stabilizes the banks,
provides woody debris), which are key parameters of river ecosystems. On
the other hand, it plays a buffer role against agricultural and urban diffuse
pollution (e.g., nutrients, sediments, pesticides). Maintaining and restoring

healthy riparian buffer conditions might therefore contribute to preserving the integrity of river ecosystems and, ultimately, to maintaining river ecosystem services.

River functional processes within a catchment interact at multiple scales in time and space, and they are closely related to human activity (Allan 2004; Wang 2006; Johnson and Host 2010). Pressures from human activities act in combination and are irregularly distributed on the catchment. Therefore, the production of reliable scientific information for prioritizing and designing efficient riparian buffers over a catchment involves a better understanding of biological responses (reflecting the integrity of river ecosystem functioning) (Hynes 1970; Karr 1993) to riparian vegetation conditions and human pressures at multiple spatial scales. However, reports on the influence of riparian vegetation are diverging in the literature; and several authors point at the limit of spatial information on Riparian Area Land Cover (RALC) in the interpretation of these findings (e.g., Frimpong et al. 2005; Roy et al. 2007; Wasson et al. 2010).

For a long time, accurately mapping RALC over broad territories remained inconceivable due to the small-scale areas on the one hand (Aguiar and Ferreira 2005) and the photointerpretation mapping methods, based on aerial photography, that are time-consuming and very expensive on the other (Coulter et al. 2000). In the 1970s, satellite images provided broader views (such as Landsat 30 m or Spot 20 m), but their moderate spatial resolution (MRS) did not capture the riparian system in great detail except for some very specific contexts such as large floodplains (e.g., a 2700-km² scene for the Amazon floodplain) (Mertes et al. 1995). Considering the spatial extension of riparian ecosystems and the diversity of land cover types in these systems, moderate spatial resolution images appear to be insufficient for a detailed characterization of RALC (Müller 1997; Hollenhorst et al. 2006; Tormos et al. 2011). Yet, most of the studies analyzing relationships between RALC and biological responses are based on MRS-derived land cover maps.

Progress in sensor and platform technologies since the middle of the 1990s has provided broad view imagery with ground cell sizes between 0.6 and 4 m, which are comparable in quality to aerial photographs but with the added advantage of digital multispectral information, stable geometry, and consistent viewing conditions (Goetz 2006). However, at the end of 1990s, RALC mapping studies, as with most high spatial resolution and very high spatial resolution (VHSR) mapping studies, encountered difficulties in processing accurately and quickly rich and detailed radiometric information contained in such images with traditional pixel-based techniques (e.g., Müller 1997; Neale 1997). During the past decade, progress in computer science and geographic information systems (GIS) technology, especially in object-based image analysis, has revolutionized the processing of high spatial resolution and VHSR images by providing effective computer-assisted classification techniques, for which results come close to the quality of manual photointerpretation, while being much faster, cheaper, and more reproducible (e.g., Durieux et al. 2007; Tiede et al. 2010; Dupuy et al. 2012). These improvements allow the production of fine-scale and reliable spatial information on riparian areas over broad territories, explored, for example, by Goetz et al. (2003) and Johansen et al. (2008). Nevertheless, these studies on large-scale systems did not take into account constraints regarding image availability for managers and did not characterize both the main human pressures (agricultural and urban areas) and vegetation conditions over the riparian areas.

In order to deal with the raised questions, this chapter introduces new approaches to (1) the mapping of RALC over broad territories from high spatial resolution and VHSR Earth observation data and (2) the analysis of biological responses to riparian vegetation conditions and human pressures at multiple spatial scales.

10.2 Study Area

This study focuses on a part of Normandy's river network (northwestern France) in the limestone shelf hydroecoregion (HER9; see Figure 10.1a). It contains all the watersheds of tributaries in the downstream part of the River

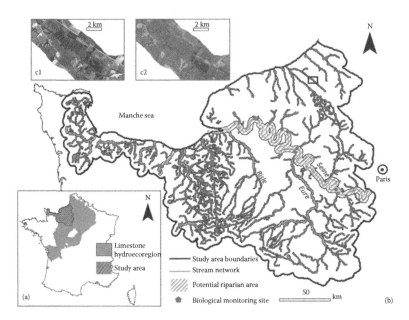

FIGURE 10.1 (See color insert.)
Presentation of the study area (the Normandy coastal region); (a) localization on the limestone shelf French hydroecoregion; (b) stream network, potential riparian area, and localization of biological monitoring sites; and (c) illustration of Earth observation data: (c1) aerial photography (0.5-m pixel) and (c2): Spot5 XS (10-m pixel).

Seine: the Eure (5935 km²), the Epte (1490 km²), the Iton (1300 km²), and the Andelle (740 km²), and several rivers flowing directly into the sea such as the Risle (2310 km²), the Touques (1605 km²), the Dives (1573 km²), and the Bethune (307 km²). It comprises a total of 297 subwatersheds that make up an area of 25,000 km² drained by a river network more than 6000 km in length, of which 48% are reaches of first-order, 20% second-order, 18% third-order, 6% fourth-order, 2% fifth-order, 1% sixth-order, and 5% seventh-order (the Seine), following the ordination of Strahler (1957). One hundred and fifty-five sites from the national monitoring network allow access to the ecological integrity of rivers in this study area.

The HER9 region is a plain with an altitude below 200 m; the basement consists of sedimentary rocks that are mostly tabular carbonate. These rocks show variable characteristics in surface permeability inducing differences in the density of the drainage network, or differences in hardness and erosion resistance, which locally produce more pronounced landforms (hills and coasts) in the study area. Complex structures of riparian strips are present with alternation of deciduous forest and grassland near the river in an area dominated by a strong agricultural activity, mainly focused on field crops (e.g., oilseed, mixed farming, forage) and livestock. This agricultural landscape is interspersed with sparsely urban areas.

10.3 Datasets and Method

10.3.1 Datasets

10.3.1.1 Remotely Sensed and Ancillary Data

Remotely sensed and ancillary data were chosen according to their overall availability and cost-effectiveness for managers in French territories. A comparative economic analysis of the different data led to the collection of two types of high and VHSR, multispectral remotely sensed data: orthophotos and Spot5 satellite images. The characteristics of these images and preprocessing data are summarized in Table 10.1. Orthophotos (0.5-m pixel) provide the textural information required to detect narrow and fragmented cover types along rivers (Müller 1997) (Figure 10.1[c1]). Orthophotos are produced by the French National Geographic Institute (IGN) and updated every 5 years for each French department. Acquisitions are generally realized in spring or summer to avoid large cloud cover. Radiometric equalizations are computed between orthophotos of the same department to reduce radiometric heterogeneities within and between aerial images (Paparoditis et al. 2006). Spot5 XS (10-m pixel) was also acquired in order to get information in the near-infrared (NIR) band essential for discriminating vegetation classes (Johansen and Phinn 2006) (Figure 10.1[c2]).

TABLE 10.1

Characteristics of Earth Observation Data and Preprocessing

Image Data	Footprint	Spatial Resolution	Spectral Resolution	Acquisition Date	No. of Blocks	Producer	Preprocessing
Orthophotos	5 km	0.5 m	B, G, R	Summer period, between 2003 and 2006	2455	IGN®	Clip into riparian area and mosaic
Spot5 XS	60 km	10 m	G, R, NIR, SWIR	Summer period, between 2003 and 2007	11	Spotimage® Archive images	TOA radiometric corrections, clip into riparian area and mosaic

Note: B = blue; G = green; R = red; NIR = near-infrared; SWIR = short wave infrared.

In addition to the images, spatial thematic data (ancillary data) with decametric and metric precision was collected (Table 10.2). We collected the road network geodatabase (BDRoute®, decametric precision) produced by IGN because roads are particularly difficult to extract automatically from remote sensing images due to their heterogeneous spectral behavior. We used the surface hydrographic entities (rivers, lakes, and reservoirs) of the French hydrography database (BDCarthage®, decametric precision) produced by IGN. The Numerical Parcel Register (NPR, metric precision) was used to improve the discrimination between grassland and crops. The NPR gives spatial information of blocks (adjacent plots) benefiting subsidies from the European common agricultural policy. The block boundaries are photointerpreted

TABLE 10.2

Correspondence between Common Land Cover Typology

Common Land Cover Typology (6 categories)	Classes Obtained from OBIA Classification Procedure (25 classes)	Data Sources
C1. Water bodies	Water courses	a & b
	Lakes and reservoirs	c
	Sea and ocean	c
C2. Agricultural areas	Ploughing	a & b
	Annual cultures (wheat, rape, corn, etc.)	a & b & e
	Forage	e
	Vineyards	a & b
	Fruit trees	e
C3. Urban areas	Continuous urban fabric	c
	Artificial soil	a & b
	Industrial or commercial units	c
	Road and rail networks and associated land	c
	Mineral extraction sites	c
	Sport and leisure facilities	c
	Local road	d
	Department road	d
	National road	d
	Freeway	d
C4. Tree vegetation	Tree vegetation	a & b
C5. Herbaceous/shrub vegetation	Permanent grassland	e
	Temporary grassland	e
	Seminatural herbaceous and shrub vegetation	a & b
	Bare soil with sparse vegetation	a & b
C6. Seminatural bare soil	Bare soil	a & b
	Sand	a & b

Note: Typology used in literature for studying impacts on river ecological integrity and land cover classes extracted from object-based image analysis (OBIA) classification procedure obtained on the riparian corridor of the study area with (a) orthophotos; (b) Spot5 XS; (c) CORINE Land Cover database; (d) BDRoute; (e) BDCarthage; and (f) Numerical Parcel Register.

annually and digitized by farmers on orthophotos. And finally, the CORINE (Coordination of information on the environment) Land Cover database (CLC, decametric precision) is derived from the visual interpretation of both Landsat and Spot satellite images acquired in 1990 and 2000. Interpretation of the images is based on transparencies overlaid on 1/100,000 hard copy prints of satellite images (Bossard et al. 2000). CLC features, characterized by a 25-ha minimum area, are either homogenous areas or combinations of land cover types with a certain recognizable structure. The classification is based on a hierarchical standard nomenclature with three levels: five broad land cover categories at level 1 (1: artificial surfaces; 2: agricultural areas; 3: forests and seminatural areas; 4: wetlands; and 5: waterbodies), 15 land cover classes (level 2), and 44 land cover types (level 3). Although CLC is not accurate enough for characterizing the narrow and fragmented objects along rivers, it is appropriate for the extraction of large objects within the river corridor, especially large artificial surfaces (Table 10.2).

10.3.1.2 Biological Data

Macroinvertebrate datasets were extracted from the national monitoring networks over the study area covering the period 1992–2002 (155 sites, 1038 samples; see Figure 10.1b). During this period, the number of samples per site was variable. The monitoring procedure was stabilized after 2002 with the implementation of the Water Framework Directive (WFD). The sites were sampled on average 6.7 times (about once every two years), the extremes being 2 times (4 stations) and 10 times (44 stations). We also noticed that the sites were sampled only once a year (mainly in spring).

In France, the biological monitoring sites were checked by the regional environmental boards using the l'Indice Biologique Global Normalisé (IBGN) index (AFNOR 1992). The IBGN is a combination of two metrics: the total number of taxa (14 classes at family level) and the faunistic indicator group representing the presence/absence of 39 indicator taxa, grouped into nine classes of sensitivity. The index is sensitive to pollution (including toxics) and to general degradation (including habitat alteration). The mean IBGN value covering the 1992–2002 period for each site was used in the models.

10.3.2 Classification of RALC

10.3.2.1 Classification Procedure Overview

We designed a typology composed of six classes (Table 10.2). "C1: water bodies" and "C6: semi-natural bare soils" are necessary to delimitate the river bed; "C2: agricultural areas" and "C3: urban areas" considered to be the main alteration causes of stream ecological status (Allan 2004); and "C4: tree vegetation" and "C5: seminatural herbaceous and shrub vegetation," constituting the main natural elements of the river corridor landscape that maintain biodiversity and regulate nonpoint source pollution (Naiman et al. 2005).

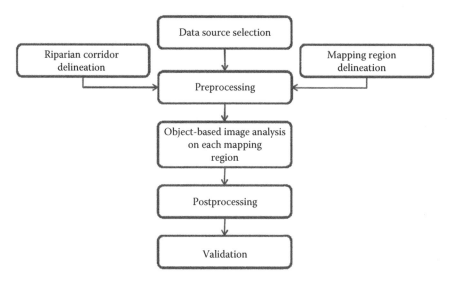

FIGURE 10.2
Flowchart of the Riparian Area Land Cover classification procedure. (From Tormos, T., P. Kosuth, et al., *International Journal of Remote Sensing*, 33, 4603–4633, 2012.)

For this purpose, the object-based image analysis (OBIA) procedure designed by Tormos et al. (2012) was used. The primary objective of this procedure is to produce a detailed and finely resolved RALC map over large riparian territories in different geographical contexts (i.e., relief, climate, and geology). This procedure can be based on information from multisource spatial data accessible for managers or stakeholders. The second objective is to validate RALC map accuracy. A flowchart showing the different stages of this procedure is presented in Figure 10.2.

The procedure is organized in seven stages. Stage 1 is dedicated to select input data. In stage 2, the potential riparian corridor area is delineated (using a buffer around the hydrographic network whose width depends on the Strahler stream order; results in our study area are presented in Figure 10.1b). In stage 3, homogeneous regions (in terms of both geographical context and image acquisition) are mapped. The following three stages are dedicated to the classification of RALC with a preprocessing (stage 4), an OBIA on each mapping region (stage 5), and a postprocessing (stage 6) stage. A final stage is dedicated to the validation of the resulting RALC map (Tormos et al. 2012).

10.3.2.2 OBIA Designed for RALC

OBIA is a suitable technique in high-resolution situation, that is, when pixels are significantly smaller than the object under consideration (Blaschke 2010). It solves the problem of low accuracy of pixel-based classification and the "salt and pepper" effect on classification results in this situation (e.g., Latty and Hoffer 1981; Kressler et al. 2003; Durieux et al. 2007). In a first step, a segmentation process identifies and builds up homogeneous regions (segments or image objects) in order to delineate objects under consideration at different scales. In a second step, a classification process is applied to these objects using spectral as well as spatial information, such as texture, shape, and context features, to increase the discrimination between spectrally similar land cover types. The majority of studies comparing OBIA and pixel classification approaches revealed a higher accuracy result with OBIA (e.g., Carleer and Wolff 2006; Cleve et al. 2008; Myint et al. 2011).

The originality of the OBIA approach designed in Tormos et al. (2012) lies in (1) the use of the thematic spatial information in the classification, (2) the hierarchical image object network, (3) the construction of the classification tree, and (4) the classification rules definition:

(1) Thematic spatial data contain fully reliable information to assess RALC. Therefore, this is the first information that is exploited by the OBIA procedure. A first level segmentation is created from this information ("thematic" level). The study area is segmented according to the boundaries of thematic data entities. (It constitutes the first level in the hierarchical image object network, see Figure 10.3.) Then, the image objects that result are attributed—or not—to a thematic class using Boolean rules.

(2) In order to extract the objects of different sizes that are present in the riparian landscapes (from narrow and fragmented objects, such as sandbanks along rivers, isolated tree vegetation, or small impervious surfaces, to large objects such as agricultural plots or continuous urban areas), a segmentation procedure composed of three levels was designed (see Figure 10.3 for a schematic presentation). In our case, the first level ("micro" level) of this segmentation procedure is based on the limits of the "thematic" level using orthophoto information (the image with the highest spatial resolution). The "macro" level used information from Spot5 XS images. Therefore, our OBIA is based on a hierarchical image object network composed of four segmentation levels.

(3) The construction of the classification tree can be categorized as top-down (i.e., low to high level) image interpretation. The tree starts from the classes that are the easiest to extract, according to available data sources, to the classes of interest (e.g., "water surface"/"land surface," and then within "land surface," "soils with high vegetation"/"soils with low vegetation"…). It stops when no more classes can be reliably separated and unambiguously defined over the segmentation level.

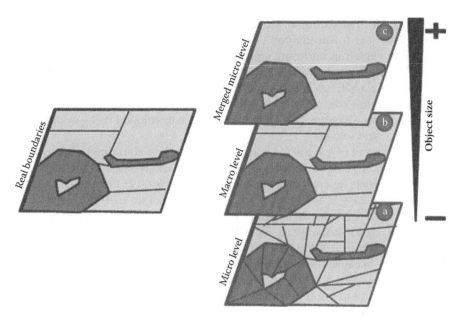

FIGURE 10.3
Schematization of the segmentation procedure designed for extracting objects of different sizes within a riparian area: (a) micro level allows detecting narrow and fragmented objects in the riparian landscapes that would be ignored if the segmentation process was realized at a coarser level, and (b and c) Macro and merged micro levels are necessary for defining medium-sized objects while keeping isolated objects that have been delineated and classified at the micro level. For that purpose, a merged micro level is created from a copy of the micro level, and the merging of neighbor objects of the same class is obtained at the micro level. Then, a macro level is created based on the limits of the micro and merged micro levels.

(4) For each decision in the classification, rules are developed using fuzzy or crisp membership function based on one or several relevant spectral, spatial, and contextual features selected by either expert knowledge, trial and error runs, or sole visual judgment.

10.3.2.3 Classification Automation and Accuracy Assessment

The study zone was first divided into homogeneous mapping regions according to the image acquisition date (stage 3 of the procedure). Then, a master rule set was designed according to the OBIA approach on a pilot mapping region. Features used were (1) radiometric value (mean of pixel values at the object scale for each band); (2) texture value and particularly entropy gray-level difference vector in all directions (Haralick et al. 1973) from the blue band of orthophotos to differentiate artificialized areas to other soils without vegetation; (3) contextual features provided by the software (e.g., distance to a class, existence of super-objects); and (4) standard indexes such as brightness index and the Normalized Difference Vegetation

Index (NDVI). For example, we used the NDVI, which was derived from the red and near-infrared Spot5 XS wavebands, well known to correctly differentiate vegetation and nonvegetation in urban areas (Carleer and Wolff 2006). Finally, the master rule set was implemented for each mapping region in adjusting classification rules, and eventually some classes were added to the classification tree after taking into account region specificities. Master rule set and classification rules adjustment were developed using eCognition 8 developer and server.

For assessment of accuracy, a confusion matrix was computed after grouping classes according to common land cover typology used in literature for studying impacts on river ecological integrity (Table 10.2). Given that the features used for classification are calculated at the object scale in OBIA, objects or polygons have been chosen as sampling units for the selection of control data (Tiede et al. 2006; Grenier et al. 2008). However, confusion matrices were computed using the area of the selected control objects (expressed as a number of 0.5-m pixels) because the land cover maps result from the implementation of the multilevel OBIA scheme and contain objects of different sizes, from fine to large. In order to build a confusion matrix dedicated to classification rules derived from remotely sensed data, the objects derived from ancillary data were not sampled. As a result, 50 samples were randomly collected for each class derived from image information (Table 10.2) obtained in a homogenous way from the study area. Considering the size of the study area (and mapping region), collecting field data for the control sample would be extremely time-consuming. As suggested by Zhu et al. (2000), selected control objects were photointerpreted using the image with the highest spatial resolution as control data. To maintain objectivity of photointerpretation, the classified maps were not viewed during the process.

10.3.3 Modeling Large-Scale Relationships between Land Cover and Stream Ecological Integrity

10.3.3.1 Model Overview

Land cover maps from Earth observation data can serve as a basis for calculating simple spatial (or landscape) indicators that provide a direct way of quantifying man-induced pressure over the different functional scales of river ecosystems (from watershed to local scale) and riparian vegetation conditions. They can be correlated with many river biological indicators currently used to characterize river ecological integrity for better understanding the impact of human pressures (e.g., Allan et al. 1997; Sponseller et al. 2001; Wasson et al. 2010).

We designed a statistical model to investigate the relationship between stream ecological integrity, assessed with a macroinvertebrate biological indicator (IBGN, see Section 10.3.1.2), and land cover spatial indicator (LCSI) estimated over the three spatial scales widely used in such studies (Allan 2004): the upstream watershed, the upstream riparian corridor (corresponding to

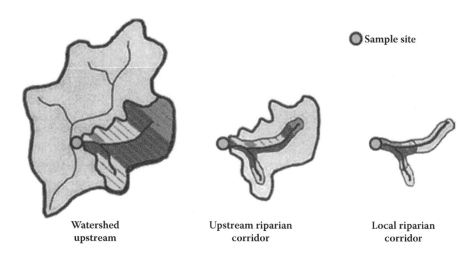

FIGURE 10.4

Illustration of the three spatial scales used for relating land cover to biological measures of stream integrity for a given biological monitoring site: the upstream watershed, the upstream riparian corridor (corresponding to riparian corridor of the whole stretch upstream), and the local riparian corridor (corresponding to a riparian corridor of san upstream stretch up to 3 km). (Modified from Morley, S. A., and J. R. Karr, *Conservation Biology*, 16, 1498–1509, 2002.)

the space along the entire network upstream of the biological monitoring site riparian corridor), and the local riparian corridor (corresponding to the space along the 3-km network upstream of the biological site) (Figure 10.4.) A wide range of buffers varying according to (1) the longitudinal distance upstream from the biological monitoring site (only for local riparian corridor scale) and (2) the lateral distance to the river (for both of riparian scales) were used in order to better localize impact sources.

10.3.3.2 River Ecological Integrity Assessment

IBGN index values were transformed into ecological quality ratio (EQR) values—a deviation from reference conditions as defined in the WFD—by dividing the observed values by reference values for each national river type. The reference values were derived from reference site data. The criteria used for the selection of reference sites complied with the criteria agreed in the WFD intercalibration process (REFCOND 2003), and the EQR values were equivalent to those used at the national level for determining WFD ecological status.

10.3.3.3 Spatial Indicators

A spatial indicator is invariably defined by aggregating a landscape structure attribute over a delimited area (Gergel et al. 2002). The proportion of each land cover type was used as an estimation of the pressure intensity of a given land cover category (Tormos et al. 2011). As a result, many LCSIs built into this study differed according to land cover category and spatial area.

For each biological monitoring site, land cover indicators were calculated at the watershed scale. Watersheds were delineated based on the analysis of a digital terrain model derived from the BDALTI® (250 m pixels) of the French National Geographic Institute. The CLC database was used to analyze the percentage of land cover at this scale. The third level of CLC was used to describe the land cover on each watershed.

For the upstream riparian corridor, 15 buffers were built that differ only by their width (lateral distance to the river on both sides): 10 buffers between 0 and 50 m with spacing of 5 m, and 5 buffers between 50 and 300 m with spacing of 50 m (Figure 10.5).

At the local riparian corridor level, 150 buffers were built that differ in their length (upstream curvilinear distance to biological monitoring site) and width (lateral distance to the river on both sides): 10 length categories were established with a spacing of 100 m, from 0 to 500 m, and a spacing of 500 m from 500 to 3000 m (Figure 10.6a). Then, for each length category, 15 width intervals were established with spacing of 5 m between 0 and 50 m and a spacing of 50 m between 50 and 300 m (Figure 10.6b).

The buffers are defined using an algorithm developed in Python® based on GIS tools of ESRI© GIS software (spatial analysis, network analysis,

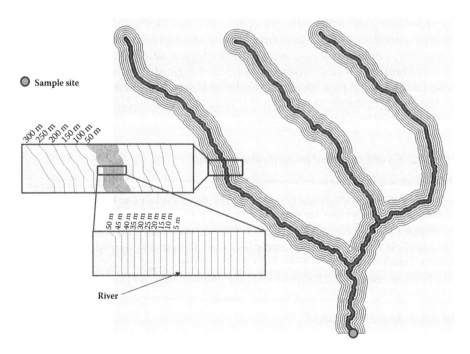

FIGURE 10.5
Schematization of buffers studied at the upstream riparian corridor scale.

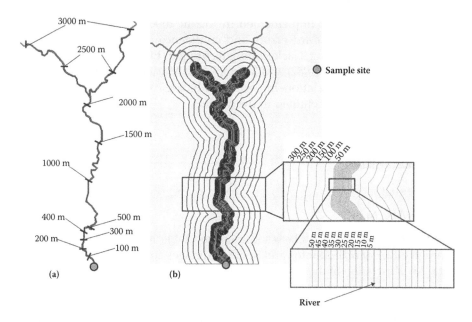

FIGURE 10.6
Schematization of buffers studied at the local riparian corridor scale: (a) the different lengths studied and (b) an example of different widths for a length (2500 m here).

and dynamic segmentation tools). The map derived from VHSR Earth observation and ancillary data was used to assess land cover at these scales. We calculated the percentage within each buffer for all land cover categories that could be extracted.

10.3.3.4 Statistical Techniques

Initially, a buffer was selected at each riparian scale where a given land cover category has the most significant influence on the EQR–IBGN. The most influential buffer for a land cover category at a given riparian scale is identified by comparing the bivariate correlations between the spatial indicators calculated on the different candidate buffers and the biological response, and choosing the buffer ensuring the highest correlation (Johnson and Covich 1997; Sponseller et al. 2001).

A second step allows identifying, among spatial indicators selected in step 1 and spatial indicators at watershed scale, the combination of LCSI exerting the greatest influence on biological response. As did Wasson et al. (2010), we used the partial least square (PLS) regression to model the influence of land cover on the invertebrate indices (EQR–IBGN). PLS is an extension of multiple linear regression (MLR) (Wold 1966, 1982). PLS regression allows investigation of complex problems by assigning unambiguously numerous and colinear predictor variables (Wold et al. 2001). We used the NIPALS algorithm to perform PLS regression implemented in XLStat 2006 software (AddinSoft, Paris, France). The standard errors and confidence intervals of the coefficients are estimated using a jack-knifing technique (Efron and Gong 1983) as recommended by Wold (1982) and Martens and Martens (2000). Only LCSIs that had a significant relationship with the biological indices according to the jack-knife test were retained in the models.

10.4 Results

10.4.1 RALC Map

A detailed RALC map (25 classes in total; see Table 10.2) has been obtained through OBIA of orthophotos and Spot5 XS images combined with ancillary data. An illustration of the resulting map can be seen in Figure 10.7; typology in six classes was used for assessment of the accuracy of the map.

According to the confusion matrix (not shown), a high accuracy was obtained: 85% of the pixels over the riparian area of the Normandy coastal

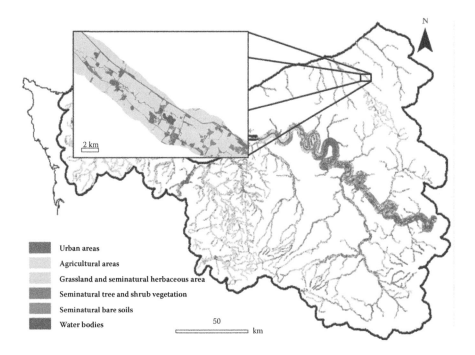

Urban areas
Agricultural areas
Grassland and seminatural herbaceous area
Seminatural tree and shrub vegetation
Seminatural bare soils
Water bodies

FIGURE 10.7 (See color insert.)
Riparian Area Land Cover map of the study area according to the typology in six classes used for the assessment of the quality of the map.

region (5600 km²) were classified correctly. Highly accurate results (both users' and producers' accuracies were greater than 90%) were obtained for "C1: water bodies," "C2: agricultural areas," and "C4: forested areas" and poorest results (both users' and producers' accuracies were less than 70%) for "C3: urban areas" and "C6: seminatural bare soils" categories.

The time for the adjustment of rule thresholds between mapping regions and processing time was estimated at 16 days to treat a 1000 km² riparian corridor.

10.4.2 Land Cover Effect

A 38% variability in the mean EQR–IBGN was explained by the combination of 14 LCSI over the three scales (upstream watershed, upstream riparian, and local riparian). Figure 10.8 provides an illustration of the contribution of significant LCSI on the mean EQR–IBGN. We can see a clear distinction between the land cover categories showing a negative relationship with the mean EQR–IBGN and those with a positive relationship over the three scales. Artificial surfaces ("discontinuous urban areas," "industrial zones," and "roads"—whatever the type) have a clear and strong negative effect at the watershed and riparian scales. Especially, the category "discontinuous urban areas" is the main factor affecting the macroinvertebrate community

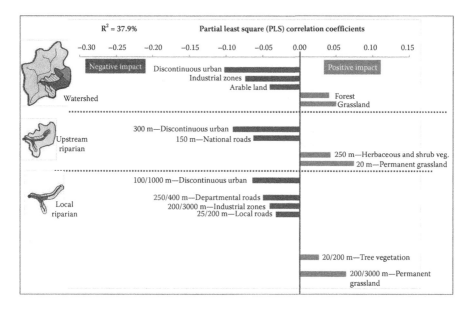

FIGURE 10.8
Standardized partial least square (PLS) regression coefficients between mean EQR–IBGN and statistically significant (according to the jack-knife test) Land Cover Spatial Indicator variables derived from CORINE Land Cover 2000 at watershed scale and from Riparian Area Land Cover map at both riparian scales (upstream and local riparian scales).

at the three scales. Agricultural areas (arable land) at watershed scale have also a negative effect on macroinvertebrate community but less marked than urban surfaces. The positive effect of grassland and forest areas is evident over the three scales. The "permanent grassland" has a benefic effect at both riparian scales, on a much narrower buffer at upstream riparian scale (a 20 m-strip on both sides of the river). Forest areas have a benefic effect at the watershed and local riparian scale on a 20 m-strip on both sides of the river near the monitoring site (upstream 200 m).

10.5 Discussion

10.5.1 Mapping RALC over Broad Territories

First, the classification rules appear efficient regarding obtained accuracies. But these results have to be taken with prudence. On the one hand, classification accuracy can be affected by training sample size (Foody et al. 2006); these results have to be validated using control sample datasets varying in numbers of parcels (as explored by Durieu et al. 2007). On the other hand, the efficiency can be overestimated by the aggregation in six classes; it would be interesting to compute confusion matrices based on the final classes derived from the remotely sensed data. Moreover, it would be useful for our analysis to quantify the confidence interval of classes using the Exact method (Sauro and Lewis 2005) or the KHAT statistic (Congalton 1991).

The proposed OBIA procedure is fully operational (i.e., reproducible, easily transferable, quickly applicable over broad areas). It allows to produce reliable fine-scale information on riparian land cover and land use over large diversified territories. Therefore it provides a tool to characterize pressures impacting river ecological integrity. Despite the diversity of landscape mosaics in the study area, it has been possible to use the same parameters of segmentation and the same class hierarchy on all mapping regions without changing classification features; the major operator task was to adjust the threshold values of the different features used to define classes. This is mainly due to the top-down approach for the construction of tree classification (that divides the feature space into finer and finer units) and the use of the "knowledge-based rules" classification technique. Top-down approaches promote the use of simple rules that are easily transposable to other mapping regions and that facilitate the appropriation of the methodology by new operators. Moreover, specific classes of a given mapping region can be integrated in the tree classification without questioning its overall construction or can be refined with the full control of the user (Lucas et al. 2007; Tiede et al. 2010). Moreover, while the knowledge-based rules classification technique can be distributed easily on a cluster machine, thus drastically reducing the computation time, the supervised classification technique cannot be distributed because it requires the processing of a mapping region as a whole in order to collect a spatially representative training sample. Thanks to fuzzy logic, a powerful technique to manage conflicts between sources (Benz et al. 2004), new data sources can easily be integrated and combined with the initial data sources. For example, in the near future, vertical information from laser-based remote sensing will be available and affordable over broad territories and might be integrated in order to (1) resolve classification errors between urban areas and seminatural bare soils using building height information; and (2) better characterize riparian vegetation classes (e.g., Antonarakis et al. 2008; Geerling et al. 2009; Johansen et al. 2010).

OBIA is obviously a suitable method for processing spatial information within riparian areas at a relevant scale level to infer relationships with river ecological integrity. However, despite the accuracy of this interpretation method, it remains a challenge to obtain a RALC map for a specific time or period with the necessary precision to relate the information to biological assessments. Obviously, it depends on the availability of imagery data over the territory.[*]

In addition, some improvements must be made in this OBIA procedure for a reliable and flexible matching—geometrically and semantically—of each thematic spatial data to the image.

10.5.2 Large-Scale Riparian Vegetation Influence on River Ecological Integrity

Our findings on the relationship between land cover (not including riparian vegetation) and biological responses do not diverge with the existing literature. The impact on macroinvertebrate communities of urbanization (discontinuous urban, industrial zones, and roads) and agricultural areas (arable land) is well recognized (e.g., Roth et al. 1996; Wang et al. 1997; Paul and Meyer 2001; Sponseller et al. 2001; Wang and Kanehl 2003; Allan 2004; Wasson et al. 2010). The multiple direct and indirect pressures generated by human occupation at watershed scale and riparian corridor scale alter the structure and functioning of the ecosystem. The presence of urbanization and agricultural practices in the riparian corridor causes a fragmentation even for the aquatic system, by limiting the recolonization process, especially for adult macroinvertebrates (Petersen et al. 1999, 2004). Similarly, the beneficial influence of seminatural areas (forest and grassland) on macroinvertebrate communities at watershed scale has been highlighted by many authors (e.g., Roth et al.1996; Wang et al. 1997; Wasson et al. 2010). These surfaces reduce diffuse pollution, as they may constitute "sinks" for the pollution depending on their position in the watershed (Gergel 2005).

---

[*] In our case, when the study started, it was difficult to obtain a mosaic of VHSR satellite images with NIR information within a reasonable cost for the study area; since 2011, the GEOSUD program (http://www.geosud.teledetection.fr/) has provided free online annual cover (for the scientific community and public and private participants in land management, addressing researchers and decision makers) of the whole French territory with Rapideye images (5 m pixels; blue, green, red, red-edge, and NIR spectral bands).

Concerning the weight of the riparian vegetation influence on biotic integrity, results are divergent in the literature. Obviously, studies are carried out in diverse natural and anthropomorphic context, but most of them are using MRS satellite images not adapted to estimate reliably the surface area of the different cover types within the riparian buffer. Without this constraint of riparian spatial information (using a VHSR-derived map), our study clearly shows the positive influence of forest and permanent grassland (with a strength equivalent to arable land, especially for grassland) at upstream riparian scale corroborating studies at local scale. Agricultural grassland along the river can play a filtering role to diffuse pollution from agricultural and urban areas (e.g., Lyons et al. 2000; Lin et al. 2004). Forest vegetation contributes to reduce diffuse pollution (plant uptake) but also exerts a key role in regulating temperature and trophic state; furthermore, it directly generates food and habitat heterogeneity for macroinvertebrates (e.g., Hachmoller et al. 1991; Maridet et al. 1997).

Moreover, thanks to the analysis of multiple buffers at the local and upstream riparian corridor scale, we have been able to identify the strip width where forest and grassland are most influential. The outcome provides important information for managers in order to prioritize riparian restoration strategies. Our results suggest that efforts should be made in restoration strategies and policy decisions governing the management of streamside vegetation to widen riparian buffers to a minimum threshold of 20-m strips on both river margins.

Therefore, fine-scale spatial information within riparian areas seems to be a condition for correctly understanding the relationship between river ecological integrity and both human pressures and riparian conditions. Further improvement in the model and its application to management strategies can be envisaged by (1) introducing new spatial indicators related to the spatial configuration of vegetation patches (width, uniformity, fragmentation) at the riparian corridor that may influence river ecological integrity (Weller et al. 2011); (2) analyzing other biological and nutriment responses; and (3) dealing with sizes of rivers and catchment constellation. For example, Potter et al. (2005) show that macroinvertebrates respond more to the presence of riparian vegetation in headwaters. These improvements are necessary to correctly quantify stream ecosystem services at the catchment scale provided by the riparian areas.

10.6 Conclusion

In this chapter, we introduce recent remote sensing processing to produce fine-scale spatial information within riparian areas over broad territories in order to analyze the relationship with river ecological integrity. This processing is based on the OBIA approach combining VHSR Earth observation data and metric ancillary data. Its implementation over the French Normandy coastal region (5600 km² of riparian area) demonstrates its efficiency, transferability, and applicability. Integration of this spatial information allowed quantifying the influence of riparian land cover over different buffers on river ecological integrity through the macroinvertebrate community response. Results show the benefic influence of the riparian forest and grassland vegetation in a 20-m buffer on both sides of the river, locally and on the whole upstream network.

From these results, we conclude that our ability to correctly understand large-scale relationships between river ecological integrity and both human pressure and riparian vegetation seems highly dependent on the availability and processing of spatial information over broad areas, as advocated by Wang (2006) and Johnson and Host (2010) in their reviews. Further investigations are required to provide fine-scale spatial information to solve current problems in river ecological research regarding nonlinearity of biological responses, model uncertainty, spatial autocorrelation, legacy effects, and natural variability (Allan 2004; Wang 2006; Johnson and Host 2010). But definitely, with this fine-scaled information, new reliable spatial indicators related to mechanistic influences of riparian vegetation (e.g., continuity, width, and uniformity; Tormos et al. 2011; Weller et al. 2011) can be explored in order to better support managers in developing effective riparian restoration strategies to maintain river ecological integrity and ecosystem services. Moreover, based on these new explorations, didactic cartographic representation of riparian area status can be performed relating river ecosystem services status at catchment scale (Pert et al. 2010).

Acknowledgments

This study has been made possible by a PhD grant from the French Ministry of Research to the first author. The work has been partly realized in the framework of a research contract between ONEMA and Irstea. Acquisition of remote sensing data benefited from the support of the ISIS program of the French national space agency CNES (Spot5 images, distribution Spot Image), while IGN's airborne orthophotos were made available by the French Ministry of Agriculture.

References

AFNOR (Association Française de NORmalisation). 1992. Essai des eaux. Détermination de l'Indice Biologique Global Normalisé (IBGN). *Association Française de Normalisation - norme homologuée T 90-350*:1–8.

Aguiar, F. C., and M. T. Ferreira. 2005. Human-disturbed landscapes: Effects on composition and integrity of riparian woody vegetation in the Tagus River basin, Portugal. *Environmental Conservation* 32:30–41.

Allan, J. D. 2004. Landscapes and riverscapes: The influence of land use on stream ecosystems. *Annual Review of Ecology, Evolution, and Systematics* 35:257–284.

Allan, J. D., D. Erickson, and J. Fay. 1997. The influence of catchment land use on stream integrity across multiple spatial scales. *Freshwater Biology* 37:149–161.

Antonarakis, A. S., K. S. Richards, and J. Brasington. 2008. Object-based land cover classification using airborne LiDAR. *Remote Sensing of Environment* 112:2988–2998.

Aylward, B., J. Bandyopadhyay, J. C. Belausteguigotia, et al. 2005. Freshwater ecosystem services. In *Ecosystems and human well-being: Policy responses*, eds. R. L. K. Chopra, R. Leemans, P. Kumar, and H Simons, 215–255. Washington, DC: Island Press.

Benz, U. C., P. Hofmann, G. Willhauck, I. Lingenfelder, and M. Heynen. 2004. Multi-resolution, object-oriented fuzzy analysis of remote sensing data for GIS-ready information. *ISPRS Journal of Photogrammetry and Remote Sensing* 58:239–258.

Blaschke, T. 2010. Object based image analysis for remote sensing. *ISPRS Journal of Photogrammetry and Remote Sensing* 65:2–16.

Bossard, M., J. Feranec, and J. Otahel. 2000. *CORINE land cover technical guide— Addendum 2000*. Technical report No 40. Copenhagen: European Environment Agency.

Carleer, A. P., and E. Wolff. 2006. Urban land cover multi-level region-based classification of VHR data by selecting relevant features. *International Journal of Remote Sensing* 27:1035–1051.

Cleve, C., M. Kelly, F. R. Kearns, and M. Moritz. 2008. Classification of the wildland-urban interface: A comparison of pixel- and object-based classifications using high-resolution aerial photography. *Computers, Environment and Urban Systems* 32:317–326.

Congalton, R. G. 1991. A review of assessing the accuracy of classifications of remotely sensed data. *Remote Sensing of Environment* 37:35–46.

Coulter, L., D. Stow, A. Hope, et al. 2000. Comparison of high spatial resolution imagery for efficient generation of GIS vegetation layers. *Photogrammetric Engineering and Remote Sensing* 66:1329–1335.

Dupuy, S., E. Barbe, and M. Balestrat. 2012. An object-based image analysis method for monitoring land conversion by artificial sprawl use of RapidEye and IRS data. *Remote Sensing* 4:404–423.

Durieux, L., E. Lagabrielle, and A. Nelson. 2007. A method for monitoring building construction in urban sprawl areas using object-based analysis of Spot 5 images and existing GIS data. *ISPRS Journal of Photogrammetry and Remote Sensing* 63:399–408.

Durrieu, S., T. Tormos, P. Kosuth, and C. Golden. 2007. Influence of training sampling protocol and of feature space optimization methods on supervised classification results. *Paper presented at the International Geoscience and Remote Sensing Symposium (IGARSS)*. 23–28 July, Barcelona, Spain.

Efron, B., and G. Gong. 1983. A leisurely look at the bootstrap, the jackknife, and cross validation. *American Statistician* 37:36–48.

Foody, G. M., A. Mathur, C. Sánchez-Hernández, and D. S. Boyd. 2006. Training set size requirements for the classification of a specific class. *Remote Sensing of Environment* 104:1–14.

Frimpong, E. A., T. M. Sutton, K. J. Lim, et al. 2005. Determination of optimal riparian forest buffer dimensions for stream biota-landscape association models using multimetric and multivariate responses. *Canadian Journal of Fisheries and Aquatic Sciences* 62:1–6.

Geerling, G. W., M. J. Vreeken-Buijs, P. Jesse, A. M. J. Ragas, and A. J. M. Smits. 2009. Mapping river floodplain ecotopes by segmentation of spectral (CASI) and structural (LiDAR) remote sensing data. *River Research and Applications* 25:795–813.

Gergel, S. E. 2005. Spatial and non-spatial factors: When do they affect landscape indicators of watershed loading? *Landscape Ecology* 20:177–189.

Gergel, S. E., M. G. Turner, J. R. Miller, J. M. Melack, and E. H. Stanley. 2002. Landscape indicators of human impacts to riverine systems. *Aquatic Sciences* 34:118–128.

Goetz, S. J. 2006. Remote sensing of riparian buffers: Past progress and future prospects. *Journal of the American Water Resources Association* 42:133–143.

Goetz, S. J., R. Wright, A. J. Smith, E. Zinecker, and E. Schaub. 2003. IKONOS imagery for resource management: Tree cover, impervious surfaces and riparian buffer analyses in the mid-Atlantic region. *Remote Sensing of Environment* 88:195–208.

Grenier, M., S. Labrecque, M. Benoit, and M. Allard. 2008. Accuracy assessment method for wetland object-based classification. *Paper presented at the Geographic Object Based Image Analysis (GEOBIA), ISPRS Commission IV*. August 5–8, 2008, Calgary, AB, Canada.

Hachmoller, B., R. A. Matthews, and D. F. Brakke. 1991. Effects of riparian community structure, sediment size, and water quality on the macroinvertebrate communities in a small, suburban stream. *Northwest Science* 65:125–132.

Haralick, R. M., K. Shanmugam, and I. Dinstein. 1973. Textural features for image classification. *IEEE Transactions on Systems, Man, and Cybernetics Society* 3:610–621.

Hollenhorst, T., G. Host, and L. Johnson. 2006. Scaling issues in mapping riparian zones with remote sensing data: Quantifying errors and sources of uncertainty. In *Scaling and uncertainty analysis in ecology*, eds. J. Wu, B. Jones, H. Li, and O. Loucks, 275–295. Dordrecht, The Netherlands: Springer.

Hynes, H. B. N. 1970. *The ecology of running waters*. Liverpool, UK: Liverpool University Press.

Johansen, K., and S. Phinn. 2006. Linking riparian vegetation spatial structure in Australian tropical savannas to ecosystem health indicators: Semi-variogram analysis of high spatial resolution satellite imagery. *Canadian Journal of Remote Sensing* 32:228–243.

Johansen, K., S. Phinn, J. Lowry, and M. Douglas. 2008. Quantifying indicators of riparian condition in Australian tropical savannas: Integrating high spatial resolution imagery and field survey data. *International Journal of Remote Sensing* 29:7003–7028.

Johansen, K., S. Phinn, and C. Witte. 2010. Mapping of riparian zone attributes using discrete return LiDAR, QuickBird and SPOT-5 imagery: Assessing accuracy and costs. *Remote Sensing of Environment* 114:2679–2691.

Johnson, L. B., and G. E. Host. 2010. Recent developments in landscape approaches for the study of aquatic ecosystems. *Journal of the North American Benthological Society* 29:41–66.

Johnson, S. L., and A. P. Covich. 1997. Scales of observation of riparian forests and distributions of suspended detritus in a prairie river. *Freshwater Biology* 37:163–175.

Karr, J. R. 1993. Defining and assessing ecological integrity: Beyond water quality. *Environmental Toxicology and Chemistry* 12:1521–1531.

Kressler, F. P., Y. S. Kim, and K. T. Steinnocher. 2003. Object-oriented land cover classification of panchromatic KOMPSAT-1 and SPOT-5 data. *Geoscience and Remote Sensing* 19:263–269.

Latty, R. S., and R. M. Hoffer. 1981. Computer-based classification accuracy due to the spatial resolution using per-point versus per-field classification techniques. *Proceedings of the 7th symposium on machine processing of remotely sensed data with special emphasis on range, forest, and wetlands assessment, IEEE Computer Society,* June 23–26, 1981, Indiana, USA: 384–393.

Lin, C. H., R. N. Lerch, H. E. Garrett, and M. F. George. 2004. Incorporating forage grasses in riparian buffers for bioremediation of atrazine, isoxaflutole and nitrate in Missouri. *Agroforestry Systems* 63:91–99.

Lucas, R., A. Rowlands, A. Brown, S. Keyworth, and P. Bunting. 2007. Rule-based classification of multi-temporal satellite imagery for habitat and agricultural land cover mapping. *ISPRS Journal of Photogrammetry and Remote Sensing* 62:165–185.

Lyons, J., S. W. Trimble, and L. K. Paine. 2000. Grass versus trees: Managing riparian areas to benefit streams of central North America. *Journal of the American Water Resources Association* 36:919–930.

Maridet, L., M. Phillippe, J. G. Wasson, and J. Mathieu. 1997. Seasonal dynamics and storage of particulate organic matter within bed sediment of three streams with contrasted riparian vegetation. In *Groundwater/surface water ecotones: Biological and hydrological interactions and management options*, eds. J. Gibert, J. Mathieu, and F. Fournier, 68–74. Cambridge, UK: Cambridge University Press.

Martens, H., and M. Martens. 2000. Modified jack-knife estimation of parameter uncertainty in bilinear modelling by partial least squares regression (PLSR). *Food Quality and Preference* 11:5–16.

Mertes, L. A. K., D. L. Daniel, J. M. Melack, B. Nelson, L. A. Martinelli, and B. R. Forsberg. 1995. Spatial patterns of hydrology, geomorphology, and vegetation on the floodplain of the Amazon River in Brazil from a remote sensing perspective. *Geomorphology* 13:215–232.

Mertz, O., H. M. Ravnborg, G. L. Lövei, I. Nielsen, and C. C. Konijnendijk. 2007. Ecosystem services and biodiversity in developing countries. *Biodiversity and Conservation* 16:2729–2737.

Morley, S. A., and J. R. Karr. 2002. Assessing and restoring the health of urban streams in the Puget Sound Basin. *Conservation Biology* 16:1498–1509.

Müller, E. 1997. Mapping riparian vegetation along rivers: Old concepts and new methods. *Aquatic Botany* 58:437.

Myint, S. W., P. Gober, A. Brazel, S. Grossman-Clarke, and Q. Weng. 2011. Per-pixel vs. object-based classification of urban land cover extraction using high spatial resolution imagery. *Remote Sensing of Environment* 115:1145–1161.

Naiman, R. J., and H. Décamps. 1997. The ecology of interfaces: Riparian zones. *Annual Review of Ecology and Systematics* 28:621–658.

Naiman, R. J., H. Décamps, and M. E. McClain. 2005. *Riparia: Ecology, conservation, and management of streamside communities.* Boston, MA: Elsevier Academic.

Neale, C. M. U. 1997. Classification and mapping of riparian systems using airborne multispectral videography. *Restoration Ecology* 5(4 Suppl.):103–112.

Paparoditis, N., J. P. Souchon, G. Martinoty, and M. Pierrot-Deseilligny. 2006. High-end aerial digital cameras and their impact on the automation and quality of the production workflow. *ISPRS Journal of Photogrammetry and Remote Sensing* 60:400–412.

Paul, M. J., and J. L. Meyer. 2001. Streams in the urban landscape. *Annual Review of Ecology and Systematics* 32:333–365.

Pert, P. L., J. R. A. Butler, J. E. Brodie, et al. 2010. A catchment-based approach to mapping hydrological ecosystem services using riparian habitat: A case study from the wet tropics, Australia. *Ecological Complexity* 7:378–388.

Petersen, I., Z. Masters, A. G. Hildrew, and S. J. Ormerod. 2004. Dispersal of adult aquatic insects in catchments of differing land use. *Journal of Applied Ecology* 41:934–950.

Petersen, I., J. H. Winterbottom, S. Orton, et al. 1999. Emergence and lateral dispersal of adult Plecoptera and Trichoptera from Broadstone Stream, UK. *Freshwater Biology* 42:401–416.

Postel, S., and S. Carpenter. 1997. Freshwater ecosystems services. In *Nature's services—Societal dependance on natural ecosystems*, ed. G. C. Daily, 195–214. Washington, DC: Island Press.

Potter, K. M., F. W. Cubbage, and R. H. Schaberg. 2005. Multiple-scale landscape predictors of benthic macroinvertebrate community structure in North Carolina. *Landscape and Urban Planning* 71:77–90.

REFCOND. 2003. Rivers and lakes—Typology, reference conditions and classification systems. Common implementation strategy for the Water Framework Directive. In *Guidance document no. 10*, ed. Office for Official Publications of the European Communities. Luxembourg: European Communities.

Roth, N. E., J. D. Allan, and D. L. Erickson. 1996. Landscape influences on stream biotics integrity assessed at multiple spatial scales. *Landscape Ecology* 11:16.

Roy, A. H., B. J. Freeman, and M. C. Freeman. 2007. Riparian influences on stream fish assemblage structure in urbanizing streams. *Landscape Ecology* 22:385–402.

Sauro, J., and J. R. Lewis. 2005. Estimating completion rates from small samples using binomial confidence intervals: Comparisons and recommendations. *Proceedings of the Human Factors and Ergonomics Society Annual Meeting. September 2005, Orlando, USA*, vol. 49 no. 24: 2100–2103.

Shearer, K. S., and W. N. Xiang. 2007. The characteristics of riparian buffer studies. *Journal of Environmental Informatics* 9:41–55.

Sponseller, R. A., E. F. Benfield, and H. M. Valett. 2001. Relationships between land use, spatial scale and stream macroinvertebrate communities. *Freshwater Biology* 46:1409–1424.

Strahler, A. N. 1957. Quantitative analysis of watershed geomorphology. *Transactions of the American Geophysical Union* 38:913–920.

Tiede, D., S. Lang, and C. Hoffmann. 2006. Supervised and forest type-specific multi-scale segmentation for a one-level-representation of single trees. *Paper presented at the 1st International Conference on Object Based Image Analysis (OBIA)*. July 4–5, 2006, Salzburg, Austria.

Tiede, D., S. Lang, D. Hölbling, and P. Füreder. 2010. Transferability of OBIA rulesets for IDP camp analysis in Darfur. *Paper presented at Geographic Object-Based Image Analysis (GEOBIA)*, Ghent, Belgium, 29 June–2 July.

Tormos, T., P. Kosuth, S. Durrieu, S. Dupuy, B. Villeneuve, and J. G. Wasson. 2012. Object-based image analysis for operational fine-scale regional mapping of land cover within river corridors from multispectral imagery and thematic data. *International Journal of Remote Sensing* 33:4603–4633.

Tormos, T., P. Kosuth, S. Durrieu, B. Villeneuve, and J. G. Wasson. 2011. Improving the quantification of land cover pressure on stream ecological status at the riparian scale using high spatial resolution imagery. *Physics and Chemistry of the Earth* 36:549–559.

Wang, L. 2006. Introduction to landscape influences on stream habitats and biological assemblages. In *Landscape influences on stream habitats and biological assemblages*, eds. R. M. Hughes, L. Wang, and P. W. Seelbach, 1–23. Bethesda, MD: American Fisheries Society.

Wang, L., and P. Kanehl. 2003. Influences of watershed urbanization and instream habitat on macroinvertebrates in cold water streams. *Journal of the American Water Resources Association* 35:1181–1196.

Wang, L., J. Lyons, P. Kanehl, and R. Gatti. 1997. Influences of watershed land use on habitat quality and biotic integrity in Wisconsin streams. *Fisheries* 22:6–12.

Wasson, J. G., B. Villeneuve, A. Iital, et al. 2010. Large scale relationships between basin and riparian land cover and ecological status of European rivers: Examples with invertebrate indices from France, Estonia, Slovakia and United Kingdom. *Freshwater Biology* 55:1465–1482.

Weller, D. E., M. E. Baker, and T. E. Jordan. 2011. Effects of riparian buffers on nitrate concentrations in watershed discharges: New models and management implications. *Ecological Applications* 21:1679–1695.

Wold, H. 1966. Estimation of principal components and related models by iterative least squares. In *Multivariate analysis*, ed. P. R. Krishnaiaah, 391–420. New York: Academic Press.

Wold, H. 1982. Soft modeling: The basic design and some extensions. In *Systems under indirect observation*, eds. K. G. Jö Reskog and H. Wold, 589–591. New York: North–Holland.

Wold, S., M. Sjostrom, and L. Eriksson. 2001. PLS-regression: A basic tool of chemometrics. *Chemometrics and Intelligent Laboratory Systems* 58:109–130.

Zhu, Z., L. Yang, S. V. Stehman, and R. L. Czaplewski. 2000. Accuracy assessment for the U.S. Geological Survey Regional Land-Cover Mapping Program: New York and New Jersey region. *Photogrammetric Engineering and Remote Sensing* 66:1425–1435.

Section IV

Ecosystem Services Related to the Water Cycle

11

Evaluation of Hydrological Ecosystem Services through Remote Sensing

C. Carvalho-Santos
University of Porto, Portugal; Wageningen University, The Netherlands

B. Marcos
University of Porto, Portugal

J. Espinha Marques
University of Porto, Portugal

D. Alcaraz-Segura
University of Granada, Spain; University of Almería, Spain

L. Hein
Wageningen University, The Netherlands

J. Honrado
University of Porto, Portugal

CONTENTS

11.1 Society and Hydrological Services...230
11.2 Hydrological Services and the Water Cycle...232
11.3 Remote Sensing of Ecosystem Functioning for Hydrological
Services Provision...234
    11.3.1 Water Supply 235
        11.3.1.1 Atmosphere...235
        11.3.1.2 Cryosphere..245
        11.3.1.3 Surface Water..246
        11.3.1.4 Soil and Ground 247
        11.3.1.5 Vegetation...248
    11.3.2 Water Damage Mitigation...248

11.4 Remote Sensing of Drivers and Pressures of Hydrological Services 249
11.5 Integrating Remote Sensing Data with Hydrological Modeling 250
    11.5.1 Hydrologic Biophysiographic Variables 251
    11.5.2 Hydrologic-State Variables 251
    11.5.3 Remote Sensing Applied in Soil and Water Assessment Tool.... 252
11.6 Conclusions and Perspectives 253
Acknowledgments 254
References 254

11.1 Society and Hydrological Services

Water plays an essential role in the functioning of ecosystems, underpinning biochemical cycles, supporting living organisms and their growth, and creating aquatic habitats on Earth (Chapin et al. 2002). In addition, humans and society rely on ecosystems to provide hydrological services and the resulting benefits (MA 2003). Two major types of hydrological services (Figure 11.1) can be identified according to the benefits they generate: (1) water supply, which includes water for household, irrigation, and industry; hydropower generation; freshwater products; transportation; and recreational and spiritual benefits and (2) water damage mitigation, which includes the reduction in the number and severity of floods, the decrease in soil erosion and sediment deposition, and the mitigation of landslides (Brauman et al. 2007). Both types of services may be evaluated according to three dimensions: (1) quantity (i.e., total amount of water), (2) timing (i.e., seasonal distribution of the water), and (3) quality (related to removal and breakdown of pollutants and trapping of sediments) (Brauman et al. 2007; Elmqvist et al. 2009).

The provision of hydrological services depends on the biophysical structures and processes involving water in ecosystems (Figure 11.1). The rate of ecosystem functioning determines the capacity to deliver a potential service for people (Haines-Young and Potschin 2010). The intrinsic capacity to provide services exists in nature independently of human options in the form of ecosystem functions, and services are only materialized when people use or feel the benefits of those functions (Fisher et al. 2009). From one service, multiple benefits can be generated that translate into a welfare gain—the subject of economic, ecological, and social valuation (Ansink et al. 2008; de Groot et al. 2010). For instance, the amount of water infiltrated will recharge groundwater reservoirs, increasing water storage capacity (properties and functions; Figure 11.1). Once people use this water, the water supply service is translated into economic benefits, such as water being available for household consumption.

Concerns over water problems have been increasing in the last decades, with special emphasis on water scarcity in arid and semiarid regions (van Beek et al. 2011). Water availability is a function of the biophysical

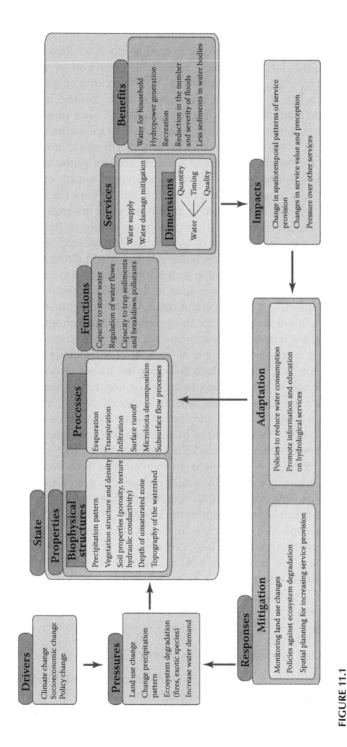

FIGURE 11.1
Framework for the provision of hydrological services, in the context of social–ecological systems. (Based on Rounsevell, M. D. A., et al., *Biodiversity and Conservation*, 19, 2823–2842, 2010; and Haines-Young, R., and M. Potschin, *Ecosystem ecology: A new synthesis*, Cambridge University Press, Cambridge, 2010.)

conditions of each region, and the implications of reduced loss of hydrological services tend to be exacerbated in a context of climate change (Brauman et al. 2007). External drivers of ecological change such as climatic, socio-economic, and political changes may affect the provision of hydrological services (Figure 11.1). These drivers may affect internal pressures that are directly influencing the state of ecosystems, such as land use change or increase in water demand (Rounsevell et al. 2010). Those pressures will impact the provision of hydrological services and ultimately affect the corresponding benefits. However, not only supply, but also demand of water must be considered for an integrated water evaluation and for planning (de Roo et al. 2012). Therefore, responses from governments and society are needed. For adaptation, policies to reduce water consumption are very important as a response to water overuse. For mitigation, monitoring the internal pressures, such as land use change or precipitation pattern, will help in the design of responses that maintain the integrity of hydrological services (Figure 11.1).

In the face of increasing pressure on water resources and stress on ecosystems regulating water flows, sustainable water management is critical. This is illustrated by the report from UN-Water for Rio+20, which stated that the success of a "green economy" depends on the sustainable management of water resources, provisioning of water supply, and adequate sanitation services (UN 2011). Su et al. (2012) recommended promoting the sharing of scientific knowledge and providing capacity building and transfer of technology regarding the water cycle among countries, which includes the use of remote sensing and especially the use of satellite technologies to improve water resource management. Hence, the use of remote sensing is crucial in supporting water management, in particular through allowing assessment and monitoring of the elements of the water balance and of the condition of ecosystems providing hydrological services over space and time (Wagner et al. 2009).

The aim of this chapter is to present an overview of how remote sensing can be used to support sustainable water management. In particular, we examine how remote sensing, especially satellite techniques, can be used to analyze and monitor ecosystem components relevant to the water cycle and the provision of hydrological services—and also, how remote sensing can support hydrological modeling. We base our analysis on a review of remote sensing and hydrological literature as well as our experience with the spatial modeling of hydrological services.

11.2 Hydrological Services and the Water Cycle

The Millennium Ecosystem Assessment (MA) framework highlighted the close relationship between human well-being and water circulation in natural systems (MA 2003). The best way to illustrate this link is by means of the

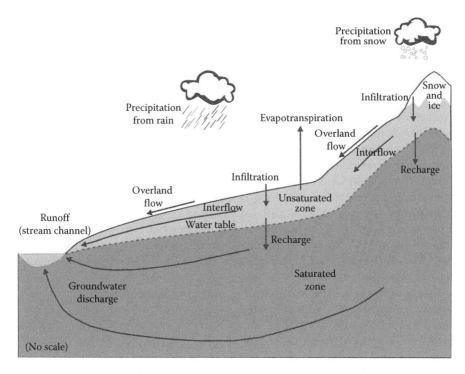

FIGURE 11.2
The water cycle in the vicinity of the land surface. (Conceptual model based on Fitts, C. R., *Groundwater science*, Academic Press, London, 2002.)

water cycle conceptual model, which reports the continuous water movement throughout the Earth's reservoirs, namely oceans, ice caps, glaciers, aquifers, rivers, lakes, soil, and atmosphere (Fetter 2001). The main physical processes involved in the water cycle are evaporation, evapotranspiration, condensation, precipitation, infiltration, overland flow, interflow, runoff, and groundwater flow (Fitts 2002).

The hydrological services described in the previous section (water supply and water damage mitigation) are closely related to the part of the water cycle that takes place on, or close to, the land surface (Figure 11.2). This phase of the global water cycle may be concisely described as follows (Fitts 2002; Van Brahana 2003): (1) water from precipitation reaches the ground surface and infiltrates or generates overland flow (depending on topographic conditions, vegetation cover, soil texture and structure, and the presence of natural or anthropic impermeable layers, among other factors); (2) once in the unsaturated zone, water may circulate subhorizontally (interflow) and emerge, after a short path, in a slope or a river bank; (3) alternatively, water may circulate downwards in the direction of the water table and recharge the aquifer; and

(4) here, groundwater circulates through a saturated medium and finally discharges to a stream (baseflow).

The water supply service depends, first of all, on the volume of precipitation and evapotranspiration. Afterwards, the balance between infiltration and overland flow determines the timing of the subsequent flows. If most of the water from precipitation infiltrates due to a gentle slope allied to a highly permeable soil and adequate vegetable cover, the aquifer recharge rate will probably increase, leading to a greater stream baseflow throughout the year. In this case, the water supply service will benefit from unsaturated zone features that favor infiltration, originating greater groundwater resources (which may be exploited through wells) as well as surface water resources that are better distributed in time (which may be exploited, e.g., through dams) (e.g., Lal 2000; Fetter 2001). Water quality greatly depends, among other factors, on the natural attenuation taking place in the unsaturated zone (Fetter 1999). In fact, the concentration of pollutants decreases as water percolates through the unsaturated zone due to the action of physicochemical and biological processes.

The hydrological services related to water damage mitigation are also closely controlled by the unsaturated zone properties. Once more, the key problem concerns the control of infiltration and aquifer recharge rates by hydropedological features, as well as by the structure of the unsaturated zone (Espinha Marques et al. 2011). The occurrence of floods and soil erosion is greatly determined by the combined effect of vegetation, soil type, and topography. For instance, sparse vegetation allied to fine textured soil (clay or silt) is likely to promote overland flow and reduce infiltration, thus increasing the risk of such natural hazards. Furthermore, a highly heterogeneous unsaturated zone (regarding hydraulic conductivity) allied to steep topography may trigger landslides as a consequence of heavy rainstorms (Fernandes et al. 2004).

11.3 Remote Sensing of Ecosystem Functioning for Hydrological Services Provision

Extensive reviews and special issues have been published, over the last decades, about remote sensing and the science of hydrology (Kite and Pietroniro 1996; Rango and Ahlam 1998; Pietroniro and Prowse 2002; Schmugge et al. 2002; van Dijk and Renzullo 2011; Fernández-Prieto et al. 2012; Su et al. 2012). More recently, some attempts to combine remote sensing and ecosystem services (all categories) have been published (Feng et al. 2010; Ayanu et al. 2012). Here, we will approach remote sensing used for the study of hydrological services, particularly satellite techniques and sensors applied to the observation and monitoring of water (liquid, solid, and gas phases) in the different Earth reservoirs: atmosphere, cryosphere, surface water, soil/ground, and vegetation. Table 11.1 describes some satellite products of interest for the assessment of each water element related to water supply and/or water damage mitigation services. This review, although comprehensive, does not describe all the satellite products available.

11.3.1 Water Supply

11.3.1.1 Atmosphere

In the atmosphere, water vapor, clouds, and precipitation (rainfall and snow) are important water elements that influence the provision of the water supply service. Water vapor and clouds absorb and emit infrared radiation (IR), and clouds also reflect visible radiation (VIS), contributing to the energy balance of the Earth (Su et al. 2012). Microwave radiometers are used to measure the emitted radiation, while optical instruments are used to measure the VIS, IR, and near-infrared radiances (NIR). The rationale behind cloud observation, and indirectly precipitation (VIS and IR), is the higher the cloud reflectance and the lower the temperature when compared to the Earth's surface (Tapiador et al. 2012).

MODIS (Moderate Resolution Imaging Spectroradiometer), on-board of Terra and Aqua satellites, which has been providing the Atmosphere product, and SEVIRI (Spinning Enhanced Visible and Infrared Imager), on-board of METEOSAT-8, are both used to retrieve cloud properties using improved algorithms at high spatial resolution (Kidd et al. 2009). Using light detection and ranging (LIDAR) and radio detection and ranging (RADAR) to retrieve cloud's physical properties, the most advanced measurements are from Cloudsat and Calypso satellites, launched in 2006 by NASA-CSA (Su et al. 2012). For the water vapor observations (total water column, water vapor profiles, and upper tropospheric humidity), some retrieval algorithms and sensors were thoroughly described by Schulz et al. (2009) and Su et al. (2012). As an example, Satellite Application Facility on Climate Monitoring (CM-SAF) uses several instruments on-board of METEOSAT and the National Oceanic and Atmospheric Administration's (NOAA) operational satellites, to provide cloud and water vapor parameters (Schulz et al. 2009).

Precipitation rates can be better measured by microwave techniques due to the strong interaction of rain particles and relative insensitivity to cloud cover (Michaelides et al. 2009; Tang et al. 2009). Precipitation satellite products were exhaustively described by Kidd et al. (2009), Michaelides et al. (2009), and Tapiador et al. (2012). Multisensor techniques, such as the Tropical Rainfall Measuring Mission (TRMM), which holds the TRMM microwave imager (TMI) on passive microwave retrievals and the first spaceborne precipitation radar (PR) on active microwaves, have provided precipitation

TABLE 11.1

Examples of Sensors and Satellites to Measure the Elements of the Water Cycle

Water Cycle	Service	Type	Product/Sensor	Satellite	Spatial Resolution	Temporal Resolution	Costs	Period	Institution/Source
Water vapor	Water supply	Optical	Total column water vapor from Global Ozone Monitoring Experiment (GOME/SCIAMACHY)	ERS	80–40 km; 60–30 km	3 days	Yes	1995–present	EUMETSAT; http://gome.aeronomie.be
			Total column water vapor from Medium-Resolution Imaging Spectrometer (MERIS)	ENVISAT	300 m	3 days	Yes	2005–present	ESA; http://www.enviport.org/meris
			Meteosat Visible and Infrared Radiation Imager (MVIRI)	METEOSAT	5 km	30 min	Yes	1977–present	EUMETSAT; http://www.eumetsat.int
			Water vapor profiles and cloud product from Atmospheric Infrared Sounder (AIRS)	Aqua	3–14 km	Daily	Free	2002–present	NASA; http://airs.jpl.nasa.gov

	Passive microwave	Total precipitable water product from Special Sensor Microwave Imager (SSM/I)	NOAA's family of polar orbiting environmental satellites (POES)	15–20 km	4 hours	Yes	1987–present	NESDIS; http://www.ncdc.noaa.gov
Clouds	Water supply quantity and timing; water damage mitigation (storm and flood prevention)	Optical	Atmosphere product (cloud temperature, height, emissivity, etc.) from Moderate Resolution Imaging Spectroradiometer (MODIS)	Terra	1 km	Once or twice a day	Free	2000–present	NASA; http://modis-atmos.gsfc.nasa.gov
			Water vapor profiles, cloud top temperature, and pressure products from Infrared Atmospheric Sounding Interferometer (IASI)	METOP-A	1–25 km	3 days	Yes	2007–present	EUMETSAT; http://www.eumetsat.int

(Continued)

TABLE 11.1 (*Continued*)

Examples of Sensors and Satellites to Measure the Elements of the Water Cycle

Water Cycle	Service	Type	Product/Sensor	Satellite	Spatial Resolution	Temporal Resolution	Costs	Period	Institution/Source
			Cloud product from Advanced Very High Resolution Radiometer (AVHRR)	NOAA's family of POES	25 km	Weekly	Free	1982–present	NESDIS; http://noaasis.noaa.gov
		Microwave radar	Cloud physical properties	Cloudsat	2.5 km	Orbital	Yes	2006–present	NASA-CSA; http://cloudsat.atmos.colostate.edu
				Cloud-Aerosol LIDAR and Infrared Pathfinder Satellite Observations (CALIPSO)	0.3–5 km	Orbital	Yes	2006–present	NASA-CSA; http://www-calipso.larc.nasa.gov
Precipitation	Water supply quantity and timing; water damage mitigation (storm and flood prevention)	Optical	Rainfall Hydro Estimator—GOES geostationary satellites	NOAA's family of POES	4 km	3 hours	Free	2002–present	NESDIS; http://www.star.nesdis.noaa.gov
			MODIS atmosphere product—36-channel VIS/IR sensor	Terra and Aqua	1 km	Daily	Free	2000–present 2002–present	NASA; http://modis-atmos.gsfc.nasa.gov
			Spinning Enhanced Visible and Infrared Imager (SEVIRI)	METEOSAT-8	1 km	15 min	Yes	2002–present	EUMETSAT; http://www.eumetsat.int

Passive microwave	Advanced Microwave Scanning Radiometer (AMSR-e)	Aqua	25 km	Subdaily	Free	2002–2011	NASA; http://wwwghcc.msfc.nasa.gov/AMSR
	TRMM microwave imager (TMI)	Tropical Rainfall Measuring Mission (TRMM) for tropical regions	250 m–5 km	3 hourly, daily, weekly, monthly	Free	1997–present	NASA + JAXA; http://trmm.gsfc.nasa.gov
Active microwave	Precipitation radar (PR) in space	TRMM	250 m–5 km	3 hourly, daily, weekly, monthly	Free	1997–present	NASA + JAXA; http://pmm.nasa.gov/node/162
Multisatellite/sensor and algorithms	PERSIANN (Artificial Neural Networks–Cloud Classification System) database	TRMM and gauged stations	4 km	3 hourly, daily, weekly, monthly	Free	2000–present	CHRS; http://chrs.web.uci.edu/persiann
	CPC Morphing Technique (Climate Prediction Center) (CMORPH)	GEO satellite IR data + passive microwave AMSR-e and TMI	8 km	30 min to daily	Free	2002–present	NWS; http://www.cpc.ncep.noaa.gov

(Continued)

TABLE 11.1 (*Continued*)

Examples of Sensors and Satellites to Measure the Elements of the Water Cycle

Water Cycle	Service	Type	Product/Sensor	Satellite	Spatial Resolution	Temporal Resolution	Costs	Period	Institution/Source
Snow and ice	Water supply quantity (glaciers); water damage mitigation (sea-level rise, flooding)	Optical	Snow product from AVHRR	NOAA's family of POES	25 km	Weekly	Free	1966–present	NESDIS; http://www.nsof.class.noaa.gov
			MODIS snow product—normalized difference snow index (NDSI)	Terra and Aqua	500 m–1 km	Daily, weekly and monthly composites	Free	2000–present	NSIDC+DAAC; http://modis-snow-ice.gsfc.nasa.gov
		Passive microwave	Snow water equivalent (SWE) from AMSR-e	Aqua	25 km	Daily	Free	2002–2011	VUA + NASA + NSIDC; http://nsidc.org
			Ice map of Arctic from CryoSat mission	CryoSat-2	5–10 km (250 m)	30 days	Yes	2010–present	ESA; http://www.esa.int
		Active microwave	RADARSAT ice monitoring	RADARSAT-1/3	500 km	Daily	Yes	1995–2013 2007–present	CSA+MDA; http://www.asc-csa.gc.ca/eng/satellites/radarsat

Multisatellite/ sensor and algorithms	The Global Land Ice Measurements from Space (GLIMS) for glaciers database	Advanced Spaceborne Thermal Emission and reflection Radiometer (ASTER) combined in a database with satellite imagery and other GIS data for glaciers monitoring	150 m	16 days	Yes	2005–present	NSIDC + USGS; http://www.glims.org
	GlobSnow	Combined ground data with SMMR, SSM/I, and AMSR-e	25 km	Daily, weekly, monthly	Yes	1979–present	ESA; http://www.globsnow.info
	Interactive Multisensor Snow and Ice Mapping System (IMS)	NOAA/ POES—VHRR; MODIS; METEOSAT	4–24 km	Daily, weekly	Free	1997–present	NIC–NOAA/ NESDIS; http://www.natice.noaa.gov/ims/ims.html
Evapo-transpiration	Water supply and drought monitoring	Multisatellite/ sensor and algorithms	MOD 16 ET product from MODIS land cover, LAI/FPAR and global surface meteorology (GMAO)	Terra	250 m–1 km	16 days	Free	2001–present	NASA; http://modis.gsfc.nasa.gov

(Continued)

TABLE 11.1 (*Continued*)

Examples of Sensors and Satellites to Measure the Elements of the Water Cycle

Water Cycle	Service	Type	Product/Sensor	Satellite	Spatial Resolution	Temporal Resolution	Costs	Period	Institution/Source
			Surface Energy Balance Algorithm for Land (SEBAL) uses different kinds of information (e.g., surface temperature, hemispherical reflectance and NDVI)	For example, METEOSAT and NOAA–AVHRR	According to the study	According to the study	Free	1998–present	Bastiaanssen et al. 1998
Surface water	Water supply (quantity, timing, and quality); water damage mitigation	Radar altimeter	Ocean topography from Jason-2	Ocean Surface Topography Mission (OSTM)	2 km	10 days	Yes	2008–present	NASA, CNES, EUMETSAT, NOAA; http://www.nasa.gov
		Optical	Surface water extent using visible band sensors (e.g., SPOT)	SPOT	5–25 m	26 days	Yes	1986–present	CNES; http://www.cnes.fr
			Water quality bands from MERIS	ENVISAT	300 m	2–3 days	Yes	2005–present	ESA; https://earth.esa.int

Soil moisture	Water supply quantity and timing; water damage mitigation (landslides and flooding)	Optical	Advanced Scatterometer (ASCAT)	METOP-A	25 km	Daily	Yes	2006–present	EUMETSAT+ESA; http://manati.star.nesdis.noaa.gov
		Active microwave	Advanced Synthetic Aperture Radar (ASAR)	ENVISAT	1–5 km	Weekly	Yes	2005–present	ESA; https://earth.esa.int
			AMSR-e	Aqua	25 km	Subdaily	Free	2002–2011	VUA+NASA+NSIDC; http://nsidc.org/data/amsre
		Multisatellite/sensor and algorithms	Land Parameter Retrieval Model (LPRM)	AMSR-e,Nimbus SMMR, TRMM TMI, SSM/I	50 km	Subdaily	Free	1978–present	VUA; http://www.falw.vu/~jeur/lprm/pubs.htm
		Passive microwave	Soil moisture from Microwave Imaging Radiometer with Aperture Synthesis (MIRAS)	Soil Moisture and Ocean Salinity (SMOS)	50 km	3 days	Free	2009–present	ESA; http://www.esa.int

(Continued)

TABLE 11.1 (Continued)

Examples of Sensors and Satellites to Measure the Elements of the Water Cycle

Water Cycle	Service	Type	Product/Sensor	Satellite	Spatial Resolution	Temporal Resolution	Costs	Period	Institution/Source
Groundwater	Water supply quantity	Optical	Visible band (vegetation identification points) and NDVI	Landsat/ IKONOS and MODIS	250 m	10–30 day composites	Free (Ikonos yes)	1972– present; 1999– present	NASA
			NIR band (temperature of groundwater discharge)	Landsat-5/7	15–60 m	16 days	Free	1999– present	NASA; http:// landsat.usgs.gov
		Gravimetry	Gravity Recovery and Climate Experiment (GRACE) Earth microgravity model	GRACE	400–500 km	30 days	Yes	2002– present	NASA+DLR; http://www.csr. utexas.edu/grace
			Gravity field and steady-state Ocean Circulation Explorer (GOCE) Earth's gravity field and geoid models	GOCE	100 km	10–30 days	Yes	2009– present	ESA; http:// www.esa.int

Note: The majority of the satellites/products described here can well serve other water cycle purposes than the ones assigned.

estimates at three-hour intervals from 1997 until the present and are often used on global climate models (Tapiador et al. 2012). Precipitation Estimation from Remotely Sensed Information using Artificial Neural Networks (PERSIANN-CCS) is a precipitation algorithm that merges low-altitude polar orbiting satellites and geostationary IR imagery. In addition, records are compared to rain-gauged observations and adjusted with passive microwave rainfall values to obtain highly accurate rainfall estimates (Sorooshian et al. 2000; Sahoo et al. 2011). To be launched in 2014, the international global precipitation measurement (GPM) will be the follower of TRMM and will utilize GPM satellite constellations, carrying advanced instruments that will set a new standard for precipitation measurements from space (Su et al. 2012; Tapiador et al. 2012).

11.3.1.2 Cryosphere

The observation of snow and ice is important because the timing of cold-water storage in a basin determines different runoff and river flows, influencing water supply timing and water damage mitigation (sea-level rise and flooding). Different physical properties compared with rainfall require different sensing techniques (Schmugge et al. 2002). Given the contrasting reflectances between snow and other types of land cover, VIS and NIR bands have been used to map snow. Even so, some limitations still exist related to the similarity between snow and clouds, as well as to the information about snow depth (Wagner et al. 2009). The most common sources on optical medium-resolution for snow and ice mapping are MODIS snow product with high temporal and spatial resolutions and Advanced Very High Resolution Radiometer (AVHRR) on-board NOAA's family of polar orbiting platforms (POES).

Microwave radiation (MR) emitted from the underlying ground is scattered in many different directions by snow grains within the snow layer. In turn, MR can be affected by some properties of the snowpack, such as the size of grains, subject to passive and active microwave observations (Schmugge et al. 2002). Snow water equivalent (SWE) is a parameter sensitive to passive microwave signals based on radiative transfer process. Similarly, GlobSnow product combines ground data with passive observations to monitor snow extent and SWE, using Scanning Multichannel Microwave Radiometer (SSMR), Special Sensor Microwave Imager (SSM/I), and Advanced Microwave Scanning Radiometer (AMSR-e; Luojus et al. 2010). Active microwave instruments offer ice and snow maps with higher spatial resolution, such as the recently launched CryoSat mission, at a spatial resolution of 5 km (Drinkwater et al. 2003).

However, passive and active microwave methods have some limitations that affect the retrievals, namely coarser temporal and spatial resolutions compared to optical observations, and a strong sensitivity to snowpack microphysical properties (Tang et al. 2009). A combination of multisensor products, such as Interactive Multisensor Snow and Ice Mapping System (IMS), which gathers information from NOAA-AVHRR, Terra/Aqua MODIS, and METEOSAT satellites, is an alternative for overcoming those limitations and generating large-scale snow data (Helfrich et al. 2007).

11.3.1.3 Surface Water

Surface water encompasses inland and ocean water, both storage and discharge. The observation of surface water contributes to evaluating water supply quantity services, as well as the quality of the water in which satellite observations have been of great importance. Water absorbs energy at NIR and middle-infrared (MIR), whereas soil and vegetation reflect at these wavelengths (Pietroniro and Prowse 2002). Similarly, optical imagery has been used to map flood plains, lakes, and reservoirs, but with constraints related to cloud presence that masks the underlying water and with limitations to water quantification (Alsdorf and Lettenmaier 2003). To overcome this, radar altimetry has been extensively used to monitor water surface level and, together with elevation, to calculate the volume of water (Alsdorf and Lettenmaier 2003; Su et al. 2012). In the early 1990s, radar altimeters, such as ERS-1 and TOPEX-Poseidon, started to provide information about very large lakes and reservoirs; this has been continued by the second generation that is currently operational, ERS-2, ENVISAT, GFO, and Jason-2 (Tang et al. 2009). In the coming future, the Surface Water and Ocean Topography (SWOT) mission will overcome the spatial and temporal resolution limitations of current altimetry instruments and include missed water bodies between tracks (Tang et al. 2009; Su et al. 2012). In fact, satellite observations alone cannot measure stream discharge and velocity (Tang et al. 2009). An approach to estimating river discharge is to use remotely sensed hydraulic information, beyond radar altimetry, combined or not with hydrological modeling (see Section 11.5) (Bjerklie et al. 2003, 2005). Finally, the space gravimetry mission Gravity Recovery and Climate Experiment (GRACE) provides measurements of water volume, which combined with precipitation and evapotranspiration data is a useful tool for estimating continental freshwater discharge at near-real-time (Syed et al. 2010).

Development of remote sensing techniques for monitoring water quality began in the early 1970s. Most of these studies evaluated empirical relationships between spectral properties and the water quality parameters (Ritchie et al. 2003). The factors that affect water quality can be grouped into (1) those that change the energy spectra of reflected solar and/or emitting thermal radiation from surface waters, which can be measured using remote sensing techniques, such as suspended sediments/turbidity (e.g., Potes et al. 2011), algae (i.e., chlorophylls, carotenoids) (Carvalho et al. 2010; Song et al. 2012), dissolved organic matter

(DOM; e.g., Del Castillo and Miller 2008; Jørgensen et al. 2011), oils (e.g., Jha et al. 2008), aquatic vascular plants (e.g., Santos et al. 2009; Ward et al. 2012), and thermal releases (e.g., Alcantara et al. 2010); and (2) those that are inferred indirectly from measurements of other water quality parameters (sensible to energy spectra), such as most chemicals (i.e., nutrients, pesticides, metals) (Hadibarata et al. 2012) and pathogens (Tran et al. 2010). Multispectral image sensors, such as the Medium Resolution Imaging Spectrometer (MERIS) instrument on-board the Envisat satellite, provide spectral bands with potential applications on suspended sediments, chlorophyll, and other water quality parameters (Ayanu et al. 2012). The advantage of using satellite observation tools to identify and monitor surface water quality problems is the spatial and temporal coverage of the parameters, compared to the nonreadily available *in situ* measurements (Ritchie et al. 2003).

11.3.1.4 *Soil and Ground*

Water under the land surface is observed as soil moisture, which is located in the unsaturated rooting zone, and as groundwater, in the saturated zone (Figure 11.2). Optical sensors operating in VIS and IR bands can indirectly infer soil moisture through partitioning of the water balance equation elements (Su et al. 2012) (see Chapter 14). However, due to the dielectric constant propriety of dry soil, which changes in the presence of water, soil moisture is more accurately measured under the microwave bands, by both active and passive retrievals (Schmugge et al. 2002). Active microwave sensors (SAR at local to regional scale and scatterometers for global monitoring) emit an electromagnetic pulse and capture the returning electromagnetic energy scattered back from Earth that is measured (Su et al. 2012). One example is the global soil moisture product derived from the Advanced Scatterometer (ASCAT), on-board the METOP-A satellite from EUMETSAT, which has been available from 2006 until the present, at a 25 km spatial resolution (Bartalis et al. 2007). In turn, passive microwave sensors measure the radiation emitted from Earth's surface. There are several algorithms to retrieve soil moisture information. One example is the Land Parameter Retrieval Model (LPRM), a global soil moisture model that combines historical datasets from 1978 to the present (Owe et al. 2008). However, passive sensors working at C-band and longer wavelengths are limited in areas with abundant vegetation cover (Tang et al. 2009). To overcome this limitation, the Soil Moisture and Ocean Salinity (SMOS) satellite mission, launched in 2009, is taking observations every three days at a 50 km spatial resolution (Albergel et al. 2011). The advantages of near-real-time observations of soil moisture are (1) better understanding of the water cycle, (2) of how it impacts climate change, and (3) the improved forecast of natural hazards, such as floods and droughts (Su et al. 2012).

Groundwater was the last component of the water cycle to benefit from satellite technologies, but its monitoring is very important due to significant seasonal and interannual variability. Satellite measurements of vegetation distribution, topography, temperature, soil moisture, and gravity have been used to collect information about groundwater presence (Becker 2006) (see Chapter 13). The gravity measurement is based on the principle of the redistribution of water in different compartments of the Earth, which changes with the gravitational field (Su et al. 2012). Similarly, groundwater is measured by GRACE, launched in 2002, a two-satellite mission to map the static and time-varying components of the Earth's gravity field (Ramillien et al. 2008). GRACE provides measurements of groundwater storage change with high accuracy, by separating the contributions of the other water compartments (soil moisture, ocean, evapotranspiration) (Rodell et al. 2006; Llovel et al. 2010). Finally, the Gravity Field and Steady-State Ocean Circulation Explorer (GOCE), launched in 2009, is mapping the Earth's gravity with unrivalled precision and is a complement to GRACE measurements.

11.3.1.5 Vegetation

Estimation of vegetation water content (VWC), from local to global scales, is central to the understanding of water flows in the environment, and it is an important variable for drought and fire monitoring. Remote sensing technologies offer an instantaneous and nondestructive method for VWC assessment (reflectance in NIR and short-wave IR), considering that *in situ* measurements can be related to spectral reflectance of VWC in a reliable way (Wu et al. 2009). However, this type of method needs further refinement to account for the observed effects of leaf structure, leaf dry matter, canopy structure, and leaf area index (LAI; Zarco-Tejada et al. 2003).

Canopy water content has been estimated through various vegetation water indices (Yilmaz et al. 2008), composed of bands in these absorption peaks—for example, normalized difference water index (NDWI), normalized difference infrared index (NDII), maximum difference water index (MDWI), and water band index (WBI)—and have been proved to be applicable in the estimation of VWC (Chen et al. 2005). Recently, researchers have explored methodologies for VWC estimation through remote sensing techniques based on radiative transfer models, such as PROSPECT and SAILH (Jacquemoud et al. 2009; Suárez et al. 2009).

11.3.2 Water Damage Mitigation

All the satellites and sensors described in Section 11.3.1 can also be used to monitor water-related hazards. Precipitation, in particular, is the triggering factor for water-related hazards. Therefore, it is important to estimate precise precipitation rates at the global scale for accurate risk assessment (Tapiador et al. 2012). The observation of meteorological phenomena from satellite platforms provides a more accurate view when compared to surface observations, with advantages for climate assessment and forecasting of extreme events (Kidd et al. 2009). Floods and related damages can be detected for different periods using optical imagery (for instance, from the Landsat series). The history of land surface dynamics can be reconstructed using microwave sensors (SRTM, MODIS, AMSR-e), contributing to predict hazards due to previous events such as flooding, coastal inundation, and landslides, especially the more devastating ones (Tralli et al. 2005; Syvitski et al. 2012). The monitoring of ice sheets and the melting process is important to follow sea-level rise and possible coastal inundations and erosion processes. Missions such as CryoSat (2010–present) and ICESat (2003–2010) provide multiyear elevation data needed to determine ice-sheet mass balance.

Landslides can be identified using optical images (e.g., QuickBird) and can be forecasted using RADAR and InSar techniques that collect information on geomorphology and soil moisture (Tralli et al. 2005; van Westen et al. 2008). High temporal and spatial resolution satellites can be used to forewarn of increased susceptibility to landslides and to help understand the processes leading to slope failures (Wasowski et al. 2010).

Droughts are caused by water deficits due to increased evapotranspiration and temperature, lack of precipitation, and reduced soil moisture. Several indices have used different elements of the water cycle to infer information about the severity of drought (vegetation, soil moisture, and evapotranspiration; Su et al. 2003). This is of particular relevance in arid and semiarid ecosystems, where droughts are a major factor controlling the inter- and intra-annual variations in the ecosystem's productivity and, consequently, the benefits provided by the ecosystem (Hein et al. 2011). Efforts have been made to collect different satellite data to support risk assessment. An example is the Dartmouth Flood Observatory, which provides a global water database and critical indices on floods and on droughts and can be used for global and regional risk assessments (e.g., global flood risk; Jongman et al. 2012).

11.4 Remote Sensing of Drivers and Pressures of Hydrological Services

Remote sensing is important to monitor the drivers and pressures that may affect the provision of hydrological services, from the security of freshwater resources to the mitigation of water hazards in the context of climate change. Climate change is a global driver of ecosystem functioning and services and is associated with the change in precipitation pattern among other factors (Figure 11.1). Climate monitoring is therefore crucial for water resource management, and highly accurate long-term datasets of precipitation have been compiled since the late 1970s from METEOSAT and NOAA series of satellites (Kidd et al. 2009). Land use change, from urban expansion to farmland abandonment or forest management intensification, is another important pressure on hydrological services. Land use dynamics can nowadays be monitored with very high spatial and temporal accuracy by satellite images, such as QuickBird and IKONOS (Rogan and Chen 2004).

11.5 Integrating Remote Sensing Data with Hydrological Modeling

The use of hydrological models enables managers to understand the response of a river catchment to atmospheric forcing conditions, which is important for more accurate water resource management and water hazards forecast and mitigation (Xie and Zhang 2010).

The increasing demand of spatial data for more complex, physically based and distributed hydrological models, together with the emergence of more sophisticated remote sensing products has increased hydrologists' interest in the use of remote sensing applications (Kite and Pietroniro 1996; Pietroniro and Prowse 2002). In particular, new opportunities have emerged from remotely sensed data to improve hydrological modeling calibration and validation (Montanari et al. 2009). The list of remote sensing products potentially useful for hydrological modeling includes the provision of data on precipitation, land use, soil moisture, discharge, and evapotranspiration (Pietroniro and Prowse 2002).

The advantages of using remotely sensed products extend to the availability of near-real-time data, with complete area coverage, which make it possible to perform hydrological modeling even in regions with spatially and temporally scarce ground observations (Grimes 2008). Focusing on near-real-time characteristics, hydrological models can be enriched with time continuity and dynamic information using a family of techniques known as data assimilation (Walker and Houser 2005). Those techniques merge models and observations accounting for uncertainties from different forcing conditions and parameterizations, improving model performance (Xie and Zhang 2010). Although remote sensing provides continuous and up-to-date measurements at varying spatial scales, it still relies on ground observations for algorithm development, calibration, and validation (Tang et al. 2009).

Remote sensing data can be used to estimate hydrologically important biophysiographic variables (terrain, land cover/use, soil data) as well as hydrologic-state variables (e.g., precipitation) that influence the water processes in a basin or region (Pietroniro and Prowse 2002).

11.5.1 Hydrologic Biophysiographic Variables

Hydrologic biophysiographic variables are the spatial input data for physical and distributed hydrological models. Terrain data are used in the delineation and discretization steps of hydrological modeling (Arnold and Fohrer 2005). This type of data is collected with high-accuracy resolution, mostly from RADAR, but also from short-wavelength sensors. The most popular and freely available sources are Shuttle Radar Topography Mission (SRTM) by NASA and Global Digital Elevation Model (GDEM) from Advanced Spaceborne Thermal Emission and Reflection Radiometer (ASTER). In addition, land cover/use and soil data are especially important for distributed hydrological models, where the hydrological response units are influenced by the spatial variability of land cover and soil characteristics (Pietroniro and Prowse 2002). There are freely available land cover datasets based on satellite imagery interpretation. For the entire world, Global Land Cover 2000 (1:5,000,000) is available at the U.S. Geological Survey website. For European countries, CORINE Land Cover (1:100,000) used satellite imagery interpretation (Landsat, SPOT-4, SOPT-5, IRS-P6, LISS III) to produce a land cover product, which is available for the years 1990, 2000, and 2006 with a common classification for all European countries (EEA 1997). Finally, for soil information, a low-resolution "Digital Soil Map of the World," from the Food and Agriculture Organization of the United Nations (FAO), is also freely available.

11.5.2 Hydrologic-State Variables

Hydrologic-state variables derived from satellite observations have been introduced to complement or even replace *in situ* model input for hydrological modeling (Tang et al. 2009). Some examples are climatological data, evapotranspiration, soil moisture, water storage, and discharge derived mainly from satellite sensors, used to calibrate and validate hydrological models (van Dijk and Renzullo 2011). Evapotranspiration is the major link between global energy budgets and the hydrological cycle (Smith and Choudhury 1990). Satellite remote sensing provides routine observations, such as vegetation, energy, and land surface temperature, used to estimate evapotranspiration (Courault et al. 2005) (see Chapter 18). Two kinds of approaches have been taken: (1) energy-balance-based physical models and (2) empirical models that relate evapotranspiration to vegetation index measurements across the growing season (Zhang et al. 2009). An example based on the first approach is the widely used model algorithm Surface Energy Balance Algorithm for Land (SEBAL; Bastiaanssen et al. 1998). SEBAL was introduced in the calibration process of hydrological modeling, with better results when compared to the use of traditional ground-based data algorithm for evapotranspiration calculation (Immerzeel and Droogers 2008).

Satellite-derived hydrologic-state variables are essential in poorly gauged catchments where availability of hydrological ground data is a challenge for the calibration process (Milzow et al. 2011). Some of these variables have been successfully used in several hydrological modeling efforts (Kite and Pietroniro 1996; Fernández-Prieto et al. 2012). Precipitation is the most important input for hydrological modeling, and several satellite missions with their derived products have estimated it, such as TRMM. In some regions, such as in the Amazon basin, the performance of satellite rainfall data from TRMM is comparable to data obtained by rain-gauge observations (Collischonn et al. 2008). However, a recent study in China has shown that the use of TRMM data is good for monthly streamflow simulation but unsuited for daily simulations, when compared to the use of rain-gauge observations. Therefore, further developments in the algorithms of satellite-based daily estimation of rainfall are needed (Li and Zhang 2012). Satellite radar altimetry over land allows data to be retrieved for small and narrow water bodies, which can be converted into discharge using a rating curve method, to calibrate and validate hydrological modeling (Leon et al. 2006; Calmant et al. 2008). In combination with ground discharge data, it may improve discharge estimates on finer time and spatial scales (Michailovsky and McEnnis 2012).

11.5.3 Remote Sensing Applied in Soil and Water Assessment Tool

A combination between ground-based measurements and remotely sensed Earth observations in a coupled model output is an interesting approach to address more accurately the water balance equation (Tang et al. 2009) (see Chapter 12). A widely used hydrologic model that combines this approach is the Soil and Water Assessment Tool (SWAT), developed in the early 1990s by the U.S. Department of Agriculture Research Service (Arnold and Fohrer 2005). SWAT was developed to predict the impact of land management on water resources, performing routines of simulated discharge on monthly or daily time steps. It is a distributed, physically based model.

Improvements to the original SWAT that introduce remote sensing data to calibrate and to validate have been emerging (Gassman et al. 2007). Data assimilation techniques to improve model reliability have also been introduced—for instance, techniques using soil moisture (Han et al. 2012). Tobin and Bennett (2012) compared the efficiency of different precipitation products (one from rain-gauged observations and the others from satellite data) in six watersheds in the United States, generating streamflows under submonthly time steps. Comparable performance between TRMM and ground precipitation data was observed. Narasimhan et al. (2005) applied Normalized Difference Vegetation Index (NDVI), derived from NOAA-AVHRR sensor, to verify soil moisture simulated by SWAT, as a complement to traditional streamflow calibration and validation. This study showed that simulated soil moisture could be a good indicator of crop stress in semi-arid conditions. In a study of a southern African river, some changes were introduced in the original SWAT code allowing the combination of radar altimetry, precipitation products, SAR surface soil moisture, and GRACE total storage changes. Although the estimates of rainfall differed among precipitation products, surface soil moisture and total water storage allowed the identification of likely errors in the periods when precipitation had higher discrepancies (Milzow et al. 2011).

11.6 Conclusions and Perspectives

This chapter provides an exhaustive review of the many advantages that Earth observation offers for the evaluation, management, and monitoring of hydrological services. First, it strongly improves the spatial quality of distributed information when compared to ground-based measurements. Similarly, it allows better access to remote regions with poor ground measurements. Second, it can offer information on a near-real-time basis, which is particularly useful in predicting natural disasters and in activating emergency plans, already highly developed for precipitation alerts (e.g., TRMM). Finally, the use of remote sensing allows the assessment of the three dimensions of hydrological services (i.e., quantity, timing, and quality). Considering the consistent and frequent time coverage information of the water cycle, it is possible to improve the study of interannual variability and seasonal behavior of the water cycle elements (Schulz et al. 2009). Landsat and other images (for more detailed analysis of vegetation and land use patterns) have been available on a planetary scale since the early 1970s.

However, there are still some limitations regarding the parameterization as well as the costs of the observations. The hydrological information gathered by satellite needs a robust parameterization and validation to improve the accuracy and consistency in hydrological studies, preferably using ground-based measurements (Tang et al. 2009; Hein et al. 2011). Although the cost of Earth observations can be high, depending on their spatial and temporal resolutions, the use of satellite observations seems to show higher cost effectiveness than conventional methods for hydrological parameters observation (Dreher et al. 2000; Pietroniro and Prowse 2002).

In the future, the increasing number of more sophisticated satellite missions will create new opportunities for observing, analyzing, and monitoring the different components of the water cycle (Fernández-Prieto et al. 2012). In addition, the integration of remote sensing applications, particularly satellite-derived products, in hydrological modeling has progressed significantly, improving the spatial and temporal resolutions of the outputs. A remaining challenge is the development/improvement of satellite algorithms, adapting them to the routines of hydrological models, and further data assimilation processes. Moreover, these satellite products should be calibrated and validated according to the environmental characteristics of each region.

Overall, the assessment of water supply and water damage mitigation services can strongly benefit from remote sensing techniques and data, especially from satellite-based products. Satellite products contribute to improving the understanding of the processes and functions behind the provision of hydrological services on a spatially explicit and near-real-time basis. Furthermore, the drivers and pressures affecting the water cycle and the provision of hydrological services can also be assessed and monitored, thereby contributing to a more robust and efficient management of hydrological resources and services.

Acknowledgments

This study was financially supported by Fundação para a Ciência e a Tecnologia (Portugal) through PhD grant SFRH/BD/66260/2009 (to Cláudia Carvalho-Santos). João Honrado received support from Fundação para a Ciência e a Tecnologia (Portugal) through project grant "ECOSENSING" (PTDC/AGR-AAM/104819/2008). Bruno Marcos received support from Fundação para a Ciência e a Tecnologia (Portugal) through project grant "MoBiA" (PTDC/AAC–AMB/114522/2009).

References

Albergel, C., E. Zakharova, J. C. Calvet, et al. 2011. A first assessment of the SMOS data in southwestern France using in situ and airborne soil moisture estimates: The CAROLS airborne campaign. *Remote Sensing of Environment* 115:2718–2728.

Alcantara, E. H., J. L. Stech, J. A. Lorenzzetti, et al. 2010. Remote sensing of water surface temperature and heat flux over a tropical hydroelectric reservoir. *Remote Sensing of Environment* 114:2651–2665.

Alsdorf, D. E., and D. P. Lettenmaier. 2003. Geophysics. Tracking fresh water from space. *Science* 301:1491–1494.

Ansink, E., L. Hein, and K. Per Hasund. 2008. To value functions or services? An analysis of ecosystem valuation approaches. *Environmental Values* 17:489–503.

Arnold, J. G., and N. Fohrer. 2005. SWAT2000: Current capabilities and research opportunities in applied watershed modelling. *Hydrological Processes* 19:563–572.

Ayanu, Y. Z., C. Conrad, T. Nauss, M. Wegmann, and T. Koellner. 2012. Quantifying and mapping ecosystem services supplies and demands: A review of remote sensing applications. *Environmental Science & Technology* 46:8529–8541.

Bartalis, Z., W. Wagner, V. Naeimi, et al. 2007. Initial soil moisture retrievals from the METOP-A Advanced Scatterometer (ASCAT). *Geophysical Research Letters* 34:L20401.

Bastiaanssen, W. G. M., H. Pelgrum, J. Wang, et al. 1998. A remote sensing surface energy balance algorithm for land (SEBAL): Part 2: Validation. *Journal of Hydrology* 212:213–229.

Becker, M. W. 2006. Potential for satellite remote sensing of ground water. *Ground Water* 44:306–318.

Bjerklie, D. M., S. L. Dingman, C. J. Vorosmarty, C. H. Bolster, and R. G. Congalton. 2003. Evaluating the potential for measuring river discharge from space. *Journal of Hydrology* 278:17–38.

Bjerklie, D. M., D. Moller, L. C. Smith, and S. L. Dingman. 2005. Estimating discharge in rivers using remotely sensed hydraulic information. *Journal of Hydrology* 309:191–209.

Brauman, K. A., G. C. Daily, T. K. Duarte, and H. A. Mooney. 2007. The nature and value of ecosystem services: An overview highlighting hydrologic services. *Annual Review of Environment and Resourses* 32:67–98.

Calmant, S., F. Seyler, and J. F. Crétaux. 2008. Monitoring continental surface waters by satellite altimetry. *Surveys in Geophysics* 29:247–269.

Carvalho, G. A., P. J. Minnett, L. E. Fleming, V. F. Banzon, and W. Baringer. 2010. Satellite remote sensing of harmful algal blooms: A new multi-algorithm method for detecting the Florida Red Tide (*Karenia brevis*). *Harmful Algae* 9:440–448.

Chapin III, F. S., P. A. Matson, and H. A. Mooney. 2002. *Principles of terrestrial ecosystem ecology.* 1st ed. New York: Springer.

Chen, D., J. Huang, and T. J. Jackson. 2005. Vegetation water content estimation for corn and soybeans using spectral indices derived from MODIS near- and shortwave infrared bands. *Remote Sensing of Environment* 98:225–236.

Collischonn, B., W. Collischonn, and C. E. M. Tucci. 2008. Daily hydrological modeling in the Amazon basin using TRMM rainfall estimates. *Journal of Hydrology* 360:207–216.

Courault, D., B. Seguin, and A. Olioso. 2005. Review on estimation of evapotranspiration from remote sensing data: From empirical to numerical modeling approaches. *Irrigation and Drainage Systems* 19:223–249.

de Groot, R. S., R. Alkemade, L. Braat, L. Hein, and L. Willemen. 2010. Challenges in integrating the concept of ecosystem services and values in landscape planning, management and decision making. *Ecological Complexity* 7:260–272.

Del Castillo, C. E., and R. L. Miller. 2008. On the use of ocean color remote sensing to measure the transport of dissolved organic carbon by the Mississippi River Plume. *Remote Sensing of Environment* 112:836–844.

de Roo, A., F. Bouraoui, P. Burek, et al. 2012. *Current water resources in Europe and Africa*. Italy: Joint Research Center.

Dreher, J., F. Gampe, M. Kirchebner, et al. 2000. Hydrological services: The need for the integration of space based information. *Proceedings of the International Symposium GEOMARK 2000*, Paris. 10–12 April 2000, 119–127.

Drinkwater, M. R., R. Francis, G. Ratier, and D. J. Wingham. 2003. The European space agency's earth explorer mission CryoSat: Measuring variability in the cryosphere. *Annals of Glaciology* 39:313–320.

EEA (European Environment Agency). 1997. *CORINE Land Cover—Technical guide, European Commission, Brussels* pp. 1–130. Retrieved from http://www.eea.europa.eu/publications/COR0-landcover.

Elmqvist, T., E. Maltby, T. Barker, et al. 2009. Biodiversity, ecosystems and ecosystem services (chapter 2). In *(TEEB) The Economics of Ecosystems and Biodiversity: Ecological and Economic Foundations*, ed. P. Kumar, 41–111. London: Earthscan.

Espinha Marques, J., J. Samper, B. Pisani, et al. 2011. Evaluation of water resources in a high-mountain basin in Serra da Estrela, Central Portugal, using a semi-distributed hydrological model. *Environmental Earth Sciences* 62:1219–1234.

Feng, X., B. Fu, X. Yang, and Y. Lü. 2010. Remote sensing of ecosystem services: An opportunity for spatially explicit assessment. *Chinese Geographical Science* 20:522–535.

Fernandes, N. F., R. F. Guimarães, R. A. T. Gomes, B. C. Vieira, D. R. Montgomery, and H. Greenberg. 2004. Topographic controls of landslides in Rio de Janeiro: Field evidence and modeling. *Catena* 55:163–181.

Fernández-Prieto, D., P. van Oevelen, Z. Su, and W. Wagner. 2012. Advances in Earth observation for water cycle science. *Hydrology and Earth System Sciences* 16:543–549.

Fetter, C. W. 1999. *Contaminant hydrogeology*. 2nd ed. Upper Saddle River, NJ: Prentice Hall.

Fetter, C. W. 2001. *Applied hydrology*. 4th ed. Upper Saddle River, NJ: Prentice Hall.

Fisher, B., R. K. Turner, and P. Morling. 2009. Defining and classifying ecosystem services for decision making. *Ecological Economics* 68:643–653.

Fitts, C. R. 2002. *Groundwater science*. London: Academic Press.

Gassman, P. W., M. R. Reyes, C. H. Green, and J. G. Arnold. 2007. The soil and water assessment tool: Historical development, applications, and future research directions. *Transactions of the ASABE* 50:1211–1250.

Grimes, D. I. F. 2008. An ensemble approach to uncertainty estimation for satellite-based rainfall estimates. *Hydrological Modelling and the Water Cycle* 63:145–162.

Hadibarata, T., F. Abdullah, A. R. M. Yusoff, R. Ismail, S. Azman, and N. Adnan. 2012. Correlation study between land use, water quality, and heavy metals (Cd, Pb, and Zn) content in water and green lipped mussels *Perna viridis* (Linnaeus.) at the Johor Strait. *Water, Air, & Soil Pollution* 223:3125–3136.

Haines-Young, R., and M. Potschin. 2010. The links between biodiversity, ecosystem services and human well-being. In *Ecosystem ecology: A new synthesis*, eds. D. Raffaelli and C. Frid, 110–139. BES Ecological Reviews Series. Cambridge, UK: Cambridge University Press.

Han, E., V. Merwade, and G. C. Heathman. 2012. Implementation of surface soil moisture data assimilation with watershed scale distributed hydrological model. *Journal of Hydrology* 416–417:98–117.

Hein, L., N. de Ridder, P. Hiernaux, R. Leemans, A. de Wit, and M. Schaepman. 2011. Desertification in the Sahel: Towards better accounting for ecosystem dynamics in the interpretation of remote sensing images. *Journal of Arid Environments* 75:1164–1172.

Helfrich, S. R., D. McNamara, B. H. Ramsay, T. Baldwin, and T. Kasheta. 2007. Enhancements to, and forthcoming developments in the Interactive Multisensor Snow and Ice Mapping System (IMS). *Hydrological Processes* 21:1576–1586.

Immerzeel, W. W., and P. Droogers. 2008. Calibration of a distributed hydrological model based on satellite evapotranspiration. *Journal of Hydrology* 349:411–424.

Jacquemoud, S., W. Verhoef, F. Baret, et al. 2009. PROSPECT+SAIL models: A review of use for vegetation characterization. *Remote Sensing of Environment* 113:S56–S66.

Jha, M. N., J. Levy, and Y. Gao. 2008. Advances in remote sensing for oil spill disaster management: State-of-the-art sensors technology for oil spill surveillance. *Sensors* 8:236–255.

Jongman, B., P. J. Ward, and J. C. J. H. Aerts. 2012. Global exposure to river and coastal flooding: Long term trends and changes. *Global Environmental Change* 22:823–835.

Jørgensen, L., C. A. Stedmon, T. Kragh, S. Markager, M. Middelboe, and M. Søndergaard. 2011. Global trends in the fluorescence characteristics and distribution of marine dissolved organic matter. *Marine Chemistry* 126:139–148.

Kidd, C., V. Levizzani, J. Turk, and R. Ferraro. 2009. Satellite precipitation measurements for water resource monitoring. *Journal of the American Water Resources Association* 45:567–579.

Kite, G. W., and A. Pietroniro. 1996. Remote sensing applications in hydrological modelling. *Hydrological Sciences Journal* 41:563–591.

Lal, R. 2000. Rationale for watershed as a basis for sustainable management of soil and water resources. In *Integrated watershed management in the global ecosystem*, ed. R. Lal, 2–16. Boca Raton, FL: CRC Press.

Leon, J. G., S. Calmant, F. Seyler, et al. 2006. Rating curves and estimation of average water depth at the Upper Negro River based on satellite altimeter data and modeled discharges. *Journal of Hydrology* 328:481–496.

Li, X. H., and Q. Zhang. 2012. Suitability of the TRMM satellite rainfalls in driving a distributed hydrological model for water balance computations in Xinjiang catchment, Poyang lake basin. *Journal of Hydrology* 426–427:28–38.

Llovel, W., M. Becker, A. Cazenave, J. F. Cretaux, and G. Ramillien. 2010. Global land water storage change from GRACE over 2002–2009. Inference on sea level. *Comptes Rendus—Geoscience* 342:179–188.

Luojus, K., J. Pulliainen, M. Takala, and J. Lemmetyinen. 2010. *Global snow monitoring for climate research*. European Space Agency Study Contract Report ESRIN Contract 21703/08/I-EC.

MA (Millenium Ecosystem Assessment). 2003. *Ecosystems and human well-being: A framework for assessment*. Washington, DC: Island Press.

Michaelides, S., V. Levizzani, E. Anagnostou, P. Bauer, T. Kasparis, and J. E. Lane. 2009. Precipitation: Measurement, remote sensing, climatology and modeling. *Atmospheric Research* 94:512–533.

Michailovsky, C. I., and S. McEnnis. 2012. River monitoring from satellite radar altimetry in the Zambezi River basin. *Hydrology and Earth System Sciences* 16:2181–2192.

Milzow, C., P. E. Krogh, and P. Bauer-Gottwein. 2011. Combining satellite radar altimetry, SAR surface soil moisture and GRACE total storage changes for hydrological model calibration in a large poorly gauged catchment. *Hydrology and Earth System Sciences* 15:1729–1743.

Montanari, M., R. Hostache, P. Matgen, G. Schumann, L. Pfister, and L. Hoffmann. 2009. Calibration and sequential updating of a coupled hydrologic–hydraulic model using remote sensing-derived water stages. *Hydrology and Earth System Sciences* 13:367–380.

Narasimhan, B., R. Srinivasan, J. G. Arnold, and M. Di Luzio. 2005. Estimation of long-term soil moisture using a distributed parameter hydrologic model and verification using remotely sensed data. *Transactions of the ASAE* 48:1101–1113.

Owe, M., R. de Jeu, and T. Holmes. 2008. Multisensor historical climatology of satellite-derived global land surface moisture. *Journal of Geophysical Research* 113:17.

Pietroniro, A., and T. D. Prowse. 2002. Applications of remote sensing in hydrology. *Hydrological Processes* 16:1537–1541.

Potes, M., M. J. Costa, J. C. B. da Silva, A. M. Silva, and M. Morais. 2011. Remote sensing of water quality parameters over Alqueva Reservoir in the south of Portugal. *International Journal of Remote Sensing* 32:3373–3388.

Ramillien, G., J. S. Famiglietti, and J. Wahr. 2008. Detection of continental hydrology and glaciology signals from GRACE: A review. *Surveys in Geophysics* 29:361–374.

Rango, A., and I. S. Ahlam. 1998. Operational applications of remote sensing in hydrology: Success, prospects and problems. *Hydrological Sciences Journal* 43:947–968.

Ritchie, J., P. Zimbra, and J. Everitt. 2003. Remote sensing techniques to assess water quality. *Photogrammetric Engineering and Remote Sensing* 69:695–704.

Rodell, M., J. Chen, H. Kato, J. S. Famiglietti, J. Nigro, and C. R. Wilson. 2006. Estimating groundwater storage changes in the Mississippi River basin (USA) using GRACE. *Hydrogeology Journal* 15:159–166.

Rogan, J., and D. M. Chen. 2004. Remote sensing technology for mapping and monitoring land-cover and land-use change. *Progress in Planning* 61:301–325.

Rounsevell, M. D. A., T. P. Dawson, and P. A. Harriuson. 2010. A conceptual framework to assess the effects of environmental change on ecosystem services. *Biodiversity and Conservation* 19:2823–2842.

Sahoo, A. K., M. Pan, T. J. Troy, R. K. Vinukollu, J. Sheffield, and E. F. Wood. 2011. Reconciling the global terrestrial water budget using satellite remote sensing. *Remote Sensing of Environment* 115:1850–1865.

Santos, M. J., S. Khanna, E. L. Hestir, et al. 2009. Use of hyperspectral remote sensing to evaluate efficacy of aquatic plant management. *Invasive Plant Science and Management* 2:216–229.

Schmugge, T. J., W. P. Kustas, J. C. Ritchie, T. J. Jackson, and A. Rango. 2002. Remote sensing in hydrology. *Advances in Water Resources* 25:1367–1385.

Schulz, J., P. Albert, H. D. Behr, et al. 2009. Operational climate monitoring from space: The EUMETSAT satellite application facility on climate monitoring (CM-SAF). *Atmospheric Chemistry and Physics* 9:1687–1709.

Smith, R. C. G., and B. J. Choudhury. 1990. Relationship of multispectral satellite data to land surface evaporation from the Australian continent. *International Journal of Remote Sensing* 11:2069–2088.

Song, K., D. Lu, L. Li, S. Li, Z. Wang, and J. Du. 2012. Remote sensing of chlorophyll-*a* concentration for drinking water source using genetic algorithms (GA)–partial least square (PLS) modeling. *Ecological Informatics* 10:25–36.

Sorooshian, S., K. L. Hsu, G. Xiaogang, H. V. Gupta, B. Imam, and D. Braithwaite. 2000. Evaluation of PERSIANN system satellite-based estimates of tropical rainfall. *Bulletin of the American Meteorological Society* 81:2035–2046.

Su, Z., R. A. Roebeling, J. Schulz, et al. 2012. Observation of hydrological processes using remote sensing. In *Treatise on water science*, ed. P. Wilderer, vol. 2:351–399. Oxford: Academic Press.

Su, Z., A. Yacob, J. Wen, et al. 2003. Assessing relative soil moisture with remote sensing data: Theory, experimental validation, and application to drought monitoring over the North China Plain. *Physics and Chemistry of the Earth, Parts A/B/C* 28:89–101.

Suárez, L., P. J. Zarco-Tejada, J. A. J. Berni, V. González-Dugo, and E. Fereres. 2009. Modelling PRI for water stress detection using radiative transfer models. *Remote Sensing of Environment* 113:730–744.

Syed, T. H., J. S. Famiglietti, D. P. Chambers, J. K. Willis, and K. Hilburn. 2010. Satellite-based global-ocean mass balance estimates of interannual variability and emerging trends in continental freshwater discharge. *Proceedings of the National Academy of Sciences of the United States of America* 107:17916–17921.

Syvitski, J. P. M., I. Overeem, G. R. Brakenridge, and M. Hannon. 2012. Floods, floodplains, delta plains—A satellite imaging approach. *Sedimentary Geology* 267–268:1–14.

Tang, Q., H. Gao, H. Lu, and D. P. Lettenmaier. 2009. Remote sensing: Hydrology. *Progress in Physical Geography* 33:490–509.

Tapiador, F. J., F. J. Turk, W. Petersen. 2012. Global precipitation measurement: Methods, datasets and applications. *Atmospheric Research* 104–105:70–97.

Tobin, K. J., and M. E. Bennett. 2012. Temporal analysis of soil and water assessment tool (SWAT) performance based on remotely sensed precipitation products. *Hydrological Processes* 27:505–514.

Tralli, D. M., R. G. Blom, V. Zlotnicki, A. Donnellan, and D. L. Evans. 2005. Satellite remote sensing of earthquake, volcano, flood, landslide and coastal inundation hazards. *ISPRS Journal of Photogrammetry and Remote Sensing* 59:185–198.

Tran, A., F. Goutard, L. Chamaille, N. Baghdadi, and D. Lo Seen. 2010. Remote sensing and avian influenza: A review of image processing methods for extracting key variables affecting avian influenza virus survival in water from Earth observation satellites. *International Journal of Applied Earth Observation and Geoinformation* 12:1–8.

UN (United Nations). 2011. *Water in a green economy*. UN-Water. Retrieved from www.unwater.org.

van Beek, L. P. H., Y. Wada, and M. F. P. Bierkens. 2011. Global monthly water stress: 1. Water balance and water availability. *Water Resources Research* 47:W07517.

Van Brahana, J. 2003. Hydrological cycle. In *Encyclopedia of water science*, eds. B. A. Stewart and T. A. Howell, 412–414. New York: Marcel Dekker.

van Dijk, A. I. J. M., and L. J. Renzullo. 2011. Water resource monitoring systems and the role of satellite observations. *Hydrology and Earth System Sciences* 15:39–55.

van Westen, C. J., E. Castellanos, and S. L. Kuriakose. 2008. Spatial data for landslide susceptibility, hazard, and vulnerability assessment: An overview. *Engineering Geology* 102:112–131.

Wagner, W., N. E. C. Verhoest, R. Ludwig, and M. Tedesco. 2009. Remote sensing in hydrological sciences. *Hydrology and Earth System Sciences* 13:813–817.

Walker, J., and P. Houser. 2005. Hydrologic data assimilation. In *Advances in water science methodologies*, ed. U. Aswathanarayana, 230. Boca Raton, FL: CRC Press.

Ward, D. P., S. K. Hamilton, T. D. Jardine, et al. 2012. Assessing the seasonal dynamics of inundation, turbidity, and aquatic vegetation in the Australian wet-dry tropics using optical remote sensing. *Ecohydrology* 6:312–323.

Wasowski, J., C. Lamanna, G. Gigante, and D. Casarano. 2010. High resolution satellite imagery analysis for inferring surface–subsurface water relationships in unstable slopes. *Remote Sensing of Environment* 124:1–14.

Wu, C., Z. Niu, Q. Tang, and W. Huang. 2009. Predicting vegetation water content in wheat using normalized difference water indices derived from ground measurements. *Journal of Plant Research* 122:317–326.

Xie, X., and D. Zhang. 2010. Data assimilation for distributed hydrological catchment modeling via ensemble Kalman filter. *Advances in Water Resources* 33:678–690.

Yilmaz, M. T., E. R. Hunt, and T. J. Jackson. 2008. Remote sensing of vegetation water content from equivalent water thickness using satellite imagery. *Remote Sensing of Environment* 112:2514–2522.

Zarco-Tejada, P. J., C. A. Rueda, and S. L. Ustin. 2003. Water content estimation in vegetation with MODIS reflectance data and model inversion methods. *Remote Sensing of Environment* 85:109–124.

Zhang, J., Y. Hu, X. Xiao, et al. 2009. Satellite-based estimation of evapotranspiration of an old-growth temperate mixed forest. *Agricultural and Forest Meteorology* 149:976–984.

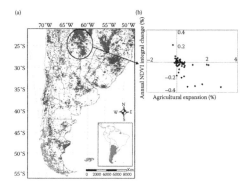

FIGURE 2.3
Refer to the text for the figure caption.

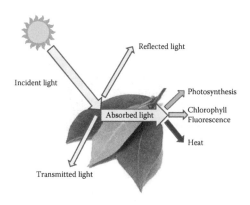

FIGURE 3.1
Refer to the text for the figure caption.

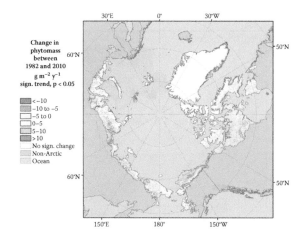

FIGURE 4.1
Refer to the text for the figure caption.

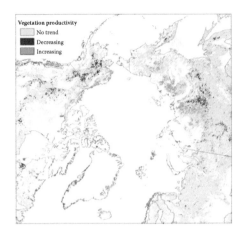

FIGURE 4.2
Refer to the text for the figure caption.

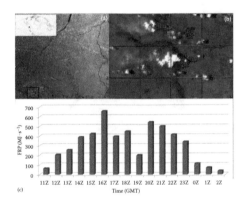

FIGURE 7.2
Refer to the text for the figure caption.

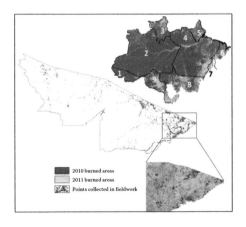

FIGURE 7.4
Refer to the text for the figure caption.

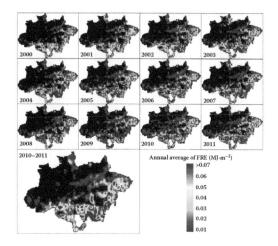

FIGURE 7.5
Refer to the text for the figure caption.

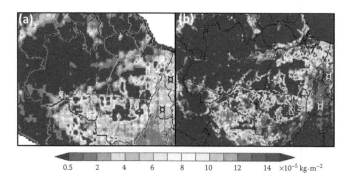

FIGURE 7.7
Refer to the text for the figure caption.

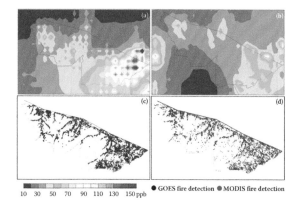

FIGURE 7.8
Refer to the text for the figure caption.

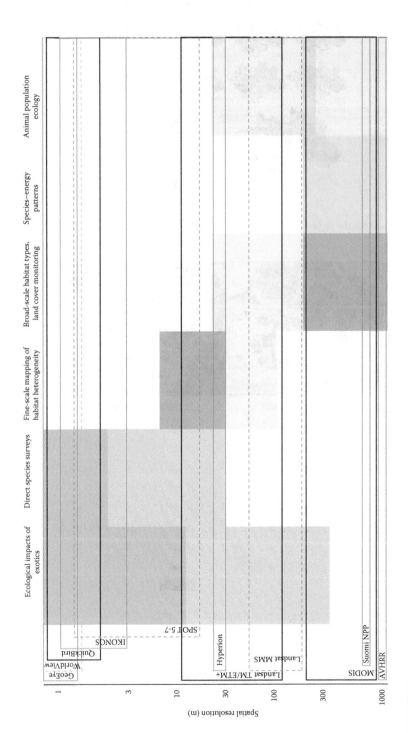

FIGURE 8.1
Refer to the text for the figure caption.

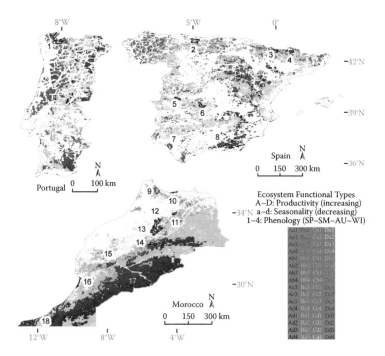

FIGURE 9.2
Refer to the text for the figure caption.

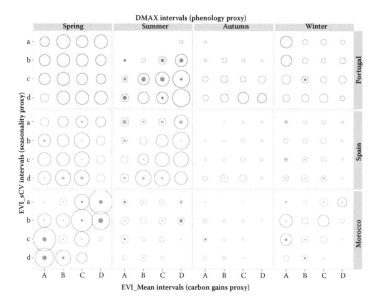

FIGURE 9.3
Refer to the text for the figure caption.

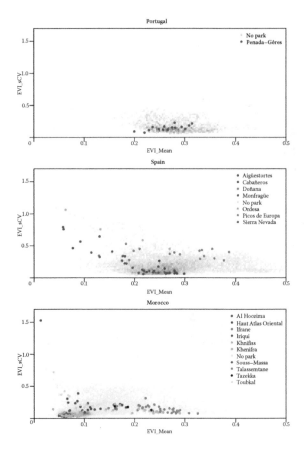

FIGURE 9.4
Refer to the text for the figure caption.

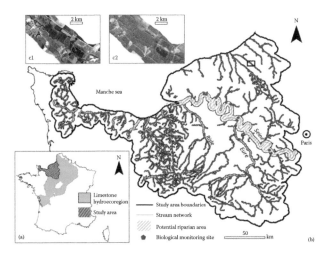

FIGURE 10.1
Refer to the text for the figure caption.

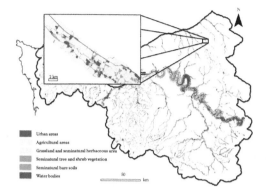

FIGURE 10.7
Refer to the text for the figure caption.

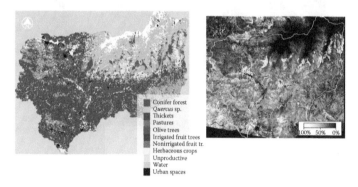

FIGURE 12.1
Refer to the text for the figure caption.

FIGURE 12.2
Refer to the text for the figure caption.

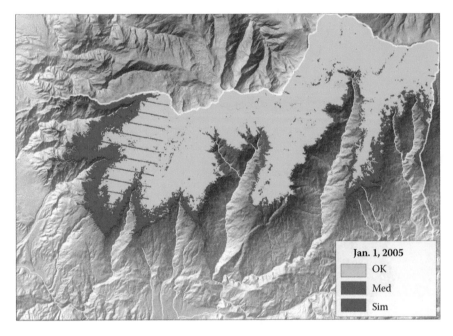

FIGURE 12.3
Refer to the text for the figure caption.

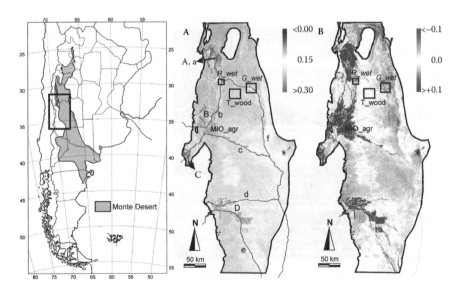

FIGURE 13.1
Refer to the text for the figure caption.

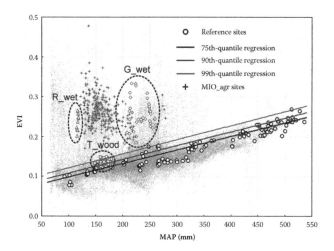

FIGURE 13.3
Refer to the text for the figure caption.

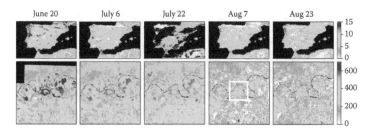

FIGURE 14.2
Refer to the text for the figure caption.

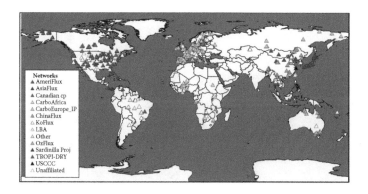

FIGURE 14.3
Refer to the text for the figure caption.

(a)

(b)

(c)

FIGURE 14.4
Refer to the text for the figure caption.

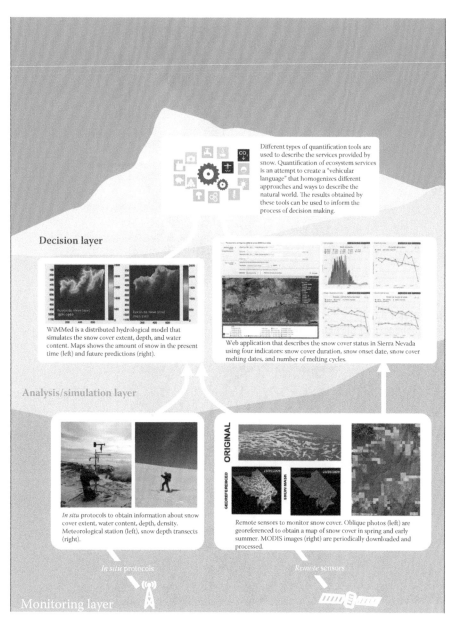

Different types of quantification tools are used to describe the services provided by snow. Quantification of ecosystem services is an attempt to create a "vehicular language" that homogenizes different approaches and ways to describe the natural world. The results obtained by these tools can be used to inform the process of decision making.

Decision layer

WiMMed is a distributed hydrological model that simulates the snow cover extent, depth, and water content. Maps shows the amount of snow in the present time (left) and future predictions (right).

Web application that describes the snow cover status in Sierra Nevada using four indicators: snow cover duration, snow onset date, snow cover melting dates, and number of melting cycles.

Analysis/simulation layer

In situ protocols to obtain information about snow cover extent, water content, depth, density. Meteorological station (left), snow depth transects (right).

Remote sensors to monitor snow cover. Oblique photos (left) are georeferenced to obtain a map of snow cover in spring and early summer. MODIS images (right) are periodically downloaded and processed.

In situ protocols

Remote sensors

Monitoring layer

FIGURE 15.2
Refer to the text for the figure caption.

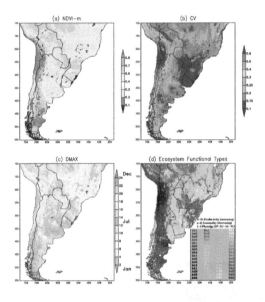

FIGURE 16.1
Refer to the text for the figure caption.

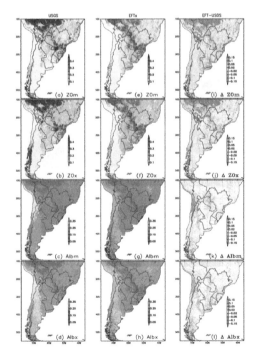

FIGURE 16.2
Refer to the text for the figure caption.

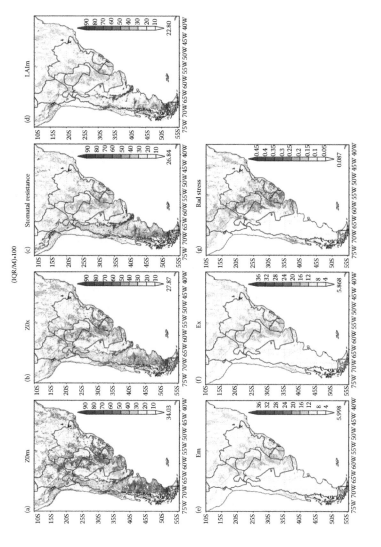

FIGURE 16.3
Refer to the text for the figure caption.

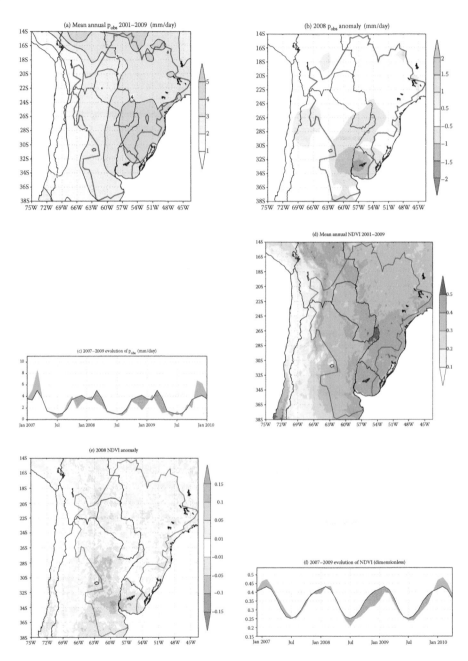

FIGURE 16.4
Refer to the text for the figure caption.

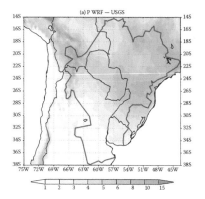
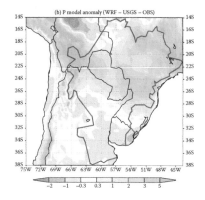
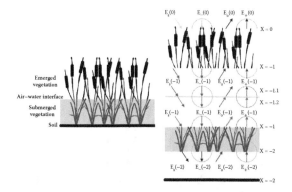

FIGURE 16.5
Refer to the text for the figure caption.

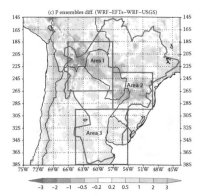

FIGURE 17.1
Refer to the text for the figure caption.

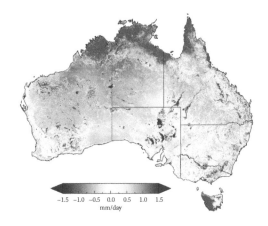

FIGURE 18.1
Refer to the text for the figure caption.

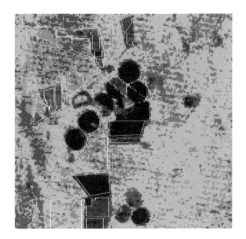

FIGURE 18.2
Refer to the text for the figure caption.

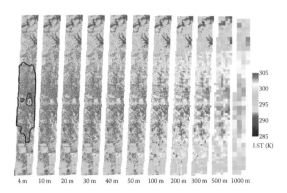

FIGURE 19.3
Refer to the text for the figure caption.

12

Assimilation of Remotely Sensed Data into Hydrologic Modeling for Ecosystem Services Assessment

J. Herrero and A. Millares
University of Granada, Spain

C. Aguilar
University of Córdoba, Spain

F. J. Bonet
University of Granada, Spain

M. J. Polo
University of Córdoba, Spain

CONTENTS

12.1 Introduction 261
12.2 Hydrologic Modeling and Ecosystem Services Quantification 262
12.3 Hydrologic Modeling and Remote Sensing 268
12.4 Water Quality Monitoring and Remote Sensing 272
12.5 Conclusions 275
References 276

12.1 Introduction

The so-called water cycle was already observed, studied, and described by the ancient civilizations (Biswas 1970). But it was during the nineteenth century that hydrology was consolidated as an individual science, when the measuring capacity reached a significant level to acquire relevant volumes of data. It led to a rationalization of hypotheses, conclusions, and modeling during the twentieth century, which finished with the development and use of complex hydrological models. It was in the last decades when observation of the Earth's surface took a leap from the ground to space, shifting the concern about the scale effects arising from the use of point measurement to characterize continuous three-dimension systems (upscaling) to the downscaling of remotely sensed data and their products. Good examples of application and state of the art of remote sensing for hydrological observation and modeling can be found in Schultz and Engman (2000), Schmugge et al. (2002), or Su et al. (2011).

The hydrologic simulation of systems, whatever their scale, is a first and significant basis for analyzing and simulating water quality and, thus, ecosystem dynamics. However, the accuracy of hydrologic simulation is one of the most significant sources of uncertainty in calculations derived for ecological variables in the complete chain of interactions and forcing of processes. Wagner et al. (2009) addressed the need for improving modeling strategies based on these new means of observation and the capture of the macroscale processes as well as the quantification of associated uncertainties, as the main challenge for hydrologists and scientists in general. In this context, data assimilation methods play a key role in fostering the application of remote sensing in hydrology and other sciences.

12.2 Hydrologic Modeling and Ecosystem Services Quantification

As the importance of ecosystem services is recognized by society, a growing interest in mathematical models as tools that provide necessary information for decision-making processes has arisen. The use of these models has increased over the last 30 years mainly due to a greater availability of information and data acquisition techniques (satellite images, aerial photography, remote data transmission, digital elevation models, etc.) and an increase in computer calculation capacity.

A huge number of models have been developed and adapted to different processes and systems. Thus, we can distinguish between biogeochemical models, such as CANDY (Franko et al. 1995) and ICBM (Andren et al. 2004); terrestrial vegetation models, such as TRIFFID model (Cox 2001) and YieldSafe (Van der Werf et al. 2007); carbon cycle models, such as Hybrid (Friend et al. 1997), CenW (Kirschbaum 1999), and ASPECTS (Rasse et al. 2001); or hydrologic models, such as TOPMODEL (Beven and Kirkby 1979) or SWAT (Arnold et al. 1998; Neitsch et al. 2005). Each of these models gives results linked to the system or subsystem for which they were developed, taking into account the spatiotemporal scales of the processes, the initial information available, and the final results desired for decision making. The complexity of all the systems and processes involved in a widespread evaluation limits even nowadays the possibility of global modeling. Recently, some works have begun to integrate different models, either in a coupled way or embedded in a single model, allowing the quantification of services corresponding to systems that, although interrelated, have traditionally been modeled separately.

The hydrologic cycle is a clear example of the conceptualization of nature as a set of connected systems (atmosphere–Earth–sea), in which concatenated physical processes take place (precipitation, snowmelt, runoff, infiltration, subsurface flow, aquifer contributions, etc.). Such processes directly condition morphologic, chemical, biological, economic, and social behavior with evident influence on the services to society. Hence, it is no coincidence that many of the works developed for assessing ecosystem services with a modeling approach are structured, in one way or another, as subroutines based on hydrologic modeling (e.g., Band et al. 1991; Tague and Band 2001; Yates et al. 2005). The processes involved in the hydrologic cycle allow a discrete quantification of their components (e.g., volume of available water, soil loss, snow, or groundwater storage) and, therefore, the possibility of measuring both accessible and nonaccessible benefits. This potentially powerful quantification of tangible resources (water, snow, sediments, nutrients, etc.) is often far removed from the actual benefit obtained by society in practice. The existence of less tangible resources to be included in the analysis, such as cultural or aesthetic values, adds a complexity to the valuation process. Some authors (Porras et al. 2008; Carpenter et al. 2009) refer to a relative success in the services quantification coming from the results obtained by hydrologic modeling, mainly due to the difficulties in estimating the benefits derived by the water cycle and by the user's needs. This aspect needs to be improved in future studies aimed at the hydrologic modeling assessment of ecosystem services.

In general, two different approaches can be identified in the use of hydrologic models for ecosystem services quantification (Bellamy et al. 2011; Vigerstol and Aukema 2011): (1) from traditional hydrologic models, which requires a second step or postprocessing from the results in order to assess the final quantification and (2) from integrated ecosystem models recently developed by the combination of different methodologies or models, which gives, as a final result, a service quantification and its spatial distribution.

Hydrologic models have been used for decades to estimate different processes related to the water cycle, such as flooding (see Chapter 17), water resource availability, soil loss valuation, and so on. Strictly speaking, they have been providing decision makers with relevant information for the valuation of processes long before the concept of ecosystem services was consolidated by the scientific community and society. These tools have evolved from lumped/aggregated and event-based models, such as HEC-1 (USACE 1982), TR-20 (USSCS 1982), to more sophisticated models with a physical basis, and/or the generation of distributed and continuous simulation, such as WMS (Dellman et al. 2002), MIKE-SHE (Abbott et al. 1986), or WiMMed (Polo et al. 2009; Herrero et al. 2010). From the point of view of ecosystem services quantification, lumped-conceptual and event-based models have a limited interest, although they require less information and less computing capacity. In contrast, as mentioned previously, physically based and distributed models allow for the estimation not only of the water balance, but also of the spatial distribution of the hydrologic variables, and they can quantify both the hydrologic processes and their interactions.

The Soil and Water Assessment Tool (SWAT) model, with a semiempirical and semidistributed basis, has been widely applied in many studies, especially those focusing on the quantification of ecosystem services. These works include the assessment of the availability of hydrologic resources (Notter et al. 2012), the pollutant distribution in basins (Prochnow et al. 2008; Schilling and Wolter 2009), the evaluation of different climate scenarios (Stone et al. 2001), the best practices in basin management (Gassman et al. 2007), water storage in snow and snowmelt contributions in mountain areas (Herrero et al. 2005), and soil loss evaluation (Shen et al. 2009), to mention some relevant works. The main limitations found, however, are related to the empirical approach of many of the processes and the spatial discretization performed because it aggregates the space in uniform hydrologic response units (HRU), which sometimes do not include the spatial distribution of many significant processes.

It is important to highlight the role of physically based and distributed models for ecosystem services quantification, although different works mention their limitations due to high computing and data requirements and the need to calibrate a large number of parameters. It is important to point out the contribution of remote sensing and its ability to provide models with distributed information in relatively short periods of time (e.g., land cover types, soil moisture, LAI (Leaf Area Index)), which allows both the initial configuration of the model and the final validation of the obtained results (snow cover evolution, flooded areas, soil loss, and so on). Furthermore, it is possible to develop relationships between remotely sensed data and different hydrologic parameters in order to create new information from those observations (Kite and Pietroniro 1996; Chen et al. 2005; Liu and Li 2008; Feng et al. 2010; Su et al. 2010; Aguilar et al. 2012). Examples of this are included in Box 12.1 and Box 12.2.

A second approach is related to models which integrate hydrologic modeling and ecosystem services quantification, related or not to water. These models are in their first developmental stages and, despite not being fully contrasted

BOX 12.1   ANNUAL AND SEASONAL VARIATION OF SURFACE ALBEDO AND COVER FRACTION OF THE VEGETATION FROM LANDSAT IMAGES: ASSIMILATION INTO A HYDROLOGIC MODEL AS INPUT VARIABLES

Energy and water budgets on the Earth's surface are coupled processes that share evaporation and transpiration terms. The available water on the surface and on the upper layer of the soil acts as a source or sink of the deficit or excess of energy in this budget; vegetation directly transports water from the root zone to the atmosphere by means of respiration and also modifies the soil evaporation regime through stoma control of the transpiration rates under water scarcity conditions. In practice, it is difficult to discriminate transpiration and evaporation rates in vegetated areas, and evapotranspiration (ET) is the term used to refer to this transport of water vapor. Moreover, the local and regional energy and water budgets in the atmospheric boundary layer and in the surface soil layer are also deeply related through the presence of vegetation: The root zone determines the effective soil depth to be considered as control volume in the soil budgets; the vegetative cover, its species, density, and structure constitute a rough three-dimensional layer with a fundamental role in the turbulent transfer of momentum, energy, and water between the atmosphere and the terrain surface; vegetation type, its vigor, and its density also influence the fraction of the incident solar energy that is reflected back to the atmosphere, its albedo; the aerial structure of the vegetation can retain a given fraction of rainfall, which is evaporated back to the atmosphere instead of infiltrating through soil, the interception term in the water budget. On a spatial basis, the surface density of the vegetation is estimated by the cover fraction, the fraction of the horizontal projection of the terrain surface that is covered by vegetation, ranging from 0 (bare soil) to 1 (completely vegetated soil). The cover fraction is, thus, the index that scales the terms in the energy and in the water budgets in which the vegetation is involved from the unit vegetation area to the unit surface area (Figure 12.1).

For medium- to large-scale hydrologic analyses, quantifying the cover fraction evolution is of great interest. Different products can be used to acquire this information through the calculation of the Normalized Difference Vegetation Index (NDVI) and its relationship to the cover fraction (Curran 1981; Sellers 1989; Bannari et al. 1995), such as Moderate Resolution Imaging Spectroradiometer (MODIS) data, although in heterogeneous areas, their spatial resolution poses a constraint for a proper quantification because significant scale effects arise. Landsat images have been widely used (Ramsey et al. 2004) instead, due to their balanced spatial resolution, to estimate NDVI values at the watershed and regional scales. To quantify interception losses in the Guadalfeo River watershed (southern Spain), a heterogeneous mountainous coastal area, Díaz-Gutiérrez (2007) coupled an interception model based on the approaches of Gash (1979) and Rutter et al. (1971, 1975) to a vegetation cover fraction map series obtained from a seasonal characterization of NDVI, by means of analyzing four to six Landsat 5 TM and seven Landsat 7 ETM+ images per year during the 2002–2005 period; a simple interpolation algorithm with a steady state and evolving periods proved to be satisfactory for simulating a continuous daily time step in the series production (Polo et al. 2011).

The assimilation of this time series into the hydrologic model, Watershed integrated Management for Mediterranean environments (WiMMed), resulted in the estimation of rainfall interception (Figure 12.2) losses along the watershed and provided managers with an efficient tool to evaluate different options of crop selection, wildfire effects, and drought consequences in terms of the intercepted fraction change at watershed scale.

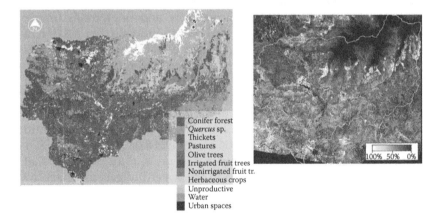

Conifer forest
Quercus sp.
Thickets
Pastures
Olive trees
Irrigated fruit trees
Nonirrigated fruit tr.
Herbaceous crops
Unproductive
Water
Urban spaces

100% 50% 0%

FIGURE 12.1 (See color insert.)
Vegetation classification (left) and example of cover fraction distribution (right), March 2005 in the Guadalfeo River watershed.

yet, they constitute highly promising tools. The Integrated Valuation of Ecosystem Services and Tradeoffs (InVEST) model (Tallis et al. 2011) allows the assessment of the quantity and value of different ecosystem services for current or future scenarios. The results are provided in distributed maps for services directly related or not to water (reservoir hydropower production, prevention of reservoir siltation, water purification, biodiversity

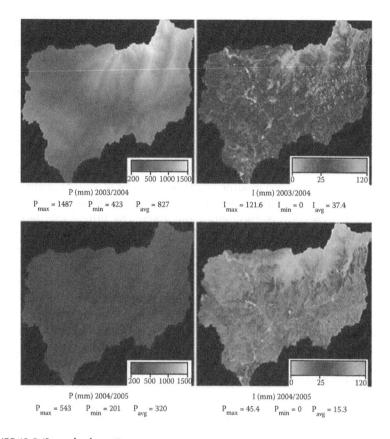

P (mm) 2003/2004

P_{max} = 1487 P_{min} = 423 P_{avg} = 827

I (mm) 2003/2004

I_{max} = 121.6 I_{min} = 0 I_{avg} = 37.4

P (mm) 2004/2005

P_{max} = 543 P_{min} = 201 P_{avg} = 320

I (mm) 2004/2005

P_{max} = 45.4 P_{min} = 0 P_{avg} = 15.3

FIGURE 12.2 (See color insert.)
Precipitation (P, left) and interception (I, right) in two consecutive hydrological years in the Guadalfeo River watershed.

BOX 12.2 DELTA COASTAL AREAS RETREAT RELATED TO SEDIMENTS SUPPLY (THE GUADALFEO STUDY CASE)

Bedload erosion processes significantly affect fluvial dynamics and frequently condition fluvial management due to their high impact on dam siltation, streambed particle stability, and estuarine dynamics, among others. The social and economic repercussions of sediment loads have different effects, sometimes opposed, which suggest the need for more complex methodologies in order to understand these processes and their associated costs.

The Guadalfeo River basin (southern Spain) exhibits an important amount of alluvial sediments stored throughout its main stream and secondary dry affluents (or *ramblas*) to its mouth in the Mediterranean Sea. The building of the Rules Dam in 2004, with 110 hm³ of capacity, posed a risk for significant environmental changes in the up- and downstream surroundings, especially in the delta.

Monitoring works in two control points located at the main channel were complemented by an analysis of remote sensing information to estimate not only the loss of storage capacity in the reservoir due to siltation, but also the impact downstream on the delta dynamics. Orthophotography and satellite images between 1956 and 2008 (Red de Información Ambiental de Andalucía 2010) were used together with field bathymetries during 2003–2011 to estimate siltation rates from 2002, with an estimated volume of 2,000,000 m^3 of sediment infilling.

This strong sediment retention in the reservoir resulted in an enhanced regression of the delta from 2005. The fluvial inputs have practically ceased, causing an unbalanced loss of beach material during intense storms on the coast, which is retreating. Under normal conditions, the coastal erosive action performed by the currents and breaking waves maintains the same direction and transports the sediment from the delta to the adjacent coastal areas, mainly beaches and harbors (Ávila 2007). The direct impact can be measured from the additional expenses in beach nourishments in this tourist area, quantified in 1,121,000 m^3 between 2004 and 2009, with an associated cost of 7,286,500 € (Ruiz de Almirón 2011). These dynamics can be found in many deltas and estuaries in regulated Mediterranean watersheds.

sustainability, carbon storage, timber production assessment). A large number of application examples are given for biodiversity loss (Goldman et al. 2012; Reyers et al. 2012) and present/future soil management scenarios (Daily et al. 2009; Nelson et al. 2009; Johnson et al. 2012). In this case, some limitations can be found related to the simplification of many of the hydrologic processes involved and the absence of some important contributors, such as groundwater resources (see Chapter 13). Other models following this philosophy are based on Bayesian network approaches to establish relationships between the input data and the different services pursued. Such is the case of the Artificial Intelligence for Ecosystem Services (ARIES) model (Villa et al. 2009), which has been applied for the valuation of different climate scenarios. Further details of both models can be found in Box 20.1.

12.3 Hydrologic Modeling and Remote Sensing

The coupled water and energy budgets on the Earth determine the primary local regime of the environmental conditions, with most of the physical, chemical, biological, and ecological regimes being highly dependent on them. Moreover, water constitutes one of the major pathways for sediment, nutrients, pollutants, and so on throughout the different ecosystem scales. The biogeochemical cycle is similar to the ecosystem dynamics of the water cycle because it defines systems and subsystems, their internal links and exchanges, their external forcing and associated responses, and their interactions. To quantify biogeochemical fluxes, the water and energy fluxes must be estimated first. The biogeochemical cycle relies both physically and mathematically on the hydrological cycle and, thus, receives and propagates the quality level achieved in the performance of the latter, as well as the uncertainty associated with its results. This fact justifies the importance and influence of the hydrologic modeling in the ecosystem services quantification.

A computer model for the simulation of the hydrologic cycle permits not only being able to predict behaviors in foreseeable future scenarios but also deepening the actual knowledge of this behavior in the present time for a particular region. To take advantage of the spatial resolution provided by remote sensing abilities nowadays, distributed models must be used. Moreover, to fully exploit the present hydrologic knowledge to describe the different processes taking place and to estimate with enough accuracy the different water paths and interactions, these models should be physically based. A model with both characteristics is capable of estimating a significant number of components and water fluxes within the hydrologic cycle, which are otherwise impossible to quantify reliably.

There are numerous distributed and physically based hydrologic models that are widely used, such as TOPMODEL, DHSVM (Wigmosta et al. 1994), or SHE, to give just a few examples. Despite these similarities, these models differ in how they solve some particular hydrologic processes. For example, some of them attach importance to the subsurface flow, or the overland flow, as part of a strategy for implementation in wetter or drier environments (such approaches usually are related to the area where the model was originally developed). The time step for continuous simulations varies from daily to hourly calculations. WiMMed is a distributed and physically based hydrologic model developed by the Universities of Granada and Córdoba, in Spain (Polo et al. 2009; Herrero et al. 2010). Initially conceived for Mediterranean mountainous environments, it is especially intended for dealing with mountainous basins where snowmelt at low latitudes is present in a context of extreme variability sources: abrupt topography, meteorological gradients, vegetation cover heterogeneity, mountainous aquifers with preferential topographic subsurface flow, torrential precipitation and dry period during the seasons, great production of sediments, and extended hyperannual droughts. The physical approach to the resolution of the different equations related to such processes is especially important in such a highly heterogeneous and time-variable environment. This is particularly relevant for the snow hydrology (see Chapter 15), which can only be correctly simulated by an energy and mass balance approach. As in many other distributed models, remote sensing plays a significant role during hydrologic simulation with WiMMed.

The use of remote sensing technologies in the modeling process allows the identification of parameters attached to some physical meaning, as well as the description of water presence over a whole region or hydrologic unit simultaneously. These parameters can be incorporated into the model as input data, which can be considered constant or variable in time. Distributed models must be able to simulate water cycles on the same resolution as that from the satellite information, to fully exploit these data. Data assimilation, as well as the comparison between model results in cells and remote measurements of state variables, will be straightforward in such a case. As an example, if Landsat multispectral data are used to obtain information to be incorporated into a distributed model, 30 m × 30 is the minimal spatial resolution that the model should reach in order to take advantage of all the available information. The high required computing capacity for that used to be claimed as a major constraint for the use of these models but is no longer true nowadays.

A distributed hydrologic model can assimilate data from different sources and approaches (Houser et al. 1998, 2012; Reichle et al. 2002; De Lannoy et al. 2011; Malik et al. 2012). As an example, WiMMed is fed by two direct results from multispectral data through the calculation of the NDVI: the vegetation cover fraction, fv, and the surface albedo, α (Díaz-Gutiérrez 2007). The first indicates the fraction of each pixel that is covered by vegetation, whereas α is the shortwave reflectivity of the terrain, both of them obtained as effective values for the spatial resolution of the multispectral sources. When Landsat Thematic Mapper data are used, a time frequency of 15 days is available, provided no clouds interfere. The resulting fv or α maps can then be interpolated to obtain daily distributed series, which constitute direct inputs to the model to estimate rainfall interception or the energy budget in the soil or in the snowpack. Among other variables, the reference evapotranspiration (see Chapter 18) and the actual evapotranspiration are derived from these calculations (Aguilar et al. 2010). This is an example of direct assimilation of physical parameters obtained from remote sensing data as inputs to the equations in a model.

On the contrary, hydrologic models can also use estimated values of their state variables by means of remotely sensed information to calibrate and validate their performance. The snow cover distribution (see Chapter 15) and the soil moisture are reference examples of this (see Chapter 14). The assimilation procedure can incorporate correction techniques to include the quality and uncertainty of the source data, by means of methods based on the use of Kalman filtering, such as the Data Assimilation Research Testbed (DART) model (Anderson et al. 2009). A simpler approach is a direct insertion, which consists of replacing the simulated variable with the "measured" values at the given states in which a satisfactory degree of adjustment is not achieved, as WiMMed does with snow data from aerial or remotely sensed sources (Pimentel et al. 2012) (Box 12.3). The presence of snow can be detected by visible images or multispectral ones, by means of the near-infrared reflectance analyses through the use of indexes such as the Normalized Difference Snow Index (NDSI; Hall et al. 1995). In a similar way, the water equivalent of the snow can be estimated from light detection and ranging (LiDAR) radar, or terrestrial gamma-ray attenuation data (DeWalle and Rango 2008), and soil moisture from microwave, multispectral analysis, or gamma-radiation measurements (Carroll 1981; Wang and Qu 2009).

BOX 12.3   CALIBRATION AND VALIDATION OF A SNOWMELT MODEL THROUGH SNOW MAPS DERIVED FROM REMOTE SENSING

Herrero et al. (2011), when applying a hydrological model in the Guadalfeo River basin (near the Sierra Nevada mountain range, Spain), used the WiMMed model to map at an hourly scale the snow cover extension and its water equivalent. With a 30-m spatial resolution, the results from the model were directly compared to the snow cover map obtained from Landsat for specific days with an adequate visibility. This comparison led to the definition of four possible different combinations for every pixel: (1) pixels with simulated and measured snow, (2) pixels free of simulated and measured snow—both (1) and (2) correct cases, (3) pixels with measured but not simulated snow, and (4) pixels free of measured snow but covered with snow in the simulation. The accuracy of the calibration–validation process was performed by means of different indexes deduced from the four possible combinations described.

Figure 12.3 shows an example of this pixel-to-pixel comparison between the simulated and measured snow presence. Measurements were obtained after the processing of a Landsat TM7 image, while simulation was run with the snow module of WiMMed model. Some correct pixels are in green and others are in a transparent color, while blue and red pixels represent underestimation and overestimation of the snow cover by the model. The identified deviations were mainly due to an incorrect assignment of the temperature and of the precipitation in every pixel during the simulation, which is usually the main source of uncertainty in hydrologic simulations in mountainous basins, where gradients are very pronounced and the coverage of the meteorological networks is usually insufficient (this being even more important than the inherent limitations of the physical modeling itself).

FIGURE 12.3 (See color insert.)
Pixel to pixel comparison between the snow cover simulated with WiMMed and measured from Landsat TM7 for Jan. 1, 2005.

12.4 Water Quality Monitoring and Remote Sensing

Surface water bodies offer very different provisioning, regulating, and cultural ecosystem services, such as water supply, fish production, transportation, recreation, and so on. They are also vital to the survival of many key species that use them to live, feed, and reproduce. As a consequence of the constant and variable interaction between human pressures and natural forces, water bodies are constantly changing. Therefore, surface water systems are at once resilient and fragile (Ji 2008).

As rivers flow throughout watersheds, they collect water, sediment, nutrients, and pollutant discharges. Other surface water systems such as reservoirs and estuaries filter the water and associated pollutants, sediments, nutrients, toxics, and so on from the upstream contributing areas. Abrupt changes in land uses, which may increase the discharges of pollutants or cause the overexploitation of water bodies, often lead to common environmental problems downstream such as eutrophication, loss of habitat, algal bloom development, decline in fish and wildlife, seawater intrusion, siltation of hydraulic infrastructures, and so on. Therefore, the management and planning of water resources require the correct assessment of not only the amount but also the quality of the water.

The first concerns about poor water quality focused on health and sanitation issues. Control measures were mainly implemented in sewage treatment plants and in industrial discharges through pipes or open channels (Engman and Gurney 1991). But, more recently, nonpoint source pollution has been the subject of both general concern and scientific investigation. Nonpoint source pollution is considered as part of storm runoff, and so the identification, measurement, and control of this type of pollution may be very complex. Once again, integrated models at watershed scale constitute an important tool for water resource management, by combining quantity and quality criteria. These models aim to characterize precisely the hydrologic and erosion processes that influence water and sediment fluxes throughout the watershed, coupled with the physical, chemical, and biological processes that affect water quality. In nonmonitored areas, physically based models allow estimation of the watershed response, although some level of calibration from field measurements is always required.

There are numerous hydrological models that reproduce nonpoint source pollution processes at watershed scale with different details such as ANSWERS, SWMM, AGNPS, HSPF, GLEAMS, SWRRBWQ, CREAMS, SWAT, and so on. However, most of them can be very complex to implement and to calibrate due to the large amount of parameters involved. In addition, the quantification of nonpoint source discharges is difficult due to the lack of available measurements or to the historical data record not being long enough to determine its dynamics and its evolution at the required temporal scale. In this way, remote sensing is a very valuable source in water quality evaluation, especially when the spatially distributed nature of nonpoint source pollution and the broad spatial scale required for these kinds of studies is considered (Engman and Gurney 1991). At the regional scale, the availability of high-frequency monitoring is only economically viable through remote sensing. Also, nowadays, the global coverage of satellites allows for the estimation of water quality in remote and inaccessible areas and provides modelers with data for historical water quality studies during periods lacking ground measurements (Hellweger et al. 2004).

Once again, remote sensing constitutes a very valuable data source at three levels in the application of nonpoint source pollution models: as input data, state variables, and measured data for the calibration and validation of the model estimates. As for input data and state variables, remote sensing data are often used for the generation of facts related to coverage and land uses, topography, and soil types, information that greatly affects the potential water quality of runoff. Regarding the model outputs or water quality indicators, remote sensing permits the derivation of surface estimations of turbidity, suspended sediments and chlorophyll concentrations, colored dissolved organic matter, and temperature in water bodies. These water quality characteristics can be used as indicators of more specific pollution problems (e.g., eutrophication levels) and be related to nonpoint source model outputs (Engman and Gurney 1991).

Turbidity is an optical effect related to the total concentration of suspended sediments and other organic matter. Chlorophyll-a is a key indicator for monitoring aquatic populations, mainly phytoplankton, and the state of aquatic ecosystems can be obtained by sensors that are able to quantify the photosynthetic process. Colored dissolved organic matter (CDOM) is a product of plant and animal decomposition processes. All these water quality variables can be measured *in situ* using conventional techniques that involve direct sampling of water (Salama et al. 2012). However, they greatly vary in both space and time affected by the loadings received in the water body and the hydrodynamics of the system. Thus, the monitoring from point samples is often inadequate as it is time consuming and only representative for a limited spatial and temporal domain. In this way, remote sensing sources provide very valuable spatial and temporal data due to their capability to monitor vast areas nearly instantaneously (Hadjimitsis et al. 2006; Budhiman et al. 2012).

Reflectance in the visible and near-infrared regions of the spectrum is used for the evaluation of water quality indicators in the surface or near surface of water. Thermal infrared is also used for estimating water quality indicators from the direct measure of emitted energy (Engman and Gurney 1991). In general, sediments present a high reflectivity in all the bands in the visible regions even though the correlations between the Secchi depth, an indirect measure of turbidity, and reflectances in the blue and green bands (450–600 nm) are much lower than those in the red ones (600–690 nm). However, the spectral response is strongly influenced by the nature of the water system. In water bodies affected by the discharge of freshwater, there is a high correlation between turbidity and the reflectance in the red band. In coastal waters with low discharges of freshwater and, therefore, nonsignificant sediment loadings, reflectance is more affected by the concentration of phytoplankton, estimated from chlorophyll-a (Hellweger et al. 2004; Lane et al. 2007).

Water quality variables can be remotely quantified following empirical approaches based on regression analysis between measurements and observations. Another option is the use of semianalytical methods that apply a hydro-optical model that describes the relationships between the observed spectrum and the concentrations of the water constituents (Salama et al. 2012).

Regarding sensors and platforms, the choice is determined by the resolution of the data required according to the variable to be estimated and the spatial extent of the study area. This is why the relatively small dimensions of rivers, lakes, and estuarine waters have often restricted the derivation of water quality variables from satellite data to the open ocean and some coastal areas (Salama and Su 2010, 2011; Shen et al. 2010; Budhiman et al. 2012). In rivers, the focus is mostly on LiDAR, altimeter, and airborne hyperspectral data as the number of satellite sensors that provide the needed spectral and spatial resolutions is limited even for large rivers (Salama et al. 2012). For instance, the spatial resolution of EnviSAT (300 m) is too coarse for capturing even the largest rivers, while in Landsat data, the limited number of visible bands and the coarse spectral resolution of these bands is the main constraint (Dekker and Peters 1993). Nevertheless, Landsat has been the dominant source of satellite images for lake water quality monitoring due to the fine spatial resolution (30 m). In the literature, there are numerous studies on inland lakes that have developed expressions to estimate suspended solids, turbidity, chlorophyll-a, salinity, and temperature from Landsat data (Lathrop 1992; Baban 1993; Mayo et al. 1995; Hadjimitsis et al. 2006; Wang et al. 2006). However, the higher spatial extent of coastal systems such as estuaries, deltas, and lagoons allows the use of data from satellite sensors with medium spatial resolutions such as MODIS (Hellweger et al. 2004; Chen et al. 2007) and the Medium Resolution Imaging Spectrometer (MERIS; Matthews et al. 2010). However, there are also a lot of studies that apply finer spatial resolution satellite data for the estimation of water quality parameters such as Landsat (Lavery et al. 1993; Hellweger et al. 2004; Kabbara et al. 2008; Wang and Xu 2008; Bustamante et al. 2009) or EO-1 ALI (Chen et al. 2009). The main limitation of these studies is that, unlike the studies that apply hydro-optical models (Salama et al. 2012), most of them are site specific; however, they allow establishing an assessment of the status of water bodies at a large scale that in many cases cannot be obtained by other means (Engman and Gurney 1991). In the future, more effort is expected to fully understand the variability of the optical properties of these water bodies (Budhiman et al. 2012).

12.5 Conclusions

Remote sensing undoubtedly constitutes a powerful data source for hydrologic modeling at the watershed, regional, and global scales—a necessary basis for ecosystem services assessment, mapping, and quantification. In fact, the increasing availability and quality of these data provide modelers with schemes to include certain assimilation techniques in the modeling environment itself. Both approaches, that is, the direct estimation of the spatial distribution of the properties of the terrain that are relevant in the hydrologic processes, and the calculation of different state variables at the frequency of the satellite, provide scientists and technicians with discrete spatial information on a given location, which can be considered to be continuous at the scale given by the spatial resolution of the sensor. Scale issues, thus, constitute a significant matter to be considered. Current and future development of satellite sources, together with the already ongoing effort to combine data from multiple satellites by taking advantage of their individually higher spatial or time resolution, constitutes a reliable horizon for technicians and scientists. However, during the past decades, the possibility of acquiring such detailed distributed information has greatly increased and broadened the human capacity not only for the observation of the Earth but also for the simulation of processes and will still do so in the future.

As for water quality modeling, remote sensing provides useful information regarding the spatial distribution of the terrain and soil properties. These data can be used as input to hydrological models that reproduce nonpoint source pollution processes at the watershed scale. Nevertheless, the main field of application of remote sensing data in water quality modeling is the estimation of water quality parameters in receiving water bodies. This information could be used in the calibration and validation of water quality models. However, once again, scale issues and the complexity of the optical properties of water bodies constitute the main lines of future research.

References

Abbott, M. B., J. C. Bathurst, J. A. Cunge, P. E. O'Connel, and J. Rasmussen. 1986. An introduction to the European Hydrological System—Systeme Hydrologique "SHE", 1: History and philosophy of a physically based distributed modelling system. *Journal of Hydrology* 87:45–59.

Aguilar, C., J. Herrero, and M. J. Polo. 2010. Topographic effects on solar radiation distribution in mountainous watersheds and their influence on reference evapotranspiration estimates at watershed scale. *Hydrology and Earth System Sciences* 14:2479–2494.

Aguilar, C., J. C. Zinnert, M. J. Polo, and D. R. Young. 2012. NDVI as an indicator for changes in water availability to woody vegetation. *Ecological Indicators* 23:290–300.

Anderson, J., T. Hoar, K. Raeder, et al. 2009. The Data Assimilation Research Testbed: A community facility. *Bulletin of the American Meteorological Society* 90:1283–1296.

Andren, O., T. Kätterer, and T. Karlsson. 2004. ICBM regional model for estimations of dynamics of agricultural soil carbon pools. *Nutrient Cycling in Agroecosystems* 70:231–239.

Arnold, J. G., R. Srinivasan, R. S. Muttiah, and J. R. Williams. 1998. Large area hydrologic modeling and assessment Part I: Model development. *Journal of American Water Resources Association* 34:73–89.

Ávila, A. 2007. *Procesos de múltiple escala en la evolución de la línea de costa* [Multiple scale processes in the coastline evolution]. PhD diss., Environmental Flow Dynamics Research Group, University of Granada, Spain.

Baban, S. M. 1993. Detecting water-quality parameters in the Norfolk broads, UK, using Landsat imagery. *International Journal of Remote Sensing* 14:1247–1267.

Band, L., D. Peterson, S. Running, et al. 1991. Forest ecosystem processes at the watershed scale: Basis for distributed simulation. *Ecological Modeling* 56:171–196.

Bannari, A., D. Morin, F. Bonn, and A. R. Huete. 1995. A review of vegetation indices. *Remote Sensing Reviews* 12:335–357.

Bellamy, P., M. Camino, J. Harris, R. Corstanje, I. Holman, and T. Mayr. 2011. Monitoring and modelling ecosystem services: A scoping study for the ecosystem services pilots. Natural England Commissioned Report NECR073, Cranfield.

Beven, K., and M. Kirkby. 1979. A physically-based variable contributing area model of basin hydrology. *Hydrologic Science Bulletin* 24:43–69.

Biswas, A. K. 1970. *History of hydrology*. Amsterdam: North Holland Publishing Co.

Budhiman, S., M. S. Salama, Z. Vekerdy, and W. Verhoef. 2012. Deriving optical properties of Mahakam Delta coastal waters, Indonesia, using in situ measurements and ocean color model inversion. *ISPRS Journal of Photogrammetry and Remote Sensing* 68:157–169.

Bustamante, J., F. Palacios, R. Díaz-Delgado, and D. Aragonés. 2009. Predictive models of turbidity and water depth in the Doñana marshes using Landsat TM and ETM+ images. *Journal of Environmental Management* 90:2219–2225.

Carpenter, S., H. Mooney, J. Agard, et al. 2009. Science for managing ecosystem services: Beyond the Millennium Ecosystem Assessment. *Proceedings of the National Academy of Sciences of the United States of America* 106:1305.

Carroll, T. R. 1981. Airborne soil moisture measurement using natural terrestrial gamma radiation. *Soil Science* 132:358–366.

Chen, J., X. Chen, W. Ju, and X. Geng. 2005. Distributed hydrological model for mapping evapotranspiration using remote sensing inputs. *Journal of Hydrology* 305:15–39.

Chen, S., L. Fang, L. Zhang, and W. Huang. 2009. Remote sensing of turbidity in seawater intrusion reaches of Pearl River Estuary—A case study in Modaomen waterway, China. *Estuarine, Coastal and Shelf Science* 82:119–127.

Chen, Z., C. Hu, and F. Muller-Karger. 2007. Monitoring turbidity in Tampa Bay using MODIS/Aqua 250-m imagery. *Remote Sensing of Environment* 109:207–220.

Cox, P. M. 2001. *Description of the "TRIFFID" dynamic global vegetation model*. Technical Note 24. Berks: Hadley Centre.

Curran, P. 1981. Multispectral remote sensing for estimating vegetation biomass and productivity. In *Plants and the daylight spectrum*, ed. H. Smith, 65–99. London: Academic Press.

Daily, S., J. Polasky, P. M. Goldstein, et al. 2009. Ecosystem services in decision making: Time to deliver. *Frontiers in Ecology and the Environment* 7:21–28.

Dekker, A., and S. Peters. 1993. The use of the Thematic Mapper for the analysis of eutrophic lakes—A case study in the Netherlands. *International Journal of Remote Sensing* 14:799–821.

De Lannoy, G. J. M., R. H. Reichle, K. Arsenault, et al. 2011. Multi-scale assimilation of AMSR-E snow water equivalent and MODIS snow cover fraction observations in northern Colorado. *Water Resources Research* 48:W01522.

Dellman, P. N., C. E. Ruiz, C. T. Manwaring, and E. J. Nelson. 2002. *Watershed modeling system hydrological simulation program; watershed model user documentation and tutorial*. Vicksburg, MS: Engineer Research and Development Center, Environmental Lab.

DeWalle, D. R., and A. Rango. 2008. *Principles of snow hydrology*. Cambridge, UK: Cambridge University Press.

Díaz-Gutiérrez, A. 2007. *Series temporales de vegetación para un modelo hidrológico distribuido* [Temporal series of vegetation for a distributed hydrological model]. Monografías 2007. Spain: Grupo de Hidrología e Hidráulica Agrícola, University of Córdoba.

Engman, E. T., and R. J. Gurney. 1991. *Remote sensing in hydrology*. London: Chapman & Hall.

Feng, B., B. Fu, X. Yang, and Y. Lü. 2010. Remote sensing of ecosystem services: An opportunity for spatially explicit assessment. *Chinese Geographical Science* 20:522–535.

Franko, U., B. Oelschlagel, and S. Schenk. 1995. Simulation of temperature, water and nitrogen dynamics using the model CANDY. *Ecological Modelling* 81:213–222.

Friend, A. K., R. G. Stevens, M. G. R. Knox, and A. Cannell. 1997. Process-based terrestrial biosphere model of ecosystem dynamics (Hybrid v3.0). *Ecological Modelling* 95:249–287.

Gash, J. H. C. 1979. An analytical model of rainfall interception by forests. *Quarterly Journal of the Royal Meteorological Society* 105:43–55.

Gassman, P. W., M. R. Reyes, C. H. Green, and J. G. Arnold. 2007. The soil and water assessment tool: Historical development, applications, and future research directions. Working Paper 07-WP 443. Center for Agricultural and Rural Development, Iowa State University, Ames, Iowa.

Goldman, R. L., S. Benitez, T. Boucher, et al. 2012. Water funds and payments for ecosystem services: Practice learns from theory and theory can learn from practice. *Oryx* 46:55–63.

Hadjimitsis, D., M. Hadjimitsis, C. Clayton, and B. Clarke. 2006. Determination of turbidity in Kourris Dam in Cyprus utilizing Landsat TM remotely sensed data. *Water Resources Management* 20:449–465.

Hall, D. K., G. A. Riggs, and V. V. Salomonson. 1995. Development of methods for mapping global snow cover using moderate resolution imaging spectroradiometer data. *Remote Sensing of Environment* 54:127–140.

Hellweger, F. L., P. Schlosser, U. Lall, and J. K. Weissel. 2004. Use of satellite imagery for water quality studies in New York Harbor. *Estuarine, Coastal and Shelf Science* 61:437–448.

Herrero, J., C. Aguilar, A. Millares, et al. 2010. *WiMMed. User Manual v1.1.* Granada, Spain: Grupo de Dinámica Fluvial e Hidrología (University of Córdoba) and Grupo de Dinámica de Flujos Ambientales (University of Granada).

Herrero, J., M. J. Polo, and M. A. Losada. 2005. Modelo SWAT aplicado a la cuenca del río Guadalfeo. Balance hidrológico dentro de un modelo de gestión [Assessment of SWAT model in the Guadalfeo River basin: hydrological balance in a management model]. *VI Simposio del Agua en Andalucía (SIAGA).* IGME. 237–248.

Herrero, J., M. J. Polo, and M. A. Losada. 2011. Snow evolution in Sierra Nevada (Spain) from an energy balance model validated with Landsat TM data. Proc. SPIE 8174, *Remote Sensing for Agriculture, Ecosystems, and Hydrology XIII*, 817403.

Houser, P. R., G. J. M. De Lannoy, and J. P. Walker. 2012. *Hydrologic data assimilation, approaches to managing disaster—Assessing hazards, emergencies and disaster impacts*, ed. J. Tiefenbacher. Rijeka, Croatia: InTech.

Houser, P. R., W. J. Shuttleworth, J. S. Famiglietti, H. V. Gupta, K. H. Syed, and C. Goodrich. 1998. Integration of soil moisture remote sensing and hydrologic modeling using data assimilation. *Water Resources Research* 34:3405–3420.

Ji, Z. G. 2008. *Hydrodynamics and water quality: Modeling rivers, lakes and estuaries.* Hoboken, NJ: John Wiley & Sons.

Johnson, K. A., S. Polasky, E. Nelson, and D. Pennington. 2012. Uncertainty in ecosystem services valuation and implications for assessing land use tradeoffs: An agricultural case study in the Minnesota River Basin. *Ecological Economics* 79:71–79.

Kabbara, N., J. Benkhelil, M. Awad, and V. Barale. 2008. Monitoring water quality in the coastal area of Tripoli (Lebanon) using high-resolution satellite data. *ISPRS Journal of Photogrammetry and Remote Sensing* 63:488–495.

Kirschbaum, M. U. F. 1999. CenW, a forest growth model with linked carbon, energy, nutrient and water cycles. *Ecological Modelling* 181:17–59.

Kite, G., and A. Pietroniro. 1996. Remote sensing applications in hydrological modeling. *Hydrological Science* 41:563–591.

Lane, R. R., J. W. Day, B. D. Marx, E. Reyes, E. Hyfield, and J. N. Day. 2007. The effects of riverine discharge on temperature, salinity, suspended sediment and chlorophyll-a in a Mississippi Delta estuary measured using a flow-through system. *Estuarine, Coastal and Shelf Science* 74:145–154.

Lathrop, R. G. 1992. Landsat Thematic Mapper monitoring of turbid inland water quality. *Photogrammetric Engineering and Remote Sensing* 58:465–470.

Lavery, P., C. Pattiaratchi, and P. Hick. 1993. Water quality monitoring in estuarine waters using the Landsat Thematic Mapper. *Remote Sensing of Environment* 46:268–280.

Liu, X., and J. Li. 2008. Application of SCS model in estimation of runoff from small watershed in loess plateau of China. *Chinese Geographical Science* 18:235–241.

Malik, M. J., R. van der Velde, Z. Vekerdy, and Z. Su. 2012. Assimilation of satellite observed snow albedo in a land surface model. *Journal of Hydrometeorology* 13:1119–1130.

Matthews, M. W., S. Bernard, and K. Winter. 2010. Remote sensing of cyanobacteria-dominant algal blooms and water quality parameters in Zeekoevlei, a small hypertrophic lake, using MERIS. *Remote Sensing of Environment* 114:2070–2087.

Mayo, M., A. Gitelson, Y. Z. Yacobi, and Z. Ben-Avraham. 1995. Chlorophyll distribution in Lake Kinneret determined from Landsat Thematic Mapper data. *International Journal of Remote Sensing* 16:175–182.

Neitsch, S. L., J. G. Arnold, J. R. Kiniry, and J. R. Williams. 2005. *Soil and water assessment tool theoretical documentation*, version 2005. Temple, TX: Grassland, Soil and Water Research Service.

Nelson, E., G. Mendoza, J. Regetz, et al. 2009. Modeling multiple ecosystem services, biodiversity conservation, commodity production, and tradeoffs at landscape scales. *Frontiers in Ecology and the Environment* 7:4–11.

Notter, B., H. Hurni, U. Wiesmann, and K. C. Abbaspour. 2012. Modelling water provision as an ecosystem service in a large East African river basin. *Hydrology and Earth System Sciences* 16:69–86.

Pimentel, R., J. Herrero, and M. J. Polo. 2012. Terrestrial photography as an alternative to satellite images to study snow cover evolution at hillslope scale. *Proceedings of SPIE - The International Society for Optical Engineering*, vol 8531, art 85310Y.

Polo, M. J., A. Díaz-Gutiérrez, and M. P. González-Dugo. 2011. Interception modeling with vegetation time series derived from Landsat TM data. *Proceedings of SPIE - The International Society for Optical Engineering*, vol 8174, art. 817403.

Polo, M. J., J. Herrero, C. Aguilar, et al. 2009. WiMMed, a distributed physically-based watershed model (I): Description and validation. In *Environmental hydraulics: Theoretical, experimental and computational solutions, IWEH09*, eds. P. A. López-Jiménez, V. S. Fuertes-Miquel, P. L. Iglesias-Rey, et al., 225–228. London: CRC Press.

Porras, I., M. Grieg-Gran, and N. Neves. 2008. *All that glitters: A review of payment for watershed services in developing countries*. London: International Institute for Environment and Development.

Prochnow, S. J., J. D. White, T. Scott, and C. D. Filstrup. 2008. Multi-scenario simulation analysis in prioritizing management options for an impacted watershed system. *Ecohydrology & Hydrobiology* 8:3–15.

Ramsey, R. D., J. R. Wright, and C. McGinty. 2004. Evaluating the use of Landsat 30m Enhanced Thematic Mapper to monitor vegetation cover in shrub-steppe environments. *Geocarto International* 19:39–47.

Rasse, D. P., L. François, M. Aubinet, et al. 2001. Modelling short-term CO_2 fluxes and long-term tree growth in temperate forests with ASPECTS. *Ecological Modelling* 141:35–52.

Red de Información Ambiental de Andalucía. 2010. Ortofotografía Digital Histórica de Andalucía 1956–2008: Medio siglo de cambios en Andalucía [Historical digital ortophotography of Andalusia 1956–2008: half a century of change in Andalusia]. Seville, Spain: Junta de Andalucía.

Reichle, R. H., B. Dennis McLaughlin, and D. Entekhabi. 2002. Hydrologic data assimilation with the ensemble Kalman filter. *Monthly Weather Review* 130:103–114.

Reyers, B., S. Polasky, H. Tallis, H. Mooney, and A. Larigauderie. 2012. Finding a common ground for biodiversity and ecosystem services. *BioScience* 62:503–507.

Ruiz de Almirón, C. 2011. *Modelo conceptual para el control de salinidad en la Charca de Suárez mediante un sistema de compuertas* [Conceptual model for the control of salinity in "Charca Suárez" using a floodgate system]. Master thesis, Environmental Flows Dynamics Research Group, University of Granada, Spain.

Rutter, A. J., K. A. Kershaw, P. C. Robins, and A. J. Morton. 1971. A predictive model of rainfall interception in forests. I. Derivation of the model from observations in a plantation of Corsican pine. *Agricultural Meteorology* 9:367–384.

Rutter, A. J., A. J. Morton, and P. C. Robins. 1975. A predictive model of rainfall interception in forests. II. Generalization of the model and comparison with observations in some coniferous and hardwood stands. *Journal of Applied Ecology* 12:367–380.

Salama, M. S., M. Radwan, and R. van der Velde. 2012. A hydro-optical model for deriving water quality variables from satellite images (HydroSat): A case study of the Nile River demonstrating the future Sentinel-2 capabilities. *Physics and Chemistry of the Earth*, Parts A/B/C:224–232.

Salama, M. S., and Z. Su. 2010. Bayesian model for matching the radiometric measurements of aerospace and field ocean color sensors. *Sensors* 10:7561–7575.

Salama, M. S., and Z. Su. 2011. Resolving the subscale spatial variability of apparent and inherent optical properties in ocean color match-up sites. *IEEE Transactions on Geoscience and Remote Sensing* 49:2612–2622.

Schilling, K. E., and C. F. Wolter. 2009. Modeling nitrate-nitrogen load reduction strategies for the Des Moines River, Iowa, using SWAT. *Journal of Environmental Management* 44:671–682.

Schmugge, T. J., W. P. Kustas, J. C. Ritchie, T. J. Jackson, and A. Rango. 2002. Remote sensing in hydrology. *Advances in Water Resources* 25:1367–1385.

Schultz, G. A., and E. T. Engman, eds. 2000. *Remote sensing in hydrology and water management*. Berlin: Springer-Verlag.

Sellers, P. J. 1989. *Theory and applications of optical remote sensing*. New York: Wiley-Interscience.

Shen, F., M. S. Salama, Y. X. Zhou, J. F. Li, Z. Su, and D. B. Kuang. 2010. Remote-sensing reflectance characteristics of highly turbid estuarine waters—A comparative experiment of the Yangtze River and the Yellow River. *International Journal of Remote Sensing* 31:2639–2654.

Shen, Z. Y., Y. W. Gong, Y. H. Li, Q. Hong, L. Xu, and R. M. Liu. 2009. A comparison of WEPP and SWAT for modeling soil erosion of the Zhangjiachong watershed in the Three Gorges reservoir area. *Agricultural Water Management* 96:1435–1442.

Stone, M. C., R. H. Hotchkiss, C. M. Hubbard, T. A. Fontaine, L. O. Mearns, and J. G. Arnold. 2001. Impacts of climate change on Missouri River basin water yield. *Journal of the American Water Resources Association* 37:1119–1129.

Su, Z., R. A. Roebeling, I. Holleman, et al. 2011. Observation of hydrological processes using remote sensing. In *Treatise on water science*, ed. P. Wilderer, vol. 2:351–399. Oxford: Academic Press.

Su, Z., J. Wen, and W. Wagner. 2010. Advances in land surface hydrological processes: Field observations, modeling and data assimilation: Preface. *Hydrology and Earth System Sciences* 14:365–367.

Tague, C. L., and L. E. Band. 2001. Evaluating explicit and implicit routing for watershed hydro-ecological models of forest hydrology at the small catchment scale. *Hydrological Processes* 15:1415–1439.

Tallis, H. T., T. Ricketts, A. D. Guerry, et al. 2011. *InVEST 2.3.0 user's guide*. Stanford, CA: The Natural Capital Project.

USACE (U.S. Army Corps of Engineers). 1982. Hydrologic analysis of ungaged watersheds with HEC-1. Davis, CA: Hydrologic Engineering Center.

USSCS (U.S. Soil Conservation Service). 1982. *Project formulation, hydrology*. Technical release No. 20. Washington, DC: USSCS.

Van der Werf, K., P. J. Keesman, A. R. Burgess, et al. 2007. Yield-SAFE: A parameter-sparse process-based dynamic model for predicting resource capture, growth and production in agroforestry systems. *Ecological Engineering* 29:419–433.

Vigerstol, K. L., and J. E. Aukema. 2011. A comparison of tools for modeling freshwater ecosystem services. *Journal of Environmental Management* 92:2403–2240.

Villa, F., M. Ceroni, K. Bagstad, G. Johnson, and S. Krivovet. 2009. ARIES (ARtificial Intelligence for Ecosystem Services): A new tool for ecosystem services assessment, planning, and valuation. *11th International BIOECON Conference on Economic Instruments to Enhance the Conservation and Sustainable Use of Biodiversity*. Venice, Italy.

Wagner, W., N. E. C. Verhoest, R. Ludwig, and M. Tedesco. 2009. Remote sensing in hydrological sciences. *Hydrology and Earth System Sciences* 13:813–817.

Wang, F., L. Han, H. T. Kung, and R. B. Van Arsdale. 2006. Applications of Landsat-5 TM imagery in assessing and mapping water quality in Reelfoot Lake, Tennessee. *International Journal of Remote Sensing* 27:5269–5283.

Wang, F., and Y. J. Xu. 2008. Development and application of a remote sensing-based salinity prediction model for a large estuarine lake in the US Gulf of Mexico coast. *Journal of Hydrology* 360:184–194.

Wang, L., and J. J. Qu. 2009. Satellite remote sensing applications for surface soil moisture monitoring: A review. *Frontiers of Earth Science in China* 3:237–247.

Wigmosta, M. S., L. Vail, and D. P. Lettenmaier. 1994. A distributed hydrology-vegetation model for complex terrain. *Water Resources Research* 30:1665–1679.

Yates, D., J. Sieber, D. R. Purkey, and A. Huber-Lee. 2005. A demand, priority and preference-driven water planning model: Part 1, model characteristics. *Water International* 30:487–500.

13

Detecting Ecosystem Reliance on Groundwater Based on Satellite-Derived Greenness Anomalies and Temporal Dynamics

S. Contreras

Centre of Pedology and Applied Biology of Segura, Spain

D. Alcaraz-Segura

University of Granada, Spain; University of Almería, Spain

B. Scanlon

The University of Texas at Austin, Texas

E. G. Jobbágy

San Luis Institute of Applied Mathematics (IMASL), Argentina

CONTENTS

13.1 Introduction 284
13.2 Methods 286
    13.2.1 Study Site 286
    13.2.2 Climate and Satellite Dataset for Greenness Anomaly
           Estimation 288
    13.2.3 Greenness Timing and Metrics 290
    13.2.4 Impact of Groundwater on Vegetation Dynamics:
           A Conceptual Model 291
13.3 Results and Discussion 292
    13.3.1 MAP-EVI Regional Function 292
    13.3.2 EVI Dynamics along a Groundwater Dependence
           Gradient at the Telteca Site 292

13.3.3 Intercomparison among Phreatophytic Woodlands,
        Wetlands, and Irrigated Crops...295
13.4 Conclusions..298
Acknowledgments...299
References..299

13.1 Introduction

Groundwater-dependent ecosystems (GDEs) play a key role in human development, and are especially relevant in regions with low rates of rainfall, by providing a broad range of ecosystem services such as physical support for wildlife habitats and biodiversity hotspots, control of floods and erosion, regulation of nutrient cycling, or provision of landscape refuges for cognitive development (de Groot et al. 2002; Chen et al. 2004; Eamus et al. 2005; Bergkamp and Katharine 2006; Ridolfi et al. 2007). During the past decade, research on ecology and functioning of GDEs has received a growing interest from the scientific community and from landscape managers. However, in spite of their high intrinsic values, many of these ecosystems have been strongly impacted as a consequence of disruption of hydrological linkages with groundwater resources. This disruption has been generally promoted by excessive rates of groundwater extraction and depletion, for example, Las Tablas de Daimiel and Doñana National Reserves in Spain (Llamas 1988; Muñoz-Reinoso and García-Novo 2005); Swan Coastal Plain in southwest Australia (Groom et al. 2000); desert springs in the Mojave and Great Basin deserts in the United States (Patten et al. 2008); San Pedro River in the United States (Stromberg et al. 1996). It has also been caused by modification of morphology of stream channels or wetlands through dredging or artificial diversions (Ellery and McCarthy 1998) or as a consequence of changes in their water balance due to climatic factors (Murray-Hudson et al. 2006). A better understanding of the functioning and water consumption of GDEs is then critically required to evaluate the ecological services provided by them (Murray et al. 2006; Brauman et al. 2007) and, for developing adaptive management frameworks that reconcile compatible human activities, ecosystem conservation, and their underlying hydrological trade-offs under future scenarios of land use and climate change (MacKay 2006; Barron et al. 2012b).

GDEs are ecosystems that require groundwater inflows to maintain their current structure and functioning and the subsequent delivery of ecosystem services (Hatton and Evans 1997; Murray et al. 2003; Eamus et al. 2006). GDEs may display an obligate reliance requiring a constant groundwater presence, or a facultative one where they adapt their functioning to fluctuating groundwater availability (Murray et al. 2003; Bertrand et al. 2012). According to the aquifer–ecosystem interface relationship, GDEs include (Eamus et al. 2006; Eamus 2009): (a) caves and subterranean-aquatic ecosystems, including karst aquifers and rock-fractured systems; (b) ecosystems dependent on permanent or temporary surface expressions of groundwater, including baseflow riverine, spring, wetland or peatland, and estuarine/marine-shoreline ecosystems; and (c) ecosystems dependent on the subsurface presence of groundwater, also termed "terrestrial GDEs" or phreatophytic ecosystems (Richardson et al. 2011). Other pedological, morphological, hydrological, and biogeochemical criteria have been proposed for classifying GDEs from a functional point of view (Bertrand et al. 2012).

To preserve their ecological integrity and service provision, GDEs require water allocation plans and adaptive management strategies rooted in knowledge about their (a) typology and spatial distribution, (b) quantitative water requirements, and (c) resistance and resilience to natural and human perturbations on their groundwater regimes. A wide range of methodological approaches and techniques are commonly employed to accomplish these three aspects, including remote sensing, water balance analysis, hydrogeological modeling, tracer and isotopic studies, ecophysiological measurements, rooting system characterization, and aquatic fauna sampling (see Richardson et al. 2011 for a synthesis).

Tracking photosynthetic/greenness activity of vegetation using satellite-based indices such as Normalized Difference Vegetation Index (NDVI) or Enhanced Vegetation Index (EVI) offers a relatively inexpensive and effective way to characterize the functioning of riparian/wetland and terrestrial GDEs (Bradley and Mustard 2008; Barron et al. 2012a). These spectral indices are well correlated with aboveground net primary productivity (ANPP) and evapotranspiration (ET) in semiarid regions (Running and Nemani 1988; Paruelo et al. 1997; Jobbágy et al. 2002; Nagler et al. 2005; Guerschman et al. 2009) (see Chapter 18). When they are not influenced by the presence of groundwater and lateral-inflow resources, annual rates of ANPP and ET in those regions are primarily controlled by precipitation and, second, by radiation forcing and its seasonal coupling with rainfall inputs (Specht 1972; Specht and Specht 1989; Ellis and Hatton 2008; Palmer et al. 2010). Because groundwater supplies a more temporally reliable water source for terrestrial ecosystems than rainfall, higher and more stable ET and ANPP rates should be expected in GDEs compared to their nongroundwater ecosystem counterparts (Contreras et al. 2011; O'Grady et al. 2011). Field observations show that leaf area, ANPP, and water availability are closely correlated, supporting the use of satellite-based vegetation indices for the identification and characterization of GDEs, including the quantification of their water requirements at different temporal scales (Nagler et al. 2005; Contreras et al. 2011; Devitt et al. 2011; O'Grady et al. 2011). Field evidence also suggests that even when access to unlimited groundwater resources exists, the productivity of GDEs could be strongly constrained by other limiting resources or processes such as incoming energy, nutrient availability, morphological constraints, or disturbances, among others (Eamus et al. 2000; Do et al. 2008). Consequently, it seems that annual primary productivity estimates retrieved from annual summaries of spectral vegetation indexes could not be sufficient to identify and characterize the water requirements of terrestrial GDEs. To solve this potential constraint, complementary assessment of seasonal greenness timing can provide additional and valuable information on the functional response of ecosystems to their environment (Morisette et al. 2009). In these studies, in addition to primary productivity estimates, seasonality and phenology traits are commonly retrieved from annual greenness dynamics of ecosystems to classify and characterize ecosystem functional types, that is, patches of land surface with similar exchanges of matter and energy between the biota and the physical environment (Paruelo et al. 2001; Alcaraz-Segura et al. 2006; Fernández et al. 2010) (see Chapters 9 and 16).

This study aims to evaluate a satellite-based approach for identifying inflow-dependent ecosystems and to detect the type and degree of groundwater reliance of wetland and phreatophytic ecosystems. The approach consists of the complementary analysis of the annual greenness anomalies computed according to Contreras et al. (2011), and land surface phenological metrics retrieved from intra-annual and interannual greenness trajectories. The performance of this approach is tested in the lowlands of the central Monte Desert (Argentina), where a potential gradient of native inflow-dependent ecosystems has been previously identified. Finally, productivity, seasonality, and phenological metrics computed for a representative sample of those types of ecosystems are compared with those extracted from a sample of sites located at an upstream irrigated oasis that exploits surface and groundwater resources for its maintenance.

13.2 Methods

13.2.1 Study Site

The study region covers an area of 87,500 km^2 of lowlands ($\leq$ 1000 m a.s.l.) and expands over the central Monte Desert in Argentina between 31° S and 36° S (Figure 13.1). The region is bounded by the Andes Cordillera to the west and by the Sierras Pampeanas to the east. Precipitation in the region ranges from 150 to 400 mm y^{-1}, most of it concentrated in the austral summer (from October to March), and mean annual temperature ranges from 13°C to 19°C. Potential evapotranspiration reaches 1400 mm y^{-1} in the driest parts of the study region. A detailed review of the main biophysical and socioeconomic characteristics of the Monte Desert is provided by Abraham et al. (2009), while Villagra et al. (2009) review some of the effects that land use and disturbance factors have had on the dynamics of the natural ecosystems of this desert.

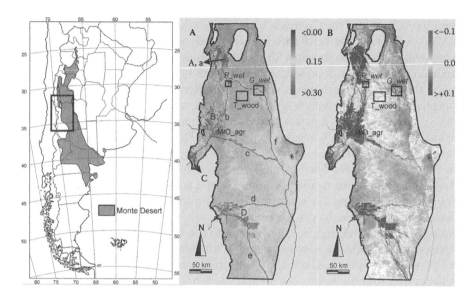

FIGURE 13.1 (See color insert.)
Study region and mean annual values of EVI (A, left) and EVI anomalies (B, right) computed from September 2000 through August 2009. Main rivers in lowercase letters: (a) San Juan River; (b) Mendoza River; (c) Tunuyán River; (d) Diamante River; (e) Atuel River; and (f) Desaguadero-Salado River. Irrigated oases in capital letters: (A) San Juan oasis; (B) Mendoza oasis; (C) Upper Tunuyan oasis; and (D) San Rafael oasis. Sample sites in control areas (black squares) were located at the open *Prosopis* woodlands of the Telteca Natural Reserve and surroundings (T_wood), the Rosario and Guanacache wetland systems (R_wet and G_wet, respectively) and irrigated crops at the Mendoza oasis (MIO_agr).

The area is crossed by five major rivers (from north to south: San Juan, Mendoza, Tunuyán, Diamante, and Atuel) with their origins in the Andes Cordillera. After crossing the mountains, these rivers reach the alluvial fans and sedimentary plains of the central Monte Desert to finally discharge into the Desaguadero-Salado river system. Andean rivers are the main sources of water for four large artificial oases located in the foothills of the region (Figure 13.1), with vineyards, olives, and fruit trees being the main crops. These oases represent approximately 90% of the economic activity in the region, and more than 1.5 million people live there. Along their route in the alluvial funs, rivers recharge large unconsolidated aquifers that extend downstream of the artificial oases to reach the lowlands of the region that are covered by sandy plains of fluvial, lacustrine, and eolian origin.

Alluvial and lowland plains at the foothills of the Andes are mainly covered by three types of ecosystems: (a) shrub-steppes dominated by *Larrea* spp. (*jarillales*), (b) open phreatophytic woodlands of *Prosopis* spp. trees (locally known as *algarrobales*), and (c) marshes and wetlands that are along the main rivers. Two of the largest wetland systems are the Rosario system— at the last section of the Mendoza River just before its confluence with the

San Juan River, and the Guanacache system at the end of the San Juan River. Because of the regulation of the river upstream of the irrigated oases and the great water diversion for agriculture, the Mendoza River has an ephemeral hydrological regime downstream from Mendoza city. Riparian vegetation along the distal section of this river and the Rosario wetlands at its end are supplied with surface waters only after intense rainfall events. Nevertheless, the San Juan River has a permanent water regime acting as a constant source of surface water to the Guanacache wetlands. However, in the past decades, similar to the occurrence in the Mendoza River, the discharge values to the wetland system have been strongly affected by hydraulic regulation and irrigation agriculture in the San Juan oasis. Lacustrine vegetation, such as *Scirpus californicus* and *Typha dominguensis*, dominates wetlands but alien species of the genus *Tamarix* are invading those areas more and more because of changes in the water regime of the river and the streams that fed them. Open *Prosopis* woodlands are mostly located in the alluvial plains on soils that are > 90% sand. These woodlands have different structures depending on their reliance on groundwater resources and show higher growth rates and health status in the Telteca National Reserve and the distal section of the Tunuyán River (phreatic level at 6–15 m depth) than at the Ñacuñan National Reserve (water table at 70–80 m depth) (Villagra et al. 2005). In the Telteca area, where an extensive dune system dominates, those open woodlands are well developed in the interdune valleys. The strong reliance of these woodlands on groundwater has been demonstrated by isotopic and hydrochemical profiling studies (Aranibar et al. 2011; Jobbágy et al. 2011). Both open woodlands and wetlands have historically provided the local settlements and economies with timber, peat, and charcoal, and food and water, and also with the physical support required for domestic livestock (Villagra et al. 2009).

In the framework of this study, satellite-based metrics of vegetation dynamics were extracted at the Rosario (R_wet) and Guanacache (G_wet) wetland systems and at the open *Prosopis* woodlands located at the Telteca National Reserve and surroundings (T_wood). Vegetation dynamics were equally characterized for a representative sample of irrigated crops located in the Mendoza irrigated oasis (MIO_agr) (Figure 13.1).

13.2.2 Climate and Satellite Dataset for Greenness Anomaly Estimation

According to Contreras et al. (2011), we define "greenness anomaly" as the absolute difference between mean annual greenness observed at any pixel of the landscape and a site-specific reference greenness value estimated depending on the local precipitation. For this study, we used the EVI as an indicator of vegetation greenness. The precipitation-based reference greenness value is assumed to be linearly related to mean annual precipitation (MAP) as follows:

$$EVI_{ref} = a\,MAP + b \tag{13.1}$$

where EVI_{ref} is precipitation-based EVI and a and b are fitted-parameters computed empirically from a quantile regression analysis developed over the observed EVI-MAP scatterplot defined for a set of reference sites. We assumed a linear relationship between EVI and evapotranspiration based on the field data support available for semiarid regions (e.g., Nagler et al. 2005; Guerschman et al. 2009; O'Grady et al. 2011). With such assumption, we are able to estimate the expected EVI value for a vegetation cover that exclusively uses local precipitation and is in equilibrium with long-term precipitation (Boer and Puigdefábregas 2003; Contreras et al. 2008). This condition, in which annual ET approaches the MAP, has been proposed to be reached at our study region for 75th quantile threshold value (Contreras et al. 2011). At an annual timescale, we defined the concept of *EVI anomaly* as follows:

$$EVI_a = EVI_{ma} - EVI_{map} \tag{13.2}$$

where EVI_{ma} is the observed annual average of EVI computed from satellite images at each pixel, and EVI_{map} is the EVI_{ref} in Equation 13.1 estimated using the 75th quantile threshold value. From a functional point of view, both metrics, EVI_a and EVI_{ma}, are considered here as surrogates of primary productivity.

A map of MAP for the region, which is required to estimate EVI_{map}, was calculated from long-term average monthly values reported in the CRU CL 2.0 dataset (New et al. 2002), which were previously corrected with data from the CLIMWAT 2.0 database (FAO 2006) and local meteorological stations. Maps of precipitation were finally resampled to a 250-m spatial resolution, which is compatible with the satellite data (Contreras et al. 2011).

The EVI MOD13Q1 land product from MODIS Collection 5 (Solano et al. 2010) was extracted for the region (tile h12v12) covering nine hydrological years from September 2001 through August 2009 (23 scenes per hydrological year). Before processing, raw EVI data at 250-m spatial resolution were filtered using a local polynomial function based on an adaptive Savitzky–Golay filter using the TIMESAT software (Jönsson and Eklundh 2004). EVI, which combines data from the blue, red, and infrared spectral bands, was preferable to NDVI because atmospheric interferences and soil background signal are more effectively removed and because of its greater sensitivity to high biomass situations (Huete et al. 2002).

Equation 13.1 was parameterized across 125 reference sites that meet the criteria of having low disturbance rates and lacking artificial or runoff water supplies (Contreras et al. 2008, 2011). From the resulting EVI-MAP scatterplot, three quantile threshold values were used here to propose a preliminary gradient of classes of groundwater reliance. First, as stated earlier, we used the 75th quantile regression of mean annual EVI versus MAP function as a conservative value to generate EVI_{map} values. Observed EVI values lower than EVI_{map} (i.e., negative greenness anomaly values) were assumed not to have any dependency on groundwater resources. For values higher than EVI_{map}, a gradient of three potential classes of low, moderate, and high degree of reliance were established using the 75th, 90th, and 99th quantile thresholds, respectively. In this study, the former quantile threshold values were arbitrarily selected in order to evaluate the potential agreement between the resulting reliance levels and the phenological metrics extracted from the greenness timing analysis.

13.2.3 Greenness Timing and Metrics

A representative sample of pixels for the four study systems (Telteca woodlands, Rosario and Guanacache wetlands, and irrigated crops at the Mendoza oasis) was selected for retrieving metrics, or phenometrics, related to vegetation traits of primary productivity, seasonality, and phenology (Figure 13.2). At the Telteca Reserve, 78 pixels were sampled to cover all the potential groundwater-reliance degrees identified by the anomaly greenness (no reliance, low, moderate, and high) although pixels

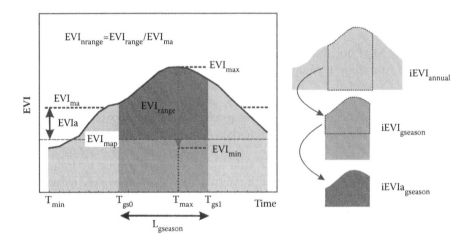

FIGURE 13.2
Vegetation metrics retrieved from Enhanced Vegetation Index (EVI) trajectories to identify groundwater-dependent ecosystems and to quantify their reliance on groundwater. All metrics were retrieved from annual (September–August) and average long-term seasonal trajectories. Metrics are related to *productivity* traits: EVI_{ma} = mean annual EVI; EVI_{gs} = mean EVI accumulated during the growing season; *seasonality* traits: EVI_{max} and EVI_{min} = maximum and minimum EVI values; EVI_{range} = annual amplitude of EVI ($EVI_{max} - EVI_{min}$), EVI_{nrange} = normalized annual amplitude (EVI_{range}/EVI_{ma}); and *phenology* traits: T_{max} and T_{min} = times at which maximum and minimum EVI values are reached; T_{gs0} and T_{gs1} = times at which growing season starts and ends, $L_{gseason}$ = growing season length. Productivity traits are estimated from integrated values of greenness at the annual ($iEVI_{annual}$) and growing season ($iEVI_{gs}$) scales. EVI_{map} is the reference EVI value expected according the local mean annual precipitation and is required to compute greenness anomalies at the annual ($EVIa_{ma}$) and growing season ($EVIa_{gs}$).

with moderate- and high-reliance degrees were finally grouped together to obtain a more robust comparison among classes. In the wetland systems, vegetation metrics were only extracted from pixels (Guanacache, n = 40; Rosario, n = 10) with a potential high degree of reliance on groundwater. Metrics related to vegetation primary productivity, seasonality, and phenology were extracted at intra-annual and interannual timescales from the average seasonal trajectory resulting from September 2000 to August 2009 (intra-annual variability) and from the annual trajectories for the nine hydrological years covered by the study (interannual variability). In this study, we extracted the following EVI metrics related to traits of: (a) vegetation productivity: EVI_{ma} and EVI_{gs}; (b) vegetation seasonality: EVI_{min}, EVI_{max}, and EVI_{nrange}; and (c) vegetation phenology: L_{gs} and T_{max} (see Figure 13.2 for more details).

13.2.4 Impact of Groundwater on Vegetation Dynamics: A Conceptual Model

From a functional point of view, a water table close to the land surface is expected to impact intra-annual (seasonal) and interannual (multiyear) variability of the EVI dynamics in several ways. The following hypothesis guided our analyses (Table 13.1).

At the intra-annual scale, we hypothesize that MAP (EVI_{ma}), the cumulated productivity during the growing season (EVI_{gs}), and maximum (EVI_{max}) and minimum (EVI_{min}) values of greenness are expected to increase

TABLE 13.1

Trends in Vegetation Traits

Vegetation Traits	Greenness Metrics	Intra-Annual Scale	Interannual Variability
Productivity	EVI_{ma}	↑	↓
	EVI_{gs}	↑	↓
Seasonality	EVI_{max}	↑	↓
	EVI_{min}	↑	↓
	EVI_{nrange}	↓	↓
Phenology	L_{gs}	↑	↓
	T_{max}	↑	↓

Note: Trends (arrows) measured by satellite-based metrics expected in terrestrial GDEs as groundwater reliance increases. Metrics are related to *productivity* traits: EVI_{ma} = mean annual EVI; EVI_{gs} = mean EVI accumulated during the growing season; *seasonality* traits: EVI_{max} and EVI_{min} = maximum and minimum EVI values; EVI_{range} = annual amplitude of EVI ($EVI_{max} - EVI_{min}$), EVI_{nrange} = normalized annual amplitude (EVI_{range}/EVI_{ma}); and *phenology* traits: T_{max} = time at which maximum value is reached.

with greater reliance of ecosystems on groundwater. Because shallow water tables represent a perennial source of water for ecosystems, we also hypothesize that vegetation with any reliance on groundwater would show less variable seasonal trajectories of productivity (EVI_{nrange}) as groundwater reliance increases. As a consequence of less variability in the greenness trajectory (less EVI_{nrange}), a longer period should be required to reach 50% of the total annual productivity, here defined as the growing season length (L_{gs}). At the interannual scale, we expect that variability of all vegetation metrics described for productivity, seasonality, and phenology traits should be lower in GDEs than in non-GDEs and should decrease as reliance on groundwater increases. The matrix of conceptual rules proposed here to evaluate ecosystem reliance on groundwater has been designed under the assumption that the access to groundwater by vegetation remained relatively constant without large changes in the water table depth. Then, changes in the water table depth or in the hydrological regime of those ecosystems are expected to be followed by modifications in their greenness dynamics and phenological patterns.

13.3 Results and Discussion

13.3.1 MAP-EVI Regional Function

According to the MAP-EVI function described for the region (Figure 13.3), positive EVI anomalies cover 26,000 km^2 (~30% of the total area) with 36% distributed over the irrigated oases of the region (Table 13.2; Figure 13.1). High positive anomalies represent almost 24% of the total positive anomalies mapped on natural ecosystems/rangelands, but almost 95% of the total area at irrigated oases, which proves the important role that irrigation has on agricultural development in the region.

13.3.2 EVI Dynamics along a Groundwater Dependence Gradient at the Telteca Site

Almost synchronous intra-annual (Figure 13.4) and interannual (Figure 13.5) trajectories of EVI were found in *Prosopis* woodlands located at the Telteca natural reserve, with annual and growing season values higher than those observed at the control sites (sites with no positive greenness anomalies) as EVI anomalies increased (Figure 13.6a). Trends in productivity metrics were also confirmed for seasonality values with higher EVI_{max} and EVI_{min} values, but lower seasonality variation (EVI_{nrange}), as EVI anomalies increased (Figure 13.6b). No significant trends were found, however, for phenological metrics, that is, growing season length (L_{gs}) and time at which maximum

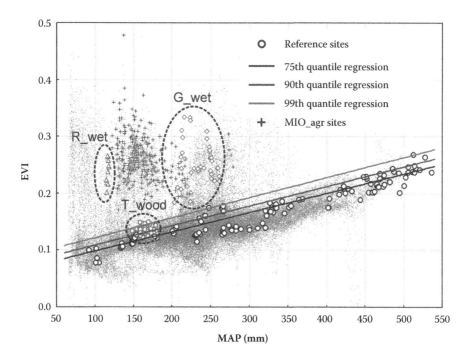

FIGURE 13.3 (See color insert.)

Mean Annual Precipitation–Enhanced Vegetation Index (MAP-EVI) quantile regression functions for the study region. Sample of pixels selected at each control area (brown symbols) are embraced by dashed lines. Functions were computed from MAP-EVI values measured at 125 reference sites (black-white circles). EVI thresholds corresponding to 75th, 90th, and 99th quantile regressions are used to classify systems into their low, moderate, and high reliance on water inputs besides local precipitation, respectively.

TABLE 13.2

Negative and Positive Enhanced Vegetation Index (EVI) Anomalies

Type of Land Cover	Negative Anomaly	Positive Anomaly				Total
		Low	Moderate	High	Total	
Natural ecosystems or rangelands	61.526	6.416	6.222	3.846	16.483	78.009
Irrigated oases	265	139	334	8.796	9.269	9.534
Total	61.791	6.555	6.556	12.642	25.752	87.543

Note: Total area (in km²) with negative and positive EVI anomalies in the study region. Positive and negative anomalies were computed from EVI_{map} values estimated from 75th quantile regression function. Positive anomalies, which represent portions of landscape with any reliance degree on water inputs besides precipitation, have been divided into low, moderate, and high levels of reliance if EVI is higher than 75th, 90th, and 99th quantile threshold values, respectively.

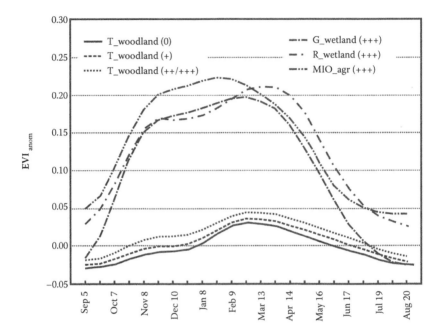

FIGURE 13.4
Mean seasonal Enhanced Vegetation Index (EVI) anomaly trajectories observed for representative sites. Trajectories were extracted for groups of sample of pixels with different levels of reliance on water inputs besides local precipitation (0: no reliance; +: low reliance; ++: moderate reliance; +++: high reliance); T_woodland = open *Prosopis* woodlands at Telteca; G_wetland = Guanacache wetland; R_wetland = Rosario wetland; MIO_agr = Mendoza irrigated oasis (agroecosystem).

EVI is reached (T_{max}), although average values for both were higher than control sites as EVI anomalies increased (Figure 13.6c). Because patterns and trends predicted by our conceptual model were matched at the Telteca woodlands, it seems that greenness anomaly may be a good surrogate for the reliance that woodland ecosystems have on groundwater resources. In addition to the metrics and trends recorded, a higher increase rate in the greenness trajectory was also observed during the late spring period, from November to December, in sites with moderately high EVI anomalies than in the control sites (Figure 13.4). This "early upraise" makes EVI differences among sites the highest during this seasonal period when energy constraints (low temperatures) start to disappear for phreatophytic woodlands, but rainfall inputs are still low for promoting vegetation growth in their nonphreatophytic counterparts.

Significant differences in EVI annual values were found throughout the entire study period between control sites and sites with moderately high EVI anomalies (Figure 13.5). Although weak absolute differences in EVI were observed, those differences were not significant between sites with

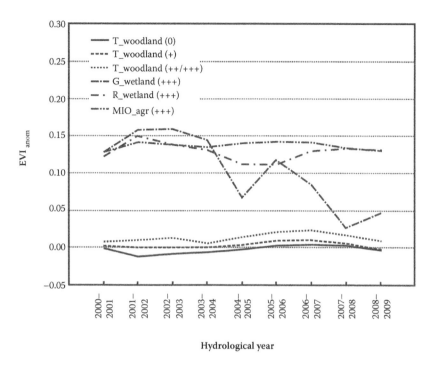

FIGURE 13.5
Interannual Enhanced Vegetation Index (EVI) anomaly trajectories at systems with different levels of reliance on water inputs besides local precipitation (0: no reliance; +: low reliance; ++: moderate reliance; +++: high reliance). T_woodland = open *Prosopis* woodlands at Telteca; G_wetland = Guanacache wetland; R_wetland = Rosario wetland; MIO_agr = Mendoza irrigated oasis (agroecosystem).

low positive EVI anomalies and control sites. As EVI anomalies increased, interannual variability (measured as the coefficient of variation among annual values of each hydrological year) decreased for growing season EVI (EVI_{gs}), minimum EVI (EVI_{min}), and the intra-annual variability (EVI_{nrange}) (Table 13.3). The remaining EVI metrics did not show any clear trends, although values for sites with moderately high EVI anomalies were slightly lower than for control sites (Table 13.3).

13.3.3 Intercomparison among Phreatophytic Woodlands, Wetlands, and Irrigated Crops

Wetlands and irrigated sites at the Mendoza oasis showed EVI trajectories clearly different from those of open woodlands of *Prosopis* (Figure 13.4). Annual productivity (EVI_{ma}) at the wetlands of Guanacache and Rosario and at irrigated sites was 8.7, 9.4, and 12.4 times higher than those measured at the Telteca woodlands, respectively (Figure 13.6a). According to

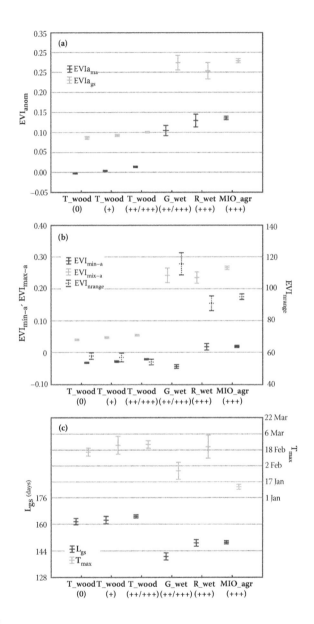

FIGURE 13.6
Whisker plots showing average and confidence intervals (at 95% level) for (a) productivity
metrics (EVI$_{ma}$ and EVI$_{gs}$), (b) seasonality metrics (EVI$_{min}$, EVI$_{max}$, EVI$_{nrange}$), and (c) phenological metrics (L$_{gs}$, T$_{max}$). Metrics were computed at *Prosopis* woodlands at Telteca (T_wood),
Guanacache and Rosario wetlands (G_wet, R_wet), and the Mendoza irrigated oasis (MIO_agr).
Samples at each system were selected for different levels of reliance on water inputs besides
local precipitation (0: no reliance; +: low reliance; ++/+++: moderate and high reliance). For
comparison purposes, all average metrics were computed in terms of their corresponding EVI
anomalies.

TABLE 13.3

Coefficients of Variation for Greenness Metrics

Sites	Productivity		Seasonality			Phenology	
	EVI_{ma}	EVI_{gs}	EVI_{max}	EVI_{min}	EVI_{nrange}	L_{gs}	T_{max}
T_wood	6.02	0.89	11.88	6.36	25.17	5.96	18.42
(0)	(0.78)	(0.21)	(2.01)	(0.85)	(3.58)	(1.39)	(5.05)
T_wood	5.61	0.69	10.67	5.86	22.56	5.05	15.67
(+)	(0.79)	(0.31)	(1.85)	(1.29)	(3.62)	(1.08)	(4.30)
T_wood	6.00	0.50	10.72	5.55	22.12	5.34	18.97
(++/+++)	(1.23)	(0.25)	(3.52)	(1.40)	(5.52)	(1.08)	(5.50)
G_wet	26.52	1.36	23.31	30.49	20.43	9.46	30.65
(+++)	(15.53)	(1.25)	(12.90)	(19.64)	(12.59)	(4.40)	(12.15)
R_wet	9.62	0.51	11.75	8.20	14.49	5.80	22.65
(+++)	(3.55)	(0.41)	(5.68)	(2.12)	(5.53)	(1.44)	(7.57)
MIO_agr	7.54	1.08	9.85	12.61	16.13	7.56	26.45
(+++)	(5.31)	(0.72)	(4.79)	(6.15)	(6.51)	(2.84)	(10.47)

Note: Values were computed from the annual metrics computed from September 2000 to August 2009 (nine hydrological years). Standard deviations of the coefficients of variation (spatial variability observed at each ecosystem type) are shown between parentheses.

their greenness anomalies, mean annual evapotranspiration rates reported for open *Prosopis* woodlands reached approximately 185 mm y^{-1}, of which approximately 25 mm y^{-1} are estimated to be supplied by shallow groundwater reserves (Contreras et al. 2011). These results agree with independent estimates computed from independent isotopic and hydrochemical evidences (Jobbágy et al. 2011). Supplementary water consumption of wetlands is even higher than in open woodlands, with rates that can reach up to 450–500 mm y^{-1} in addition to rainfall inputs. No accurate data exist on the relative contribution of groundwater supplies to the average productivity of the Guanacache and Rosario wetlands, but the observation of different average seasonal EVI trajectories suggests two patterns of ecological functioning: vegetation at the Guanacache wetland is characterized by higher intra-annual variability (EVI_{nrange}), lower minimum EVI values (EVI_{min}; Figure 13.6b), and a shorter growing season (L_{gs}; Figure 13.6c) than at the Rosario wetlands. Although no significant differences in maximum greenness were found between both wetland systems (Figure 13.6b), the time at which they were reached was approximately 16 days earlier at the Guanacache system than at the Rosario system (Figure 13.6c). The lower interannual variability found for all EVI metrics at the Rosario wetlands compared to the Guanacache wetlands would suggest that ecological functioning of the Rosario system relies more on groundwater resources than the Guanacache system. This fact is confirmed by the interannual greenness dynamics at the Guanacache system (Figure 13.5), where abrupt rises and falls in the mean annual EVI values suggest a higher dependence on the water discharges supplied by the San Juan River and, consequently, by the water abstractions accounted for irrigation at the upstream San Juan oasis.

Greenness seasonal dynamics observed at the Mendoza irrigation oasis are characterized by its lack of coupling with the rest of the ecosystem types. Irrigated oases showed similar values for productivity metrics (Figure 13.6a) and seasonality (Figure 13.6b) compared to those observed in both wetland systems. However, average intra-annual trajectory of EVI (Figure 13.4) in the Mendoza oasis highlights an earlier maximum vegetation activity and a higher activity during the growing season. The existence of a phase difference between the seasonal EVI trajectories of the irrigated oasis and the natural wetland systems would suggest a competitive process for water resources. This fact was stressed earlier between the San Juan oasis and the Guanacache wetland but would be equally expected for the Mendoza oasis and the Rosario system. In the Guanacache-San Juan case, consequences of water abstraction on the wetland productivity are more clearly depicted because of limited reliance of wetlands on groundwater resources. In the Rosario-Mendoza case, where the Rosario wetlands seem to rely more on groundwater resources, it is expected that consequences of agricultural development on the wetland productivity are less evident during wet or average-rainfall hydrological years than during dry years. A more detailed study identifying those differences during the driest periods would help to identify GDEs and the effects that irrigation development could have on their ecological functioning and the services they provide.

13.4 Conclusions

GDEs offer an outstanding example of the dependence of human well-being on ecosystem services. In this study, we demonstrated the usefulness of the annual greenness anomaly concept (Contreras et al. 2011) to identify landscape systems where vegetation activity depends on abnormally high inputs of water apart from precipitation: that is, riparian ecosystems, wetlands (see Chapter 17), phreatophytic woodlands, and irrigated oases. Particularly in low-precipitation regions, provisioning services (such as biomass or water availability), regulating services (such as the maintenance of lifecycles, habitats, and gene pools, or the local climate regulation), and cultural services (intellectual, spiritual, or recreational interactions with distinctive landscapes) are locally concentrated in these ecosystems, which are tightly coupled to the groundwater dynamics. Annual greenness anomaly estimated from satellite data was demonstrated to be a simple yet robust measurement for mapping those inflow-dependent systems at vast regions with limited field data and for providing a first estimate of their water requirements. Additional information on the reliance of those ecosystems on groundwater can be obtained from complementary analysis of their EVI intra-annual and interannual trajectories and from the extraction of metrics related to productivity, seasonality, and phenology of carbon gains in those ecosystems (see Chapter 9). In this study, we showed how the average greenness during the growing season, the annual minimum greenness, and the intra-annual variability (normalized range) were higher in phreatophytic woodlands and wetlands than in their nonphreatophytic counterparts. Hence, we suggest the use of these metrics to quantify and map the ecosystem's reliance on groundwater resources and the degree of dependence of ecosystem services on the groundwater dynamics.

Satellite-based approaches based on the spatial analysis of vegetation greenness anomalies and the tracking of their phenologies during time periods explicitly selected to cover dry rainfall conditions provide an initial characterization of natural ecosystems that show any reliance on groundwater resources. Both methods are extremely useful as a first step in building conceptual models on the functioning of GDEs, to quantify their water requirements, and to evaluate the ecosystem services trade-offs that can emerge between their conservation and agricultural development options.

Acknowledgments

S. Contreras acknowledges the support given by a Juan de la Cierva postdoctoral fellowship (JCI-2009-04927) funded by the Spanish Ministry of Science and Innovation. Pablo E. Villagra, Erica Cesca, and three anonymous reviewers are acknowledged for their help in supplying technical data and for their insightful comments.

References

Abraham, E., H. F. del Valle, F. Roig, et al. 2009. Overview of the geography of the Monte Desert biome (Argentina). *Journal of Arid Environments* 73:144–153.

Alcaraz-Segura, D., J. M. Paruelo, and J. Cabello. 2006. Identification of current ecosystem functional types in the Iberian Peninsula. *Global Ecology and Biogeography* 15:200–212.

Aranibar, J. N., P. E. Villagra, M. L. Gómez, et al. 2011. Nitrate dynamics in the soil and unconfined aquifer in arid groundwater coupled ecosystems of the Monte desert, Argentina. *Journal of Geophysical Research* 116:G04015.

Barron, O., I. Emelyanova, T. G. Van Niel, D. Pollock, and G. Hodgson. 2012a. Mapping groundwater-dependent ecosystems using remote sensing measures of vegetation and moisture dynamics. *Hydrological Processes*. doi:10.1002/hyp.9609.

Barron, O., R. Silberstein, R. Ali, et al. 2012b. Climate change effects on water-dependent ecosystems in south-western Australia. *Journal of Hydrology* 434–435:95–109.

Bergkamp, G., and C. Katharine. 2006. Groundwater and ecosystem services: Towards their sustainable use. *Proceedings of the International Symposium on Groundwater Sustainability.* Alicante, Spain: Instituto Geológico y Minero de España, 24–27 January, 177–193.

Bertrand, G., N. Goldscheider, J. M. Gobat, and D. Hunkeler. 2012. Review: From multi-scale conceptualization to a classification system for inland groundwater-dependent ecosystems. *Hydrogeology Journal* 20:5–25.

Boer, M. M., and J. Puigdefábregas. 2003. Predicting potential vegetation index values as a reference for the assessment and monitoring of dryland condition. *International Journal of Remote Sensing* 24:1135–1141.

Bradley, B. A., and J. F. Mustard. 2008. Comparison of phenology trends by land cover class: A case study in the Great Basin, USA. *Global Change Biology* 14:334–346.

Brauman, K. A., G. C. Daily, T. K. Duarte, and H. A. Mooney. 2007. The nature and value of ecosystem services: An overview highlighting hydrologic services. *Annual Review of Environment and Resources* 32:67–98.

Chen, J. S., L. Li, J. Y. Wang, et al. 2004. Groundwater maintains dune landscape. *Nature* 432:459–460.

Contreras, S., M. M. Boer, F. J. Alcalá, et al. 2008. An ecohydrological modelling approach for assessing long-term recharge rates in semiarid karstic landscapes. *Journal of Hydrology* 351:42–57.

Contreras, S., E. G. Jobbágy, P. E. Villagra, M. D. Nosetto, and J. Puigdefábregas. 2011. Remote sensing estimates of supplementary water consumption by arid ecosystems of central Argentina. *Journal of Hydrology* 397:10–22.

de Groot, R. S., M. A. Wilson, and R. M. J. Boumans. 2002. A typology for the classification, description and valuation of ecosystem functions, goods and services. *Ecological Economics* 41:393–408.

Devitt, D. A., L. F. Fenstermaker, M. H. Young, B. Conrad, M. Baghzouz, and B. M. Bird. 2011. Evapotranspiration of mixed shrub communities in phreatophytic zones of the Great Basin region of Nevada (USA). *Ecohydrology* 4:807–822.

Do, F. C., A. Rocheteau, A. L. Diagne, V. Goudiaby, A. Granier, and J. P. Homme. 2008. Stable annual pattern of water use by *Acacia tortilis* in Sahelian Africa. *Tree Physiology* 28:95–104.

Eamus, D. 2009. *Identifying groundwater dependent ecosystems: A guide for land and water managers.* Sydney: Land & Water Australia.

Eamus, D., R. Froend, R. Loomes, G. Hose, and B. R. Murray. 2006. A functional methodology for determining the groundwater regime needed to maintain the health of groundwater-dependent vegetation. *Australian Journal of Botany* 54:97–114.

Eamus, D., C. M. O. Macinnis-Ng, G. C. Hose, M. J. B. Zeppel, D. T. Taylor, and B. R. Murray. 2005. Ecosystem services: An ecophysiological examination. *Australian Journal of Botany* 53:1–19.

Eamus, D., A. P. O'Grady, and L. Hutley. 2000. Dry season conditions determine wet season water use in the wet-dry tropical savannas of northern Australia. *Tree Physiology* 20:1219–1226.

Ellery, W. N., and T. S. McCarthy. 1998. Environmental change over two decades since dredging and excavation of the lower Boro River, Okavango Delta, Botswana. *Journal of Biogeography* 25:361–378.

Ellis, T. W., and T. J. Hatton. 2008. Relating leaf area index of natural eucalypt vegetation to climate variables in southern Australia. *Agricultural Water Management* 95:743–747.

FAO (Food and Agriculture Organization). 2006. *ClimWat 2.0 for CropWat*. Rome: FAO, United Nations.

Fernández, N., J. M. Paruelo, and M. Delibes. 2010. Ecosystem functioning of protected and altered Mediterranean environments: A remote sensing classification in Doñana, Spain. *Remote Sensing of Environment* 114:211–220.

Groom, P. K., R. H. Froend, and E. M. Mattiske. 2000. Impact of groundwater abstraction on a *Banksia* woodland, Swan Coastal Plain, Western Australia. *Ecological Management and Restoration* 1:117–124.

Guerschman, J. P., A. I. J. M. Van Dijk, G. Mattersdorf, et al. 2009. Scaling of potential evapotranspiration with MODIS data reproduces flux observations and catchment water balance observations across Australia. *Journal of Hydrology* 369:107–119.

Hatton, T. J., and R. Evans. 1997. *Dependence of ecosystems on groundwater and its significance to Australia.* Canberra: Land and Water Resources Research and Development Corporation.

Huete, A. K. Didan, T. Miura, E. P. Rodriguez, X. Gao, and L. G. Ferreira. 2002. Overview of the radiometric and biophysical performance of the MODIS vegetation indices. *Remote Sensing of Environment* 83:195–213.

Jobbágy, E. G., M. D. Nosetto, P. E. Villagra, and R. B. Jackson. 2011. Water subsidies from mountains to deserts: Their role in sustaining groundwater-fed oases in a sandy landscape. *Ecological Applications* 21:678–694.

Jobbágy, E. G., O. E. Sala, and J. M. Paruelo. 2002. Patterns and controls of primary production in the Patagonian steppe: A remote sensing approach. *Ecology* 83:307–319.

Jönsson, P., and L. Eklundh. 2004. TIMESAT—A program for analyzing time-series of satellite sensor data. *Computers and Geosciences* 30:833–845.

Llamas, M. R. 1988. Conflicts between wetland conservation and groundwater exploitation: Two case histories in Spain. *Environmental Geology and Water Sciences* 11:241–251.

MacKay, H. 2006. Protection and management of groundwater-dependent ecosystems: Emerging challenges and potential approaches for policy and management. *Australian Journal of Botany* 54:231–237.

Morisette, J. T., A. D. Richardson, A. K. Knapp, et al. 2009. Tracking the rhythm of the seasons in the face of global change: Phenological research in the 21st century. *Frontiers in Ecology and the Environment* 7:253–260.

Muñoz-Reinoso, J. C., and F. García-Novo. 2005. Multiscale control of vegetation patterns: The case of Doñana (SW Spain). *Landscape Ecology* 20:51–61.

Murray, B. R., G. C. Hose, D. Eamus, and D. Licari. 2006. Valuation of groundwater-dependent ecosystems: A functional methodology incorporating ecosystem services. *Australian Journal of Botany* 54:221–229.

Murray, B. R., M. J. B. Zeppel, G. C. Hose, and D. Eamus. 2003. Groundwater-dependent ecosystems in Australia: It's more than just water for rivers. *Ecological Management and Restoration* 4:110–113.

Murray-Hudson, M., P. Wolski, and S. Ringrose. 2006. Scenarios of the impact of local upstream changes in climate and water use on hydro-ecology in the Okavango Delta, Botswana. *Journal of Hydrology* 311:73–84.

Nagler, P. L., J. Cleverly, E. P. Glenn, D. Lampkin, A. R. Huete, and Z. Wan. 2005. Predicting riparian evapotranspiration from MODIS vegetation indices and meteorological data. *Remote Sensing of Environment* 94:17–30.

New, M., D. Lister, M. Hulme, and I. Makin. 2002. A high-resolution data set of surface climate over global land areas. *Climate Research* 21:1–25.

O'Grady, A. P., J. L. Carter, and J. Bruce. 2011. Can we predict groundwater discharge from terrestrial ecosystems using existing eco-hydrological concepts? *Hydrology and Earth System Sciences* 15:3731–3739.

Palmer, A. R., S. Fuentes, D. Taylor, et al. 2010. Towards a spatial understanding of water use of several land-cover classes: An examination of relationships amongst pre-drawn leaf water potential, vegetation water use, aridity and MODIS LAI. *Ecohydrology* 3:1–10.

Paruelo, J. M., H. E. Epstein, W. K. Lauenroth, and I. C. Burke. 1997. ANPP estimates from NDVI for the central grassland region of the United States. *Ecology* 78:953–958.

Paruelo, J. M., E. G. Jobbágy, and O. E. Sala. 2001. Current distribution of ecosystem functional types in temperate South America. *Ecosystems* 4:683–698.

Patten, D. T., L. Rouse, and J. C. Stromberg. 2008. Isolated spring wetlands in the Great Basin and Mojave Deserts, USA: Potential response of vegetation to groundwater withdrawal. *Environmental Management* 41:398–413.

Richardson, S., E. Irvine, R. Froend, P. Boon, S. Barber, and B. Bonneville. 2011. *Australian groundwater-dependent ecosystem toolbox. Part 1: Assessment framework.* Canberra: National Water Commission.

Ridolfi, L., P. Odorico, and F. Laio. 2007. Vegetation dynamics induced by phreatophyte-aquifer interactions. *Journal of Theoretical Biology* 248:301–310.

Running, S. W., and R. R. Nemani. 1988. Relating seasonal patterns of the AVHRR vegetation index to simulated photosynthesis and transpiration of forests in different climates. *Remote Sensing of Environment* 24:347–367.

Solano, R., K. Didan, A. Jacobson, and A. Huete. 2010. *MODIS vegetation indices (MOD13) C5—User's guide.* Tucson, AZ: The University of Arizona.

Specht, R. L. 1972. Water use by perennial evergreen plant communities in Australia and Papua New Guinea. *Australian Journal of Botany* 20:273–299.

Specht, R. L., and A. Specht. 1989. Canopy structure in eucalyptus-dominated communities in Australia along climatic gradients. *Oecologia Plantarum* 10:191–202.

Stromberg, J. C., R. Tiller, and B. Richter. 1996. Effects of groundwater decline on riparian vegetation of semiarid regions: The San Pedro, Arizona. *Ecological Applications* 6:113–131.

Villagra, P. E., G. E. Defossé, H. F. del Valle, et al. 2009. Land use and disturbance effects on the dynamics of natural ecosystems of the Monte Desert: Implication for their management. *Journal of Arid Environments* 73:202–211.

Villagra, P. E., R. Villalba, and J. A. Boninsegna. 2005. Structure and dynamics of *P. flexuosa* woodlands in two contrasting environments of the central Monte Desert. *Journal of Arid Environment* 60:187–199.

14

Surface Soil Moisture Monitoring by Remote Sensing: Applications to Ecosystem Processes and Scale Effects

M. J. Polo
University of Córdoba, Spain

M. P. González-Dugo
Andalusian Institute for Agricultural and Fisheries Research and Training (IFAPA), Spain

C. Aguilar
University of Córdoba, Spain

A. Andreu
Andalusian Institute for Agricultural and Fisheries Research and Training (IFAPA), Spain

CONTENTS

14.1 Introduction 304
14.2 Soil Moisture Monitoring by Remote Sensing Sources 305
14.3 Assimilation of Remote Sensing Data into Hydrological
      Modeling 311
14.4 Estimating Evapotranspiration as an Indirect Valuation for Soil
      Moisture Stress on Vegetation 312
14.5 Final Remarks/Conclusions 317
Acknowledgments 319
References 319

14.1 Introduction

Surface soil moisture monitoring plays an important role not only for quantifying water content variability and calibrating/validating hydrological models but also in determining limit states for ecosystems and providing information to value different ecosystem services. The soil moisture regime is the result of the energy and water budgets, coupled with evaporative fluxes from both soil and vegetation. Moisture gradients determine the direction and intensity of water fluxes through the soil (e.g., Brutsaert 1982, 2005; Jury and Horton 2004) and, among others, on the medium- and long-term distribution of vegetation, dominant species, and rooting depth. The latter affects the local regime of wetness through the transpiration process, the shadowing of the soil surface, soil structure and composition modifications, the increase in the surface effective roughness, and a long list of aspects involved in the different terms of the energy and water balance. Nutrient, and also pollutant, adsorption/desorption and movement through the soil are also greatly influenced by soil moisture and vegetation (Sposito 1989), for both advective and diffusive transport conditions (Jury and Horton 2004). Thus, not only provisioning services, such as water–nutrient–pollutant availability for plants and animals, are influenced by the changes in soil moisture, but also regulatory and maintenance services are affected (runoff generation by rainfall, flood risk, soil maintenance, surface and groundwater quality, etc.). Cultural services of ecosystems are also involved, provided the impact of vegetation and wetlands on landscape, and their effects in man's physical and intellectual activities in relation to ecosystems.

The fact that soil is a porous medium adds complexity to the characterization of the spatiotemporal evolution of the water that is in it. Soil moisture range in a given soil mainly depends on porosity, which is related to soil structure, texture, and content of certain components (such as organic matter, carbonates, etc.). The same water content in absolute values can be found in soils with a different porosity and, thus, under different flow conditions. Under saturated conditions, soil water can be considered as a continuum that completely fills the pore volume in soil, and hydrodynamic laws can be applied under this assumption. For unsaturated conditions, though, liquid water and water vapor coexist with air in the pore volume; thus, multiphase transport occurs. Because this classification of saturated/unsaturated flow is a key condition for modeling and quantifying water movement and exchange through the soil, soil moisture is mainly expressed in relative terms (referring to the moisture range that can be found in a given soil) to point out the proximity to saturated (maximum soil moisture) conditions. Water and energy balance equations, as well as the transport equations for nutrients, salts, and pollutants, are thus applied along the soil profile by using this relative moisture as a scale factor, which allows comparison among different types of soil and states.

Heterogeneity is an intrinsic feature of any soil. Uniformity of any soil property is a desirable condition for simplifying soil process modeling, and soil moisture is not an exception. Sampling soil moisture at field conditions requires an adequate experimental design if nonsignificance of results is to be avoided, which is not usually feasible at large scales. However, soil moisture is one of the main state variables in hydrological models, and it usually poses a constraint for the direct calibration of results from field measurements in medium- to large-scale studies.

Our increasing capacity to observe the Earth from space triggered our ability to apply and calibrate physical, chemical, and environmental models throughout large areas. Surface soil moisture, similarly to many other surface features, can be traced locally through the analysis of the terrain surface reflectance spectrum during a given period. Remote sensing provides distributed reflectance in huge areas with different frequencies/spatial resolutions depending on the sensor. Different sources are available for soil moisture monitoring; passive and active microwave being among those most widely used, with emerging alternatives such as Soil Moisture and Ocean Salinity (SMOS; e.g., Dente et al. 2012) and future Soil Moisture Active and Passive (SMAP) instruments for large-scale analyses. However, contrasting them with properly designed field measurements is needed to calibrate the spectral signals and assimilate these data. The scale effects arising from the different spatiotemporal resolutions of the remote sensing sources and the model equations (e.g., Su et al. 2010, 2011) must be specially addressed, due to the large nonlinearity of soil processes behavior.

The result of this data processing is of great interest for correlating multivariate sets of environmental variables, related to vegetation, water stress, soil temperature, and so on. Moreover, the final time series of soil moisture maps for a given area are highly valuable for hydrological model calibration at watershed and regional scales, for which uniform and frequent field measurements are restricted by time and cost. Further work on the scale effects due to the modeling "environment" definition of the soil, that is, the effective depth in soil where moisture is calculated as uniform value, is required to match the results of the hydrological simulation to the data obtained from remote sensing because this provides us with surface information, whereas models deal with an *a priori* defined surface layer consisting of some centimeters (10–100 cm).

14.2 Soil Moisture Monitoring by Remote Sensing Sources

The spatial heterogeneity of land cover, topography, and soil properties, as influenced by fluctuating atmospheric effects, results in high soil moisture variability in space and time. These variations need to be quantified and taken into account to properly address environmental applications, either directly or through the use of hydrological models. Traditionally, soil moisture has been measured using invasive ground-based methods providing high-quality profile data at a point; spatially distributed estimations for an area being generally obtained by interpolation using statistical methods. However, this approach is often insufficient to adequately characterize most hydrological processes in large areas, in which a further assessment of soil moisture spatial variations is required. In this sense, remote sensing represents a valuable data source for providing spatially distributed estimates of soil moisture rather than point measurements at high temporal frequency with uniform measurements (Engman and Gurney 1991; Laguardia and Niemeyer 2008).

Radiation reflected and emitted from the land surface in several regions of the electromagnetic spectrum can be used to monitor soil moisture. Over the last decades, most attempts have been based on the microwave spectral range, using either passive or active systems (Table 14.1). A thorough revision of fundamental theory and first applications of microwave remote sensing can be found in Ulaby et al. (1981) and Fung (1994). One of the most important reasons for using microwaves is their ability to penetrate clouds and, to some extent, rain. At the land surface, microwaves can penetrate more deeply

TABLE 14.1

Selected Characteristics of Representative Sensors and Recent Application Examples

Sensor	Band	Active/Passive	Spatial Resolution	Application Examples
AMSR-E	C-band (6.9 GHz), X-band (10.7 GHz), 18.7 GHz, 23.8 GHz, 36.5 GHz, and 89 GHz	P	25–50 km	Gruhier et al. (2010) Liu et al. (2011) Su et al. (2013)
TRMM-TMI	10.7, 19.4, 21.3, 37, and 85.5 GHz	P	6–50 km	Bindlish et al. (2003) Gao et al. (2003) Gruhier et al. (2010)
SMOS	L-band (1.4 GHz)	P	50 km	Kerr et al. (2001) Su et al. (2013)
RADARSAT	C-band (5.3 GHz)	A	10–100 m	Baghdadi et al. (2002) Zribi and Dechambre (2002)
ENVISAT-ASAR	C-band (5.3 GHz)	A	30 m	Zribi et al. (2005) Manninen et al. (2005)
JERS-1	L-band (1.27 GHz)	A	18 m	Narayanan and Hegde (1995) Kasischke et al. (2011)
ERS-1/ERS-2	C-band (5.3 GHz)	A	30 m	Wagner et al. (1999a, 1999b) Moran et al. (2001) Ceballos et al. (2005) Parajka et al. (2006)

into vegetation and soils than optical sensors can. The extent of penetration depends on the moisture content, vegetation density, and the wavelength of the microwaves, with longer wavelengths penetrating deeper than shorter ones. Other reasons for the use of microwaves are their independence from solar illumination, allowing day and night observations, and their ability to provide information different from, and thus complementary to, that coming from the visible and infrared regions (VNIR). At microwave frequencies, the geometric and bulk-dielectric properties of the surface studied determine its response, in contrast to the molecular resonance of the surface layers of vegetation or soils responsible for the reflections in VNIR regions. The large contrast between water and dry soil dielectric constants (80 and <5, respectively) causes an emissivity contrast in the microwave region providing a mechanism for the remote sensing of moisture content of soils. Since the 1970s, many studies have used this approach, some examples of which can be found in Schmugge and Jackson (1994), Wigneron et al. (1998, 2003), Schmugge et al. (2002), and Juglea et al. (2010). A general constraint is that the depth of microwave measurement is only 0.1–0.2 times the wavelength, and it decreases with moisture.

Passive microwaves present the lowest spatial resolutions, typically of the order of tens of kilometers, and are more suited to atmospheric and oceanic applications than to land studies. In addition, land targets generally have complicated dielectric and geometric properties. However, an important international effort has been made in recent years to develop advanced instruments with multiconfiguration capabilities (multifrequency, dual-polarization or polarimetric, multiangular observations) to improve the quantitative characterization of soil moisture at a regional and global scale. Among these instruments is the Advanced Microwave Scanning Radiometer on Earth Observing System (AMSR-E) on the AQUA satellite and the SMOS satellite of the European Space Agency (ESA) launched in 2009. This mission uses an L-band interferometer that has shown itself to be optimal for capturing soil moisture information from space (Kerr et al. 2001). In 2014, NASA plans to launch the SMAP satellite. It combines active and passive sensors to provide accurate soil moisture products with unprecedented resolution, sensitivity, area coverage, and revisit frequency.

Active microwave or radar systems can improve the spatial resolution of observation to tens of meters, particularly with the use of a virtual antenna that is artificially synthesized, the so-called Synthetic Aperture Radars (SAR). However, radar systems have specific constraints such as interference due to soil roughness and topography (Lievens et al. 2011). Some authors (Engman and Chauhan 1995) proposed the use of multitemporal information to monitor relative changes of soil moisture and minimize the effects of stationary variables over given time periods. Four main types of algorithms are suggested by Wigneron et al. (2003) to retrieve soil moisture from microwave remote sensing, taking into account various effects contributing to the surface emission at these frequencies including the amount of vegetation, soil temperature, snow cover (see Chapter 15), topography, and soil surface roughness. These algorithms are based on (1) the use of land cover classification maps, (2) of ancillary remote sensing indexes, and (3) two-parameter or (4) three-parameter retrievals (in this case, soil moisture, vegetation optical depth, and effective surface temperature are retrieved simultaneously from the microwave observations). Methods (3) and (4) are based on multiconfiguration observations, in terms of frequency, polarization, or view angle. The authors addressed the advantages and disadvantages of the different algorithms, highlighting key issues for soil moisture retrievals based on the new multiangular microwave signatures.

A further review of different active and passive microwave products can be found in Gruhier et al. (2010). Several ongoing efforts aim at producing temporal soil moisture series combining a variety of sensors, including both passive and active instruments (Fernández-Prieto et al. 2012; Liu et al. 2012).

Some authors (Carlson 2007; Crow et al. 2008; Sobrino et al. 2012) have addressed the estimation of soil moisture through indirect techniques using measurements in the optical and thermal infrared ranges. These approaches are based on the assimilation of remote sensing observations into soil–vegetation–atmosphere transfer (SVAT) models, and they rely on the basic idea that the surface radiometric temperature (TR, directly measured by the sensor) and the surface turbulent energy fluxes are sensitive to surface soil water variations. Indirect techniques can also be used to obtain qualitative classifications (Box 14.1) or assess soil wetness through other vegetation indexes (Box 14.2).

BOX 14.1   SOIL STATE CLASSIFICATION BY SPECTRAL MIXTURE ANALYSIS ON LANDSAT 5 TM DATA

The combined use of airborne and field multispectral data to identify the soil moisture state in a degraded saltmarsh area in Cadiz, southern Spain, allowed both the quantification of the restorative potential of the system (Polo et al. 2009) and the calibration of the flooding limits in a tidal creek area, where direct measurement and monitoring would have been impractical because of the low water depth values in the tidal plains. From a first classification of the features of significance for the study (wet soil, dry soil, green vegetation, dry vegetation, and shadow), a spectral mixture analysis (SMA) was performed to assess the relative importance of each one of these components within each pixel of the study area. SMA consists of a deconvolution technique that quantifies these individual fraction covers in each pixel, considering the pixel reflectance of each band in a spectrum as the linear combination of the contributions of its components (Roberts et al. 1998). Nine selected images from Landsat 5 thematic mapper (TM) covering different seasons during the period 2004–2007 were analyzed, and the SMA was calibrated with two different field campaigns performed in the creek network area in 2007, coinciding with the sensor data time. Measurements with a GER 3700 radiometer were obtained in 30 selected control points covering the previously identified potential flooding area from map and hydrodynamic analysis of the tidal cycle.

The results of the SMA, corrected for a reclassification into three significant components (dry soil, wet soil, and vegetation) led to the clear identification of changes in the tidal flood plain extension and the vegetation development, up to a 0.91 correlation value for the vegetation fraction and 0.71 for dry soil fraction. Moreover, the results at each date were adequately related to the different factors that can jointly influence soil moisture evolution in the surface, such as rainfall occurrence, diurnal tidal cycles, and seasonal tidal cycles, when combined with tidal and meteorological data. The estimation of the maximum limits of the flooding area from the tidal creek network was performed from a further analysis by a numerical one-dimensional hydrodynamic model, calibrated for the time of the image.

BOX 14.2   NATURAL VEGETATION STATE AS INDICATOR OF THE SOIL WATER DYNAMICS

The application of the Normalized Difference Vegetation Index (NDVI; Rouse et al. 1974) in ecological studies has enabled quantification and mapping of green vegetation. NDVI can be a useful tool for coupling climate and vegetation distribution and performance at large spatial and temporal scales (Pettorelli et al. 2005). Because vegetation vigor and productivity are related to hydrological variables (rainfall, evapotranspiration, etc.), NDVI serves as a surrogate measure of these factors at the landscape scale (Wang et al. 2003; Groeneveld and Baugh 2007).

Aguilar et al. (2012) quantified the responses to seasonal changes in precipitation and water availability through the groundwater lens by linking landscape-level variations in relative greenness of woody vegetation to past precipitation and hydrology in a barrier island in the eastern United States. Annual changes in NDVI calculated from Landsat TM data were coupled with changes in precipitation. Lower mean NDVI values (0.75 over 0.9 in Figure 14.1) were obtained in the dry years (2007 and 2008), and linear relationships between accumulated rainfall and mean water table depth with maximum and mean NDVI values as indicators of overall and mean productivity and biomass were observed.

Monthly variations revealed the different behavior of the woody community within the growing season according to hydrology; in the dry year (2007), water table depths fall and the histogram of NDVI classes gets narrower in the middle of the season. Linear fits with r^2 close to 1 were found between NDVI values and accumulated rainfall in the hydrological year. This shifted during dry years to rainfall accumulated in the previous month showing the quick response of vegetation to less availability of the water in the system from the recent past. Monthly relationships between NDVI and groundwater were seen only during dry years (e.g., 2007 in Figure 14.1) when rainfall was scarce and evapotranspiration was high and the freshwater lens became the major source of water for plants.

These results demonstrated the important feedback between woody vegetation responses to changes in the freshwater lens as shown in the Shao et al. (1995) model, using empirical data. Because of this dependence of NDVI on water availability, it is important to consider timing of remote sensing when applying indices to determine productivity.

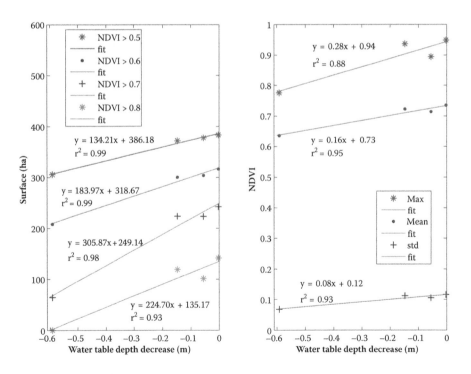

FIGURE 14.1
Linear adjustments among the surface with Normalized Difference Vegetation Index (NDVI) values above a threshold and NDVI statistics of the sample of pixels with NDVI > 0.5 with the monthly water table depth decrease (m/month) in 2007.

14.3 Assimilation of Remote Sensing Data into Hydrological Modeling

The first hydrological models developed in history were analytical ones. These models solve the fundamental physical equations that describe the water flow and generation and transport of sediments and associated substances. Theoretically, the parameters involved in these models are measurable. However, the numerous parameters involved together with the heterogeneity of the environment's characteristics make it necessary to calibrate the results with observed data. Errors in the measurement process and differences between the scale of application of the model's algorithms and scale of measurements add uncertainty to the results (Mertens 2003).

The development of Geographic Information Systems (GIS) and the increase in processing speed and storage capacity of computers, together with the opportunity to acquire spatially distributed data through remote sensing techniques, have led to the application of semidistributed models. These models are applied on each spatial unit into which the system is divided, and they route the response of each unit toward the outlet, combining these responses in time and in space to produce watershed scale outputs (Beven 1989, 2000). In this way, analytical models can be applied in highly extensive systems, but the rigorousness and accuracy of the results will depend on the quality of the spatially distributed inputs available to the model and the suitability of the global and/or analytical model for the spatial and temporal scale of the research (Gupta et al. 1986; Blöschl and Sivapalan 1995). The derivation of the expressions that describe individual processes in physically based distributed hydrological models is subject to several assumptions that simplify the process so that it can be mathematically expressed. In general, the equations that govern the processes in these models were derived at a small scale in homogeneous systems under controlled conditions, but in practice these equations are used at broader scales: cell, subwatershed, and so on and under different physical conditions. However, these equations were theoretically derived for their use with spatially and temporally continuous data although, in practice, point values representing a cell or even a region in the watershed are used. The viability of joining the small scale of physical processes to the cell scale used in most physical models is questionable according to some authors because the parameters derived in this way only represent adjusted coefficients that distort any physical meaning (Beven 1989). Therefore, it is necessary to consider the implications of certain initial assumptions about the nature of the system to be modeled. Thus, the potential offered by a suitable combination of remote sensing data, models, and *in situ* data augurs a major opportunity for the future (Fernández-Prieto et al. 2012).

Data assimilation allows integrating remotely sensed products into hydrological models as a means of merging uncertain observations with uncertain model outputs to improve estimation accuracy. Therefore, data assimilation is used not only to update the hydrological model states that optimally combine model outputs with observations but also to quantify observational and hydrological model errors (Moradkhani 2008). Remotely sensed data can be assimilated as input data into the models (e.g., rainfall, cloudiness, albedo, antecedent soil moisture conditions, etc.), as state variables (e.g., soil moisture or snow water equivalent), or for the calibration of the outputs of the model. Hydrological models at a watershed scale require the calibration of the parameters involved in the computation of hydrological processes. Because soil moisture data are often missing, the calibration process is usually carried out with water flow data recorded at certain control stations in the watershed. However, soil surface moisture estimates can be derived from remote sensing data, and so the assimilation of soil moisture information within hydrological models can make a great improvement in evaporation and runoff predictions (Brocca et al. 2010).

The explanation of the few improvements obtained so far by assimilating soil moisture observations could be related to several main factors. First, there is a spatial mismatch as the measurement extension is low for *in situ* data and coarse for satellite sensors when compared with model estimates. Second, the relevant layer depth differs because remote sensors are only able to investigate a thin surface layer (2–5 cm) that does not match the soil depth (1–2 m) usually simulated within rainfall runoff models. Finally, soil moisture data availability is limited, being determined by the recent creation of *in situ* soil moisture networks and satellite data with the temporal resolution required for hydrological applications (Vereecken et al. 2008; Brocca et al. 2010).

14.4 Estimating Evapotranspiration as an Indirect Valuation for Soil Moisture Stress on Vegetation

The predominant forcing agents for soil moisture dynamics are the evaporation and transpiration rates from the soil to the atmosphere. In practice, it is a complex matter to separate both fluxes in vegetated areas, unless scientific field research is conducted on small and highly monitored plots. For this reason, evapotranspiration (ET) is defined as the combined loss of water from both sources. Whenever soil moisture is lower than its maximum value (i.e., saturated conditions), the vegetation and also (although to a lesser extent) the soil matrix exert a limiting control on the actual rates of ET, if these are compared to the potential (maximum) ET rate, which corresponds to an unlimited water supply. The actual ET distribution in relation to potential ET values is thus usually analyzed in terms of estimating soil moisture state and water stress on vegetation. The state of natural and crop vegetated areas is traced by means of estimating ET evolution at different time scales for short-term management of the vegetation, and for medium- to long-term planning of soil uses and water resource management. In agricultural areas, this assessment is crucial.

ET is a key variable in hydrological processes, and multiple methods have been developed for its assessment. The integration of remotely sensed data into ET models has broadened the area of application of these models from point to basin and regional scales (see Chapter 18). However, significant scale effects arise from using field datasets together with these distributed data.

Spatially distributed remotely sensed data have been used at regional and continental scales for estimating ET and land surface moisture, predicting water demand and monitoring drought and climate change (Bastiaanssen et al. 2002; Chandrapala and Wimalasuriya 2003; Anderson et al. 2007), for estimating water consumption over irrigated areas and planning the irrigation schedule (Garatuza-Payan and Watts 2005; Rossi et al. 2010), and for analyzing irrigation and productivity performance indicators (Bastiaanssen et al. 1999; Akbari et al. 2007; Hamid et al. 2011). Moisture stress is often quantified in models as the reduction that the actual ET exhibits from its potential value (PET; Moran 2003).

Two different approaches for monitoring ET by using remotely sensed data have been applied with success in agricultural water use studies. In the first approach, an energy balance is applied over the surface by using the so-called radiometric surface temperature (T_R), derived from thermal data (8–14 μm), to calculate the sensible heat flux, and obtaining the latent heat flux as the residual of the balance (e.g., Moran et al. 1994; Kustas and Norman 1996; Gillies et al. 1997; Bastiaanssen et al. 1998). This approach must deal with the difference between T_R and the aerodynamic temperature (T_o), which is needed to compute sensible heat, particularly for surfaces partially covered with vegetation (Kustas 1990). Several schemes of varying levels of complexity and input requirements have been set up to solve this problem. Some of them employ empirical or semiempirical relationships to adjust T_R to T_o (e.g., Kustas et al. 1989; Lhomme et al. 1994; Chehbouni et al. 1996; Mahrt and Vickers 2004). When calibration is performed with field data, these methods provide accurate results (Chavez et al. 2005). Another scheme to avoid the determination of T_o consists of applying an internal calibration on the surface temperature (Bastiaanssen et al. 1998). This procedure also reduces the need for atmospheric correction of T_R, which is a complex process that could introduce additional errors. The methodology that best models the effects of a partial canopy cover is a two-source approach (Shuttelworth and Wallace 1985; Norman et al. 1995; Kustas and Norman 1999). In this method, the different surface fluxes are divided into soil and canopy components. Previous works (Timmermans et al. 2007; González-Dugo et al. 2009) show the advantages of these two-source energy-balance models when compared to one-source models. Furthermore, energy balance applications result in instantaneous flux estimated at the time of image acquisition; therefore, it is necessary to transform this estimation into daily values (Box 14.3).

The second approach uses vegetation indices (VI), derived from surface reflectance data, to estimate crop coefficients (Kc). This method relies on the generally close correspondence between crop growth rates and transpiration rates (Tasumi et al. 2005; Tasumi and Allen 2007; Singh and Irmak 2009). Kc relates the ET of a given crop to that of a reference vegetated surface, calculated from ground meteorological data. This index-based approach directly obtains daily values, and most high-frequency satellites provide optical input data detailed enough to feed the method (Hunsaker et al. 2005a, 2005b; Duchemin et al. 2006; Er-Raki et al. 2007, 2010; González-Dugo and Mateos 2008; Campos et al. 2010; Sánchez et al. 2010; Padilla et al. 2011).

**BOX 14.3   MAPPING EVAPOTRANSPIRATION
USING THERMAL SATELLITE IMAGERY**

The study of Anderson et al. (2011) is an example of daily evapotranspiration (ET) mapping by using thermal satellite imagery. Multiscale ET maps were generated with a physically based inverse model of Atmosphere-Land Exchange (ALEXI; Anderson et al. 1997, 2007; Mecikalski et al. 1999), which coupled a two-source land-surface model with an atmospheric boundary layer model and an associated flux disaggregation technique (DisALEXI; Norman et al. 2003). The study describes how the resulting information is being used for drought monitoring purposes, agricultural water resources, and hydrological management over different locations at the United States, Europe, and Africa. The ALEXI model was applied over the Guadalquivir River Basin in Spain (Figure 14.2), the largest watershed in southern Spain (57,527 km^2). It supports extensive agricultural production (around 8000 km^2 of irrigated land) and suffers from a chronic water deficit. Daily remote sensing estimations can provide accurate information to water managers—at field, irrigation district, and basin scales—with different temporal patterns about the actual amount of water being used by different crop types to improve decisions concerning water management.

The planning and operational tool for monitoring crop water consumption in the irrigated lands of the Guadalquivir basin, MINARET (MonitorINg irrigated AgricultuRe ET), is given as an example of index-based estimation of ET (González-Dugo et al. 2012). Vegetation indices were obtained from a series of high-spatial-resolution satellite images for 2007, 2008, and 2009, and supported the assessment of crop type and daily to seasonal ET of individual fields, enabling the analysis of crop water use. To estimate actual ET, reference ET was calculated using the Penman–Monteith combination equation (Allen et al. 1998).

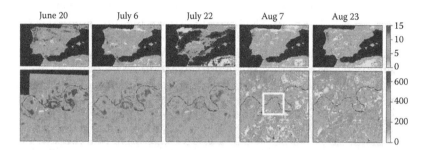

FIGURE 14.2 (See color insert.)
Maps of daytime ET (quantified as latent heat flux in MJ m^{-2} d^{-1}) from ALEXI over Spain (top row) and instantaneous ET (shortly before local noon; W m^{-2}) from DisALEXI over an irrigated agricultural area along the Guadalquivir River in southern Spain (bottom row) for five days during the 2009 growing season. White box on the 7 August DisALEXI map indicates the 3-km MSG pixel size. (From Anderson, M. C., et al., *Hydrology and Earth System Sciences*, 15, 223–239, 2011; and http://www.hydrol-earth-syst-sci.net/15/223/2011/doi:10.5194/hess-15-223-2011© Author(s) 2011. This work is distributed under the Creative Commons Attribution 3.0 License.)

To validate ET estimations, ET measurements are required (Allen et al. 2011). ET measuring systems include lysimeters, scintillometry, Bowen ratio, and eddy covariance methods. Lysimeters have been commonly used to provide information for the calibration and validation steps (Makkink 1957; Jensen 1974; Doorenbos and Pruitt 1977; Wright 1981, 1982; Allen et al. 1989; Jensen et al. 1990). The Bowen ratio energy balance (BREB) is a micrometeorological method (Bowen 1926) that solves the energy balance equation by measuring air temperature and vapor pressure gradients in the near-surface layer above the evaporating surface. A scintillometer is an optical device that measures small fluctuations of the air refractive index caused by temperature, humidity, and pressure-induced variations in density. Current scintillometers measure sensible heat flux, and to obtain ET, it is necessary to also measure the net radiation (Rn) and soil heat (G) fluxes (descriptions available in Meijninger and De Bruin 2000; Meijninger et al. 2002; Hartogensis et al. 2003; De Bruin 2008).

Eddy covariance systems measure sensible and latent heat fluxes, momentum flux, and CO$_2$ or other fluxes depending on their configuration. The method is based on the covariance between fluctuations in temperature and humidity, and upward and downward turbulent eddies. This requires the measurement of temperature, wind velocity, and humidity at high speed, with frequencies ranging between 5 and 20 Hz. The instrumentation is relatively fragile and expensive, requiring periodically maintenance, but the methodology is highly reliable. The vertical (and horizontal) wind component is generally measured by using a sonic anemometer; the temperature is measured by using ultra-fine wire thermocouples or, otherwise, it is determined sonically and is corrected later for humidity effects (Munger and Loescher 2004). The specific humidity is measured by means of quick response hygrometers (Buck 1976; Campbell and Tanner 1985; Tanner 1988; Burba and Anderson 2008). To characterize the same eddy scales, the measurements must be made at the same point, or at least in the closest vicinity. Thus, corrections are required because of instrument separation, different frequency response, coordinate rotation, and the type of hygrometer (Tanner et al. 1993; Villalobos 1997; Aubinet et al. 2000; Horst 2000; Massman 2000, 2001; Paw et al. 2000; Twine et al. 2000; Rannik 2001; Sakai et al. 2001; Wilson et al. 2002; Moncrieff et al. 2010; Mauder et al. 2013). Many software products are available for processing and correcting raw data (EdiRE, Clement 1999; TK3, Mauder and Foken 2004; EddySoft, Kolle and Rebmann 2007; ECPack, Van Dijk et al. 2004).

Turbulent heat fluxes often appear underestimated when their sum is compared to the available energy, Rn − G, net radiation minus soil heat flux. An average closure error is around 20% to 30% (Twine et al. 2000; Wilson et al. 2002; Foken 2008; Hendricks Franssen et al. 2010). Possible reasons can be found in the influence of the horizontal advection, the storage of heat in canopies, flux divergences, photosynthesis, errors in the measurement of Rn or G, frequency response of the sensors, measurement errors of turbulent fluxes, and separation of instruments. Field tower sites must fulfill some conditions: to be almost flat, with an extensive fetch, and to have uniform and homogeneous vegetation. The height at which the sensors must be placed depends on the height of the vegetation, the extent of the fetch, and the frequency response of the instruments. Box 14.4 provides two application examples at the global and local scales, respectively.

BOX 14.4   MICROMETEOROLOGICAL TOWER SITES DESCRIPTION FOR FLUXNET DATABASE

To respond to the needs of the scientific community for CO_2, water vapor, and energy flux data, a worldwide network database called FluxNet (Baldocchi et al. 2001), with more than 500 long-term micrometeorological tower sites, was developed (Figure 14.3). Different canopy covers including temperate conifer and broadleaved (deciduous and evergreen) forests, tropical and boreal forests, crops, grasslands, chaparral, wetlands, and tundra are monitored. The towers are maintained by either regional networks or individual projects.

A two-source energy-balance surface model based on Norman et al. (1995) and Kustas and Norman (1999) is being applied to a sub-basin of the Guadalquivir River, in the Martin Gonzalo watershed. To calibrate the model and validate ET estimations, an eddy covariance tower was set up on the typical area ecosystem, an oak savannah called *dehesa*. The tower is 18-m high, due to the average height of the trees (7 m), the area being nearly flat, and they having a homogeneous canopy cover (Figure 14.4). The tower is equipped with a CSAT three-dimensional sonic anemometer, a KH20 krypton hygrometer, a temperature and relative humidity probe, and a net radiometer. Two soil heat flux plates have been installed with wire thermocouples inside the soil. With this system configuration, four energy fluxes are being measured hourly.

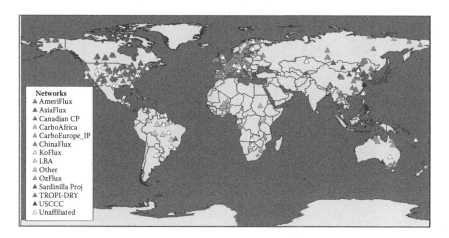

FIGURE 14.3 (See color insert.)
Distribution of tower sites within the global network of networks. (From FLUXNET, Integrating worldwide CO_2, water and energy flux measurements, http://fluxnet.ornl.gov.)

14.5 Final Remarks/Conclusions

Soil moisture is the final result of the energy and water balance in a soil volume, and at the same time, its gradient and evolution play a key role in the different hydrological cycle processes, forcing water movement through the soil, controlling water losses by evaporation from the soil and transpiration by plants, and providing nutrients and pollutants with different conditions for their absorption by organisms, transport through the soil profile, and retention by the soil components. Thus, soil moisture is determinant for provisioning ecosystem services, first to maintain vegetation and adequate

(a)

(b)

(c)

FIGURE 14.4 (See color insert.)
Eddy covariance tower (a) placed on Cardeña (Martín Gonzalo River Basin); "dehesa" landscape (b); soil heat flux plates system and complementary weather station (c).

conditions for microorganisms in soil, and second to guarantee a balanced trophic chain. But it also affects water availability for regulatory and maintenance services and has a clear impact on cultural services. Soil heterogeneity imposes a naturally high variability in soil moisture, in time and in space, which makes characterizing its regime from ground monitoring systems in medium- to large-sized areas practically unviable.

Remote sensing provides us with an alternative, robust, and high-frequency source of data by means of direct and indirect information. This chapter has revisited the most recent sources of soil moisture information, and examples of different applications have been referenced. Special focus has been directed toward the need to consider the significant scale effects arising from using soil moisture spatial distribution obtained from direct and indirect sources. Also, further explanation has been given toward applications dealing with vegetation covers, which are most usually studied with field and airborne data combined. Vegetation cover is the largest scale live component reflecting soil moisture variations, and influencing them, and it itself provides a provisioning service at upper scales.

Specific and further information can be found in the works referenced throughout the text, and a rapid advance in this topic is expected in the next few years, particularly with regard to sensor development, data analysis, and scaling, among others. Soil moisture evolution and its scaling are still challenging issues in hydrology, for which remote sensing has widened all the previous expectations for medium- and large-scale applications. The valuation of ecosystem services will benefit from further approaches increasing the accuracy in soil moisture estimation. Moreover, the evolution of ecosystem services under the current increasing trend of temperature will depend strongly on how the soil moisture regime is affected. Remote sensing provides society with a regional assessment of this change and the required information to estimate impacts on the associated ecosystem provisioning, maintenance, and cultural services.

Acknowledgments

This work was supported by the CERESS project (AGL2011-30498-02, Ministerio de Economía y Competitividad of Spain, co-funded FEDER).

References

Aguilar, C., J. C. Zinnert, M. J. Polo, and D. R. Young. 2012. NDVI as an indicator for changes in water availability to woody vegetation. *Ecological Indicators* 23:290–300.

Akbari, M., N. Toomanian, P. Droogers, W. Bastiaanssen, and A. Gieske. 2007. Monitoring irrigation performance in Esfahan, Iran, using NOAA satellite imagery. *Agricultural Water Management* 88:99–109.

Allen, R. G., M. E. Jensen, J. L. Wright, and R. D. Burman. 1989. Operational estimates of reference evapotranspiration. *Agronomy Journal* 81:650–662.

Allen, R. G., L. S. Pereira, T. A. Howell, and M. E. Jensen. 2011. Evapotranspiration information reporting: I. Factors governing measurement accuracy. *Agricultural Water Management* 98:899–920.

Allen, R. G., L. S. Pereira, D. Raes, and M. Smith. 1998. *Crop evapotranspiration. Guidelines for computing crop water requirements.* Irrigation and Drainage Paper No. 56. Rome: FAO.

Anderson, M. C., W. P. Kustas, J. M. Norman, et al. 2011. Mapping daily evapotranspiration at field to continental scales using geostationary and polar orbiting satellite imagery. *Hydrology and Earth System Sciences* 15:223–239.

Anderson, M. C., J. M. Norman, G. R. Diak, W. P. Kustas, and J. R. Mecikalski. 1997. A two-source time-integrated model for estimating surface fluxes using thermal infrared remote sensing. *Remote Sensing of Environment* 60:195–216.

Anderson, M. C., J. M. Norman, J. R. Mecikalski, J. A. Otkin, and W. P. Kustas. 2007. A climatological study of evapotranspiration and moisture stress across the continental United States based on thermal remote sensing: 2. Surface moisture climatology. *Journal of Geophysical Research: Atmospheres* 112:D11112.

Aubinet, M., A. Grelle, A. Ibrom, et al. 2000. Estimates of the annual net carbon and water exchange of forests: The EUROFLUX methodology. *Advances in Ecological Research* 30:113–175.

Baghdadi, N., C. King, A. Bourguignon, and A. Remond. 2002. Potential of ERS and RADARSAT data for surface roughness monitoring over bare agricultural fields: Application to catchments in Northern France. *International Journal of Remote Sensing* 23:3427–3442.

Baldocchi, D. D., E. Falge, L. Gu, et al. 2001. FLUXNET: A new tool to study the temporal and spatial variability of ecosystems-scale carbon dioxide, water vapor, and energy flux densities [Review]. *Bulletin of the American Meteorological Society* 82:2415–2434.

Bastiaanssen, W. G. M., M. D. Ahmad, and Y. Chemin. 2002. Satellite surveillance of evaporative depletion across the Indus 1 Basin. *Water Resources Research* 38:1273.

Bastiaanssen, W. G. M., M. Menenti, R. A. Feddes, and A. A. M. Holstlag. 1998. A remote sensing surface energy balance algorithm for land (SEBAL). 1. Formulation. *Journal of Hydrology* 212–213:198–212.

Bastiaanssen, W. G. M., S. Thiruvengadachari, R. Sakthivadivel, and D. J. Molden. 1999. Satellite remote sensing for estimating productivities of land and water. *Water Resources Development* 15:181–194.

Beven, K. J. 1989. Changing ideas in hydrology—The case of physically-based models. *Journal of Hydrology* 105:157–172.

Beven, K. J. 2000. On the future of distributed modelling in hydrology. *Hydrological Processes* 14:3183–3184.

Bindlish, R., T. J. Jackson, E. F. Wood, et al. 2003. Soil moisture estimates from TRMM Microwave Imager observations over the southern United States. *Remote Sensing of Environment* 85:507–515.

Blöschl, G., and M. Sivapalan. 1995. Scale issues in hydrological modeling—A review. *Hydrological Processes* 9:251–290.

Bowen, I. S. 1926. The ratio of heat losses by conduction and by evaporation from any water surface. *Physical Review* 27:779–787.

Brocca, L., F. Melone, T. Moramarco, et al. 2010. Improving runoff prediction through the assimilation of the ASCAT soil moisture product. *Hydrology and Earth System Science*. 14:1881–1893.

Brutsaert, W. 1982. *Evaporation into the atmosphere*. Dordrecht, The Netherlands: Kluwer Academic Publishers.

Brutsaert, W. 2005. *Hydrology. An introduction*. Cambridge, UK: Cambridge University Press.

Buck, A. 1976. The variable path Lyman-alpha hygrometer and its operating characteristics. *Bulletin of the American Meteorological Society* 51:1113–1118.

Burba, G., and D. Anderson. 2008. *Introduction to the eddy covariance method: General guidelines and conventional workflow*. Lincoln, NE: LiCor Corporation.

Campbell, G. S., and B. D. Tanner. 1985. A krypton hygrometer for measurement of atmospheric water vapor concentration. In *Moisture and humidity*, 609–612. Triangle Research Park, NC: Instrument Society of America.

Campos, I., C. M. U. Neale, A. Calera, C. Balbontin, and J. González-Piqueras. 2010. Assessing satellite-based basal crop coefficients for irrigated grapes (*Vitis vinifera L.*). *Agricultural Water Management* 97:1760–1768.

Carlson, T. 2007. An overview of the "Triangle Method" for estimating surface evapotranspiration and soil moisture from satellite imagery. *Sensors* 7:1612–1629.

Ceballos, A., K. Scipal, W. Wagner, and J. Martinez-Fernandez. 2005. Validation of ERS scatterometer-derived soil moisture data in the central part of the Duero Basin, Spain. *Hydrological Processes* 19:1549–1566.

Chandrapala, L., and M. Wimalasuriya. 2003. Satellite measurements supplemented with meteorological data to operationally estimate evaporation in Sri Lanka. *Agricultural Water Management* 58:89–107.

Chavez, J. L., C. M. U. Neale, L. E. Hipps, J. H. Prueger, and W. P. Kustas. 2005. Comparing aircraft-based remotely sensed energy balance fluxes with eddy covariance tower data using heat flux source area functions. *Journal of Hydrometeorology* 6:923–940.

Chehbouni, A., D. Lo Seen, E. G. Njoku, and B. M. Monteney. 1996. Examination of difference between radiometric and aerodynamic surface temperature over sparsely vegetated surfaces. *Remote Sensing of Environment* 58:177–186.

Clement, R. 1999. EdiRE. Available from: http://www.geos.ed.ac.uk/abs/research/micromet/EdiRe.

Crow, W. T., W. P. Kustas, and J. H. Prueger. 2008. Monitoring root-zone soil moisture through the assimilation of a thermal remote sensing-based soil moisture proxy into a water balance model. *Remote Sensing of Environment* 112:1268–1281.

De Bruin, H. A. R. 2008. Theory and application of large aperture scintillometers. Short course notes. Scintec Corp.

Dente, L., Z. Su, and J. Wen. 2012. Validation of SMOS soil moisture products over the Maqu and Twente regions. *Sensors* 12:9965–9986.

Doorenbos, J., and W. O. Pruitt. 1977. Guidelines for predicting crop-water requirements. In *FAO Irrigation and drainage*. Paper No. 24. 2nd ed. Rome: FAO. 156 pp.

Duchemin, B., R. Hadria, S. Er-Raki, et al. 2006. Monitoring wheat phenology and irrigation in Central Morocco: On the use of relationships between evapotranspiration, crop coefficients, leaf area index and remotely-sensed vegetation indices. *Agricultural Water Management* 79:1–27.

Engman, E. T., and N. Chauhan. 1995. Status of microwave soil moisture measurements with remote sensing. *Remote Sensing of Environment* 51:189–198.

Engman, E. T., and R. J. Gurney. 1991. *Remote sensing in hydrology.* London: Chapman & Hall.

Er-Raki, S., A. Chehbouni, and B. Duchemin. 2010. Combining satellite remote sensing data with the FAO-56 dual approach for water use mapping in irrigated wheat fields of a semi-arid region. *Remote Sensing* 2:375–387.

Er-Raki, S., A. Chehbouni, N. Guemouria, B. Duchemin, J. Ezzahar, and R. Hadria. 2007. Combining FAO-56 model and ground-based remote sensing to estimate water consumptions of wheat crops in a semi-arid region. *Agricultural Water Management* 87:41–54.

Fernández-Prieto, D., P. van Oevelen, Z. Su, and W. Wagner. 2012. Advances in Earth observation for water science (Editorial). *Hydrology and Earth System Science* 16:543–549.

FLUXNET. Integrating worldwide CO_2, water and energy flux measurements. Available from: http://fluxnet.ornl.gov. Accessed July 7, 2012

Foken, T. 2008. The energy balance closure problem—An overview. *Ecological Applications* 18:1351–1367.

Fung, A. K. 1994. *Microwave scattering and emission models and their applications.* Norwood, MA: Artech House, Inc.

Gao, H., E. Wood, M. Drusch, M. McCabe, T. J. Jackson, and R. Bindlish. 2003. Using TRMM/TMI to retrieve soil moisture over southern United States from 1998 to 2002: Results and validation. *EOS Transactions* 84:618.

Garatuza-Payan, J., and C. J. Watts. 2005. The use of remote sensing for estimating ET of irrigated wheat and cotton in northwest Mexico. *Irrigation and Drainage Systems* 19:301–320.

Gillies, R. T., T. N. Carlson, J. Cui, W. P. Kustas, and K. S. Humes. 1997. A verification of the "triangle" method for obtaining surface soil water content and energy fluxes from remote measurements of the Normalized Difference Vegetation Index (NDVI) and surface radiant temperatures. *International Journal of Remote Sensing* 18:3145–3166.

González-Dugo, M. P., S. Escuin, L. Mateos, et al. 2012. Monitoring evapotranspiration of irrigated crops using crop coefficients derived from time series of satellite images. II. Application on basin scale. *Agricultural Water Management* 125:92–104.

González-Dugo, M. P., and L. Mateos. 2008. Spectral vegetation indices for benchmarking water productivity of irrigated cotton and sugarbeet crops. *Agricultural Water Management* 95:48–58.

González-Dugo, M. P., C. M. U. Neale, L. Mateos, et al. 2009. A comparison of operational remote-sensing-based models for estimating crop evapotranspiration. *Agricultural and Forest Meteorology* 149:1843–1853.

Groeneveld, D. P., and W. M. Baugh. 2007. Correcting satellite data to detect vegetation signal for eco-hydrologic analyses. *Journal of Hydrology* 344:135–145.

Gruhier, C., P. de Rosnay, S. Hasenauer, et al. 2010. Soil moisture active and passive microwave products: Intercomparison and evaluation over a Sahelian site. *Hydrology and Earth System Science* 14:141–156.

Gupta, V. K., I. Rodríguez-Iturbe, and E. F. Wood (eds.). 1986. *Scale problems in hydrology.* Dordrecht, The Netherlands: Reidel.

Hamid, S. H., A. A. Mohamed, and Y. A. Mohamed. 2011. Towards a performance-oriented management for large-scale irrigation systems: Case study, Rahad scheme, Sudan. *Irrigation and Drainage* 60:20–34.

Hartogensis, O. K., C. J. Watts, J. C. Rodriguez, and H. A. R. De Bruin. 2003. Derivation of an effective height for scintillometers: La Poza experiment in northwest Mexico. *Journal of Hydrometeorology* 4:915–928.

Hendricks Franssen, H. J., R. Stöckli, I. Lehner, E. Rotenberg, and S. I. Seneviratne. 2010. Energy balance closure of eddy-covariance data: A multisite analysis for European FLUXNET stations. *Agricultural and Forest Meteorology* 150:1553–1567.

Horst, T. W. 2000. On frequency response corrections for eddy covariance flux measurements. *Boundary-Layer Meteorology* 95:517–520.

Hunsaker, D. J., E. M. Barnes, T. R. Clarke, G. J. Fitzgerald, and P. J. Pinter. 2005a. Cotton irrigation scheduling using remotely sensed and FAO-56 basal crop coefficients. *Transactions of the ASAE* 48:1395–1407.

Hunsaker, D. J., P. J. Pinter, and B. A. Kimbal. 2005b. Wheat basal crop coefficients determined by normalized difference vegetation index. *Irrigation Science* 24:1–14.

Jensen, M. E. (ed.). 1974. *Consumptive use of water and irrigation water requirements.* New York: American Society of Civil Engineers.

Jensen, M. E., R. D. Burman, R. G. Allen (eds.). 1990. *Evapotranspiration and irrigation water requirements.* New York: American Society of Civil Engineers.

Juglea, S., Y. Kerr, A. Mialon, E. López-Baeza, D. Braithwaite, and K. Hsu. 2010. Soil moisture modelling of a SMOS pixel: Interest of using the PERSIANN database over the Valencia Anchor Station. *Hydrology and Earth System Science* 14:1509–1525.

Jury, W. A., and R. Horton. 2004. Chemical transport in soil. In *Soil physics*, 225–276. Hoboken, NJ: Wiley.

Kasischke, E. S., M. A. Tanase, L. L. Bourgeau-Chavez, and M. Borr. 2011. Soil moisture limitations on monitoring boreal forest regrowth using spaceborne L-band SAR data. *Remote Sensing of Environment* 115:227–232.

Kerr, Y., P. Waldteufel, J. P. Wigneron, J. M. Martinuzzi, J. Font, and M. Berger. 2001. Soil moisture retrieval from space. The Soil Moisture and Ocean Salinity (SMOS) mission. *IEEE Transaction on Geoscience and Remote Sensing* 39:1729–1735.

Kolle, O., and C. Rebmann. 2007. EddySoft—Documentation of a software package to acquire and process Eddy covariance data. Jena, Germany: Max Planck Institute for Biogeochemistry.

Kustas, W. P. 1990. Estimates of evapotranspiration with a one- and two-layer model of heat transfer over partial cover. *Journal of Applied Meteorology and Climatology* 29:704–715.

Kustas, W. P., B. J. Choudhury, M. S. Moran, et al. 1989. Determination of sensible heat flux over sparse canopy using thermal infrared data. *Agricultural and Forest Meteorology* 44:197–216.

Kustas, W. P., and J. M. Norman. 1996. Use of remote sensing for evapotranspiration monitoring over land surfaces. *Hydrological Sciences* 41:495–516.

Kustas, W. P., and J. M. Norman. 1999. Evaluation of soil and vegetation heat flux predictions using a simple two source model with radiometric temperatures for partial canopy cover. *Agricultural and Forest Meteorology* 94:13–29.

Laguardia, G., and S. Niemeyer. 2008. On the comparison between the LISFLOOD modelled and the ERS/SCAT derived soil moisture estimates. *Hydrology and Earth System Science* 12:1339–1351.

Lhomme, J. P., B. Monteny, and M. Amadou. 1994. Estimating sensible heat flux from radiometric temperature over sparse millet. *Agricultural and Forest Meteorology* 68:77–91.

Lievens, H., N. E. C. Verhoest, E. De Keyser, et al. 2011. Effective roughness modelling as a tool for soil moisture retrieval from C- and L-band SAR. *Hydrology and Earth System Science* 15:151–162.

Liu, Y. Y., W. A. Dorigo, R. M. Parinussa, et al. 2012. Trend-preserving blending of passive and active microwave soil moisture retrievals. *Remote Sensing of Environment* 123:280–297.

Liu, Y. Y., R. M. Parinussa, W. A. Dorigo, et al. 2011. Developing an improved soil moisture dataset by blending passive and active microwave satellite-based retrievals. *Hydrology and Earth System Science* 15:425–436.

Mahrt, L., and D. Vickers. 2004. Bulk formulation of the surface heat flux. *Boundary-Layer Meteorology* 110:357–379.

Makkink, G. P. 1957. Testing the Penman formula by means of lysimeters. *Journal of the Institution of Water Engineers* 11:277–288.

Manninen, T., P. Stenberg, M. Rautiainen, P. Voipio, and H. Smolander. 2005. Leaf area index estimation of boreal forest using ENVISAT ASAR. *IEEE Transactions on Geoscience and Remote Sensing* 43:2627–2635.

Massman, W. J. 2000. A simple method for estimating frequency response corrections for eddy covariance systems. *Agricultural and Forest Meteorology* 104:185–198.

Massman, W. J. 2001. Reply to comment by Rannik on: A simple method for estimating frequency response corrections for eddy covariance systems. *Agricultural and Forest Meteorology* 107:247–251.

Mauder, M., M. Cuntz, C. Drüe, et al. 2013. A strategy for quality and uncertainty assessment of long-term eddy-covariance measurements. *Agricultural and Forest Meteorology* 169:122–135.

Mauder, M., and T. Foken. 2004. *Documentation and instruction manual of the eddy covariance software package TK2*. Arbeitsergebnisse 26. Bayreuth, Germany: Universitaet Bayreuth, Abt. Mikrometeorologie.

Mecikalski, J. M., G. R. Diak, M. C. Anderson, and J. M. Norman. 1999. Estimating fluxes on continental scales using remotely sensed data in an atmosphere-land exchange model. *Journal of Applied Meteorology and Climatology* 38:1352–1369.

Meijninger, W. M. L., and H. A. R. De Bruin. 2000. The sensible heat fluxes over irrigated areas in western Turkey determined with a large aperture scintillometer. *Journal of Hydrology* 229:42–49.

Meijninger, W. M. L., O. K. Hartogensis, W. Kohsiek, J. C. B. Hoedjes, R. M. Zuurbier, and H. A. R. De Bruin. 2002. Determination of area-averaged sensible heat fluxes with a large aperture scintillometer over a heterogeneous surface—Flevoland field experiment. *Boundary-Layer Meteorology* 105:37–62.

Mertens, J. 2003. Parameter estimation strategies in unsaturated zone modelling. PhD diss., Leuven Catholic University.

Moncrieff, J. B., R. Clement, J. Finnigan, and T. Meyers. 2010. Averaging, detrending, and filtering of eddy covariance time series. In *Handbook of micrometeorology: A guide for surface flux measurement and analysis,* eds. X. Lee, W. Massman, and B. Law, 7–31. Dordrecht, The Netherlands: Kluwer Academic Publishers.

Moradkhani, H. 2008. Hydrologic remote sensing and land surface data assimilation. *Sensors* 8:2986–3004.

Moran, M. S. 2003. Thermal infrared measurement as an indicator of plant ecosystem health. In *Thermal remote sensing in land surface processes*, eds. D. A. Quattrochi and J. Luvall, 257–282. Philadelphia, PA: Taylor & Francis.

Moran, M. S., T. R. Clarke, Y. Inoue, and A. Vidal. 1994. Estimating cropwater deficit using the relation between surface-air temperature and spectral vegetation index. *Remote Sensing of Environment* 49:246–263.

Moran, M. S., D. C. Hymer, J. Qi, and Y. Kerr. 2001. Comparison of ERS-2 SAR and Landsat TM imagery for monitoring agricultural crop and soil conditions. *Remote Sensing of Environment* 79:243–252.

Munger, J. W., and H. W. Loescher. 2004. Guidelines for making eddy covariance flux measurements. Available from: http://public.ornl.gov/ameriflux/measurement standards 4.doc. Accessed May 12, 2012.

Narayanan, R. M., and M. S. Hegde. 1995. Soil moisture inversion algorithms using ERS-1, JERS-1, and ALMAZ SAR data. Geoscience and Remote Sensing Symposium. IGARSS '95. *Quantitative Remote Sensing for Science and Applications* 1:504–506.

Norman, J. M., M. C. Anderson, W. P. Kustas, et al. 2003. Remote sensing of surface energy fluxes at 101-m pixel resolutions. *Water Resources Research* 39:1221.

Norman, J. M., W. P. Kustas, and K. S. Humes. 1995. A two-source approach for estimating soil and vegetation energy fluxes from observations of directional radiometric surface temperature. *Agricultural and Forest Meteorology* 77:263–293.

Padilla, F. L. M., M. P. González-Dugo, P. Gavilán, and J. Domínguez. 2011. Integration of vegetation indices into a water balance model to estimate evapotranspiration of wheat and corn. *Hydrology and Earth System Science* 15:1213–1225.

Parajka, J., V. Naeimi, G. Blöschl, W. Wagner, R. Merz, and K. Scipal. 2006. Assimilating scatterometer soil moisture data into conceptual hydrologic models at the regional scale. *Hydrology and Earth System Science* 10:353–368.

Paw, U. K. T., D. D. Baldocchi, T. P. Meyers, and K. B. Wilson. 2000. Correction of eddy covariance measurements incorporating both advective effects and density fluxes. *Boundary-Layer Meteorology* 97:487–511.

Pettorelli, N., J. O. Vik, A. Mysterud, J. M. Gaillard, C. J. Tucker, and N. C. Stenseth. 2005. Using the satellite-derived NDVI to assess ecological responses to environmental change. *Trends in Ecology and Evolution* 20:503–510.

Polo, M. J., J. Regodón, and M. P. González-Dugo. 2009. Tidal flood monitoring in marsh estuary areas from Landsat TM data. In *Remote sensing for agriculture, ecosystems, and hydrology XI*, eds. C. M. U. Neale and A. Maltese. Washington, DC: SPIE.

Rannik, U. 2001. A comment on the paper by W. J. Massman: A simple method for estimating frequency response corrections for eddy covariance systems. *Agricultural and Forest Meteorology* 107:241–245.

Roberts, D. A., G. T. Batista, J. L. G. Pereira, E. K. Waller, and B. W. Nelson. 1998. Change identification using multitemporal spectral mixture analysis: Applications in eastern Amazonia. In *Remote sensing change detection: Environmental monitoring methods and applications*, eds. R. S. Lunetta and C. D. Elvidge, 137–161. Ann Arbor, MI: Ann Arbor Press.

Rossi, S., A. Rampini, S. Bocchi, and M. Boschetti. 2010. Operational monitoring of daily crop water requirements at the regional scale with time series of satellite data. *Journal of Irrigation and Drainage Engineering* 136:225–231.

Rouse, J. W., Jr., R. H. Haas, J. A. Schell, and D. W. Deering. 1974. Monitoring vegetation systems in the Great Plains with ERTS. *Proceedings of the Third ERTS Symposium* 1:309–317.

Sakai, R. K., D. R. Fitzjarrald, and K. E. Moore. 2001. Importance of low-frequency contributions to eddy fluxes observed over rough surfaces. *Journal of Applied Meteorology and Climatology* 40:2178–2192.

Sánchez, N., J. Martínez-Fernández, A. Calera, E. Torres, and C. Pérez-Gutiérrez. 2010. Combining remote sensing and in situ soil moisture 1 data for the application and validation of a distributed water balance model (HIDROMORE). *Agricultural Water Management* 98:69–78.

Schmugge, T. J., and T. J. Jackson. 1994. Mapping surface soil moisture with microwave radiometers. *Meteorology and Atmospheric Physics* 54:213–223.

Schmugge, T. J., W. P. Kustas, J. C. Ritchie, T. J. Jackson, and A. Rango. 2002. Remote sensing in hydrology. *Advances in Water Resources* 25:1367–1385.

Shao, G., H. H. Shugart, and D. R. Young. 1995. Simulation of transpiration sensitivity to environmental changes for shrub (*Myrica cerifera*) thickets on a Virginia barrier island. *Ecological Modelling* 78:235–248.

Shuttelworth, W., and J. Wallace. 1985. Evaporation from sparse crops: An energy combination theory. *Quarterly Journal of the Royal Meteorological Society* 111:1143–1162.

Singh, R., and A. Irmak. 2009. Estimation of crop coefficients using satellite remote sensing. *Journal of Irrigation and Drainage Engineering* 135:597–608.

Sobrino J. A., B. Franch, C. Mattar, J. C. Jiménez-Muñoz, and C. Corbari. 2012. A method to estimate soil moisture from Airborne Hyperspectral Scanner (AHS) and ASTER data: Application to SEN2FLEX and SEN3EXP campaigns. *Remote Sensing of Environment* 117:415–428.

Sposito, G. 1989. *The chemistry of soils*. New York: Oxford University Press.

Su, C. H., D. Ryu, R. I. Young, A. W. Western, and W. Wagner. 2013. Inter-comparison of microwave satellite soil moisture retrievals over the Murrumbidgee basin, southeast Australia. *Remote Sensing of Environment* 134:1–11.

Su, Z., J. Wen, L. Dente, et al. 2011. The Tibetan Plateau observatory of plateau scale soil moisture and soil temperature (Tibet-Obs) for quantifying uncertainties in coarse resolution satellite and model products. *Hydrology and Earth System Science* 15:2303–2316.

Su, Z., J. Wen, and W. Wagner. 2010. Advances in land surface hydrological processes: field observations, modeling and data assimilation: Preface. *Hydrology and Earth System Science* 14:365–367.

Tanner, B. D. 1988. Use requirements for Bowen ratio and eddy correlation determination of evapotranspiration. In *Planning now for irrigation and drainage in the 21st century*, ed. D. R. Hay, 605–616. Lincoln, NE: ASCE.

Tanner, B. D., E. Swiatek, and J. P. Green. 1993. Density fluctuations and use of the krypton hygrometer in surface flux measurements. In *Management of irrigation and drainage systems: Integrated perspectives*, eds. R. G. Allen and C. M. U. Neale, 945–952. Lincoln, NE: ASCE.

Tasumi, M., and R. G. Allen. 2007. Satellite-based ET mapping to assess variation in ET with timing of crop development. *Agricultural Water Management* 88:54–62.

Tasumi, M., R. Trezza, R. G. Allen, and J. L. Wright. 2005. Operational aspects of satellite-based energy balance models for irrigated crops in the semi-arid U.S. *Irrigation and Drainage Systems* 19:355–376.

Timmermans, W. J., W. P. Kustas, M. C. Anderson, and A. N. French. 2007. An inter-comparison of the surface energy balance algorithm for land (SEBAL) and the two source energy balance (TSEB) modeling schemes. *Remote Sensing of Environment* 108:284–369.

Twine, T. E., W. P. Kustas, J. M. Norman, et al. 2000. Correcting eddy-covariance flux underestimates over a grassland. *Agricultural and Forest Meteorology* 103:279–300.

Ulaby, F. T., R. K. Moore, and A. K. Fung. 1981. *Microwave remote sensing: Active and passive, Vol. I—Microwave remote sensing fundamentals and radiometry.* Reading, MA: Addison-Wesley, Advanced Book Program.

Van Dijk, A., A. F. Moene, and H. A. R. De Bruin. 2004. *The principles of surface flux physics: Theory, practice and description of the ECPACK library.* Internal Report 2004/1, Meteorology and Air Quality Group. Wageningen, The Netherlands: Wageningen University.

Vereecken, H., J. A. Huisman, H. Bogena, J. Vanderborght, J. A. Vrugt, and J. W. Hopmans. 2008. On the value of soil moisture measurements in vadose zone hydrology: A review. *Water Resources Research* 44:W00D06.

Villalobos, F. J. 1997. Correction of eddy covariance water vapor flux using additional measurements of temperature. *Agricultural and Forest Meteorology* 88:77–83.

Wagner, W., G. Lemoine, M. Borgeaud, and H. Rott. 1999a. A study of vegetation cover effects on ERS scatterometer data. *IEEE Transactions on Geoscience and Remote Sensing* 37:938–948.

Wagner, W., G. Lemoine, and H. Rott. 1999b. A method for estimating soil moisture from ERS scatterometer and soil data. *Remote Sensing of Environment* 70:191–207.

Wang, J., P. M. Rich, and K. P. Price. 2003. Temporal responses of NDVI to precipitation and temperature in the central Great Plains, USA. *International Journal of Remote Sensing* 24:2345–2364.

Wigneron, J. P., J. C. Calvet, T. Pellarin, A. A. V. D. Griend, M. Berger, and P. Ferrazzoli. 2003. Retrieving near-surface soil moisture from microwave radiometric observations: Current status and future plans. *Remote Sensing of Environment* 85:489–506.

Wigneron, J. P., T. Schmugge, A. Chanzy, J. C. Calvet, and Y. Kerr. 1998. Use of passive microwave remote sensing to monitor soil moisture. *Agronomie* 18:27–43.

Wilson, K. B., A. H. Goldstein, and F. Falge. 2002. Energy balance closure at FLUXNET sites. *Agricultural and Forest Meteorology* 113:223–243.

Wright, J. L. 1981. *Crop coefficients irrigation scheduling for water and energy conservation in the 80s.* St. Joseph, MO: American Society of Agricultural Engineers.

Wright, J. L. 1982. New evapotranspiration crop coefficients. *Journal of the Irrigation and Drainage Division ASCE* 108:57–74.

Zribi, M., N. Baghdadi, N. Holah, O. Fafin, and C. Guérin. 2005. Evaluation of a rough soil surface description with ASAR-ENVISAT radar data. *Remote Sensing of Environment* 95:67–76.

Zribi, M., and M. Dechambre. 2002. A new empirical model to inverse soil moisture and roughness using two radar configurations. *Proceedings of the IEEE International Geoscience and Remote Sensing Symposium and 24th Canadian Symposium on Remote Sensing,* Toronto, Canada: Institute of Electrical and Electronics Engineers. 2223–2225.

15

Snowpack as a Key Element in Mountain Ecosystem Services: Some Clues for Designing Useful Monitoring Programs

F. J. Bonet
University of Granada, Spain

A. Millares and J. Herrero
University of Granada, Spain

CONTENTS

15.1 Introduction: From Monitoring Snow Cover to Quantifying
Ecosystem Services 330
15.2 Design and Implementation of Methodologies to Monitor the
Services Provided by Snow Cover 332
    15.2.1 *In Situ* Measurements 332
            15.2.1.1 Transects, Snow Surveys, and Automatic Stations 332
            15.2.1.2 Oblique Photographs to Monitor Snow Cover............333
    15.2.2 Satellite Measurements 334
            15.2.2.1 Moderate Resolution Imaging Spectroradiometer.....334
            15.2.2.2 Landsat 335
            15.2.2.3 Advanced Very High Resolution Radiometer 336
    15.2.3 Airborne Sensors 336
15.3 Final Services Provided by Snow in Mountain Areas.........................337
    15.3.1 Provisioning Services..337
    15.3.2 Regulating and Maintenance Services 338
    15.3.3 Cultural Services..340
15.4 A Case Study of Monitoring Snow Cover to Quantify Ecosystem
Services: Sierra Nevada Biosphere Reserve (Spain)..............................341
15.5 Conclusion 343
Acknowledgments...344
References..344

15.1 Introduction: From Monitoring Snow Cover to Quantifying Ecosystem Services

Snow cover is the layer of snow accumulated over a terrain that can suffer a grouping and rearranging process yielding to the production of ice. It remains for a given period of time with variable degrees of metamorphism. It is composed of a solid matrix that hosts a hollow component that can store gases and liquids. Thanks to this structure, snow cover has some physical characteristics such as density, porosity, or hydraulic and thermal conductivity. A peculiarity of snow cover is that its matrix is frozen water. Ice cannot exist at temperatures higher than 0°C. In these situations, the matrix melts and becomes part of the pores. This liquid water can become ice (matrix) again if the temperature is below 0°C. It can also leach into the terrain or become part of superficial runoff. Snow cover, a porous system where the three possible phases of water coexist (gas, liquid, and solid), is a unique element on the Earth's surface. Also, it is very peculiar because it can experience huge fluctuations, both spatially and temporally. A cyclonic event can increase the extent of snow cover to the order of 1000 km² (Cohen and Rind 1991).

Thanks to these features, snow cover is a key element capable of shaping the landscape on those mountains where it is present. Its capacity for storing water, for example, explains its role in the hydrological cycle. Its weight and thermal conductivity could explain its role in some ecological processes. According to the Common International Classification of Ecosystem Services (CICES; Haines-Young and Potschin 2013), ecosystem services are outputs fundamentally dependent on living processes. Essentially, abiotic outputs from nature are not regarded as ecosystem services. Thus, snow cover should not be considered as an element capable of providing ecosystem services. The final services are provided by ecosystems, although snow contributes toward creating an abiotic context for those services.

The aforementioned reasoning serves as a guide for the conceptual framework that we are following in this chapter: (1) Snow cover can be described as having several structural and functional features. (2) These features can be measured using several monitoring protocols. (3) The measured properties contribute to the creation of services by mountain ecosystems. This last step is due to quantification methodologies ranging from hydrological models (see Chapter 12) to (for example) ecophysiological models that are capable of simulating biomass production using information from the water content of snow. Figure 15.1 shows our conceptual model for this chapter.

Following the aforementioned conceptual framework, the main objectives of this chapter are (1) to describe different methodologies that could help to implement a monitoring program to gather information about snow cover; (2) to outline the final services provided by snowpack; and (3) to present a case study of a monitoring program designed to quantify the final services.

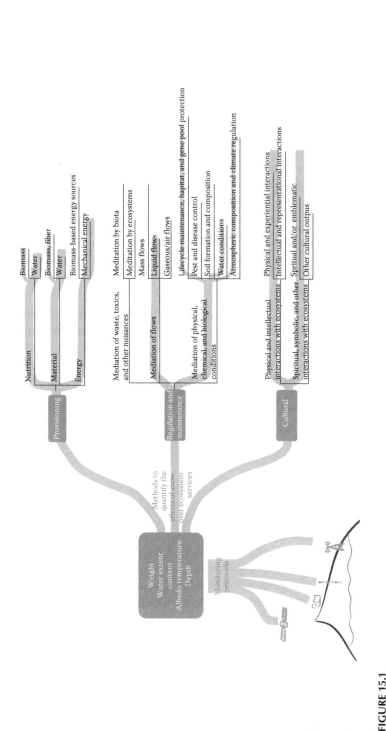

FIGURE 15.1
Schema showing the conceptual framework that we have followed in this chapter. Snow cover has several functional and structural characteristics (temperature, water content, depth, etc.) that can be evaluated using monitoring protocols. The obtained information can be used for quantifing different types of ecosystem services. (Classified according to Haines-Young, R., and Potschin, M., CICES V4.3–Revised report prepared following consultation on CICES Version 4, August–December 2012, EEA Framework Contract No EEA/IEA/09/003, 2013.)

We have selected mountain areas as the spatial context for this chapter. Although snow covers huge flat areas around the world, we have chosen mountains because they are very peculiar areas among terrestrial systems. They condense different climate types, a very rich biodiversity, and a huge human population in a not very large area. These features make them little "Earth systems" where it is especially interesting to analyze the impact of global changes on the biosphere. They could be considered as relevant observing systems that act as "canaries in a coal mine" (Diaz et al. 2003) thereby alerting us to the important changes affecting the Earth. In addition, mountains can be considered as cost-effective places to conserve ecosystem services and biodiversity (Sanderson et al. 2002).

15.2 Design and Implementation of Methodologies to Monitor the Services Provided by Snow Cover

We will describe an ideal monitoring program for snow cover in mountain regions. This program will take into account the peculiar characteristics of mountain areas when defining monitoring methodologies. As shown in the conceptual framework, these protocols are capable of quantitatively describing the structure and functions of snow cover. This is the first step toward quantifying services provided by snow. The proposed monitoring program will also consider the spatiotemporal hierarchy of elements that are present in mountain areas and monitoring procedures.

15.2.1 *In Situ* Measurements

This section contains monitoring procedures that allow us to gather information in the field. Thanks to these methods, we can obtain information on snow depth and also on the snow water equivalent (SWE). It is also possible to get information about the structure of the snow layer. The implementation of these methodologies implies the installation of sensors and other devices in the field and/or the gathering of information by humans. These methods are especially useful to collect detailed information on the water content of the snow. Thus, they are used to quantify provisioning and regulating services provided by snow cover. They are also helpful for calibrating hydrological models, the results of which can be used to quantify the regulating and provisioning services provided by snow.

15.2.1.1 Transects, Snow Surveys, and Automatic Stations

In order to obtain very detailed information about the volume of snow and water content, it is necessary to create closed enclosures with lysimeters, depth sensors (using ultrasonics), pressure sensors, and so on (Dingman 2002). This intensive methodology is difficult to accomplish, especially in mountain regions where snow depth is very heterogeneous. Creating transects is a much more efficient way of obtaining information about SWE in a mountain area. The methodology is easier and involves designing a set of transects over the snow cover with periodic sample points. Using a metallic cylinder, it is possible to extract a column of snow and measure depth, weight, and water content (Chow 1964; Rallison 1981). Both methods are difficult to implement in remote areas. The spatial extent of the measures is very short, and the periodicity of collecting information is very coarse (it is not realistic to visit transects very often). To avoid these problems, it is possible to install automatic stations that are capable of measuring SWEs. A snow telemetry (SNOTEL; Rallison 1981; Watson et al. 2008) system is probably the most commonly used program worldwide to monitor the SWE. This system, which is a network operated by the U.S. Department of Agriculture, is able to monitor SWE, precipitation, temperature, and snow depth. There are 600 SNOTEL stations in the western United States. Because of systems such as SNOTEL, it is possible to overcome the problems of transects and punctual snow surveys.

One of the most important drawbacks of these methodologies is their limited spatial extent. It is very expensive and time consuming to design, implement, and maintain a dense monitoring system based on transects or automatic field stations.

Citizen science (Silvertown 2009) could help to overcome this lack of spatial extent of the previous methodologies. Some public administrations, such as the Canadian government, have realized that there could be thousands of people willing to gather information on snow and ice cover with their only motivation being an urge to be of use in monitoring global change. Canada has created a program called Icewatch (http://www.naturewatch. ca/english/icewatch) that allows its citizens to participate in discovering how and why snowpack and ice are changing. This network is helping scientists to understand the freeze–thaw cycles of ice and snow in Canada.

15.2.1.2 Oblique Photographs to Monitor Snow Cover

Point sample methods are not useful for obtaining information on the extent of snowpack. They are not capable of detecting significant snow cover variations across areas (Hinkler et al. 2002). Standard digital cameras can be of help in overcoming this problem, allowing the monitoring the extent of snow cover from a single *in situ* measurement point (Hinkler et al. 2002; Laffly et al. 2011). This technique combines methods from field monitoring and remote sensing and is especially interesting at a small catchment scale (Parajka et al. 2012). Oblique photographs allow a very high spatiotemporal resolution at a relatively low cost. A standard digital camera is connected to an automatic trigger that can be controlled by an electronic timer. The energy supply could possibly come from solar panels. The information is stored in a flash memory card and can be sent automatically via general packet radio service (GPRS), where available. The next step is to georeference the obtained oblique photographs to a digital elevation model (DEM) in order to obtain a layer where snow surface can be measured. To obtain such a layer, it is necessary to create a function that relates two-dimensional pixels in the photograph to three-dimensional points in the DEM (Corripio 2004; Laffly et al. 2011). It is also mandatory to correct the topographic and atmospheric influences that affect the reflectance values existing in the image. Finally, it is possible to detect the snow automatically using either visible wavelength or a combination of red and infrared wavelengths (multispectral radiation). This last option allows the calculation of the Normalized Difference Snow Index (NDSI; Hall et al. 1995; Dozier and Painter 2004), which can distinguish snow from other white elements, such as clouds or rocks.

15.2.2 Satellite Measurements

Monitoring of snow cover using satellites has a long history (since the 1960s) (Dietz et al. 2012). This is due to the importance of snow cover in the hydrological cycle and in the climate system, and its role in providing useful services to human societies. The first satellite that started to monitor snow cover was the Television and InfraRed Observation Satellite (TIROS-1) in 1960. From that date to the present, many spatial agencies have designed sensors with different spectral channels and spatiotemporal resolutions. This section provides a brief review of the most relevant methods used to monitor snow cover with remote sensing devices. Most satellites are capable of measuring the extent of snow. Some of them can obtain information on the SWE and the snow depth. These two variables are very useful for quantifying the water provisioning services provided by snow cover.

The major advantage of these techniques is the aerial information gathered by the satellite. Thanks to these devices, it is possible to monitor snow cover at a global scale. All the obtained images can be processed following similar procedures, and it is possible to obtain long time series showing the extent of snow cover. These procedures include techniques to calibrate sensors, which suffer degradation due to the harsh space environment (Wang et al. 2012). There are different techniques to calibrate sensors using onboard apparatus (Green and Shimada 1997), vicarious calibration (using natural or artificial sites on the surface of the Earth) (Martiny et al. 2005), or postlaunch methods (flying an aircraft with a calibrated radiometer and simultaneously measuring the radiance).

15.2.2.1 Moderate Resolution Imaging Spectroradiometer

The Moderate Resolution Imaging Spectroradiometer (MODIS) sensor was installed on two satellites: Terra (launched in December 1999) and Aqua (launched in May 2002). This sensor captures data in 36 spectral bands (from 0.4 to 14.4 µm). These bands allow the creation of relevant geophysical information about land use, cloud cover, aerosols, biological productivity in oceans and land, surface temperature, and snow cover. Regarding this last element, MODIS is able to provide information about the extent of snow cover and fractional snow cover (percentage of snow per pixel). Snow cover is obtained due to an algorithm called Snowmap (Hall et al. 1995). It uses the reflectance of visible (VIS) and infrared (IR) channels to obtain NDSI. MODIS provides MOD10A2 as a product obtained using the Snowmap algorithm. This product has a periodicity of eight days and a spatial resolution of 500 m. In each MOD10A2 image, each pixel is labeled with snow if it has had snow in one of the previous eight days. Thus, MOD10A2 shows the maximum extent of snow cover in eight days. Fractional snow cover is also provided by MODIS in a daily product called MOD10A1. The percentage of surface covered by snow in each pixel is obtained by an algorithm developed by Salomonson and Appel in 2006. These two products are very useful for calibrating and validating hydrological models (Parajka and Blöschl 2008). Powell et al. (2011) used MODIS snow cover area as input data for a hydrological daily model that predicts stream flow in a California river. Andreadis and Lettenmaier (2006) successfully assimilated MODIS products into the variable infiltration capacity (VIC) of a hydrologic model in the United States. The MODIS product has been improved to address specific objectives. Thirel et al. (2012) created a real-time snow cover product from MODIS information. This map was used to validate the output obtained by a hydrological model.

One of the most important advantages of MODIS products is the availability of data through versatile Web applications. National Aeronautics and Space Administration (NASA) has developed several procedures to easily download raw data from all the MODIS products. The most advanced at the moment is ECHO Reverb (http://reverb.echo.nasa.gov/reverb)—a powerful Web application that allows the downloading of MODIS snow products worldwide. Users can select an area to obtain all the existing images. It is also possible to subscribe to several products and receive new images via file transfer protocol (FTP) making it possible to create third-party applications that automatically download, process, and display information about snow cover using MODIS.

15.2.2.2 Landsat

Landsat satellites have been in operation since 1972. The first three satellites had four spectral bands. Their spatial resolution was 79 m, and they gathered images every 18 days. Landsat 4–5 (launched in 1982) and Landsat 7 (launched in 1999) have seven to eight spectral bands, 30 m of spatial resolution, and a periodicity of 16 days. Landsat and MODIS provide similar snow products: snow cover extent and fractional snow cover. However, Landsat should not be used for monitoring because of its poor temporal resolution (16 days).

However, Landsat products are useful for validating outputs from hydrological models (Herrero et al. 2011). Both products can be obtained using algorithms similar to those of MODIS (Rosenthal and Dozier 1996; Vikhamar and Solberg 2003). Unfortunately, Landsat 5 is no longer operational, and Landsat 7 is has developed severe problems (a scan line corrector failure). The Landsat Data Continuity Mission (LDCM) was launched in February 2013, and it is considered the next generation of Landsat satellites (Irons et al. 2012). It has a spatial resolution of 15 m for panchromatic, 30 m for multispectral, and 100 m for thermal channels. It will have a 16-day ground track repeat cycle with an equatorial crossing at 10:00 a.m.

15.2.2.3 Advanced Very High Resolution Radiometer

The first Advanced Very High Resolution Radiometer (AVHRR) sensor was launched by the National Oceanic and Atmospheric Administration (NOAA) in 1978. Currently, there are at least two polar orbiting satellites in orbit at all times. They have a spatial resolution of 1,090 m and take images on a daily basis. AVHRR hosts five spectral bands ranging from 0.58 to 12.5 μm. Information provided by this sensor has been frequently used to monitor snow cover worldwide. Fernandes and Zhao (2008) designed an algorithm capable of deriving maps showing snow cover from AVHRR images. This algorithm uses reflectance of AVHRR channels 1 and 2, plus NDVI, albedo, and skin temperature. More recently, Hüsler et al. (2012) developed a more accurate algorithm to map snow cover with AVHRR. Although AVHRR supplies the longest time series for snow cover monitoring, it has some drawbacks. The most important one is that its spatial resolution is too coarse to carry out local detailed studies determining the quantification of services provided by snow. In addition, its spectral resolution only allows the creation of snow cover products, creating a fractional snow cover layer is not possible.

15.2.3 Airborne Sensors

Although there is a wide variety of sensors onboard several Earth observation satellites, their spatial resolution is not always accurate enough to monitor snow status and water content. This is the main reason why airborne sensors could be a good strategy to improve the quality of information collected about snow cover. Some sensors have been developed in the past decades to improve the quality of information regarding SWE, snow depth, and snow density. These airborne sensors can be deployed almost anywhere with weather conditions and budget being the only limitations. We will describe two of the most important methodologies: microwaves and Laser Imaging Detection and Ranging (LIDAR). There are some other interesting techniques such as hyperspectral data (Airborne Visible/Infrared Spectrometer, or AVIRIS) and gamma sensors capable of detecting this natural radiation.

Microwave sensors are among the most interesting airborne devices. Microwaves penetrate the snow, allowing collection of information on snow water content (Chang et al. 1982), snow extent, depth, and wet/dry state (Sokol et al. 2003). Depending on the origin of the microwave radiation, we can distinguish two types: (1) Passive microwave—the intensity of microwaves emitted by the ground below the snowpack depends on its structural features (grain size, density, vertical heterogeneity, crystallization status of snow, and temperature). Using these microwave devices, it is possible to estimate the mass of water per unit area in the snowpack, the SWE. However, this method has several drawbacks. One major limitation is the presence of liquid water: its microwave emission masks the snow signal. This makes it difficult for microwave sensors to detect this type of wet snow and (2) Active microwave—these sensors emit their own radiation to record the backscattering response from the terrain. This backscattering depends on the snow cover properties (water content, grain size, etc.). The response of the terrain when there is dry snow is similar to that of bare ground. This makes it difficult to separate both substances (Guneriussen 1997). Therefore, active microwave is not useful for dry snow cover.

Airborne LIDAR is another airborne sensor that can provide information about snow depth and volume. This technique (also called laser altimetry) is based on the emission of laser pulses by a device located in the flight platform. Light is reflected from the ground to the sensor, allowing the creation of a high-resolution digital terrain model of the area. To use LIDAR to monitor snow cover, it is necessary to perform two surveys: the first one without snow (to obtain a sort of baseline DEM) and another one with the snowpack whose depth is going to be measured (Hopkinson et al. 2004). However, it is also possible to perform terrestrial laser scanning using a terrestrial instrument that emits pulses and also detects its reflections with a photodiode (Prokop 2008).

15.3 Final Services Provided by Snow in Mountain Areas

15.3.1 Provisioning Services

The most evident service offered by snow cover is its capacity for storing water. Because of its porosity, snow cover can store huge amounts of water. This storing capacity depends on three factors: the amount of water per volume of snow, the total amount of snow (depth and extent), and its permanence over the landscape.

Therefore, when present, snow cover is an important water reservoir in mountains. It is estimated that snow can cover more than 224,000 km^2 worldwide (Shiklomanov 2009). This area can store more than 24 million km^3 of fresh water. Most of this water comes from rivers and is stored in the soil, acting as a dynamic engine of ecological and human systems. Agriculture is the most important consumer of the water provided by snow in mountain areas. According to Oki and Kanae (2006), this economic sector needs more than 12,000 km^3 of water per year.

Melting water also "carries" some of the potential energy created by altitudinal gradients in mountains. Humans use this energy to create electricity. Some developed countries (Austria, Switzerland) meet between 60% and 70% of their electric demands with hydropower plants that use melting water (Lehner et al. 2005).

15.3.2 Regulating and Maintenance Services

The high albedo and low thermal conductivity of snow are the main characteristics that explain the effect of snow cover on local and global climate (Rittger et al. 2012). Snow cover can change the radiation of the lower atmosphere and of the surface. This physical effect has important consequences for local, regional, and global climate. Snow cover exerts a strong influence on climate (and vice versa), but this effect is very difficult to quantify. The impact of snow on climate can be studied using climatic models (Vavrus 2007). These models forecast that snow cover significantly cools the air of the troposphere at most locations. This cooling effect is one-third the magnitude of the expected warming in a $2 \times CO_2$ scenario.

The local effects of snow on climate have been widely studied in the past decades. Walsh (1984) showed that there is a strong relationship between local temperatures and snow cover. He concluded that colder temperatures were associated with above-normal snow cover. Actually, his studies found that presence of snow could be a good tool for predicting temperature. The regional effects have also been well studied. Global circulation models have helped to disentangle the complex relationships between snow cover and climate. A good example is the effect of snow cover in the Himalayas on the monsoons in south Asia. There is a proven reverse relationship between the extent of Eurasian snow cover in winter and the amount of rainfall in the Indian monsoons in the following summer (Barnett et al. 1989; Yasunari et al. 1991; Douville and Royer 1996). The explanation for lies with albedo: The high albedo of snow cover reduces the total amount of solar radiation, which in turn reduces the heating of the ground surface of the continent. This heating is the most important driver of heavy rains in monsoon areas (Yasunari et al. 1991).

In the previous section, we noted the quantitative effect of snow cover on the hydrological cycle. But the importance of snow is not only about quantity. The solid status of snow gives it a very important role as a buffer for the hydrological cycle in mountain regions. This means that snow cover can retain water and discharge it at a given rate depending on the local temperature. This delay in discharge, provoked by progressive snowmelt, enhances the regulating capacity of aquifers. They are optimally recharged when snowmelt is progressive during the spring. The combination of snowmelt and aquifer recharge explains why basins with most of their water coming from snow display a more constant discharge during the hydrological year (Manga 1999). Besides these natural processes, humans have developed some managerial actions that attempt to regulate the hydrological cycle. From building small dams to store water to using irrigation channels to distribute it (Kamash 2012), humans have tried to regulate hydrological cycles in the mountains across the centuries (Liu and Yamanaka 2012). Box 15.1 provides an example of these human actions.

BOX 15.1 ACEQUIAS: A NETWORK OF IRRIGATION CHANNELS TO REGULATE HYDROLOGICAL CYCLE IN MEDITERRANEAN MOUNTAINS

The presence of snow in high mountains provokes a particular hydrological regime. Flows and processes are gentler. This feature make easier the use of water resources by human settlements. In semiarid environments, these resources—temporally held in the highest parts of the basins—represent essential water storage. At the Guadalfeo watershed, located in southeastern Iberian Peninsula, the amount of water stored as snow in the Sierra Nevada represents more than 70% of the annual water resources. In this area, the historical use of snowmelt (through an intricate system of channels or *acequias* dating from the twelfth century) is one of the more complex water regulating systems seen in a Mediterranean environment. Here, the redistribution of snowmelt occurs in two ways: (1) in a surface manner for downstream irrigation areas, mainly in terraced crops and high-mountain pastures and (2) as artificial groundwater recharge throughout local fractures, or *simas*, that carry the water to the aquifer system. The latter is one of the first systems of its kind on the Iberian Peninsula (Díaz-Marta 1989). Here, the delay of the snowmelt is achieved through the storage–discharge relationship of the aquifer. Afterward, this water is returned downstream through different sources, allowing greater availability during the dry season (Castillo and Fedeli 2002; Millares et al. 2009). Today, thanks to the aesthetic and cultural values, new vegetation and fauna have appeared that are associated with this artificial drainage network. The value of this drainage network in Sierra Nevada National Park has led to numerous investments in inventory, conservation, and restoration of these channels by public managers.

The maintenance role of snow cover is important because it allows the creation of habitat for various life forms. Some of these organisms contribute to creating other ecosystem services. The physical and chemical properties of snow create a habitat suitable for microorganisms, animals, and plants (Jones et al. 2001).

Glaciers and other environments with permanent ice or snow harbor ecosystems populated by unicellular algae, bacteria, and fungi (Segawa et al. 2005). These organisms have adapted to extreme conditions of temperature, pH, irradiation levels, and lack of nutrients (Jones et al. 2001). They can even suffer from drought due to the absence of liquid water. However, snow cover can be a suitable habitat for small animals and plants due to the thermal regime of the subnivean environment. The snow cover acts as a sort of ecotone between the dry, windy, and cold atmosphere and the wet, warm air underneath. Invertebrates and small mammals take advantage of this thermal service to survive the rigors of winter. Snow cover properties also have a deep impact on plant life. Several studies have indicated that phenologies (Ostler et al. 1982) and distribution patterns (Kudo and Ito 1992) of mountain plants are controlled mainly by the snowmelt date and snow cover duration.

Snow cover also has an important impact on the structure of the landscape. Actually, it has been recognized as the most important variable affecting the pattern of vegetation in alpine regions (Jones et al. 2001). Snow cover duration and snow accumulation are key factors explaining the structure and composition of vegetal communities in alpine regions. In addition, snow has a deep effect on vegetation functioning. Changes in maximum snow accumulation can explain more than 50% of the interannual variability in forest greenness in the U.S. Sierra Nevada region (Trujillo et al. 2012). The status of some snow cover services affects the capability of some ecosystems (forest) to supply services (primary production).

15.3.3 Cultural Services

Cultural services supplied by snow cover can be easily described. The most challenging issue is their quantification, due to the high degree of subjectivity involved in the process. In this section, we will outline some of the most important cultural services provided by snow cover. Skiing is probably the most obvious cultural service provided by snow. Mountains and snow cover attract a sizable number of tourists worldwide. There are more than 3500 ski resorts throughout the world, with most located in North America and Europe. A conservative estimation states that there are more than 65 million skiers worldwide. Thus, the economic impact of this service is very important, reaching over US $621 billion by the year 2000 (Hudson 2000). This is an important factor affecting the local economy of some countries. In Austria, for example, snow tourism contributes approximately 18% of the total gross domestic product (GDP; Amelung and Moreno 2009).

Another important service provided by snow cover (and mountain areas) is the scenic value. Mountains are among the most revered landscapes of the world (Beza 2010), and snow cover is an inherent part of many mountains. Although the role of snow in the human perception of beauty is difficult to quantify, some studies note that it is among the five most important characteristics in the mountain landscape (Clay and Daniel 2000).

15.4 A Case Study of Monitoring Snow Cover to Quantify Ecosystem Services: Sierra Nevada Biosphere Reserve (Spain)

Sierra Nevada is a Mediterranean mountain located at the southernmost part of Spain. It reaches 3400 m a.s.l., and it is considered one of the most important hotspots of biodiversity in the Mediterranean region. More than 50 urban areas surround this mountain with a total population of approximately 500,000 inhabitants. Most of this population depends on Sierra Nevada for irrigation and drinking water. This means that services provided by Sierra Nevada are very important for human well-being. Snow cover acts as a fundamental water reservoir that becomes ever more important due to the frequent droughts typical of the Mediterranean climate. Sierra Nevada hosts a Long Term Ecological Research node (http://www.ilternet.edu) that gathers information on the effects of global change in ecosystems and fosters their resilience in coping with those impacts.

One of the first tasks accomplished by this observatory was the creation of a comprehensive system to monitor snow cover at different spatiotemporal scales, to simulate the hydrological effects of snow, and to quantify the services provided by snow to ecosystems and human communities (Figure 15.2). This system can be described using a multilayered approach.

The first layer is a set of monitoring procedures that collects information about snow on Sierra Nevada. There are six protocols covering a wide combination of spatial, temporal, and thematic resolutions. There are three monitoring procedures that harvest information *in situ*: (1) poles, (2) automatic meteorological stations with digital cameras and ultrasonic sensors to measure snow depth, and (3) transects useful to quantify SWE and other physical properties. Three other protocols collect information on snow cover using remote devices: (1) an automatic system that is able to download, process, and analyze all the images created by MODIS snow cover products from 2000 to the present time; (2) a more detailed monitoring protocol based on Landsat images that are periodically processed to extract the extent of snow cover; and (3) oblique photographs are taken daily from the city of Granada (located 30 km from Sierra Nevada) to the hills of the Sierra Nevada. The objective is to monitor the process of snow melt in spring and early summer.

The raw data obtained by the first layer are transferred to the second layer whose main function is to create useful knowledge by aggregating and

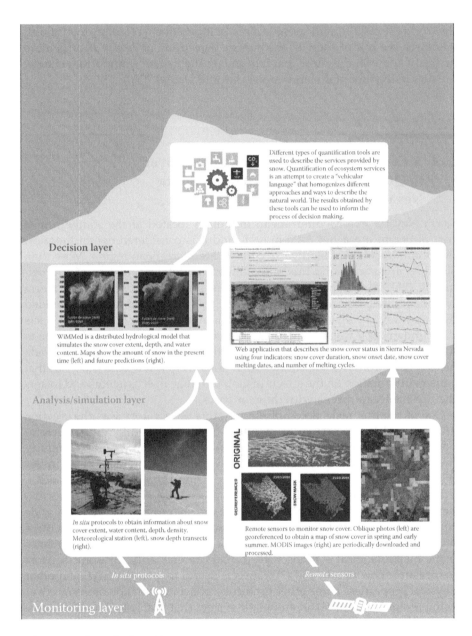

FIGURE 15.2 (See color insert.)
Schema showing the structure of a comprehensive system capable of monitoring snow cover using different methodologies, analyzing raw data to obtain snow cover indicators, and finally quantifying the services provided by snow. This system is being implemented in the Sierra Nevada Mountains (Spain).

analyzing information. This analytical layer is conformed by a set of complex tools capable of simulating processes related to snow formation. We have developed two analytical tools: (1) The most powerful is WiMMed (Polo et al. 2009). It is a distributed hydrological model that simulates water flow in rivers, snow water content, the extent of snow cover, and even soil moisture (see Chapter 14). WiMMed uses Landsat and oblique photographs to validate the snow cover dynamics. SWE and snow depth provided by snow transects, poles, and meteorological stations are used to calibrate the model; and (2) *Linaria* is another set of analytical procedures that download, process, and analyze all the information provided by MODIS snow products. This system relies on Kepler (Altintas et al. 2004) scientific workflows to process all Hierarchical Data Format (HDF) files provided by NASA, and it obtains graphs and tables showing four important indicators of snow status. These indicators are snow duration, snow onset dates, snow cover melt dates, and the number of melting cycles per hydrological year (Wang and Xie 2009). The resulting information is accessible through a Web portal (free registration at http://linaria.obsnev.es).

The third layer comprises a set of methodologies that should quantify the services provided by snow cover by using the aforementioned raw and processed information. To do this, it is necessary to create mixed ecohydrological models that simulate quantity, quality, and temporal evolution of different types of ecosystem services (supporting, regulating, provisioning). A good example of models that could be used in this layer is Integrated Valuation of Ecosystem Services and Tradeoff (InVEST; Tallis et al. 2011). This software is designed to quantify ecosystem services and to evaluate trade-offs among multiple human uses.

15.5 Conclusion

Snow cover is a very important abiotic element that shapes mountain landscape. Although it cannot be considered as an ecosystem services provider (because it is not an ecosystem), snow cover is implicated in several final ecosystem services.

Remote sensing and other techniques are very useful for monitoring the extent of snow cover, water content, depth, and so on. A well-designed monitoring program should include a combination of different monitoring techniques, from field campaigns for gathering information about water content to remote sensing information that assesses the extent of snow cover. This combination of methods is especially important in mountain areas, where topography explains the distribution of snow over the terrain.

Obtaining information about snow cover structure is only the first step in the process of ecosystem services quantification. It is necessary to establish links between the snow cover structure and the services provided by mountain ecosystems. Snow is related to several final ecosystem services: production of water is probably the most evident one, but not the only one. Another provisioning service is the capacity to generate electric power. Regulatory and maintenance services are also very important. Snow's albedo explains its important role in climate regulation at local and regional scales (see Chapter 16). The solid state of snow gives it a very important role as a buffer of the hydrological cycle in mountain areas. Snow can retain water and discharge it at a given rate depending on the local temperature. Finally, snow cover is considered as a favorable habitat for several species and biological communities. Some plants, invertebrates, and even mammals take advantage of the thermal regime of the subnivean environment. Snow also affects some ecological functions such as phenology and the distribution of several plant species.

But the real challenge that should be addressed is the quantification of the ecosystem services where snow cover is implicated. From the aforementioned properties of snow, it is possible to quantify their effects over ecosystem services. This quantification could be addressed using both hydrological and ecological models. The raw data obtained via monitoring protocols could be used as input data by these distributed simulation models.

Acknowledgments

This chapter has been written in the collaborative framework of the Sierra Nevada Global Change Observatory (http://obsnev.es). Funding was provided by the Fundación Biodiversidad (Spanish government) and CEI Biotic from the University of Granada.

References

Altintas, I., C. Berkley, E. Jaeger, M. Jones, B. Ludascher, and S. Mock. 2004. Kepler: An extensible system for design and execution of scientific workflows. *Proceedings of the 16th International Conference on Scientific and Statistical Database Management, Greece.* Vol. I. doi:10.1109/SSDM.2004.1311241.

Amelung, B., and A. Moreno. 2009. *Impacts of climate change in tourism in Europe. PESETA-Tourism Study, Luxembourg: Office for Official Publications of the European Communities.* doi:10.2791/33218.

Andreadis, K. M., and D. P. Lettenmaier. 2006. Assimilating remotely sensed snow observations into a macroscale hydrology model. *Advances in Water Resources* 29:872–886.

Barnett, T. P., L. Dümenil, U. Schlese, E. Roeckner, and M. Latif. 1989. The effect of Eurasian snow cover on regional and global climate variations. *Journal of the Atmospheric Sciences* 46:661–685.

Beza, B. B. 2010. The aesthetic value of a mountain landscape: A study of the Mt. Everest trek. *Landscape and Urban Planning* 97:306–317.

Castillo, A., and B. Fideli. 2002. Algunas pautas del comportamiento hidrogeológico de rocas duras afectadas por glaciarismo y periglaciarismo en Sierra Nevada (España). *Geogaceta* 32:189–191.

Chang, A. T. C., J. L. Foster, and D. K. Hall. 1982. Snow water equivalent estimation by microwave radiometry. *Cold Regions Science and Technology* 5:259–267.

Chow, V. T. 1964. *Handbook of applied hydrology*. New York: McGraw-Hill.

Clay, G. R., and T. C. Daniel. 2000. Scenic landscape assessment: The effects of land management jurisdiction on public perception of scenic beauty. *Landscape and Urban Planning* 49:1–13.

Cohen, J., and D. Rind. 1991. The effect of snow cover on the climate. *Journal of Climate* 4:689–706.

Corripio, J. G. 2004. Snow surface albedo estimation using terrestrial photography. *International Journal of Remote Sensing* 25:5705–5729.

Diaz, H. F., M. Grosjean, and L. Graumlich. 2003. Climate variability and change in high elevation regions: Past, present and future. *Climatic Change* 2001:1–4.

Díaz-Marta, M. 1989. Esquema histórico de la ingeniería y la gestión del agua en España. *Revista de obras públicas* 13:8–23.

Dietz, A. J., C. Kuenzer, U. Gessner, and S. Dech. 2012. Remote sensing of snow—A review of available methods. *International Journal of Remote Sensing* 33:4094–4137.

Dingman, L. 2002. *Physical hydrology*. Upper Saddle River, NJ: Prentice Hall.

Douville, H., and J. F. Royer. 1996. Sensitivity of the Asian summer monsoon to an anomalous Eurasian snow cover within the Météo-France GCM. *Climate Dynamics* 12:449–466.

Dozier, J., and T. H. Painter. 2004. Multispectral and hyperspectral remote sensing of alpine snow properties. *Annual Review of Earth and Planetary Sciences* 32:465–494.

Fernandes, R., and H. Zhao. 2008. Mapping daily snow cover extent over land surfaces using NOAA AVHRR imaginery. In *Remote sensing of land ice and snow*, 11–13. The Netherlands: IOS Press.

Green, R. O., and M. Shimada. 1997. On-orbit calibration of a multi-spectral satellite sensor using a high altitude airborne imaging spectrometer. *Advances in Space Research* 19:1387–1398.

Guneriussen, T. 1997. Backscattering properties of a wet snow cover derived from DEM corrected ERS-1 SAR data. *International Journal of Remote Sensing* 18:375–392.

Haines-Young, R., and M. Potschin. 2013. CICES V4.3–Revised report prepared following consultation on CICES Version 4, August–December 2012. EEA Framework Contract No EEA/IEA/09/003.

Hall, D. K., G. A. Riggs, and V. V. Salomonson. 1995. Development of methods for mapping global snow cover using moderate resolution imaging spectroradiometer data. *Remote Sensing of Environment* 54:127–140.

Herrero, J., M. J. Polo, and M. A. Losada. 2011. Snow evolution in Sierra Nevada (Spain) from an energy balance model validated with Landsat TM data. Proc. SPIE 8174, *Remote Sensing for Agriculture, Ecosystems, and Hydrology* XIII, 817403.

Hinkler, J., S. B. Pedersen, M. Rasch, and B. U. Hansen. 2002. Automatic snow cover monitoring at high temporal and spatial resolution, using images taken by a standard digital camera. *International Journal of Remote Sensing* 2:4669–4682.

Hopkinson, C., M. Sitar, L. Chasmer, and P. Treitz. 2004. Mapping snowpack depth beneath forest canopies using airborne LIDAR. *Photogrammetric Engineering & Remote Sensing* 70:323–330.

Hudson, S. 2000. *Snow business: A study of the international ski industry*. London: Cassell.

Hüsler, F., T. Jonas, S. Wunderle, and S. Albrecht. 2012. Validation of a modified snow cover retrieval algorithm from historical 1-km AVHRR data over the European Alps. *Remote Sensing of Environment* 121:497–515.

Irons, J. R., J. L. Dwyer, and J. A. Barsi. 2012. The next Landsat satellite: The Landsat Data Continuity Mission. *Remote Sensing of Environment* 122:11–21.

Jones, H. G., J. W. Pomeroy, D. A. Walker, and R. W. Hoham. (eds.). 2001. *Snow ecology: An interdisciplinary examination of snow-covered ecosystems*. Cambridge, UK: Cambridge University Press.

Kamash, Z. 2012. Irrigation technology, society and environment in the Roman near east. *Journal of Arid Environments* 86:65–74.

Kudo, G., and K. Ito. 1992. Plant distribution in relation to the length of the growing season in a snow-bed in the Taisetsu Mountains, northern Japan. *Vegetatio* 98:165–174.

Laffly, D., E. Bernard, M. Griselin, et al. 2011. High temporal resolution monitoring of snow cover using oblique view ground-based pictures. *Polar Record* 48:11–16.

Lehner, B., G. Czisch, and S. Vassolo. 2005. The impact of global change on the hydropower potential of Europe: A model-based analysis. *Energy Policy* 33:839–855.

Liu, Y., and T. Yamanaka. 2012. Tracing groundwater recharge sources in a mountain-plain transitional area using stable isotopes and hydrochemistry. *Journal of Hydrology* 464–465:116–126.

Manga, M. 1999. On the timescales characterizing groundwater discharge at springs. *Journal of Hydrology* 219:56–69.

Martiny, N., R. Santer, and I. Smolskaia. 2005. Vicarious calibration of MERIS over dark waters in the near infrared. *Remote Sensing of Environment* 94:475–490.

Millares, A., M. J. Polo, and M. A. Losada. 2009. The hydrological response of baseflow in fractured mountain areas. *Hydrology and Earth System Science* 13:1261–1271.

Oki, T., and S. Kanae. 2006. Global hydrological cycles and world water resources. *Science* 313:1068–1072.

Ostler, W. K., K. T. Harper, K. T. McKnight, and D. C. Anderson. 1982. The effects of increasing snowpack on a subalpine meadow in the Uinta Mountains, Utah, USA. *Arctic and Alpine Research* 14:203–214.

Parajka, J., and G. Blöschl. 2008. The value of MODIS snow cover data in validating and calibrating conceptual hydrologic models. *Journal of Hydrology* 358:240–258.

Parajka, J., P. Haas, R. Kirnbauer, J. Jansa, and G. Blöschl. 2012. Potential of time-lapse photography of snow for hydrological purposes at the small catchment scale. *Hydrological Processes* 26:3327–3337.

Polo, M. J., J. Herrero, and C. Aguilar. 2009. WiMMed, a distributed physically-based watershed model (I): Description and validation. In *Environmental hydraulics: Theoretical, experimental and computational solutions, IWEH09*, ed. P. A. López-Jiménez, 225–228. London: CRC Press.

Powell, C., L. Blesius, J. Davis, and F. Schuetzenmeister. 2011. Using MODIS snow cover and precipitation data to model water runoff for the Mokelumne River Basin in the Sierra Nevada, California (2000–2009). *Global and Planetary Change* 77:77–84.

Prokop, A. 2008. Assessing the applicability of terrestrial laser scanning for spatial snow depth measurements. *Cold Regions Science and Technology* 54:155–163.

Rallison, R. E. 1981. Automated system for collecting snow and related hydrological data in mountains of the Western United States. *Hydrological Sciences* 26:83–89.

Rittger, K., T. H. Painter, and J. Dozier. 2012. Assessment of methods for mapping snow cover from MODIS. *Advances in Water Resources* 51:367–380.

Rosenthal, W., and J. Dozier. 1996. Automated mapping of montane snow cover at subpixel resolution from the Landsat Thematic Mapper. *Water Resources Research* 32:115–130.

Salomonson, V. V., and I. Appel. 2006. Development of the Aqua MODIS NDSI fractional snow cover algorithm and validation results. *IEEE Transactions on Geoscience and Remote Sensing* 44:1747–1756.

Sanderson, E. W., M. Jaiteh, M. A. Levy, K. H. Redford, A. V. Wannebo, and G. Woolmer. 2002. The human footprint and the last of the wild. *BioScience* 52:891–904.

Segawa, T., K. Miyamoto, and K. Ushida. 2005. Seasonal change in bacterial flora and biomass in mountain snow from the Tateyama Mountains, Japan, Analyzed by 16S rRNA gene sequencing and real-time PCR. *Applied and Environmental Microbiology* 71:123–130.

Shiklomanov, I. A. 2009. *The hydrological cycle*. Vol. I. St. Petersburg, Russia: EOLSS Publishers.

Silvertown, J. 2009. A new dawn for citizen science. *Trends in Ecology and Evolution* 24:467–471.

Sokol, J., T. J. Pultz, and A. E. Walker. 2003. Passive and active airborne microwave remote sensing of snow cover. *International Journal of Remote Sensing* 24:5327–5344.

Tallis, H. T., T. Ricketts, A. D. Guerry, et al. 2011. InVEST 2.4.3 user's guide: Integrated valuation of environmental services and tradeoffs. A modeling suite developed by the Natural Capital Project to support environmental decision-making. Stanford, CA: The Natural Capital Project.

Thirel, G., C. Notarnicola, M. Kalas, et al. 2012. Assessing the quality of a real-time snow cover area product for hydrological applications. *Remote Sensing of Environment* 127:271–287.

Trujillo, E., N. P. Molotch, M. L. Goulden, A. E. Kelly, and R. C. Bales. 2012. Elevation-dependent influence of snow accumulation on forest greening. *Nature Geoscience* 5:705–709.

Vavrus, S. 2007. The role of terrestrial snow cover in the climate system. *Climate Dynamics* 29:73–88.

Vikhamar, D., and R. Solberg. 2003. Subpixel mapping of snow cover in forests by optical remote sensing. *Remote Sensing of Environment* 84:69–82.

Walsh, J. E. 1984. Snow cover and atmospheric variability. *American Scientist* 72:50–57.

Wang, D., D. Morton, J. Masek, et al. 2012. Impact of sensor degradation on the MODIS NDVI time series. *Remote Sensing of Environment* 119:55–61.

Wang, X., and H. Xie. 2009. New methods for studying the spatiotemporal variation of snow cover based on combination products of MODIS Terra and Aqua. *Journal of Hydrology* 371:192–200.

Watson, F. G. R., T. N. Anderson, and W. B. Newman. 2008. Modeling spatial snow pack dynamics. *Terrestrial Ecology* 7961:18–23.

Yasunari, T., A. Kitoh, and T. Tokioka. 1991. Local and remote responses to excessive snow mass over Eurasia appearing in the northern spring and summer climate. A study with the MRI-GCM. *Journal of the Meteorological Society of Japan* 69:473–487.

Section V

Ecosystem Services Related to the Land-Surface Energy Balance

16

Characterizing and Monitoring Climate Regulation Services

D. Alcaraz-Segura
University of Granada, Spain; University of Almería, Spain

E. H. Berbery
University of Maryland, Maryland

O. V. Müller
National University of Litoral, Argentina

J. M. Paruelo
University of Buenos Aires, Argentina

CONTENTS

16.1 Introduction 352
    16.1.1 The Ecosystem Service of Climate Regulation 352
    16.1.2 Ecosystem–Climate Feedbacks 353
    16.1.3 Modeling of Ecosystem–Atmosphere Interactions 354
16.2 Identification of EFTs 357
    16.2.1 Satellite Data Record 357
    16.2.2 Definition of EFTs 357
    16.2.3 EFTs of Southern South America 359
16.3 Biophysical Properties from EFTs 360
    16.3.1 Land Surface Parameterization of EFTs 360
    16.3.2 U.S. Geological Survey versus EFT-Derived Biophysical
           Properties 361
    16.3.3 Interannual Variability of Vegetation Properties 363
16.4 Climate Regulation Services in Regional Modeling 363
    16.4.1 Accounting for Changes in Biophysical Properties in a
           Regional Climate Model 363
    16.4.2 A Case Study for a Drought Episode 366

16.5 Discussion and Conclusions 373
Acknowledgments 374
References 374

16.1 Introduction

16.1.1 The Ecosystem Service of Climate Regulation

Ecosystem services related to the atmospheric composition and climate regulation group (Haines-Young and Potschin 2013) are associated with the maintenance of both global and local climate conditions that are favorable for health, crop production, and other human activities. At the global scale, ecosystem biogeochemical processes influence the climate by emitting/absorbing greenhouse gases and aerosols to/from the atmosphere. Forests capture and store carbon dioxide, while marshlands, lakes, rice paddies, and cattle-raising rangelands emit methane. Ecosystems' biophysical properties, such as albedo, latent heat, and sensible heat, affect the local or regional temperature, precipitation, and other climatic factors (Pielke et al. 2002; Bonan 2008; Oki et al. 2013). For instance (see Bonan, 2008), part of the incoming solar radiation in a tropical forest is used for water evapotranspiration (latent heat flux), which decreases surface temperature. Furthermore, the forest's evapotranspiration may favor cloud formation as part of the local climate but also help maintain air quality. In a tropical desert, radiation heats up the soil, which then heats the air (sensible heat flux).

Despite ecosystems regulate climate through biogeochemistry (e.g., greenhouse gas exchanges) and biophysics (e.g., water and energy balance), current policies only focus on biogeochemical influences (i.e., CO_2 emissions). Recently, Anderson-Teixeira et al. (2012) proposed a climate regulation value (CRV) index that accounts for the biogeochemical and biogeophysical ecosystem properties that affect the value of ecosystem–climate services. The CRV converts the biophysical effects into biogeochemical units. Hence, the CRV offers the possibility of expanding the suite of climate regulation services considered in the current global policies and carbon markets. The biophysical part of the CRV is estimated from ecosystem's surface net radiation and latent heat flux, which are simulated using land surface models such as Integrated Biosphere Simulator (IBIS) (Foley et al. 1996; Kucharik et al. 2000) or Noah Land Surface Model (LSM) (Chen et al. 1996; Chen and Dudhia 2001; Ek et al. 2003). In general, these simulations involve the use of variables related to vegetation such as leaf area index, stomatal resistance, rooting depth, albedo and transmittance in the visible and near infrared, heat, water and snow capacities, and others. This chapter presents an original method to estimate and monitor such variables using satellite information and coupled climate regional models.

16.1.2 Ecosystem–Climate Feedbacks

Climate is the main regional driver of ecosystem structure and functioning that determines the timing and amount of energy (both heat and solar radiation) and water that are available in the system (Stephenson 1990). Conversely, ecosystems also influence climate through multiple pathways, primarily by determining the energy, momentum, water, and chemical balance between the land surface and the atmosphere through changes in albedo, longwave radiation, surface roughness, evapotranspiration, greenhouse gases, or aerosols (Chapin et al. 2008). Hence, impacts on ecosystems, of natural and human origin, may alter one or several pathways of the ecosystem–climate feedbacks that may end up affecting the regional and global climate. Indeed, several studies (e.g., Weaver and Avissar 2001; Pielke et al. 2002; Werth and Avissar 2002; Kalnay and Cai 2003) have concluded that the contribution of land use changes to climate change might be about 10% of the total global change but that regionally the relative contribution of land use change may be notably larger, even larger than that from greenhouse gas emissions. There are known cases showing how land use changes may alter the regional climate, such as the aridification of the Mediterranean basin during the Roman period (Reale and Dirmeyer 2000; Reale and Shukla 2000) or changes in the hydrometeorology of Amazonia after deforestation (Gedney and Valdes 2000; Roy and Avissar 2002).

Interannual variability in climate conditions significantly affects vegetation structural and functional properties (Brando et al. 2010; Zhao and Running 2010), possibly influencing the regional climate. For example, the decline in vegetation density produced by droughts increases albedo and may reduce convective uplift and moisture advection (Bonan 2008). Insect outbreaks (Maness et al. 2013) and overgrazing can aggravate these effects by further reducing vegetation density, which strengthens the albedo-induced decline in the convective uplift. On the other hand, vast areas of South America are also suffering from human-made changes in land cover and management practices of crop-systems that may affect ecosystem–climate feedbacks (Foley et al. 2003). From these, deforestation and land clearing for agriculture and cattle ranging are the most important ones (Foley et al. 2007; Volante et al. 2012). Land clearing produces (1) an increase in albedo, which reduces energy transfer to the ecosystem (and, subsequently, to the atmosphere); (2) a reduction of transpiration, which reduces moisture transport from the soil and surface aquifer to the atmosphere; and (3) a net release of CO_2, which increases the heat-trapping capacity of the atmosphere. Conversely, other extensive land use changes in South America, such as grassland afforestation (Nosetto et al. 2005, and references therein), produce (1) a decrease in albedo, which increases the energy transfer to the ecosystem; (2) a rise of evapotranspiration, which increases moisture transport from the soil and surface aquifer to the atmosphere; and (3) greater surface roughness (Beltrán-Przekurat et al. 2012). Yet other examples of ecosystem–climate feedback derive from the extensive practice of no-tillage agriculture or the extensive expansion of irrigated agriculture over drylands (De Oliveira et al. 2009), which increases evapotranspiration and decreases albedo.

The ecosystem–climate feedbacks are a central problem not only for modeling the land–atmosphere interactions of the climate system (Mahmood et al. 2010) but also for many other biological and environmental issues (Oki et al. 2013). As stated, ecosystem–atmosphere interactions and feedbacks depend on the physical properties of the underlying land cover. Changes in the structure and functioning of the ecosystems will thus have an impact on those properties and, consequently, will affect the radiation balance at the surface as well as the exchange of momentum, heat, moisture, and other gaseous/aerosol materials.

The exchanges of mass and energy may be modified by at least two mechanisms: by human activity changing the land surface conditions (land cover and land use changes) or by natural variability in climate that affects the ecosystem's health, performance, and biophysical properties. Some typical cases are low-frequency modulation of regional climate and the development of droughts and wet spells.

As much as climate affects ecosystems, the inverse path is also common. The size, geographic location, and patchiness of an area where land cover changes take place may determine the extent to which they affect local, regional, and global climate (Marland et al. 2003; Pielke et al. 2007). Small areas (e.g., of the order 10 km) of land cover change can result in changes in the local pattern and intensity of precipitation (Pielke et al. 2007). In tropical regions, where large thunderstorms are frequent, the effect will be larger with possible impacts even escalating up to the global scales (Pielke 2001; Werth and Avissar 2002). In the United States, numerical and observational studies show that changes in land cover and increased use of irrigation can influence regional climate and vegetation (Stohlgren et al. 1998; Baidya Roy et al. 2003; Diffenbaugh 2009).

16.1.3 Modeling of Ecosystem–Atmosphere Interactions

Ecosystem–climate feedbacks are a central problem for modeling the land–atmosphere interactions of the climate system. The incorporation of land surface–atmosphere interactions and feedbacks in current regional and global circulation models is not straightforward. Some approaches use land cover maps to estimate maps of biophysical properties (West et al. 2011).

Such estimates rely on the relationship between particular plant functional traits (e.g., such as having an evergreen, deciduous, or annual life form) and different ecosystem functioning properties (Smith et al. 1997). However, the understanding of how these functional traits determine upscale effects on ecosystems and biogeochemical cycles is not fully understood (Lavorel et al. 2007). In fact, several works have shown how plant functional type classifications were not reliable for predicting ecosystem functioning (Wright et al. 2006; Bret-Harte et al. 2008). The functional traits and properties are usually determined from limited observations and models usually assume them to be constant within a plant functional type or land cover despite the fact that in the real world they vary within and between plant functional types (e.g., a needleleaf forest will be assumed to have the same constant properties anywhere in the world even though the observations may vary and have been taken at few locations) (Reich and Oleksyn 2004; Wright et al. 2004, 2005, 2006; Reich et al. 2006). In addition, these land cover maps are difficult to update on a yearly basis and some structural features of vegetation (such as leaf life span) have little sensitivity to environmental changes (McNaughton et al. 1989). Overall, a simplified representation of vegetation properties may result in a delayed response and reduces the ability of models to represent rapid changes including land use shifts, fires, floods (see Chapter 17), droughts, and insect outbreaks. Hence, improving the way spatial and interannual variability of vegetation dynamics is considered in land surface models is necessary to account for land use/cover change effects on circulation models.

Dynamic vegetation models that include the carbon cycle add a significant improvement in the area of ecosystem–atmosphere interactions because they allow for vegetation changes and have advanced assumptions regarding surface processes. However, many land surface models do not consider the concept of ecosystems. This is the case of models of intermediate complexity used on operational forecasting (such as the Noah LSM model) that have constant (or static) vegetation classes with look-up tables to identify the value or the annual cycle of their corresponding biophysical properties. More complex models employ land cover classifications that identify patches of the land surface that are homogeneous in terms of their plant functional type composition. Plant functional types are groups of species that share similar functional features such as leaf life, metabolic route, or nitrogen fixation (Smith et al. 1997). Land cover types that are assumed to remain constant in their composition in reality may experience important changes. For instance, the biophysical properties of a typical vegetation type during a wet period should be very different during a drought. The same is true during anomalous periods of intense rain that can create numerous ponds or flooding. Interestingly, for a model that prescribes the same annual cycle of the surface properties for any year, all these cases will behave similarly in terms of land–atmosphere interactions, the radiation budget, and the surface water, energy, and carbon cycles.

Functional attributes of vegetation describing the energy and matter exchange between the biota and the atmosphere (Valentini et al. 1999;

Virginia et al. 2001) at the ecosystem scale, may help to fulfill these needs since they show a quicker response to environmental changes than structural ones (McNaughton et al. 1989). In addition, many ecosystem functional properties are relatively easy to monitor using satellite-derived spectral indices. The most widely used spectral index to characterize and monitor ecosystem functioning is the Normalized Difference Vegetation Index (NDVI). This spectral index is strongly related to the fraction of the incoming photosynthetically active radiation that is absorbed by green vegetation (Sellers et al. 1986; Tucker and Sellers 1986) and, hence, with primary production (Monteith 1972), which is the most integrative indicator of ecosystem functioning (Virginia et al. 2001). The NDVI has been widely and satisfactorily used for describing intra- and interannual variability of vegetation dynamics (Alcaraz-Segura et al. 2009), monitoring ecosystem changes in structure and function (Pettorelli et al. 2005), detecting long-term trends in vegetation growth and phenology (Kathuroju et al. 2007), providing inputs for primary production models (Cao et al. 2004), and providing a reference to model the carbon balance worldwide (Potter et al. 2005).

NDVI-derived functional descriptors of the interception of radiation by vegetation have also been used to produce classifications of ecosystem functional types (EFTs) that capture the spatial heterogeneity of ecosystem functioning (Paruelo et al. 2001). EFTs are defined as patches of the land surface that exchange mass and energy with the atmosphere in a common way, showing a coordinated and specific response to environmental factors (Soriano and Paruelo 1992; Valentini et al. 1999; Paruelo et al. 2001). EFTs can thus be considered a top-down approach to classify ecosystems at a higher level of classification than the more traditional bottom-up land cover types approach based on plant functional types. Alcaraz-Segura et al. (2006) modified the approach by Paruelo et al. (2001) to define EFTs based on fixed limits between classes, which opened the possibility to monitor and compare ecosystem functioning through time.

This chapter discusses the interannual changes in EFTs and their corresponding biophysical properties. Then, a method is proposed to replace the traditional land cover types in regional models by time-varying EFTs with corresponding properties. We focused our analysis in southern South America because it is strongly affected by interannual changes in climatic conditions due to El Niño southern oscillation, and it suffered vast land cover changes during the last few decades, which presents a unique opportunity to assess the impact of climate variability and land transformations on the biophysical properties of vegetation using satellite information. For this, we first produced annual EFTs maps from 1982 to 1999 using three metrics of the NDVI dynamics. Then, we obtained the biophysical properties of each EFT based on the Noah LSM parameterization for the U.S. Geological Survey (USGS) land cover classes. Finally, we present a case study of the implications of implementing a more realistic representation of lower boundary conditions with EFTs on regional climate simulations.

16.2 Identification of EFTs

16.2.1 Satellite Data Record

The identification of EFTs was based on the NDVI dataset produced by the Land Long-Term Data Record team (Pedelty et al. 2007). The NDVI is calculated from the reflectance in the red and near-infrared wavelengths (Tucker and Sellers 1986): NDVI = (NIR − R)/(NIR + R), NIR and R stand for the spectral reflectance in the near-infrared and red regions, respectively. LTDR is the most recent NDVI dataset derived from the Advanced Very High Resolution Radiometer (AVHRR) archive. LTDR is a NASA-funded REASoN (Research, Education and Applications Solutions Network) project that aims to produce a consistent long-term dataset from AVHRR, Moderate Resolution Imaging Spectroradiometer (MODIS), and Visible/Infrared Imager/Radiometer Suite (VIIRS) sensors. The LTDR project is reprocessing Global Area Coverage (GAC) archive from 1981 to the present by applying the preprocessing improvements identified in the Pathfinder AVHRR Land II (PAL-II) project and the atmospheric and BRDF corrections used in MODIS preprocessing steps (http://ltdr.nascom.nasa.gov) (Pedelty et al. 2007). Version 2 of the LTDR dataset consists of daily global images at a spatial resolution of $0.05° \times 0.05°$ (~25 km^2 at the equator) that expands from 1981 to 1999. The images have been corrected for sensor degradation, water and ozone vapor absorptions, Raleigh scattering, and intersensor differences. In this study, we calculated 15-day maximum value composites (Holben 1986) to minimize the noise due to cloud cover, cloud shadow, and aerosol contamination of the daily images (though residual noise may remain in the time series; Nagol et al. 2009). We also used the quality assessment flag information to discard NDVI values with low quality (i.e., invalid NDVI channels, high solar zenith angle, and presence of sun glint, cloud shadow, and cloudy or partially cloudy) (further details in Alcaraz-Segura et al. 2010, 2013).

16.2.2 Definition of EFTs

First, we classified as water bodies all pixels that were masked as water in the LTDR quality assessment flags (in more than 80% of the 15-day composites of all years). The remaining pixels were classified into EFTs following the approach introduced by Alcaraz-Segura et al. (2006) where EFTs were identified using fixed limits between classes, which allows for interannual comparisons. For this, we used three metrics of the NDVI seasonal dynamics of each year that are known to capture most of the variability of the NDVI time series (Paruelo et al. 2001; Alcaraz-Segura et al. 2009): NDVI annual mean (NDVI-m), seasonal coefficient of variation (CVseas, i.e., intra-annual standard deviation divided by annual mean) and date of the maximum NDVI value (DMAX). NDVI-m is a linear estimator of primary production, CVseas is a descriptor of seasonality (i.e., intra-annual differences in carbon gains between growing and nongrowing seasons), and DMAX is a phenological indicator of the growing season. Then, the range of values of each NDVI metric was divided into four fixed intervals, giving a potential number of $4 \times 4 \times 4 = 64$ EFTs. In the case of DMAX, the four intervals agreed with the four seasons of the year that occur in temperate ecosystems. In the case of NDVI-m and CVseas, the three fixed limits between intervals were calculated as the 18-year median of the first, second, and third quartiles of the 18 histograms obtained for each variable. We assigned codes to each EFT following the terminology suggested by Paruelo et al. (2001) based on two letters and a number (three characters). The first letter of the code (capital) corresponded to the NDVI-m level, ranging from "A" to "D" for low to high NDVI-m. The second letter (small) showed the seasonal CV, ranging similarly from "a" to "d" for low to high CV. The numbers indicated the season of maximum NDVI (see Table 16.1). The definition and coding of EFTs were based only on descriptors of ecosystem functioning and allow for a straightforward ecological interpretation of the legend. Finally, we applied the fixed limits of the former procedure to each year, so we could compare how the EFT distribution varied across years. To produce a map with the average distribution of EFTs

TABLE 16.1

Range of LTDR/NDVI-Derived Traits

	Functional Code	Lower Limit	Upper Limit
NDVI-m	A	0.0001	0.3476
	B	0.3476	0.5186
	C	0.5186	0.5834
	D	0.5834	0.8891
CVseas	d	0.0000	0.1388
	c	0.1388	0.1813
	b	0.1813	0.2307
	a	0.2307	3.8974
DMAX	1	Spring	
	2	Summer	
	3	Autumn	
	4	Winter	

Note: Range traits used in the definition of ecosystem functional types in South America for the 1982–1999 period: NDVI annual mean (NDVI-m), seasonal coefficient of variation NDVI (CVseas), and date of the maximum NDVI (DMAX). Capital letters correspond to the NDVI-m level, ranging from "A" to "D" for low to high NDVI-m. Lowercase letters show the seasonal coefficient of variation, ranging similarly from "a" to "d" for high to low CVseas. The numbers indicate the season of maximum NDVI.

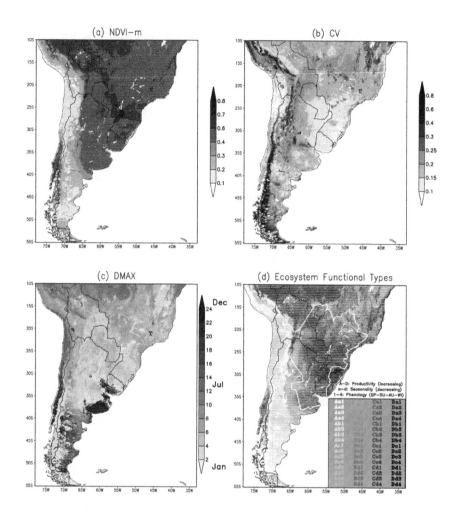

FIGURE 16.1 (See color insert.)
Distribution and relative extension of ecosystem functional types in South America based on the LTDR/NDVI dynamics. The maps show (a) the NDVI annual mean (NDVI-m), (b) seasonal coefficient of variation NDVI (CVseas), (c) date of the maximum NDVI (DMAX), and (d) ecosystem functional type (EFTs) corresponding to the median of the 17-year 1982–1999 period. See Table 20.1 for the interpretation of the EFT naming codes.

for the 1982–1999 period, we calculated the EFT corresponding to the 18-year median of the EFTs histogram for each pixel. Almost all possible combinations of NDVI-m, CVseas, and DMAX were identified in the study area (Figure 16.1).

16.2.3 EFTs of Southern South America

The EFT map presented in Figure 16.1d shows a concise characterization of spatial patterns of ecosystem functioning in temperate South America for the 1982–1999 period. The EFT map captures in a synthetic way the spatial variability of three integrative descriptors of ecosystem functioning: productivity (NDVI-m, Figure 16.1a), seasonality (CVseas, Figure 16.1b), and phenology (DMAX, Figure 16.1c). Almost all possible combinations of NDVI-m, CVseas, and DMAX were identified in the study area. On average, most ecosystems showed NDVI maxima in autumn and summer (Figure 16.1c). EFTs with summer maxima tended to have medium to low productivity and high seasonality, while EFTs with autumn and spring maxima showed most of the possible combinations of productivity and seasonality. EFTs with NDVI maxima during winter tended to exhibit either very low or very high productivity with very low seasonality values. The definition and coding of EFTs allow for an ecological interpretation of the legend in terms of the three NDVI metrics related to productivity (NDVI-m), seasonality (CVseas), and phenology (DMAX). The greatest NDVI-m (D) was reached in the Alto Paraná Atlantic Forests and in the Southwestern Amazonian Moist Forests (Figure 16.1a). All spatial references in this article are based upon the World Wildlife Fund (WWF) Ecoregions Map of the World (Olson et al. 2001). The lowest NDVI-m (A) occurred in the Atacama-Sechura Deserts, in the Central Andean Dry Puna, and in some central areas of the Patagonian Steppe (Figure 16.1a). The CVseas was relatively low (c, d) throughout the study area (Figure 16.1b). The greatest seasonality (a) occurred in the highest altitudes of (1) the southern Andes (highest parts of the Valdivian Temperate Forests) and (2) the northern Andes (Bolivian Montane Dry Forest and Central Andean Wet Puna) (Figure 16.1b). High seasonality (a, b) is also found in agricultural lands of central and northwestern Argentina (across the Espinal and Dry Chaco) and eastern Brazil (across the Caatinga and Atlantic Dry Forests) (Figure 16.1b). The lowest seasonality (d) was observed in both (1) the driest ecoregions, such as the Atacama Desert and the Patagonian Steppe, and (2) in very humid ecoregions, such as the Alto Paraná Atlantic Forests, and the Southwestern Amazonian Moist Forests (Figure 16.1b). The phenological indicator of the growing season, DMAX, showed that most areas of extratropical South America have summer (2) and autumn (3) NDVI maxima (Figure 16.1c). Spring (1) maxima were observed just in the southeastern half of the Humid Pampas and the Uruguayan Savanna. Winter (4) maxima are scarce and mainly occur in the northernmost Chilean Matorral and in the Southwestern Amazonian Moist Forests (Figure 16.1c).

16.3 Biophysical Properties from EFTs

16.3.1 Land Surface Parameterization of EFTs

To obtain the values of land surface properties of each EFT, first, we used the 1992 USGS global land cover map at 1-km pixel resolution (see http://edc2. usgs.gov/1KM/1kmhomepage.php; Eidenshink and Faundeen 1994) and the Noah land surface model table of physical properties to produce maps of 15 land surface parameters for the year 1992. Then, the former 15 USGS-derived maps were spatially overlapped to the EFT classification for the year 1992 to calculate the spatial mean of each land surface property for each EFT. The definition of EFTs is able to capture differences in ecosystem functioning within the same land cover type. For instance, it distinguishes dense from open scrublands, or irrigated from rainfed croplands, by means of their differences in primary production dynamics. To show the spatial variability introduced in the land surface properties by the EFT approach, we calculated the relative difference between USGS-derived and EFT-derived property maps calculated as: [(USGS − EFT)/USGS] (Figure 16.2).

16.3.2 U.S. Geological Survey versus EFT-Derived Biophysical Properties

The large-scale regional patterns of all surface properties were similar between the USGS and the EFT-derived maps (Figure 16.2). Spatial autocorrelation, a intrinsic property of ecological variables (Legendre 1993), was slightly greater in the EFTs than in the USGS-derived maps (the global Moran's I mean across all properties for the EFTs and USGS-derived maps was 0.82 and 0.72, respectively, P-value < 0.05, n = 335, 534). This is related to the more gradual changes in the EFT-derived property maps compared to the more abrupt changes in the USGS-derived maps. The spatial differences between the two approaches were greater for some properties (such as minimum (Z0m) and maximum (Z0x) surface roughness with an averaged absolute difference of 69% and 53%, respectively; Figure 16.2a, b, e, f, i, and j) than others (such as minimum and maximum albedo with an averaged absolute difference of 16% and 17%, respectively, Figure 16.2c, d, g, h, k, and l). Since the EFT properties were calculated as spatial means, a side effect was that the EFT approach tended to reduce the range of values of the biophysical properties. For instance, because of the spatial smoothing, the EFTs approach increased values close to zero for variables, such as minimum and maximum surface roughness in nonvegetated and scarcely vegetated areas (e.g., Figure 16.2i and j). It also caused a less intense increase of low values in green vegetation fraction, leaf area index, and root depth in these same areas (not shown), and of maximum and minimum albedo in densely vegetated subtropical areas (Figure 16.2k and l). On the other hand, the EFT approach tended to decrease very high values such as minimum and maximum surface roughness in densely vegetated areas (e.g., Figure 16.2i and j), and of minimum and maximum albedo in slightly vegetated areas (Figure 16.2k and l). Most regions showed an average difference lower than 30%, but particular regions repeatedly concentrated large differences (greater than 70%) in most properties (not shown), such as shrublands of southern Patagonian steppe, arid and semiarid areas of the southern and central Andean Cordillera, the Valdivian temperate forests, and agricultural areas in the cerrado ecoregion. Average differences were found in the transition between the Espinal and Low Monte

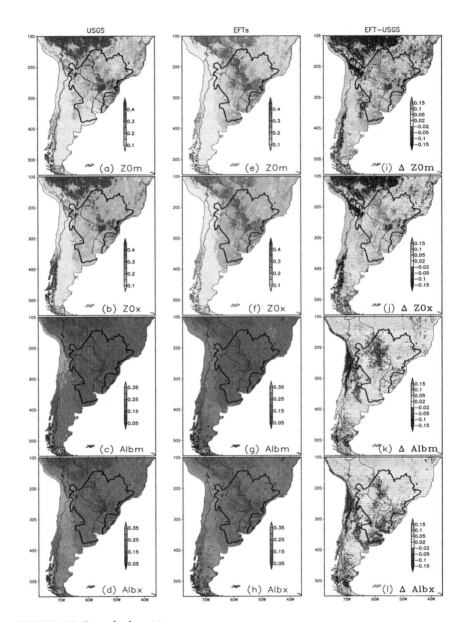

FIGURE 16.2 (See color insert.)
Selected biophysical properties: minimum surface roughness (Z0m), maximum surface roughness (Z0x); minimum albedo (Albm); and maximum albedo (Albx) estimated from the 1982–1999 median EFT maps (left column, a–d); as obtained from the USGS land cover field (middle column, e–h) and their relative difference [(USGS – EFT)/USGS] (right column, i–l).

ecoregions, and the southwestern Amazonian moist forest (ecoregion names follow Olson et al. 2001).

16.3.3 Interannual Variability of Vegetation Properties

We also characterized the interannual variability of the land surface properties based on the year-to-year changes in the EFT distribution from 1982 to 1999 and the former land surface parameterization of each EFT (Figure 16.3). Both the coefficient of variation (standard deviation divided by mean; not shown) and the interquartile range divided by median [(3rd quartile – 1st quartile)/median × 100] were calculated for each land surface property as relative descriptors of the interannual differences.

Some land surface properties showed greater interannual variability than others across the entire study area (Figure 16.3). Great interannual variability was found for minimum and maximum surface roughness, stomatal resistance, and minimum leaf area index (34%, 28%, 27%, and 23%, respectively) (Figure 16.3a through d). Low interannual variability was observed for maximum and minimum emissivity, and radiation stress (lower than 6%) (Figure 16.3e through g). Rooting depth, minimum and maximum background albedo, green vegetation fraction, and maximum leaf area index showed intermediate variability (not shown). On average, the interannual coefficient of variation of the entire study area across all biophysical properties was relatively low (13%) (not shown). However, some regions repeatedly showed high interannual variability across all properties. The frontiers (ecotones) between the semiarid areas of the Patagonian steppe with wetter regions to the south, west, and northeast showed the greatest interannual variability (greater than 60%). Relatively high interannual variability (between 30% and 60%) was observed in (1) semiarid areas of the Patagonian steppe and the Andes Cordillera, (2) the northwest–southeast transect from southeastern Bolivia to Uruguay, and (3) the Brazilian Atlantic Plateau (a mountain range in eastern Brazil in the north–south direction between the São Francisco River and the Atlantic coast).

16.4 Climate Regulation Services in Regional Modeling

16.4.1 Accounting for Changes in Biophysical Properties in a Regional Climate Model

To study the effect of the interannual changes of ecosystem biophysical properties on the regional climate, model simulations with conventional fixed land cover types were compared against simulations using EFTs as lower boundary conditions for the 2008 drought episode in the La Plata

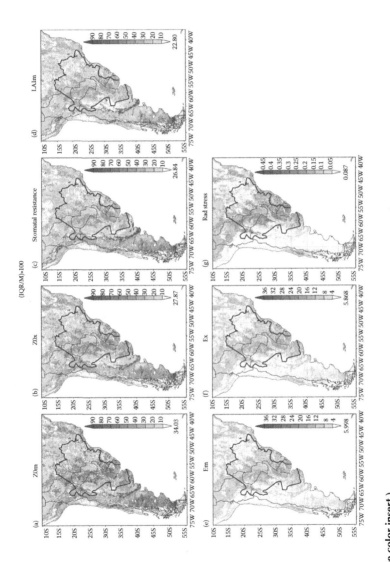

FIGURE 16.3 (See color insert.)
Interannual variability (IQR/M) of selected biophysical properties in South America based on the interannual variation of the EFT distribution for the 1982–1999 period. IQR/M = (3rd quartile – 1st quartile)/median × 100. (a) Minimum surface roughness (Z0m); (b) maximum surface roughness (Z0x); (c) stomatal resistance; (d) minimum leaf area index (LAIm); (e) minimum emissivity (Em); (f) maximum emissivity (Ex); (g) radiation stress (Rad Stress). The numbers at the bottom of each panel indicate the (spatial) mean interannual variability for the whole region.

Basin of South America as a case study. Details of this event are given in the next section.

Long-term simulations were carried out using the Weather Research and Forecasting (WRF) model coupled to the Noah LSM. The WRF model is a mesoscale numerical weather prediction system designed to serve both atmospheric research and operational forecasting needs. Also it serves for regional and global applications through the use of scales ranging from meters to thousands of kilometers. WRF solves the fully compressible non-hydrostatic equations to simulate the atmosphere's behavior. The Noah LSM has four soil layers with a corresponding thickness from the top down of 10, 30, 60 and 100 cm (2 m total depth), and includes representations of the root zone, vegetation categories, monthly vegetation fraction, and soil texture. The Noah LSM that solves the surface energy and water balances to provide surface conditions, like moisture and heat fluxes, to the boundary-layer.

The coupled model was used over the southern part of South America on a domain limited by 40°S and 12°S in latitude and 77°W and 42°W in longitude. This domain includes important topographic features such as the Andes Mountains along the Pacific coast and the Brazilian Plateau along the Brazilian coast. Lowlands and plains complete the topography. In terms of hydrology, the major feature is the presence of the La Plata Basin, second in river discharge in South America and of great economical value for the region. The model was configured to use a spatial resolution of 18 km × 18 km with 28 vertical levels for the atmosphere and four levels for soil depth, with a time step of 90 s. The National Centers for Environmental Prediction/National Center for Atmospheric Research (NCEP/NCAR) Reanalysis Project (NNRP) dataset (Kalnay and Cai 2003) was used as initial conditions and 6-hour lateral boundary conditions.

The definition of land cover types in the coupled model is part of the Noah LSM. Specifically, Noah LSM uses a map where each pixel represents a land cover type. Then, a look-up table assigns 15 biophysical properties associated with each land cover type: green vegetation fraction, rooting depth, stomatal resistance, a parameter used in radiation stress function, a parameter used in vapor pressure deficit function, threshold water-equivalent snow depth that implies 100% snow cover, upper bound on maximum albedo over deep snow, minimum and maximum leaf area index through the year, minimum and maximum background emissivity through the year, minimum and maximum background albedo through the year, and minimum and maximum background roughness length through the year. The model offers two land cover classifications: USGS and International Geosphere-Biosphere Program (IGBP). USGS has 27 categories determined by information from 1992/93 of the AVHRR (Eidenshink and Faundeen 1994). IGBP has 20 land cover types assigned by information from 2001 of the MODIS Land Cover (MOD12Q1) Product (Friedl et al. 2010).

To evaluate the performance of the new biophysical properties dataset and its influence on surface conditions, model simulations with the conventional

USGS land cover map were compared against simulations using the EFT map of 2008 with its associated properties as lower boundary conditions. The EFT map for 2008 was derived from the MODIS MOD13C1 product for the 2001–2009 period (full details are provided in Alcaraz-Segura et al. 2013). Such procedure is equivalent to the one explained earlier (in Section 16.2.3) for the 1982–1999 period using LTDR NDVI.

Two sets of five simulations starting on consecutive days were performed to generate two ensembles identified as: WRF–USGS and WRF–EFTs. The WRF–USGS ensemble uses the conventional USGS land cover categories to represent vegetation. The WRF–EFTs ensemble uses the EFT map of 2008 with its associated properties. This experiment is designed to help to understand the possible advantages of using realistic vegetation data and the model sensitivity to this information. The hypothesis is that land use changes and vegetation variability due to extreme events or human activities modify vegetation properties that may affect heat fluxes at the planet boundary layer and, therefore, may act as a forcing over the overlying atmospheric states. The results and discussion will be presented in the next sections (full details are provided in Müller et al. 2013).

16.4.2 A Case Study for a Drought Episode

The severe drought of 2008 affecting southeastern South America was mainly forced by the combination of a La Niña event with a warm tropical North Atlantic. In addition, as the drought progressed, surface processes driven by land cover changes may have had a local effect over heat fluxes, which in turn may act as forcing in the lower atmosphere. Figure 16.4 shows the similarity between precipitation and NDVI anomalies in terms of spatial distribution and evolution in time for the 2008 episode. The precipitation anomaly map (Figure 16.4b) and the NDVI anomaly map (Figure 16.4e) exhibit a similar pattern with negative values over Uruguay and northeastern Argentina. Around the tripartite border (Argentina, Paraguay, and Brazil), the dense subtropical forest has the ability to withstand the deficit in precipitation without loss of vegetation cover. Figure 16.4c and f presents the evolution of the drought in the box (63°W–55°W, 38°S–28°S), which is mostly dominated by crops and pastures. The precipitation time series demonstrates the persistence of the drought with an almost continuous negative trend of rainfall anomaly of two years starting in June 2007. The temporal evolution of NDVI seems to follow the deficit of precipitation with a lag of about one month. Values below normal start on August 2007 and finish on November 2009. Only two exceptions on November 2007 and March 2008 showed weak positive anomalies. The anomaly maps and both time-series show visually the direct influence of precipitation deficit on vegetation, particularly in regions of crops and pastures without artificial irrigation. As we will further explain below the vegetation may have different feedbacks with the atmosphere depending on the type of land cover and its current state.

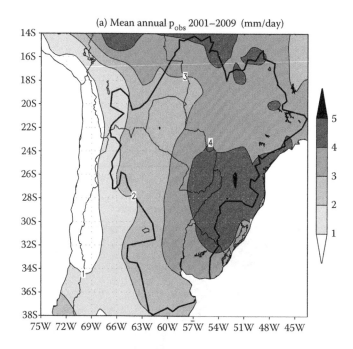

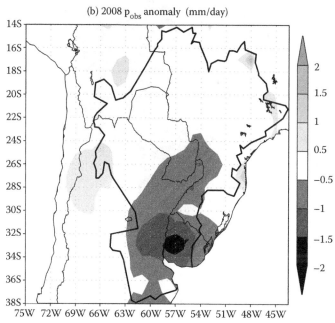

FIGURE 16.4 (See color insert.)
(a) Observed mean precipitation (Pobs) (2001–2009) and (b) observed precipitation anomalies
(Pobs anomaly) for the year 2008.

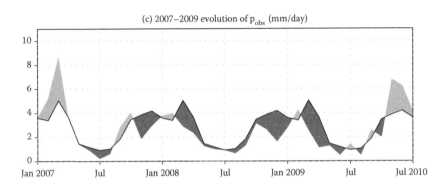

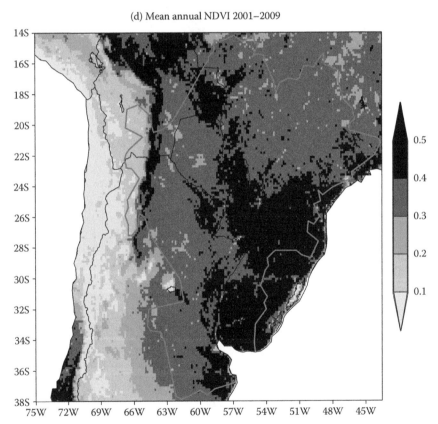

FIGURE 16.4 (See color insert.) (*Continued*)
(c) Time series of precipitation averaged for the drought region (black line, mean annual cycle; green color area over the black line indicates excess of precipitation and brown color area under the black line indicates deficit of precipitation).

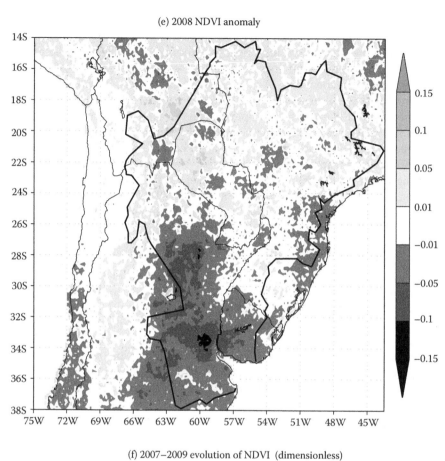

(e) 2008 NDVI anomaly

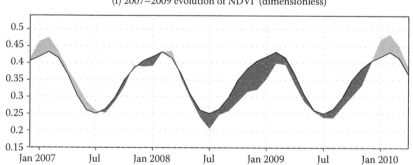

(f) 2007–2009 evolution of NDVI (dimensionless)

FIGURE 16.4 (See color insert.) (*Continued*)
(d–f) as (a–c) but for NDVI.

As shown in Figure 16.4, extreme events such as droughts have direct effects on certain supporting ecosystem services such as the reduction of annual primary production (e.g., negative anomalies in Figure 16.4e and f, reduction of NDVI-m) and increases of the seasonal variability of carbon gains (not shown, increases in CVseas). The 2008 drought also ended up affecting biophysical properties of the land surface. Focusing on the differences between EFTs and USGS properties, on the same drought area selected to plot the time-series in Figure 16.4, i.e. [63° W – 55° W, 38° S – 28° S], we observed that during the drought period, green vegetation fraction (–13.2%), maximum and minimum leaf area index (–13.9% and –20.7% respectively) were significantly lower in the EFTs dataset than in the USGS dataset. In contrast, maximum and minimum albedo (0.7% and 9.0% respectively) and stomatal resistance (105.8%) were larger (more results shown in Müller et al. 2013).

Figure 16.5 compares the results of simulated rainfall for the WRF–USGS and WRF–EFT ensembles. The spatial distribution of the WRF–USGS in Figure 16.5a presents high values to the northeast of the domain that decreases toward the southwest. The anomaly map (Figure 16.5b) illustrates the quantitative error of the WRF–USGS ensemble with positive values (green) for wet biases over Brazil, Paraguay, and southern Bolivia, and negative anomalies (brown) in the southeastern part of the domain. The implication is that the model simulates precipitation below observations in most parts of the drought area (i.e., the drought was exaggerated by the model), with the exception of the northern part of the domain where the model has biases towards greater values.

To analyze the performance of the WRF–EFTs ensemble, the map of the difference between both ensembles is plotted in Figure 16.5c (further analyses are provided in Section 5 of Müller et al. 2013). When using EFTs as lower boundary conditions, areas 1 and 2 exhibit a reduction in the wet bias (brown shades in Figure 16.5 b), but also a reduction of the dry bias over area 3 (green shades in area 3 in Figure 16.5 b). Both ensembles tend to overestimate the precipitation over area 1, which was not affected by the drought (not shown for WRF-EFTs ensemble). However the WRF-EFTs ensemble reduces the wet bias as it can be inferred graphically from the negative values (brown shades in area 1). The reduction of the model error when using EFTs is about 19%. The bias reduction is related to a general reduction of all components of the water balance caused mainly by a decrease of surface roughness that reduces the degree of turbulence near the surface and then, strengthens the stability of the boundary layer. Also, the greater stomatal resistance introduced by the EFTs (not shown) compared to USGS, leads to a reduction in evapotranspiration. In area 2, although both ensembles failed to depict the northern sector of the drought region (not shown for WRF-EFTs ensemble), EFTs reduces the wet bias of USGS by about 7%. This region has suffered the replacement of natural forests by crops in recent years (Izquierdo et al. 2008). As the EFTs account for inter-annual changes (see section 16.3), their properties represent

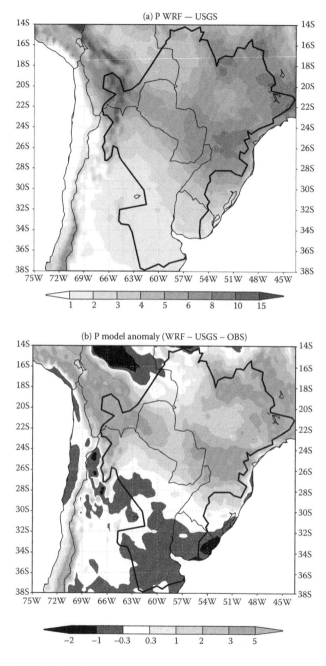

FIGURE 16.5 (See color insert.)
Time-average precipitation during the drought period: (a) WRF–USGS ensemble, (b) model anomaly compared to observations.

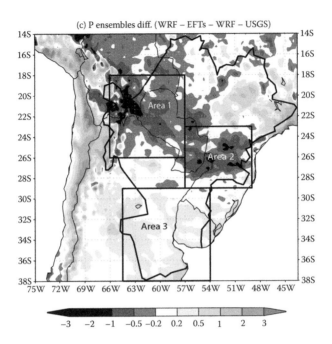

FIGURE 16.5 (See color insert.) (*Continued*)
Time-average precipitation during the drought period: (c) differences between WRF–USGS ensemble and WRF–EFTs ensemble. Red lines define three selected areas called Area 1 (top rectangle, normal precipitation), Area 2 (central rectangle, drought), and Area 3 (rectangle at bottom, drought).

these interannual changes with a large reduction in roughness length but also with a decrease in vegetation fraction and leaf area index (not shown). The loss of vegetation reduces evapotranspiration and then precipitation (soil moisture and runoff are also reduced). Finally, the WRF–EFTs ensemble slightly increases the rain over the core region of the drought (area 3), hence, correcting the excessive drought bias of the WRF–USGS ensemble. On this area, crops and pastures (that dominate the region) exposed to dry conditions do not have the capacity to retain water on their leaves. Then, the rainfall directly infiltrates into the soil layers and leads to an increase in the direct soil evaporation, favoring more rainfall.

In summary, though the model using the conventional USGS land covers is able to represent the spatial distribution and temporal evolution of precipitation, it tends to magnify positive and negative anomalies. However, the use of the EFTs vegetation map for the year in course contributes to a large extent to reduce these errors in several regions by representing the realistic conditions of ecosystem biophysical properties, and it offers a better connection between ecosystem biophysical properties and climate services.

16.5 Discussion and Conclusions

The biophysical part of the climate regulation services is estimated from ecosystem's surface net radiation and latent heat flux, which are simulated using land surface models. In general, these simulations involve the use of ecosystem biophysical properties such as leaf area index; stomatal resistance; rooting depth; albedo and transmittance in the visible and near infrared; heat, water, and snow capacities; and others. Here we offer an original method based on satellite mapping of EFTs to estimate and monitor such variables. We also present results of climate model simulations of a drought case study showing a reduction in precipitation biases when this method is used.

Our approach, based on the definition of EFTs from three simple NDVI-derived surrogates for productivity, seasonality, and phenology of carbon gains (Figure 16.1), opens a straightforward way to incorporate the year-to-year variations of vegetation biophysical properties into land surface and climate modeling. Currently, climate models that use land surface schemes of intermediate complexity do not account for year-to-year changes in the biophysical properties of vegetation or the land cover type (Oki et al. 2013). Our study showed how some biophysical properties vary between years across the entire study area and how these interannual changes are noticeable in several regions (Figure 16.3). Our EFTs classification was entirely based on just three descriptors of the seasonal dynamics of NDVI, which is used as a surrogate of primary production and, hence, as an integrative indicator of overall ecosystem functioning (Virginia et al. 2001). Since EFTs can be defined in a year-to-year basis, they can represent time-varying land surface properties that reflect the actual characteristics of vegetation functioning and not just time-fixed vegetation types. In this sense, the use of time-varying EFTs captures the effect of human-driven changes in land use and management. In addition, the NDVI dynamics of a particular year not only reflects the vegetation response to the environmental conditions of that particular year but also exhibits the memory of the system to the climatic conditions and disturbance effects from previous years (Wiegand et al. 2004).

The approach presented here could be further improved by setting a decision rule for each year to classify nonvegetated pixels (water bodies, snow, and absolute deserts) before the histogram calculation of the NDVI descriptors. For instance, water bodies and pixels covered by snow could be identified using the LTDR or MODIS quality assessment flags. Deserts could also be classified as those pixels with extremely low NDVI values (e.g., NDVI values always lower than 0.15 throughout the year). Finally, the remaining pixels can be classified into EFTs as described here. As stated in the introduction, the definition of EFTs is dependent on the variables included. Remote sensing can provide additional functional variables to improve ecosystem descriptions (Nemani and Running 1997). Evapotranspiration (see Chapter 18), shortwave albedo (see Chapter 17), and surface temperature (see Chapter 19), three variables closely linked to the exchange of water and energy of the ecosystem, may provide a complementary description to the one presented here, based on the seasonal dynamics of carbon gains (e.g., Piñeiro et al. 2002; Garbulsky and Paruelo 2004; Fernández et al. 2010). One major advantage of using satellite information is the near-real-time distribution and the global availability of the information. In the near future, we will make available an EFTs map operational product based on the Global Vegetation Index (GVI) derived from MetOp AVHRR global 1-km data from the National Environmental Satellite, Data, and Information Service (NESDIS) website.

Acknowledgments

Financial support was provided by the Inter-American Institute for Global Change Research (IAI, CRN II 2031 and 2094) under the U.S. National Science Foundation (Grant GEO-0452325), University of Maryland, NASA Grant NNX08AE50G, Proyecto Estratégico of the University of Buenos Aires, CONICET, FONCYT, FEDER Funds, the Andalusian Government (projects GLOCHARID and SEGALERT P09–RNM-5048), the Spanish National Parks Agency (project 066/2007), the Spanish Ministry of Science and Innovation (project CGL2010-22314, Plan Nacional I+D+I 2010), and the Ecología de Zonas Áridas Research Group of the University of Almería. LTDR data were obtained from the LTDR team website.

References

Alcaraz-Segura, D., J. Cabello, and J. Paruelo. 2009. Baseline characterization of major Iberian vegetation types based on the NDVI dynamics. *Plant Ecology* 202:13–29.

Alcaraz-Segura, D., E. Liras, S. Tabik, J. M. Paruelo, and J. Cabello. 2010. Evaluating the consistency of the 1982–1999 NDVI trends in the Iberian Peninsula across four time-series derived from the AVHRR sensor: LTDR, GIMMS, FASIR, and PAL-II. *Sensors* 10:1291–1314.

Alcaraz-Segura, D., J. Paruelo, and J. Cabello. 2006. Identification of current ecosystem functional types in the Iberian Peninsula. *Global Ecology and Biogeography* 15:200–212.

Alcaraz-Segura, D., J. Paruelo, H. Epstein, and J. Cabello. 2013. Environmental and human controls of ecosystem functional diversity in temperate South America. *Remote Sensing* 5:127–154.

Anderson-Teixeira, K. J., P. K. Snyder, T. E. Twine, S. V. Cuadra, M. H. Costa, and E. H. DeLucia. 2012. Climate-regulation services of natural and agricultural ecoregions of the Americas. *Nature Climate Change* 2:177–181.

Baidya Roy, S., G. C. Hurtt, C. P. Weaver, and S. W. Pacala. 2003. Impact of historical land cover change on the July climate of the United States. *Journal of Geophysical Research: Atmospheres* 108: ACL 11-1–ACL 11-14.

Beltrán-Przekurat, A., R. A. Pielke Sr., J. L. Eastman, and M. B. Coughenour. 2012. Modelling the effects of land-use/land-cover changes on the near-surface atmosphere in southern South America. *International Journal of Climatology* 32:1206–1225.

Bonan, G. B. (2008). *Ecological climatology: concepts and applications.* Second Edition. Cambridge University Press.

Bonan, G. B. 2008. Forests and climate change: Forcings, feedbacks, and the climate benefits of forests. *Science* 320:1444–1449.

Brando, P. M., S. J. Goetz, A. Baccini, D. C. Nepstad, P. S. A. Beck, and M. C. Christman. 2010. Seasonal and interannual variability of climate and vegetation indices across the Amazon. *Proceedings of the National Academy of Sciences of the United States of America* 107:14685.

Bret-Harte, M. S., M. C. Mack, G. R. Goldsmith, et al. 2008. Plant functional types do not predict biomass responses to removal and fertilization in Alaskan tussock tundra. *Journal of Ecology* 96:713–726.

Cao, M., S. D. Prince, J. Small, and S. J. Goetz. 2004. Remotely sensed interannual variations and trends in terrestrial net primary productivity 1981–2000. *Ecosystems* 7:233–242.

Chapin, F. S., J. T. Randerson, A. D. McGuire, J. A. Foley, and C. B. Field. 2008. Changing feedbacks in the climate-biosphere system. *Frontiers in Ecology and the Environment* 6:313–320.

Chen, F., and J. Dudhia. 2001. Coupling an advanced land surface-hydrology model with the Penn State-NCAR MM5 modeling system. Part I: Model implementation and sensitivity. *Monthly Weather Review* 129:569–585.

Chen, F., K. Mitchell, J. Schaake, et al. 1996. Modeling of land surface evaporation by four schemes and comparison with FIFE observations. *Journal of Geophysical Research: Atmospheres* 101:7251–7268.

De Oliveira, A. S., R. Trezza, E. Holzapfel, I. Lorite, and V. P. S. Paz. 2009. Irrigation water management in Latin America. *Chilean Journal of Agricultural Research* 69:7–16.

Diffenbaugh, N. S. 2009. Influence of modern land cover on the climate of the United States. *Climate Dynamics* 33:945–958.

Eidenshink, J. C., and J. L. Faundeen. 1994. The 1 km AVHRR global land data set: First stages in implementation. *International Journal of Remote Sensing* 15:3443–3462.

Ek, M. B., K. E. Mitchell, Y. Lin, et al. 2003. Implementation of Noah land surface model advances in the National Centers for Environmental Prediction operational mesoscale Eta model. *Journal of Geophysical Research: Atmospheres* 108:GCP 12-1–GCP 12-16.

Fernández, N., J. Paruelo, and M. Delibes. 2010. Ecosystem functioning of protected and altered Mediterranean environments: A remote sensing classification in Doñana, Spain. *Remote Sensing of Environment* 114:211–220.

Foley, J. A., G. P. Asner, M. H. Costa, et al. 2007. Amazonia revealed: Forest degradation and loss of ecosystem goods and services in the Amazon Basin. *Frontiers in Ecology and the Environment* 5:25–32.

Foley, J. A., M. H. Costa, C. Delire, N. Ramankutty, and P. Snyder. 2003. Green surprise? How terrestrial ecosystems could affect Earth's climate. *Frontiers in Ecology and the Environment* 1:38–44.

Foley, J. A., I. C. Prentice, N. Ramankutty, et al. 1996. An integrated biosphere model of land surface processes, terrestrial carbon balance, and vegetation dynamics. *Global Biogeochemical Cycles* 10:603–628.

Friedl, M. A., D. Sulla-Menashe, B. Tan, et al. 2010. MODIS Collection 5 global land cover: Algorithm refinements and characterization of new datasets. *Remote Sensing of Environment* 114:168–182.

Garbulsky, M. F., and J. M. Paruelo. 2004. Remote sensing of protected areas to derive baseline vegetation functioning characteristics. *Journal of Vegetation Science* 15:711–720.

Gedney, N., and P. J. Valdes. 2000. The effect of Amazonian deforestation on the northern hemisphere circulation and climate. *Geophysical Research Letters* 27:3053–3056.

Haines-Young, R., and M. Potschin. 2013. *Common International Classification of Ecosystem Services (CICES): Consultation on Version 4.* Nottingham, UK: University of Nottingham.

Holben, B. N. 1986. Characteristics of maximum-value composite images from temporal AVHRR data. *International Journal of Remote Sensing* 7:1417–1434.

Izquierdo, A. E., C. D. De Angelo, and T. M. Aide. 2008. Thirty years of human demography and land-use change in the Atlantic forest of Misiones, Argentina: An evaluation of the forest transition model. *Ecology and Society* 13:3.

Kalnay, E., and M. Cai. 2003. Impact of urbanization and land-use change on climate. *Nature* 423:528–531.

Kathuroju, N., M. A. White, J. Symanzik, M. D. Schwartz, J. A. Powell, and R. R. Nemani. 2007. On the use of the Advanced Very High Resolution Radiometer for development of prognostic land surface phenology models. *Ecological Modelling* 201:144–156.

Kucharik, C. J., J. A. Foley, C. Delire, et al. 2000. Testing the performance of a dynamic global ecosystem model: Water balance, carbon balance, and vegetation structure. *Global Biogeochemical Cycles* 14:795–825.

Lavorel, S., S. Díaz, J. Cornelissen, et al. 2007. Plant functional types: Are we getting any closer to the Holy Grail? In *Terrestrial ecosystems in a changing world*, eds. J. G. Canadell, D. E. Pataki, and L. F. Pitelka, 149–164. New York: Springer.

Legendre, P. 1993. Spatial autocorrelation: trouble or new paradigm? *Ecology* 74:1659–1673.

Mahmood, R., R. A. Pielke, K. G. Hubbard, et al. 2010. Impacts of land use/land lover change on climate and future research priorities. *Bulletin of the American Meteorological Society* 91:37–46.

Maness, H., P. J. Kushner, and I. Fung. 2013. Summertime climate response to mountain pine beetle disturbance in British Columbia. *Nature Geoscience* 6:65–70.

Marland, G., R. A. Pielke Sr., M. Apps, et al. 2003. The climatic impacts of land surface change and carbon management, and the implications for climate-change mitigation policy. *Climate Policy* 3:149–157.

McNaughton, S. J., M. Oesterheld, D. A. Frank, and K. J. Williams. 1989. Ecosystem-level patterns of primary productivity and herbivory in terrestrial habitats. *Nature* 341:142–144.

Monteith, J. 1972. Solar radiation and productivity in tropical ecosystems. *Journal of Applied Ecology* 9:747–766.

Müller, O. V., E. H. Berbery, and D. Alcaraz-Segura, M. B. Ek. 2013. Regional model simulations of the 2008 drought in Southern South America using a consistent set of land surface properties. *Journal of Climate*. (In revision.)

Nagol, J. R., E. F. Vermote, and S. D. Prince. 2009. Effects of atmospheric variation on AVHRR NDVI data. *Remote Sensing of Environment* 113:392–397.

Nemani, R., and S. Running. 1997. Land cover characterization using multitemporal red, near-IR, and thermal-IR data from NOAA/AVHRR. *Ecological Applications* 7:79–90.

Nosetto, M. D., E. G. Jobbágy, and J. M. Paruelo. 2005. Land-use change and water losses: The case of grassland afforestation across a soil textural gradient in central Argentina. *Global Change Biology* 11:1101–1117.

Oki, T., E. M. Blyth, E. H. Berbery, and D. Alcaraz-Segura. 2013. Land cover and land use changes and their impacts on hydroclimate, ecosystems and society. In *Climate science for serving society: Research, modeling and prediction priorities*, eds. G. R. Asrar and J. W. Hurrel. Dordrecht, The Netherlands: Springer Science+Business Media.

Olson, D. M., E. Dinerstein, E. D. Wikramanayake, et al. 2001. Terrestrial ecoregions of the worlds: A new map of life on Earth. *BioScience* 51:933–938.

Paruelo, J. M., E. G. Jobbagy, and O. E. Sala. 2001. Current distribution of ecosystem functional types in temperate South America. *Ecosystems* 4:683–698.

Pedelty, J., S. Devadiga, E. Masuoka, et al. 2007. Generating a long-term land data record from the AVHRR and MODIS instruments. *IGARSS 2007 IEEE International Geoscience and Remote Sensing Symposium* 1021–1025.

Pettorelli, N., J. O. Vik, A. Mysterud, J. M. Gaillard, C. J. Tucker, and N. C. Stenseth. 2005. Using the satellite-derived NDVI to assess ecological responses to environmental change. *Trends in Ecology & Evolution* 20:503–510.

Pielke, R. A., Sr. 2001. Influence of the spatial distribution of vegetation and soils on the prediction of cumulus convective rainfall. *Reviews of Geophysics* 39:151–177.

Pielke, R. A., Sr., J. Adegoke, A. BeltráN-Przekurat, et al. 2007. An overview of regional land-use and land-cover impacts on rainfall. *Tellus B* 59:587–601.

Pielke, R. A., Sr., G. Marland, R. A. Betts, et al. 2002. The influence of land-use change and landscape dynamics on the climate system: Relevance to climate-change policy beyond the radiative effect of greenhouse gases. *Philosophical Transactions of the Royal Society A: Mathematical, Physical and Engineering Sciences* 360:1705–1719.

Piñeiro, G., D. Alcaraz, J. Paruelo, et al. 2002. A functional classification of natural and human-modified areas of 'Cabo de Gata,' Spain, based on Landsat TM data. *Presented at the 29th International Symposium of Remote Sensing of Environment*, Buenos Aires, Argentina, 8–12.

Potter, C., S. Klooster, P. Tan, M. Steinbach, V. Kumar, and V. Genovese. 2005. Variability in terrestrial carbon sinks over two decades. Part III: South America, Africa, and Asia. *Earth Interactions* 9:1–15.

Reale, O., and P. Dirmeyer. 2000. Modeling the effects of vegetation on Mediterranean climate during the Roman Classical Period: Part I: Climate history and model sensitivity. *Global and Planetary Change* 25:163–184.

Reale, O., and J. Shukla. 2000. Modeling the effects of vegetation on Mediterranean climate during the Roman Classical Period: Part II. Model simulation. *Global and Planetary Change* 25:185–214.

Reich, P. B., and J. Oleksyn. 2004. Global patterns of plant leaf N and P in relation to temperature and latitude. *Proceedings of the National Academy of Sciences of the United States of America* 101:11001.

Reich, P. B., M. G. Tjoelker, J. L. Machado, and J. Oleksyn. 2006. Universal scaling of respiratory metabolism, size and nitrogen in plants. *Nature* 439:457–461.

Roy, S. B., and R. Avissar. 2002. Impact of land use/land cover change on regional hydrometeorology in Amazonia. *Journal of Geophysical Research* 107:8037.

Sellers, P. J., Y. Mintz, Y. C. Sud, and A. Dalcher. 1986. A simple biosphere model (SIB) for the use within general circulation models. *Journal of the Atmospheric Sciences* 43:505–531.

Smith, T. M., H. H. Shugart, and F. I. Woodward. (eds.). 1997. *Plant functional types: Their relevance to ecosystem properties and global change.* Cambridge, UK: Cambridge University Press.

Soriano, A., and J. M. Paruelo. 1992. Biozones: Vegetation units defined by functional characters identifiable with the aid of satellite sensor images. *Global Ecology & Biogeography Letters* 2:82–89.

Stephenson, N. L. 1990. Climatic control of vegetation distribution—The role of the water-balance. *American Naturalist* 135:649–670.

Stohlgren, T. J., T. N. Chase, R. A. Pielke, T. G. F. Kittel, and J. S. Baron. 1998. Evidence that local land use practices influence regional climate, vegetation, and stream flow patterns in adjacent natural areas. *Global Change Biology* 4:495–504.

Tucker, C. J., and P. J. Sellers. 1986. Satellite remote-sensing of primary production. *International Journal of Remote Sensing* 7:1395–1416.

Valentini, R., D. D. Baldocchi, J. D. Tenhunen, and P. Kabat. 1999. Ecological controls on land-surface atmospheric interactions. In *Integrating hydrology, ecosystem dynamics and biogeochemistry in complex landscapes*, eds. D. Tenhunen and P. Kabat, 105–116. Berlin: John Wiley & Sons.

Virginia, R. A., D. H. Wall, and S. A. Levin. 2001. Principles of ecosystem function. In *Encyclopedia of biodiversity*, ed. S. Levin, 345–352. San Diego, CA: Academic Press.

Volante, J. N., D. Alcaraz-Segura, M. J. Mosciaro, E. F. Viglizzo, and J. M. Paruelo. 2012. Ecosystem functional changes associated with land clearing in NW Argentina. *Ecosystem Services and Land-Use Policy* 154:12–22.

Weaver, C. P., and R. Avissar. 2001. Atmospheric disturbances caused by human modification of the landscape. *Bulletin of the American Meteorological Society* 82:269–281.

Werth, D., and R. Avissar. 2002. The local and global effects of Amazon deforestation. *Journal of Geophysical Research* 107:8087.

West, P. C., G. T. Narisma, C. C. Barford, C. J. Kucharik, and J. A. Foley. 2011. An alternative approach for quantifying climate regulation by ecosystems. *Frontiers in Ecology and the Environment* 9:126–133.

Wiegand, T., H. A. Snyman, K. Kellner, and J. M. Paruelo. 2004. Do grasslands have a memory: Modeling phytomass production of a semiarid South African grassland. *Ecosystems* 7:243–258.

Wright, I. J., P. B. Reich, J. H. C. Cornelissen, et al. 2005. Assessing the generality of global leaf trait relationships. *New Phytologist* 166:485–496.

Wright, I. J., P. B. Reich, M. Westoby, et al. 2004. The worldwide leaf economics spectrum. *Nature* 428:821–827.

Wright, J. P., S. Naeem, A. Hector, et al. 2006. Conventional functional classification schemes underestimate the relationship with ecosystem functioning. *Ecology Letters* 9:111–120.

Zhao, M., and S. W. Running. 2010. Drought-induced reduction in global terrestrial net primary production from 2000 through 2009. *Science* 329:940–943.

17

Ecosystem Services Related to Energy Balance: A Case Study of Wetlands Reflected Energy

C. M. Di Bella and M. E. Beget

National Institute of Agricultural Technology (INTA), Argentina

CONTENTS

17.1 Introduction 379
17.2 Wetlands Reflected Energy 383
    17.2.1 Canopy Reflectance Simulation Model 384
    17.2.2 Simulated Case Studies 385
    17.2.3 Simulated Reflectance 387
17.3 Conclusions 394
Acknowledgments 394
References 395

17.1 Introduction

Ecosystems play an essential role providing goods and services to humans. Directly or indirectly, their benefits to humans include provisioning (e.g., food, water, fuel, fiber, and genetic resources), supporting (e.g., ecosystems primary production, soil formation, oxygen production), cultural developments (e.g., recreation, cognitive development, reflection, and spiritual enrichment), and regulating services (e.g., air maintenance, water purification, regulation of human diseases, erosion control, and climate regulation) (MA 2005). Studying, quantification, and mapping of ecosystem services have gained increasing interest in the scientific community in recent years.

At the ecosystem level, several processes regulate water, mass, and energy fluxes toward the atmosphere (e.g., Baldocchi and Wilson 2001; Noe et al. 2011): photosynthesis—affecting the level of carbon dioxide in the atmosphere; evapotranspiration—controlling latent heat and water released from the soil and plants toward the atmosphere; aerosol production—altering the radiative heating of the atmosphere; and solar radiation reflection (albedo)—modifying the available energy and thus changing the temperature of the land surface (e.g., House et al. 2005; Smith et al. 2011).

The surface albedo plays a major role in energy balance and climate regulation because it determines the quantity of energy entering the system. As explained by Alcaraz-Segura (see Chapter 16), a change in albedo, due to land use change, modifies the amount of energy transfer to the ecosystem and then to the atmosphere. For example, desertification in tropical and subtropical drylands produces an increase in albedo and a reduction in evapotranspiration leading to a reduction in regional rainfall (House et al. 2005) (see Chapter 18). In seasonally snow-covered forests, land clearing leads to regional cooling during snow season and warming during summer due to an increase in albedo and due to a reduction in evapotranspiration, respectively. Depending on season and geographic location, ecosystem service of climate regulation is provided through the biogeophysical process of surface albedo (House et al. 2005).

Surface albedo, defined as the fraction of incident diffuse and direct radiation that is reflected in all directions over solar spectrum (Pinty and Verstraete 1992), integrates reflected radiation over the electromagnetic spectrum band comprised within 0.3 and 3 μm. Nevertheless, this range can be further divided into "visible albedo" (0.3–0.7 μm) and "near-infrared albedo" (0.7–3 μm) (e.g., Bsaibes et al. 2009).

In order to measure the albedo over any piece of land, a specific instrument is used: the pyranometer. This instrument is a radiation broadband sensor with a 180° field of view. Albedo is calculated as the ratio of reflected solar radiation (downward orientation of the pyranometer) and incident solar radiation (upward orientation of the pyranometer). Even though a series of field measurement networks are installed (e.g., Ameriflux, Asiaflux, Fluxnet, CarboEurope), mapping albedo becomes a difficult task due to wide spatial and temporal variability.

In this context, satellite observation arises as a favorable way to map the albedo of the land surface. During the 1980s, two instruments were launched at the time of the Earth Radiation Budget Experiment (ERBE), designed specifically to measure broadband surface albedo. Those instruments were assembled in two satellite platforms: Earth Radiation Budget Satellite (ERBS) and National Oceanic and Atmospheric Administration (NOAA). From 1984 to 1999, these instruments created monthly products of 2.5° of spatial resolution (S-9 and S-10 products), thus allowing an important improvement in the global observation of this variable (e.g., Hatzianastassiou et al. 2004).

Nowadays, many of the current satellite platforms get spectral data that allow approximation of land surface albedo (Table 17.1), thereby contributing

TABLE 17.1

Albedo Calculation from Different Satellite Sensors

Sensor/Platform	Bands	Spatial Resolution	References
ASTER/TERRA	0.52–0.6 0.63–0.69 0.78–0.86 1.6–1.7 2.15–2.18 2.18–2.22 2.23–2.28 2.29–2.36 2.36–2.43	15 m (visible and NIR)/30 m (IR)/90 m	Nasipuri et al. (2006)
AVHRR/NOAA	0.57–0.71 0.72–1.01	1.25 km/5 km	Csiszar and Gutman (1999); Key et al. (2001)
MISR/TERRA	0.42–0.45 0.54–0.55 0.66–0.67 0.85–0.87	275 m	Diner et al. (1998)
MODIS/TERRA	0.62–0.67 0.84–0.87 0.46–0.48 0.54–0.56 1.23–1.25 1.63–1.65 2.11–2.15	1 km	Wang and Zender (2010)
POLDER/ADEOS	0.43–0.46 0.66–0.68 0.74–0.79 0.84–0.88	6 km × 7 km	Deschamps et al. (1994)
VEGETATION/ SPOT	0.43–0.47 0.61–0.68 0.78–0.89 1.58–1.75	1 km (NADIR)	Stroppiana et al. (2002)
LANDSAT/ETM+	0.45–0.51 0.63–0.69 0.75–0.9 1.55–1.75 2.09–2.35	30 m	Shuai et al. (2011)

Note: NIR = near infrared; IR = infrared.

to the study of ecosystem energy balance (Table 17.2). One of the most popular products in remote sensing—used to estimate albedo at a global scale—is the one provided by the Moderate Resolution Imaging Spectroradiometer (MODIS) (Lucht et al. 2000). It is based on the narrow band to broadband conversion coefficients produced by Liang et al. (1999). Prior to this calculation, atmospheric and angular corrections through the Bidirectional Reflectance Distribution Function (BRDF) model were applied to MODIS data.

TABLE 17.2

Remote Sensing Studies in Relation to Energy Balance and Albedo Ecosystem Services

References	Application
Consoli et al. (2006); Mariotto and Gutschick (2010)	Evapotranspiration estimates
Mueller et al. (2011); Posselt et al. (2012)	Climate monitoring and analysis
Alton (2009)	Simulation of carbon, water, and energy exchange in land surface
Beck et al. (2011)	Forest fire impact
Bright et al. (2012)	Bioenergy climate impact
Kuusinen et al. (2012)	Seasonal variation on boreal forests
Puzachenko et al. (2011)	Estimation of thermodynamic parameters of the biosphere
Allen et al. (2013)	Land cover dynamics
Maes et al. (2011)	Ecosystem succession and energy dissipation
Yang et al. (2012)	Water management on irrigated lands
Zwart et al. (2010)	Water productivity
Ju et al. (2010)	Monitoring of soil water content

Since March 2000, this product has been available every 16 days through the integration of daily sequential multiangle observations (over that precise period) as well as a spatial resolution of 1 km^2 (MOD43; http://modis. gsfc.nasa.gov/data/dataprod) (Lucht et al. 2000; Schaaf et al. 2002). Liang (2000) then improved the algorithm that converts directional/hemispherical reflectance—for each solar zenith angle—and bihemispherical reflectance into albedo. Total shortwave albedo (α, 0.25–2.5 μm) conversion is given by Equation 17.1:

$$\alpha = 0.160\alpha_1 + 0.291\alpha_2 + 0.243\alpha_3 + 0.116\alpha_4 + 0.112\alpha_5 + 0.081\alpha_7 - 0.0015 \quad (17.1)$$

where α_n denotes albedo for MODIS bands n (1–7).

Artificial or natural wetlands are an important source of goods and services. Several times, politics and decisions related to wetlands use and conservation are undervalued because of the open-access nature and public good characteristics of ecosystems (Brander et al. 2006). Wetlands use assessment has become a very important issue and has led the research community to focus on services valuation. In our case, the interest is centered on the effect of wetland conditions on the energy balance, particularly on surface albedo.

In wetlands ecosystems, water modifies spectral behavior of vegetation canopies (Beget and Di Bella 2007); albedo estimation would depend on water levels and leaf area index (LAI) above and below water (e.g., Beget and

Di Bella 2007; Sumner et al. 2011; Beget et al. 2013). Despite albedo's importance in terms of surface or goods and services it provides as a whole, few studies have been carried out considering those changes that are produced in the albedo due to the replacement of wetlands with other land covers (e.g., by forest, pastures, or rice crops) or because of changes in growing conditions (water level or LAI). In order to answer these questions, a radiative transfer model (SAILHFlood, Beget et al. 2013) was used to simulate canopy reflectance over multiple conditions (e.g., land use type, LAI, and water level). We assumed that simulated reflectance, as an intermediate variable to calculate spectral albedo (Liang 2000), will help to evaluate—in a first approach—all impacts produced by the change of land use and cover, over the regional scale energy assessment.

17.2 Wetlands Reflected Energy

Wetlands are among the most productive terrestrial ecosystems covering around 7 million km^2 to 9 million km^2 on all continents and across a wide range of climate conditions (Mitsch and Gosselink 2000). Even natural or artificial wetlands (e.g., rice plantations) are an important source of goods and services for humanity. In addition to the fact that they provide goods that can be appropriated by humans, these ecosystems provide essential services such as discharge and recharge of aquifers, flood control, storm control, improvement in water quality, sources of biological diversity and phylogenetic material, opportunities for transportation, recreation and tourism, and climate regulation (Lambert 2003).

Regarding climate regulation, particularly, water plays an important role in the fraction of reflected energy determined by wetlands (surface albedo). In reflective terms, these environments are highly variable. Among those factors that modify the spectral behavior in these cover lands water is—with no doubt—the one that plays a major role, as much for its reflective characteristics associated with turbidity, sediment types, or algae presence, as well as for the changes produced in water levels in each of the systems. Land cover, proportion of bare soil, and LAI (submerged and emerged) are other factors that strongly influence the spectral response of the surface. If we consider that in wetlands these features can combine in infinite ways, and that responses are often the result of multiple interactions among these factors, it is not easy to know—far less to understand, the effect that these changes can have on the spectral response of wetlands. In addition, wetlands are continuously changing as a result of fluctuating water levels and also as a consequence of being replaced with crops (Kingsford 2000; Bunn and Arthington 2002; Schottler et al. 2013).

Albedo product MOD43 has shown very good approximations in the calculation of such variables for different ecosystems and under very different land use conditions (Disney et al. 2004; Stroeve et al. 2005; Román et al. 2010; Wang and Zender 2010; Cescatti et al. 2012). Nevertheless, there exists a short bibliography that has presented the effects of the change of land cover/use on the albedo for wetlands, far less about water-level changes on those ecosystems. This study gets even more difficult if we take into account the spatial resolution of 1 km² for most satellite products. In this sense, it is interesting to assess the spectral behavior of these systems and, thus, to prevent possible changes in the surface albedo. In this work, reflected energy is simulated for different land covers, water levels, and LAI attained to extrapolate these variations for surface albedo estimation. Reflectance models provide the possibility to study the spectral behavior of land covers, which would be very difficult to carry out in experimental conditions (e.g., Jacquemoud et al. 2009). Spectral reflectance could be integrated in any broadband sensor, such as in the case of MODIS (Liang 2000). A comparison of simulated reflectance for different land covers allowed us to study indirectly the effect of land use/cover change, or the differences in water level, on spectral albedo.

17.2.1 Canopy Reflectance Simulation Model

Radiative transfer models are very useful tools to study the effects of land cover characteristics on spectral behavior (e.g., Jacquemoud et al. 2000, 2009; Weiss et al. 2000). Considering the optical properties of leaves, these models calculate canopy reflectance through the description of dispersion/extinction radiation fluxes.

In the case of wetland ecosystems, the model should consider water because, unlike air, it absorbs and scatters energy transmitted and reflected by leaves. On the other hand, water–air interface interferes in energy fluxes changing their direction due to the difference in refractive rates of both media (Beget et al. 2010). The radiative transfer model SAILHFlood was developed by Beget et al. (2010, 2013) to simulate partially submerged canopy reflectance. SAILHFlood incorporates to SAILH (Scattering by Arbitrarily Inclined Leaves) model (Verhoef 1984, 1985) a submerged vegetation layer in which the propagation medium is water (Figure 17.1). SAILHFlood model calculates the reflectance of partially submerged canopy from variables related to the geometry of illumination and observation, the optical properties of leaves and soil, and also from canopy characteristics (Table 17.3).

SAILHFlood model incorporates three new inputs to the ones of SAILH model. These inputs are submerged LAI, emerged LAI, and water level (Table 17.3). SAILHFlood model was tested under laboratory conditions using artificial flooded canopies (Beget et al. 2013). The testing model included three illumination zenith angles, nine observation zenith angles, five LAI levels, and three levels of water. Root mean square error (RMSE) between

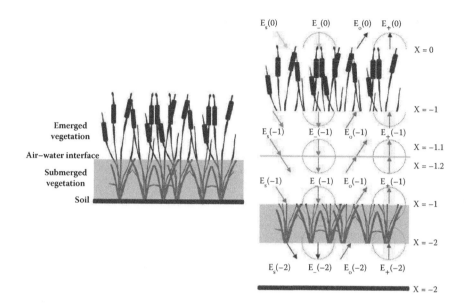

FIGURE 17.1 (See color insert.)
Illustration of SAILHFlood model. X represents vertical dimension with upward orientation. Arrows represent fluxes (direct flux is denoted by an arrow and diffuse flux is denoted by an arrow plus the hemisphere). E_s represents the direct solar irradiance; E_- and E_+ are diffuse downward and upward irradiance, respectively; E_o is π times the radiance in the observer direction.

simulated and measured reflectance was 0.0355. Test results showed a satisfactory model performance.

17.2.2 Simulated case studies

Spectral reflectance was simulated (using SAILHFlood model) between 0.4 and 2.5 μm within a spectral resolution of 0.01 μm for three land covers (wetland, rice crop, and forest) and different LAI (between 1 and 12) and water (between 0 and 1.4 m) levels (Table 17.4). Forest land cover was chosen because in some regions forests are planted in wetland areas (e.g., Baigún et al. 2008). Although paddy rice is an artificial wetland because of its own flood characteristic, water has different seasonalities and levels than common reed wetlands. Spectral reflectances were averaged to MODIS spectral bands for each of 90 simulations (Tables 17.1 and 17.4).

Common reed (*Phragmites australis* [Cav.] Trin. ex Steud.) was chosen to represent the spectral behavior of a wetland. *P. australis* is widely distributed around the world (Clevering and Lissner 1999) and shows large plasticity of LAI (ranging between 2.2 and 12.2) (Mal and Narine 2004). For reflectance simulations, *P. australis* canopy characteristics were taken from the description made by Burba et al. (1999) and Mal and Narine (2004). Six LAI and eight water levels were combined to run the model (Table 17.4). A canopy vertical

TABLE 17.3

Input Variables for SAILHFlood Model

SAILHFlood Input Variables	Symbol	Units/ Range
Water level	h	Meters
Emerged LAI	LAI_{em}	–
Submerged LAI	LAI_{sum}	–
Leaf angle inclination average	θ_l	Degrees
Hotspot	hot	–
Leaf reflectance	refl	0–1
Leaf transmittance	tran	0–1
Solar zenith angle	θ_s	Degrees
Viewing zenith angle	θ_o	Degrees
Azimuth relative angle	ψ	Degrees
Diffuse radiation factor	skyl	0–1
Soil bidirectional reflectance	r_{os}	0–1
Soil bihemispherical reflectance	r_{dd}	0–1
Soil directional-hemispherical reflectance	r_{sd}	0–1
Soil hemispherical-directional reflectance	r_{do}	0–1

Source: Modified from Beget, M. E., et al., *Ecol Model*, 257, 25–35, 2013.

Note: $\theta_s = 8°$, $\theta_o = 30°$, $\psi = 0°$, skyl = 0.

TABLE 17.4

Values of Input Parameters Used in SAILHFlood Simulations

Land Cover	Species	LAI	Water Levels (m)	$\overline{\theta}_l$ (°)	Hot	Number of Simulations
Wetland (Common reed)	*Phragmites australis*	LAI = [2, 4, 6, 8, 10, 12]	h = [0, 0.2, 0.4, 0.6, 0.8, 1, 1.2, 1.4]	70	0.2	48
Cropland (Paddy rice)	*Oryza sativa*	LAI = [2, 3, 4, 5, 6, 7]	h = [0, 0.1, 0.2, 0.3, 0.4, 0.5]	70	0.1	36
Forest plantation (Poplar)	*Populus canadensis*	LAI = [1, 2, 3, 4, 5, 6]	No water	56.5	0.005	6

distribution was assumed to be similar to other grasses (Zheng et al. 2007). The vertical distribution of LAI values enabled calculation of emerged and submerged LAI at each water level, assuming no changes as a consequence of the submersion. The hotspot parameter and the mean foliar inclination angle were fixed over all simulations and were set following the study made by Bacour et al. (2002).

In the case of cropland, an irrigated rice plantation was used for the case study. Most of rice production (75% of the world production per International Rice Research Institute) is developed in lowland irrigated systems. Irrigation implies the application of 5–25 cm of water over field. Rice crops could reach up to an LAI value of seven (Casanova et al. 1998; Stroppiana et al. 2006). Simulations were carried out using combinations of six LAI and six water levels (Table 17.4). The canopy vertical distribution and the mean foliar inclination angle used were simulated by Zheng et al. (2007) with a 3D model of rice canopy architecture for Shanyou63 variety. The hotspot parameter considered was similar to the one taken by Moulin et al. (2002) for a wheat crop.

To represent a forest plantation, reflectance canopy was simulated for a poplar canopy (*Populus canadensis*). In this case, different simulations were carried out for different LAI levels that ranged between one for a young plantation, to seven for a mature one (Meroni et al. 2004). The hotspot parameter and the mean foliar inclination angle were considered fixed over all simulations and were taken from Meroni et al. (2004). Because poplar is planted upland, no water was considered for simulations.

For each species, soil and leaf optical characteristics stayed constant between simulations. Leaf optical properties (reflectance and transmittance) of all species were taken from the LOPEX93 (Leaf Optical Properties EXperiment 93) database at 0.005-μm resolution (Hosgood et al. 1994). Spectral resolution was 0.001-μm applying a cubic spline smoothing. Soil spectral reflectance was measured over a silty clay loam using a handheld spectroradiometer Analytical Spectral Devices© FieldSpecPro (0.4–2.5 μm, ASD) and was assumed to be Lambertian.

17.2.3 Simulated Reflectance

Reflectance was simulated over the spectral range from 0.4 to 2.4 μm with 0.001-μm spectral resolution. For canopies with the same LAI (four), the forest showed a more pronounced red edge (Gitelson et al. 1996), higher reflectance in the near-infrared portion, and more pronounced absorption peaks in shortwave infrared than paddy rice and even more than wetland (Figure 17.2). In comparative terms, wetland showed the lowest reflectance values in near-infrared bands and the lowest absorption in visible and shortwave infrared bands (Figure 17.2). Reflectance decreased from 0.6 μm when canopies of wetland and paddy rice were flooded (h = 0.4 m). The decrease in reflectance of flooded canopies was larger as water level increased according to the results from Beget et al. (2013).

Wetlands and rice plantations showed a similar spectral behavior for most of the seven MODIS bands (Figure 17.3) and a greater variability between simulations. This kind of response occurred mainly because of the effects of water levels, low LAI levels (Beget and Di Bella 2007), or high proportions of bare soil (Oguro et al. 2003) (Figure 17.3). Water introduces variability because

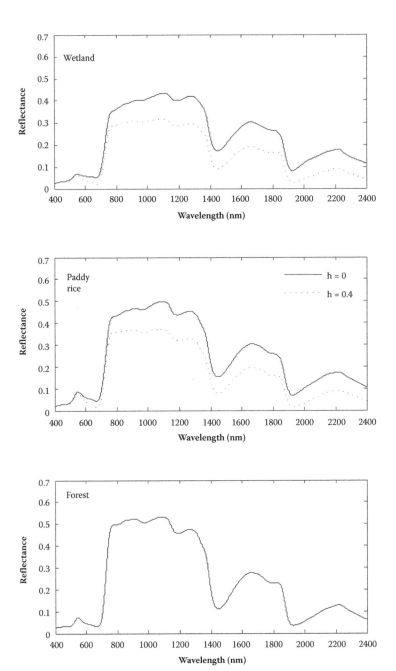

FIGURE 17.2
Simulated reflectance for hyperspectral bands carried out on wetland, rice paddy, and forest plantation. LAI is four for all cases. Filled lines represent cases without water and dotted lines represent flooded cases, h = 0.4 m.

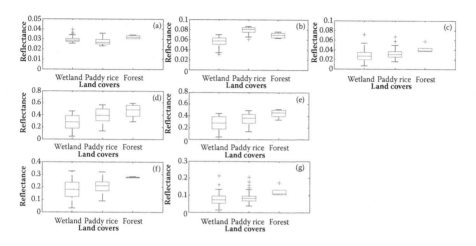

FIGURE 17.3
Boxplot of simulated reflectance at each MODIS band for studied covers ($n_{wetland}$ = 48, n_{rice} = 36, n_{forest} = 6). (a) Band 459–479 nm; (b) band 545–565 nm; (c) band 620–670 nm; (d) band 841–876 nm; (e) band 1230–1250 nm; (f) band 1628–1652 nm; (g) band 2005–2155 nm. Reflectance axis scale is different at each band. In each box, the central mark is the median, the edges of the box are the 25th and 75th percentiles, the whiskers extend to the most extreme data points, and outliers are plotted individually as a "+" sign. An outlier is a value that is more than 1.5 times the interquartile range away from the top or bottom of the box.

(as mentioned earlier) reflectance decreases as water level increases. Forest plantations differed from wetlands and rice plantations in almost all spectral bands—except in the blue band (459–479 nm). Nevertheless, the bands that stood out were medium-infrared bands (1628–1652 nm and 2005–2155 nm) (Figure 17.3f and g, respectively). The effect of water in canopy reflectance was emphasized in the infrared range (Figure 17.3d and g) (Beget et al. 2013). Even though LAI was expected to affect (at a greater extent) infrared reflectance (e.g., Jacquemoud et al. 2009), reflectance variability on the near-infrared bands was similar among wetlands, rice plantations, and forested lands (Figure 17.3d) regardless of the simulated LAI differences (up to 50%) among land covers.

Albedo was estimated from Liang's (2000) algorithm (see Equation 17.1) for simulated cases of wetlands, rice paddies, and forests (Figure 17.4). In this case, forests showed a 0.22 albedo value, which was 17% and 59% higher than those for rice plantations and wetlands, respectively. Even though forest plantations in general showed low albedo values with regard to other covers such as meadows and bushes (e.g., Houldcroft et al. 2009; Cescatti et al. 2012), they are known to show comparatively higher albedo values than water bodies (Houldcroft et al. 2009). This is the reason why the presence of surface water in wetlands and paddy rice crops decreases albedo. Variability (assessed as the difference between the maximum and

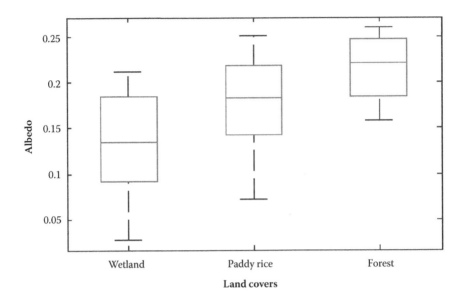

FIGURE 17.4
Boxplot of albedo for studied land covers ($n_{wetland} = 48$, $n_{rice} = 36$, $n_{forest} = 6$). In each box, the central mark is the median, the edges of the box are the 25th and 75th percentiles, the whiskers extend to the most extreme data points not considered. An outlier is a value that is more than 1.5 times the interquartile range away from the top or bottom of the box.

minimum albedo) was very high and similar between wetlands and rice plantations (between approximately 0.185 and 0.179, respectively). Bsaibes et al. (2009) found that when calculating albedo in different land covers (rice, wheat, corn, bare soil, and meadows) rice showed more difficulties mainly because of the complexity of the combined water and vegetation system. In the case of forest plantations, variability on albedo was only about 0.1 (Figure 17.4).

When albedo simulations for different LAI were analyzed, a positive relationship was found for wetlands and rice plantations (Figure 17.5a). These differences in albedo values were significant only between low and high LAI values (0.058 vs. 0.183 for low and high wetlands LAI, respectively; 0.106 vs. 0.233 for low- and high-rice LAI, respectively). Rautiainen et al. (2011) did not find any measurable effect over the LAI variation on calculated forest albedo. In this case, the relation found between LAI and albedo is connected to the simulations that were carried out. Input variables used in the model, except for LAI, were constant for all simulations. However, it is difficult to find these situations in nature because canopy characteristics (input model variables) are related between themselves and they covary simultaneously. Simulations showed a considerable increase of reflectance

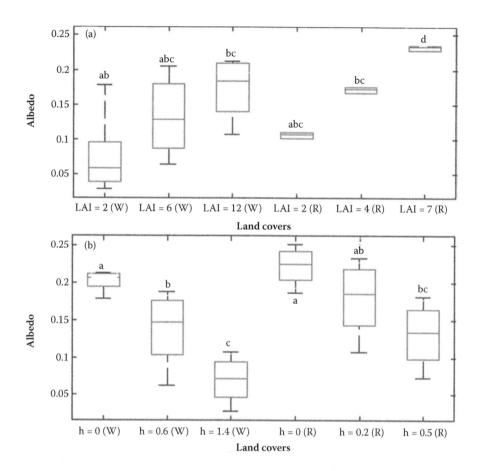

FIGURE 17.5
(a) Boxplot albedo for different LAI in wetlands (W) and rice paddies (R). (b) Boxplot of integrated reflectance (integration of MODIS bands 1 to 7 simulated reflectance) for different water levels (h) in wetlands (W) and rice paddies (R). In each box, the central mark is the median, the edges of the box are the 25th and 75th percentiles, and the whiskers extend to the most extreme data points. Differences between means were compared by T-student's t-test ($n_{wetlands} = 48$, $n_{rice} = 36$, $p = 0.05$). Means with the same letter do not differ significantly (Tukeys significant difference criterion).

in the near-infrared band, just at the moment LAI increased, with no important changes in the reflectance of other bands. Therefore, albedo increased according to LAI influence, contrary to what was expected. In the case of wetlands, regardless of LAI value, water level causes a great variability in albedo (Figure 17.5a). This variability does not seem to decrease with higher LAI values—probably because of high water levels and very variable proportions of submerged and emerged LAI. Variability differences found in wetland and rice plantations come from the fact that in wetland simulations, water ranges between 0 and 1.4 m, whereas in rice plantations it ranges between 0 and 0.5 m.

Albedo significantly decreased with higher water levels in wetlands and in rice paddies (Figure 17.5b). This result was expected because of the low albedo values shown in water (e.g., Rautiainen et al. 2011). In the case of wetlands, differences in albedo values were significant ($P < 0.5$) for all three measured water levels (0.207, 0.147, and 0.071 for 0, 0.6, and 1.4 m of water, respectively). Rice plantations showed significant differences only in extreme water values (0 and 0.5 m). In both cases, water level increased variability as an effect of LAI (submerged and emerged). When there was water in the system, LAI contribution to canopy reflectance—above water level (emerged LAI)—was very important (Beget et al. 2013). Similarly, small differences in total LAI resulted in large differences in emerged LAI; this, in turn, resulted in large differences in canopy reflectance. In the case of rice, particularly, we noticed an important effect of bare soil over variability in the integrated value, apart from emerged and submerged LAI (Figure 17.5b).

Apart from instant value changes coming from albedo (Figures 17.4 and 17.5), we expected very important temporal changes in albedo due to LAI changes or water levels, and also to changes made through the replacement of natural covers with new crops. Such changes could come from land use/land cover changes or from severe droughts and floods. To answer these questions, we analyzed the evolution of the albedo values in simulations during a growing season (Figure 17.6). During the months of July, common reed wetlands maintain (with small growing rates) an important green cover and an accumulation of senescent materials with high water levels (Mal and Narine 2004). In the case of rice, however, soil is fallow and mostly there is low humidity. Under these conditions, rice fallow land albedo is 40% higher than that of the wetland (0.150 and 0.091, respectively). In October, a month after rice sowing and with an LAI low value (LAI = 2), albedo showed very low values (0.109)—even lower than that of the wetland (0.159). For January (the beginning of summer), which is a preflowering and flowering period with maximum wetland LAI, albedo value is very similar in both cases (0.230 and 0.194). Finally, in April (autumn), a period very near rice harvest and the highest point of wetland species growth, albedo value once again is slightly higher than that of the new crop (0.150 for rice and 0.159 for wetlands).

In general, forests are very efficient at trapping radiation (House et al. 2005). However, compared to forests, wetland ecosystems are more efficient systems for trapping radiation through very low albedo values. This fact alters surface energy balance—lower values of albedo imply increasing energy transfer to the ecosystem and, subsequently, to the atmosphere, which alters local climate (see Chapter 16). Integrating a radiative transfer model, such as SAILHFlood, into the ecosystem service models would allow scientists to evaluate—in a basic way—albedo components of energy balance and their trade-offs.

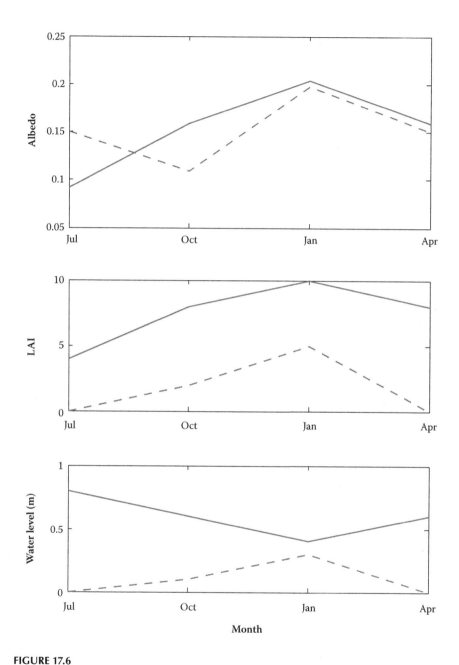

FIGURE 17.6
Albedo, LAI, and water level (m) for temporal evolutions of a wetland situation (filled lines) and a paddy rice situation (dotted lines). July (winter), October (spring), January (summer), and April (autumn) were considered to simulate temporal evolution in the Southern Hemisphere.

17.3 Conclusions

We used the SAILHFlood model to simulate reflectance of canopies (partially submerged and nonflooded). Thanks to these examples, we were able to indirectly study how land use changes would affect surface albedo. Wetland replacement by rice paddies or forest plantation or changes in land cover characteristics, such as LAI and water level, were also considered.

Reflectance simulation models could be used as interesting tools to evaluate changes in reflectance due to land cover/land use change. Spectral albedo could be estimated from narrow spectral bands after broadband conversion and after atmospheric and angular corrections.

Flooding would decrease albedo in all cases of wetland and paddy rice. Decreases in albedo would depend on LAI and water levels. Land use changes, as in the case of wetland replacement or a decrease in flooding level, would increase surface albedo and therefore would alter surface energy balance.

The use of a radiative transfer model (SAILHFlood) allowed us to study situations that are difficult to copy in real land covers, not only because of water complexity in wetland systems and rice plantations but also because of complexity associated with spatial and temporal variability.

Radiative transfer models make possible the evaluation of trade-offs between different land uses; for example, a land use change would capture more CO_2 (see Chapter 16) but would also warm more of the surface (lower albedo). In the case of wetlands, lower albedo values are not always associated with higher CO_2 capture because radiation is also being absorbed by water. The use of radiative transfer models became a valuable approximation for the quantification of ecosystems services related to biogeophysical processes as albedo. These models could complement satellite albedo assessment through the comprehension of factors involved in the determination of reflected energy. The SAILHFlood model in particular includes flooding water; therefore, it allows evaluation of energy balance and, subsequently, climate regulation in wetland ecosystems for services quantification.

Acknowledgments

This work was funded by grants from the International Research Development Center (IDRC-Canada, project 106601-001) and INTA (projects AERN4-AERN4642).

References

Allen, T. R., Y. Wang, and T. W. Crawford. 2013. Remote sensing of land cover dynamics. In *Treatise on geomorphology*, ed. J. F. Shroder, 80–102. San Diego, CA: Academic Press.

Alton, P. 2009. A simple retrieval of ground albedo and vegetation absorptance from MODIS satellite data for parameterisation of global land-surface models. *Agricultural and Forest Meteorology* 149:1769–1775.

Bacour, C., S. Jacquemoud, M. Leroy, et al. 2002. Reliability of the estimation of vegetation characteristics by inversion of three canopy reflectance models on airborne POLDER data. *Agronomie* 22:555–565.

Baigún, C. R. M., A. Puig, P. G. Minotti, et al. 2008. Resource use in the Parana River Delta (Argentina): Moving away from an ecohydrological approach? *Ecohydrology and Hydrobiology* 8:245–262.

Baldocchi, D. D., and K. B. Wilson. 2001. Modeling CO_2 and water vapor exchange of a temperate broadleaved forest across hourly to decadal time scales. *Ecological Modelling* 142:155–184.

Beck, P. S. A., S. J. Goetz, M. C. Mack, et al. 2011. The impacts and implications of an intensifying fire regime on Alaskan boreal forest composition and albedo. *Global Change Biology* 17:2853–2866.

Beget, M. E., V. A. Bettachini, C. M. Di Bella, and F. Baret. 2013. SAILHFlood: A radiative transfer model for flooded vegetation. *Ecological Modelling* 257:25–35.

Beget, M. E., and C. M. Di Bella. 2007. Flooding: The effect of water depth on the spectral response of grass canopies. *Journal of Hydrology* 335:285–294.

Beget, M. E., C. M. Di Bella, F. Baret, and J. F. Hanocq. 2010. Modeling reflectance of partially submerged canopies. *III International Symposium on Recent Advances in Quantitative Remote Sensing*. Valencia, Spain.

Brander, L. M., R. J. Florax, and J. E. Vermaat. 2006. The empirics of wetland valuation: A comprehensive summary and a meta-analysis of the literature. *Environmental and Resource Economics* 33:223–250.

Bright, R. M., F. Cherubini, and A. H. Strømman. 2012. Climate impacts of bioenergy: Inclusion of carbon cycle and albedo dynamics in life cycle impact assessment. *Environmental Impact Assessment Review* 37:2–11.

Bsaibes, A., D. Courault, F. Baret, et al. 2009. Albedo and LAI estimates from FORMOSAT-2 data for crop monitoring. *Remote Sensing of Environment* 113:716–729.

Bunn, S. E., and A. H. Arthington. 2002. Basic principles and ecological consequences of altered flow regimes for aquatic biodiversity. *Environmental Management* 30:492–507.

Burba, G. G., S. B. Verma, and J. Kim. 1999. Surface energy fluxes of *Phragmites australis* in a prairie wetland. *Agricultural and Forest Meteorology* 94:31–51.

Casanova, D., G. F. Epema, and J. Goudriaan. 1998. Monitoring rice reflectance at field level for estimating biomass and LAI. *Field Crops Research* 55:83–92.

Cescatti, A., B. Marcolla, S. K. Santhana Vannan, et al. 2012. Intercomparison of MODIS albedo retrievals and in situ measurements across the global FLUXNET network. *Remote Sensing of Environment* 121:323–334.

Clevering, O., and J. Lissner. 1999. Taxonomy, chromosome numbers, clonal diversity and population dynamics of *Phragmites australis*. *Aquatic Botany* 64:185–208.

Consoli, S., G. D'Urso, and A. Toscano. 2006. Remote sensing to estimate ET-fluxes and the performance of an irrigation district in southern Italy. *Agricultural Water Management* 81:295–314.

Csiszar, I., and G. Gutman. 1999. Mapping global land surface albedo from NOAA AVHRR. *Journal of Geophysical Research—Atmospheres* 104:6215–6228.

Deschamps, P. Y., F. Brion, M. Leroy, et al. 1994. The POLDER mission: Instrument characteristics and scientific objectives. *IEEE Transactions on Geoscience and Remote Sensing* 32:598–615.

Diner, D. J., J. C. Beckert, T. H. Reilly, et al. 1998. Multi-angle Imaging Spectroradiometer (MISR) instrument description and experiment overview. *IEEE Transactions on Geoscience and Remote Sensing* 36:1072–1087.

Disney, M., P. Lewis, G. Thackrah, T. Quaife, and M. Barnsley. 2004. Comparison of MODIS broadband albedo over an agricultural site with ground measurements and values derived from Earth observation data at a range of spatial scales. *International Journal of Remote Sensing* 25:5297–5317.

Gitelson, A. A., M. N. Merzlyak, and H. K. Lichtenthaler. 1996. Detection of red edge position and chlorophyll content by reflectance measurements near 700 nm. *Journal of Plant Physiology* 148:501–508.

Hatzianastassiou, N., C. Matsoukas, D. Hatzidimitriou, C. Pavlakis, M. Drakakis, and I. Vardavas. 2004. Ten year radiation budget of the Earth: 1984–93. *International Journal of Climatology* 24:1785–1802.

Hosgood, B., S. Jacquemoud, G. Andreoli, J. Verdebout, A. Pedrini, and G. Schmuck. 1994. Leaf Optical Properties EXperiment 93 (LOPEX93). *Report EUR 16095 EN*. Ispra, Italy: European Commission, Joint Research Centre, Institute for Remote Sensing Applications.

Houldcroft, C. J., W. M. F. Grey, M. Barnsley, et al. 2009. New vegetation albedo parameters and global fields of soil background albedo derived from MODIS for use in a climate model. *Journal of Hydrometeorogy* 10:183–198.

House, J., V. Brovkin, R. Betts, et al. 2005. Climate and air quality. In *Millennium ecosystem assessment. Ecosystems and human well-being: Current states and trends*, eds. P. Kabat, S. Nishioka, 355–390. Washington, DC: Island Press.

Jacquemoud, S., C. Bacour, H. Poilvé, and J. P. Frangi. 2000. Comparison of four radiative transfer models to simulate plant canopies reflectance—Direct and inverse mode. *Remote Sensing of Environment* 74:471–481.

Jacquemoud, S., W. Verhoef, F. Baret, et al. 2009. PROSPECT+SAIL models: A review of use for vegetation characterization. *Remote Sensing of Environment* 113:S56–S66.

Ju, W., P. Gao, J. Wang, Y. Zhou, and X. Zhang. 2010. Combining an ecological model with remote sensing and GIS techniques to monitor soil water content of croplands with a monsoon climate. *Agricultural Water Management* 97:1221–1231.

Key, J. R., X. Wang, J. C. Stoeve, and C. Fowler. 2001. Estimating the cloudy-sky albedo of sea ice and snow from space. *Journal of Geophysical Research—Atmospheres* 106:12489–12497.

Kingsford, R. T. 2000. Ecological impacts of dams, water diversions and river management on floodplain wetlands in Australia. *Austral Ecology* 25:109–127.

Kuusinen, N., P. Kolari, J. Levula, A. Porcar-Castell, P. Stenberg, and F. Berninger. 2012. Seasonal variation in boreal pine forest albedo and effects of canopy snow on forest reflectance. *Agricultural and Forest Meteorology* 164:53–60.

Lambert, A. 2003. *Economic valuation of wetlands: An important component of wetland management strategies at the river basin scale.* Ramsar: Gland.

Liang, S. 2000. Narrowband to broadband conversions of land surface albedo I: Algorithms. *Remote Sensing of Environment* 76:213–238.

Liang, S., A. H. Strahler, and C. W. Walthall. 1999. Retrieval of land surface albedo from satellite observations: A simulation study. *Journal of Applied Meteorology* 38:712–725.

Lucht, W., C. B. Schaaf, and A. H. Strahler. 2000. An algorithm for the retrieval of albedo from space using semiempirical BRDF models. *IEEE Transactions on Geoscience and Remote Sensing* 38:977–998.

MA (Millennium Ecosystem Assessment). 2005. *Ecosystems and human well-being: Current state and trends.* Washington, DC: Island Press.

Maes, W. H., T. Pashuysen, A. Trabucco, F. Veroustraete, and B. Muys. 2011. Does energy dissipation increase with ecosystem succession? Testing the ecosystem energy theory combining theoretical simulations and thermal remote sensing observations. *Ecological Modelling* 222:3917–3941.

Mal, T. K., and L. Narine. 2004. The biology of Canadian weeds. 129. Phragmites australis (Cav.) Trin. ex Steud. *Canadian Journal of Plant Science* 84:365–396.

Mariotto, I., and V. P. Gutschick. 2010. Non-Lambertian Corrected Albedo and Vegetation Index for estimating land evapotranspiration in a heterogeneous semi-arid landscape. *Remote Sensing* 2:926–938.

Meroni, M., R. Colombo, and C. Panigada. 2004. Inversion of a radiative transfer model with hyperspectral observations for LAI mapping in poplar plantations. *Remote Sensing of Environment* 92:195–206.

Mitsch, W. J., and J. G. Gosselink. 2000. *Wetlands.* 3rd ed. New York: John Wiley & Sons.

Moulin, S., L. Kergoa, P. Cayrol, G. Dedieu, and L. Prévot. 2002. Calibration of a coupled canopy functioning and SVAT model in the ReSeDA experiment. Towards the assimilation of SPOT/HRV observations into the model. *Agronomy* 22:681–686.

Mueller, R., J. Trentmann, C. Träger-Chatterjee, R. Posselt, and R. Stöckli. 2011. The role of the effective cloud albedo for climate monitoring and analysis. *Remote Sensing* 3:2305–2320.

Nasipuri, P., T. J. Majumdar, and D. S. Mitra. 2006. Study of high-resolution thermal inertia over western India oil fields using ASTER data. *Acta Astronautica* 58:270–278.

Noe, S. M., V. Kimmel, K. Hüve, et al. 2011. Ecosystem-scale biosphere–atmosphere interactions of a hemiboreal mixed forest stand at Järvselja, Estonia. *Forest Ecology and Management* 262:71–81.

Oguro, Y., Y. Suga, S. Takeuchi, H. Ogawa, and K. Tsuchiya. 2003. Monitoring of a rice field using Landsat-5 TM and Landsat-7 ETM+ data. *Advanced Space Research* 32:2223–2228.

Pinty, B., and M. Verstraete. 1992. On the design and validation of surface bidirectional reflectance and albedo model. *Remote Sensing of Environment* 41:155–167.

Posselt, R., R. W. Mueller, R. Stöckli, and J. Trentmann. 2012. Remote sensing of solar surface radiation for climate monitoring—The CM-SAF retrieval in international comparison. *Remote Sensing of Environment* 118:186–198.

Puzachenko, Y. G., R. B. Sandlersky, and A. Svirejeva-Hopkins. 2011. Estimation of thermodynamic parameters of the biosphere, based on remote sensing. *Ecological Modelling* 222:2913–2923.

Rautiainen, M., P. Stenberg, M. Mottus, and T. Manninen. 2011. Radiative transfer simulations link boreal forest structure and shortwave albedo. *Boreal Environment Research* 16:91–100.

Román, M. O., C. B. Schaaf, P. Lewis, et al. 2010. Assessing the coupling between surface albedo derived from MODIS and the fraction of diffuse skylight over spatially-characterized landscapes. *Remote Sensing of Environment* 114:738–760.

Schaaf, C. B., F. Gao, A. H. Strahler, et al. 2002. First operational BRDF, albedo nadir reflectance products from MODIS. *Remote Sensing of Environment* 83:135–148.

Schottler, S. P., J. Ulrich, P. Belmont, et al. 2013. Twentieth century agricultural drainage creates more erosive rivers. *Hydrological Processes.* doi:10.1002/hyp.9738.

Shuai, Y., J. G. Masek, F. Gao, and C. B. Schaaf. 2011. An algorithm for the retrieval of 30-m snow-free albedo from Landsat surface reflectance and MODIS BRDF. *Remote Sensing of Environment* 115:2204–2216.

Smith, P., H. Black, C. Evans, et al. 2011. Regulating services. In *UK National Ecosystem Assessment. Understanding nature's value to society.* Technical Report, 535–596. Cambridge, UK: UNEP-WCMC.

Stroeve, J., J. E. Box, F. Gao, S. Liang, A. Nolin, and C. Schaaf. 2005. Accuracy assessment of the MODIS 16-day albedo product for snow: Comparisons with Greenland in situ measurements. *Remote Sensing of Environment* 94:46–60.

Stroppiana, D., M. Boschetti, R. Confalonieri, S. Bocchi, and P. A. Brivio. 2006. Evaluation of LAI-2000 for leaf area index monitoring in paddy rice. *Field Crops Research* 99:167–170.

Stroppiana, D., S. Pinnock, J. M. C. Pereira, and J. M. Grégoire. 2002. Radiometric analysis of SPOT-VEGETATION images for burnt area detection in northern Australia. *Remote Sensing of Environment* 82:21–37.

Sumner, D. M., Q. Wu, and C. S. Pathak. 2011. Variability of albedo and utility of the MODIS albedo product in forested wetlands. *Wetlands* 31:229–237.

Verhoef, W. 1984. Light scattering by leaf layers with application to canopy reflectance modeling: The SAIL model. *Remote Sensing of Environment* 16:125–141.

Verhoef, W. 1985. Earth observation modeling based on layer scattering matrices. *Remote Sensing of Environment* 17:165–178.

Wang, X., and C. S. Zender. 2010. MODIS snow albedo bias at high solar zenith angles relative to theory and to in situ observations in Greenland. *Remote Sensing of Environment* 114:563–575.

Weiss, M., F. Baret, R. Myneni, A. Pragnère, and Y. Knyazikhin. 2000. Investigation of a model inversion technique for the estimation of crop characteristics from spectral and directional reflectance data. *Agronomie* 20:3–22.

Yang, Y., S. Shang, and L. Jiang. 2012. Remote sensing temporal and spatial patterns of evapotranspiration and the responses to water management in a large irrigation district of north China. *Agricultural and Forest Meteorology* 164:112–122.

Zheng, B., L. Shi, Y. Ma, Q. Deng, B. Li, B. and Y. Guo. 2007. Canopy architecture quantification and spatial direct light interception modeling of hybrid rice. *5th Workshop Functional Structural Plant Models Proceedings*, Napier, New Zealand, November 4–9, 37.1–37.3.

Zwart, S. J., W. G. M. Bastiaanssen, C. de Fraiture, and D. J. Molden. 2010. WATPRO: A remote sensing based model for mapping water productivity of wheat. *Agricultural Water Management* 97:1628–1636.

18

Energy Balance and Evapotranspiration: A Remote Sensing Approach to Assess Ecosystem Services

V. A. Marchesini

University of Buenos Aires, Argentina; The University of Western Australia, Australia

J. P. Guerschman

Commonwealth Scientific and Industrial Research Organisation (CSIRO) Land and Water, Australia

J. A. Sobrino

University of València, Spain

CONTENTS

18.1 Ecosystem Services, Energy Balance, and Evapotranspiration 400
18.2 Relating ET and Energy Balance to Ecosystem Services 401
18.3 The Use of Remote Sensing to Estimate ET and Its Relationship
      with Ecosystem Services...402
18.4 ET, Albedo, LST, and Their Relation with Ecosystem Services 404
18.5 A Case Study: The Impact of Large-Scale Deforestation on
      Dry Forests of Central Argentina: Changes in ET, Surface
      Temperature, and Albedo at Landscape Scale.....................................407
18.6 Conclusions.. 410
Acknowledgments 411
References.. 412

18.1 Ecosystem Services, Energy Balance, and Evapotranspiration

In the past four decades, humans have altered ecosystem functions and ecosystems' ability to provide basic services to humankind such as provision of water and local and regional weather control (MA 2005). The term "ecosystem services" refers to all the processes occurring in ecosystems that support and sustain human life (Daily 1997; Kremen and Ostfeld 2005; Kinzig et al. 2011). From the maintenance of biodiversity through moderation of weather and climate mitigation, natural ecosystems play an irreplaceable role in maintaining human existence. An integrated measure of ecosystem function that contributes to regulate local and regional climate is evapotranspiration (ET; Carlson et al. 1995; Anderson and Kustas 2008; Jung et al. 2010).

ET is defined as the process in which water is released into the atmosphere from vegetation, soil, and water bodies (Allen et al. 1998). ET mediates the interaction between water dynamics and below- and aboveground surface processes by regulating water vapor and heat through the vegetation canopy. ET is usually expressed as volume of water per unit of time (e.g., mm day^{-1}) and also can be expressed in energy units (e.g., W m^{-2}) considering the energy required to evaporate water at a specific temperature. The main variables that affect ET are solar radiation, air humidity, wind speed, air temperature, and soil moisture (Carlson et al. 1990; Allen et al. 2006b; Jung et al. 2010). Local factors such as soil texture and salinity also influence ET (Fernandez-Illescas et al. 2001; El-Nahry and Hammad 2009). The main biotic factor in controlling ET is vegetation, where functional group and species composition regulate its magnitude (Nosetto et al. 2005; Miao et al. 2009). Alterations in vegetation cover and structure can significantly impact the reflection and transmission of energy and, in consequence, impact ET and the hydrological cycle, with serious implications for climate and ecosystem services (Chapin et al. 2002; Chen et al. 2009).

Because the process of water evaporation from soil and vegetation requires energy, indeed globally more than half of the energy used on the Earth's surface is used to evaporate water (Jung et al. 2010), ET can be estimated and quantified by considering the energy balance. In simple terms, the surface energy balance equation (which will be described in detail in the next section) assumes that the available energy at ground level is the total amount of energy that reaches the ground surface minus the flux of energy transferred to the soil surface by conduction. This amount of energy can be partitioned into two turbulent fluxes: the sensible and the latent heat flux. The sensible heat is referred to as the flux of heat from conduction and convection processes that results in changes in temperature (Jackson et al. 1977). The latent heat is related to the energy necessary to evaporate water from a surface (e.g., vegetation cover). The partition between the latent and the sensible heat (the ET fraction) determines the balance between water vapor and heat.

Thus, estimation of this evaporative fraction using the energy balance has profound implications in the evaluation of many hydrological and biogeo-chemical processes (Wang et al. 2012).

ET can also be estimated using other methods that vary according to the spatial scale: At patch scale (10^0–10^3 of m^2), the most common approaches include pan evaporation (Brutsaert and Parlange 1998; Blight 2002), lysimeters (Paruelo et al. 1991), and sap flow sensors (Burgess et al. 2001; Doody and Benyon 2011). One of the strongest limitations of these methods is the high number of replicates required to account for soil spatial heterogeneity. Another issue is that scaling up data obtained at patch scale to the entire canopy can give mis-leading results because changes in scale could result in nonlinear ecosystem responses (Wang et al. 2012). ET at intermediate spatial scales (10^4–10^6 of m^{-2}) is best obtained by flux towers: A variety of sensors are mounted on platforms to estimate mainly energy and carbon and water dynamics (Baldocchi 2003; Cleugh et al. 2007) (see Chapter 2). The use of flux towers has become common in the past years because they measure ET at a scale that makes it suitable to obtain ground truth data for satellite observations. Although these techniques present good temporal resolution, they are limited given the number of devices necessary to cover large extensions and the lack of accuracy in fetch areas.

Despite the efforts to improve accuracy in estimations of carbon, water, and energy at different scales, remote sensing technology supplemented with ground observations remains the best approach to cover large scales (10^6 m^2 to global) and to integrate long time series data because some satellite plat-forms and remote sensing products now have been available for more than four decades. Landsat satellite data, for example, have been available since the early 1970s and the Advance Very High Resolution Radiometer (AVHRR) has been collecting information since 1980. The estimation of ET by remote sens-ing is mostly based on two approaches: (1) linking ET to vegetation cover and greenness by spectral indices such as the Normalized Difference Vegetation Index (NDVI; Caparrini et al. 2003; Mu et al. 2007; Contreras et al. 2011; Yebra et al. 2013); and (2) relating ET to land surface temperature (Norman et al. 1995; Su 2002; Sobrino et al. 2005; Kalma et al. 2008). Some of these models employ the energy balance equation and the partition of the energy available between sensible and the latent heat fluxes to connect ET and the energy balance.

18.2 Relating ET and Energy Balance to Ecosystem Services

Life on Earth is governed by energy and it is through the combination of energy and water that processes such as photosynthesis support ecosystem life. As mentioned earlier, the energy that reaches the Earth at ground level is mainly the result of the total incoming solar radiation minus the total ongoing radiation. In simple terms, net radiation (Rn) can be considered as the sum of three main terms:

$$Rn = G + H + LET \tag{18.1}$$

where Rn is the net radiation, G is the ground heat flux, LET is the heat latent flux, and H is the sensible heat flux. All terms are expressed in units of energy per time and area such as W s m^{-2} (Jackson et al. 1977; Hurtado and Sobrino 2000).

This simplified balance equation does not take into account other minor energy components such as the heat stored in soil and vegetation or fluxes as a consequence of metabolic processes. Moreover, this equation only considers vertical fluxes and not the energy that also flows in horizontal directions as a consequence of advection processes (Allen et al. 2006b). From all these terms, LET and H are those closely related to ecosystem services because ET emerges from the estimation of latent heat, a direct link between vegetation and the water cycle. The partition between LET and H, the ET fraction (λET), determines the balance between vapor and heat at surface level and can be used as an indicator of land cover changes, with a particular ecological meaning. Land use changes such as the replacement of forests by crops or landscape fragmentation and deforestation for urban expansion can have significant impact on the redistribution of LET and H fluxes and, in consequence, on ET and the water balance (Anderson et al. 2008). Changes in ET, precipitation, runoff, and deep drainage have a profound influence on ecosystem services providing climate and soil stabilization, regulating local rainfall, affecting nutrients dynamics, and groundwater recharge rate. ET also can account for more than 50% of terrestrial water input especially in arid and semiarid areas (Wilcox 2002; Reynolds et al. 2004).

18.3 The Use of Remote Sensing to Estimate ET and Its Relationship with Ecosystem Services

In recent years, progress in technology has facilitated the evaluation of human impact on ecosystem services at the same scale that the changes are occurring (Kerr and Ostrovsky 2003). The use of remote sensors attached to satellite platforms is one of the few alternatives available to assess the impact of these changes over large areas. Many methods used to evaluate flux exchange between the surface and the atmosphere obtain sensible heat flux as a residual of the energy balance equation. Disadvantages of these models are that many of them require local adaptation and, depending on the spatial scale, large extensions are required to cover a "pure pixel." However, these flaws can be corrected combining products offered by different satellite platforms at different spatial and temporal scales. The Surface Energy Balance System (SEBS; Su 2002), for example, combines variables such as albedo, land surface temperature (LST), vegetation cover, air humidity, wind speed, and downward solar radiation to estimate ET. An advantage of this method is that most of the input variables can be obtained directly from remote sensing data. NDVI, for example, is used as a surrogate when vegetation cover data are not available (Su 2002).

A pioneer model used to estimate fluxes of sensible, latent heat and soil heat flux (G) is the two source energy balance (TSEB) developed by Norman et al. (1995). This two-layer model, based mainly in thermal remote sensing data, requires as input net radiation, directional angle of view, air temperature, wind speed, and vegetation parameters such as leaf area index, vegetation height, or leaf size. The model employs a fractional vegetation cover to estimate soil and canopy temperature. The separation between vegetation and soil components is an important attribute of this model because it can account for the variability at subpixel level (Timmermans et al. 2011). Water loss predictions from this model were subsequently improved by Anderson et al. (2008), by incorporating a light-use efficiency (LUE) model of canopy resistance that considers stomatal conductance and the effects of stomatal closure as a consequence of vapor pressure deficit.

Another remote sensing model used to estimate ET is the Simplified Surface Energy Balance Index (S-SEBI) developed by Roerink et al. (2000). This method combines reflectance and surface temperature under dry and wet conditions for determining an evaporative fraction and obtaining the sensible and latent heat from it. The reasoning behind the model is as follows: under high water availability (e.g., irrigation crops), low albedo values determine a steady LST because the available energy in the system is used to evaporate water. If albedo values rise, LST increases up to a defined point called "evaporation controlled" where ET and latent heat is maximum and sensible heat is null ($Rn - G = H = 0 = LET_{max}$). As soon as soil moisture availability decreases and ET reaches values close to zero, all the available energy in the system is used to heat the soil, and as a consequence, LST increases. Beyond a reflectance threshold, LST reaches a point called "radiation controlled" where ET is null and heat flux is maximum ($LET = 0$, $H_{max} = Rn - G$). Given these two temperature points ($LET = 0$ and $LET = max$), it is possible to estimate ET as the fraction between these two temperature lines. LET is calculated as the evaporative fraction times the difference between the net available energy and G. Net available energy is calculated by considering the incoming solar radiation, atmospheric transmissivity, and surface albedo, while G is estimated empirically by a relationship between surface reflectance, surface temperature, and NDVI or other vegetation indices such as the Modified Soil-Adjusted Vegetation Index (MSAVI) or the Leaf Area Index (LAI). This model has been validated for areas in southwestern Spain (Sobrino et al. 2005) and Italy (Roerink et al. 2000), and in central Argentina (Marchesini et al. 2012), among others.

A comparison between S-SEBI and TSEB was done by Timmermans et al. (2011) in agricultural sites of northeast Germany. The authors did not find major differences in radiative fluxes (e.g., net radiation or soil heat flux), but they found considerable differences in turbulent fluxes (H, LET), especially in areas with high roughness such as tall crops exposed to dry conditions, suggesting a better agreement between these two models in humid and uniform areas.

While SEBS, S-SEBI, and other related methods use remote sensing to resolve the terms of the energy balance equation, other approaches make a more empirical connection of vegetation (and other derived) indices with ET. The LAI product derived from the Moderate Resolution Imaging Spectroradiometer (MODIS) was used as a proxy for canopy-level stomatal conductance and combined with the Penman–Monteith model of potential ET for estimating actual ET (Cleugh et al. 2007). Based on the same method, Mu et al. (2007, 2011) applied the model to global data to produce eight-day estimates of ET at the global scale. Guerschman et al. (2009) used the Enhanced Vegetation Index (EVI) derived also from the MODIS sensor to scale potential ET (ET_p) into actual (ET_a). ET_p was derived from meteorological data using the Priestley–Taylor formulation. In addition to EVI, which quantifies the vegetation cover, the Global Vegetation Moisture Index (GVMI) was used. The GVMI allows the separation between open water surfaces and bare soil when the vegetation cover (and hence EVI) is low and, therefore, provides a way to better represent open water evaporation. The method was calibrated using flux tower measurements from seven sites across Australia and evaluated using ~200 catchment-derived ET (as the long-term difference between rainfall and streamflow). Figure 18.1 shows the long-term water balance across the Australian continent derived from the method of Guerschman et al. (2009). By estimating ET dynamically across large regions and combining it with interpolated rainfall data, it is possible to identify areas that evaporate water from sources other than rainfall (Figure 18.1, in red). Those areas correspond to groundwater-dependent ecosystems, irrigated crops, wetlands, and combinations of them. Such areas tend to be important sources of ecosystem goods and services such as food, fiber, timber production, and maintenance of biodiversity.

18.4 ET, Albedo, LST, and Their Relation with Ecosystem Services

Deterioration of ecosystem services as a result of vegetation structure simplification (e.g., from more to less LAI), a decline in species richness, or a reduction in soil cover can modify the energy budget by increasing the proportion of energy that is reflected and, in consequence, affect LST and

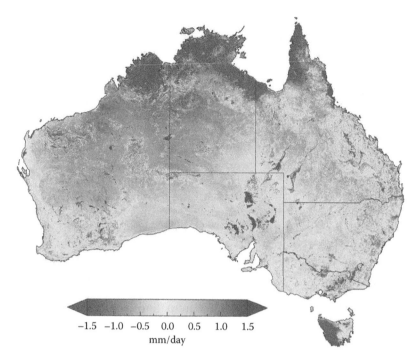

FIGURE 18.1 (See color insert.)
Difference between mean precipitation and mean monthly actual evapotranspiration across Australia for the period 2000–2006. (Adapted from Guerschman, J. P., et al., *J. Hydrol.*, 369:107–119, 2009.)

albedo (Dirmeyer and Shukla 1994). LST is the variable that best integrates the relationship between atmospheric phenomena and processes occurring at the surface. Anderson and Kustas (2008) have described LST as a thermometer in which increments in land surface temperature can be an indicator of an anomalous situation such as stress by water deficit. LST can be retrieved from the thermal infrared data measured by sensors onboard Earth observation platforms. Examples of these sensors are Thematic Mapper (TM) or Enhanced TM (ETM+) onboard the Landsat platforms and the Advanced Space borne Thermal Emission and Reflection Radiometer (ASTER) onboard the Terra platform, which provides data at high spatial resolution (around 100 m), and the AVHRR onboard the National Oceanic and Atmospheric Administration (NOAA) platforms, the MODIS onboard Terra, and the Spinning Enhanced Visible and InfraRed Imager (SEVIRI) onboard the Meteosat Second Generation (MSG) providing data at low spatial resolution (1–5 km). Estimation of LST is required for a wide variety of applications such as in the study of natural ecosystems, agricultural and urban planning (e.g., urban heat island ecology), climate mitigation, forestry, meteorological predictions, and hydrology, among others. In Jornada, New Mexico, studies with

Landsat images have revealed that shrub encroachment increased radiance and LST by 5°C in areas with lower ground vegetation cover (Ritchie et al. 2000). Thermal infrared data have also been used to evaluate the spatial and temporal precipitation trends by discriminating cloud from nonraining cloud on the basis of cloud temperature (Feidas et al. 2009).

Another integrative variable that can be used to reflect changes in ecosystem properties is albedo, the ratio between the solar radiative flux that is reflected from a surface to the flux intercepted by this same surface (Liang 2000). In arid and semiarid regions, increments in albedo lead to losses in the energy absorbed by the surface. Land clearing and increment in bare soil cover can enhance albedo and reduce ET and local precipitation by changes in surface roughness (Dirmeyer and Shukla 1994) (see Chapter 17). Changes in albedo and emissivity with consequences on near-surface climate have been reported for the Chihuahuan Desert as a result of shrub encroachment (Beltrán-Przekurat et al. 2008). The term "radiative forcing" has been largely used to indicate possible global warming effects as changes in the radiation balance, including albedo changes, at the tropopause level (Forster and Ramaswamy 2007). Houspanossian et al. (2012) used the MODIS albedo product MCD43A3 to evaluate how the transition from woody to a crop-dominated system changes albedo and LST in semiarid areas of Argentina. They found that crops showed 2.5°C diurnal surface temperature and 50% higher albedo than forests. Decreases in precipitation after clearing have been observed in Amazonian forests in areas where albedo increased from 0.09 to 0.3 (Dirmeyer and Shukla 1994). Rises in albedo and increments in the latent heat imply less water in the system and increments in LST. However, a "counterintuitive" cooling effect has been demonstrated for middle to high latitudes where a reduction in the proportion of forest cover increases snow exposure and albedo resulting in larger losses of energy to the atmosphere in both day- and nighttimes of the diurnal cycle (Lee et al. 2011).

Albedo estimation by remote sensing can be performed using either broadband or narrowband observations; however, accurate albedo values are obtained if atmospheric surface conditions at the top of the atmosphere are known. Due to the better resolution characterizing the heterogeneity at surface and atmospheric levels, narrowband multispectral sensor are more effective (Liang 2000). Albedo values obtained from narrowband observations require angular and atmospheric corrections. Many algorithms have been proposed to transform albedo values obtained from broad- to narrowband (Brest and Goward 1987; Liang 2000). In general terms, depending on the aerosol optical depth, albedo can be estimated as a value extrapolated between complete direct illumination (black-sky albedo) and complete diffuse illumination (white-sky albedo). An accurate estimation of albedo requires the computation of the bidirectional reflectance distribution function (BRDF). This function considers the radiance from a surface scattered to any direction above the hemisphere of the surface. BRDF, thus, is used to standardize observations with variable sun-view geometries into a common standard geometry. Because land surface structure influence BRDF, this function can also be used as a source of information about the biophysical characteristics of the land surface observed. BRDF estimation requires the acquisition of visible near-infrared (VNIR) and shortwave near-infrared (SWIR) data at different observational geometries.

When VNIR and SWIR data are not available, then a simple approach such as the one proposed by Saunders (1990) could be used. This simple approach obtains albedo by using visible (RED) and near-infrared (NIR) reflectance channels as:

$$\text{Albedo } (\alpha) = 0.5 \, (\rho \text{RED} + \rho \text{NIR}) \tag{18.2}$$

Here, ρRED and ρNIR are the RED and NIR reflectance, respectively (Saunders 1990). Estimation of albedo is recognized as one of the main radiative uncertainties. Many models used to predict this variable present errors larger than 15% (Liang 2000).

A final point to highlight in this section is the connection between LST, albedo, and ET. These three variables are linked via the instantaneous net radiation flux (Rni), the parameter used in the energy balance (Equation 18.1) to estimate H and λET and consequently deduce ET:

$$\text{Rni} = (1 - \alpha) \, \text{Rc } \lambda \downarrow + \varepsilon \, \text{Rg } \lambda \downarrow - \varepsilon \sigma \, \text{T4S} \tag{18.3}$$

where α is the albedo, Rc $\lambda \downarrow$ is the incoming shortwave radiation (W m^{-2}), ε Rg $\lambda \downarrow$ is the incoming longwave radiation (W m^{-2}), σ is the Stefan Boltzmann constant, ε is the surface emissivity, and TS is the surface temperature (K) (Sobrino et al. 2004, 2005).

Currently, abundant literature exists on the use of thermal infrared (associated with LST) and albedo data to estimate changes in ET (Carlson et al. 1995; Anderson and Kustas 2008; Kalma et al. 2008) with potential applications in hydrology, urban planning (see Chapter 19), weather forecasting, agriculture, and natural resources management.

18.5 A Case Study: The Impact of Large-Scale Deforestation on Dry Forests of Central Argentina: Changes in ET, Surface Temperature, and Albedo at Landscape Scale

Deforestation represents one of the major environmental issues that humanity faces in this century (Malhi et al. 2008). Land cover changes or changes in vegetation patterns (e.g., from dominant woody vegetation to grasslands or crops) can affect the magnitude of ET. Forest replacement can increase ecosystem services of provision such as food production, timber, and housing, but at the same time, deforestation can have a negative impact on ecosystem services regulation such as drought and flood mitigation (Carpenter and Folke 2006). The phenomenon of dryland salinity in Australia is one of the best known examples in which the replacement of native forest by croplands (from high to less ET) induced raises in the water tables causing salinization and the loss of thousands of hectares of productive land (Barrett-Lennard 2003; Lambers 2003). The opposite phenomenon has been observed in the Argentinean Pampas where tree establishment in native grassland reduced water tables and stream flows by almost 50% (Nosetto et al. 2012).

South America has the highest percentage of original forest and the largest carbon stocks in forest biomass (>100 gigatons C) in the world and also shows the highest deforestation rates. According to the Food and Agricultural Organization of the United Nations (FAO 2010), the percentage of primary forest losses in South America is still extremely high, with almost 4 million hectares of forest lost every year (Hamilton 2008). These forests offer a wide range of ecosystem services: soil and water conservation, desertification control, and climate change mitigation. Forested catchment areas also provide an important proportion of the water for industrial and domestic uses. Forests and the energy balance and water cycle are connected by the modulation of processes such as local precipitation, direct soil and canopy evaporation, transpiration, soil infiltration (stem flow and channel roots), fog water capture, and radiation reflectance among others (Dirmeyer and Shukla 1994; Miles et al. 2006; Hansen et al. 2010).

ET estimations using the SEBAL model for Bolivian forests showed that more than 50% of the annual rainfall depends on *in situ* ET produced by transpiration and by the evaporation of wet leaves. Massive deforestation in these areas implies a reduction in total local water input, affecting processes such as forest regeneration and increasing extreme temperatures. Seiler and Moene (2011) obtained similar results for Amazonian forests in which deforestation reduced ET by 22%.

This section describes how the conversion of dry forest areas into grasslands reduces ET and increases LST and albedo. Satellite images and a model based on the energy balance equation were used to evaluate the impact of this event over a dry forest in central Argentina. The use of satellite images allowed us to compare the same place before, during, and after the disturbance occurrence.

In San Luis Province, in central Argentina, shrubs and trees dominate forest overstory (Aguilera et al. 2003; Marchesini et al. 2012). In the past decades, agricultural expansion has intensified—especially in areas where soybean became the dominant crop. As a consequence of soybean expansion, intensive ranching production was pushed toward semiarid areas, historically used for free ranching and selective logging (Gasparri and Grau 2009). One of the most common practices for eliminating shrubby vegetation is roller chopping, a technique that uses a heavy metal cylinder moved by a bulldozer that selectively chooses trees and shrubs to eventually remove them by chopping and crushing (Blanco et al. 2005). As a result of this practice, cattle can easily access forage that ultimately increases in cover several times after tree and shrub removal.

Removal of woody vegetation on a large scale may affect the ecosystem's ability to use water integrally through ET. Some physical characteristics of forests, such as canopy height and deep root distribution, can promote higher water losses (Kelliher et al. 1993; Moore and Heilman 2011). Air turbulence, one of the variables that most influences ET, is usually elevated in tall-canopy systems due to the higher roughness, which favors a strong coupling between atmospheric water demand and water losses from the canopy.

During 2006–2009, eight plots from 60 to 300 hectares immersed in a large dry forest matrix in central Argentina were selectively cleared for the purpose of increasing forage production. The remaining undisturbed dry forest control plots were used for comparing ET, LST, and albedo. Disturbed and undisturbed plots were analyzed using remote sensing data from Landsat/TM5 platform and included at least four images per year from January 2004 to December 2009.

ET was estimated using the Simplified Surface Energy Balance Model (S-SEBI; Roerink et al. 2000), details of which were explained in the previous section. In order to compare and validate ET satellite estimations, an indirect and simple method based on air temperature (obtained from a local weather station) (Blaney and Criddle 1950) was used. Although the Blaney–Criddle method can be considered inaccurate because it is based on standard conditions, it can be used as a standard parameter to compare the magnitude and the trends observed in ET between disturbed and undisturbed areas.

Deforestation over dry forests caused a reduction in ET by 30% and an increment in LST up to 4°C. Albedo also increased more than 60% after clearing trees and shrubs. Differences between undisturbed and disturbed plots were observed immediately after the disturbance, but it was during the summer and the rainy season that these areas showed more disparities. Cleared areas also showed a larger spatial variability in ET distribution (Figure 18.2). Many factors could explain the significant drop in ET: a reduction in roots and aboveground biomass, changes in foliar index, and complementary and positive niche interactions between species. Although deforestation can have a cooling effect on the Earth's surface due to a reduction of the energy income (by albedo increments) (Lee 2010), the warming effect observed in open areas in this study seems to be more the result of changes in the ratio between latent and sensible heat (Davin et al. 2007). The forest seems to perform a better "job" in refreshing the surface and removing latent heat than the new open grasslands.

This section shows an example of remote sensing application for studying changes in water and energy balance as a consequence of changes in vegetation cover. In terms of ecosystem services, this study showed that woody vegetation in dry forests of central Argentina exerts a strong control on the dynamics of water even when grass cover and biomass significantly increased after deforestation. This reduction in tree and shrub biomass also impacted local climate by increasing LST and the proportion of longwave radiation

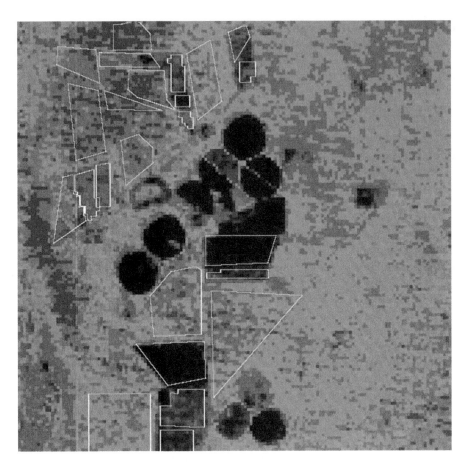

FIGURE 18.2 (See color insert.)
Evapotranspiration in mm day^{-1} estimated with Landsat after tree removal, October 2008. Black circles represent sites cleared for agriculture. Red-filled polygons represent areas where trees were removed; and blue-green areas correspond to undisturbed forest.

that was emitted by land surface. The impact of this large-scale deforestation on ET may have profound practical implications because water is a crucial factor modulating most of the biological processes in semiarid regions.

18.6 Conclusions

ET is a term closely related to ecosystem services: as a component of the water cycle, ET determines the balance between heat and water vapor and has a strong influence on climate mitigation, control of precipitation, soil equilibrium, nutrients fluxes, and on groundwater recharge. Due to the close relationship between ET and vegetation, changes in the proportion of vegetation cover and LAI could directly affect the magnitude of ET and consequently modify ecosystem services.

During the past century, land use changes have been particularly intense in some areas, affecting ET patterns and thus the capacity of many ecosystems to evaporate water. The disruption of thousands of years of a delicate equilibrium in which ecosystems have integrally used water via ET has led to drastic land degradation including reduction of local precipitation and large-scale salinization, among others. These large-scale environmental problems as well as the loss of ecosystem services can only be assessed using techniques with the appropriate temporal and spatial resolutions. Earth observation sensors on satellite platforms remain the best approach to achieve this goal at regional to global scales.

The models proposed in this chapter to estimate ET via remote sensing use different input variables—either vegetation or physical parameters, such as albedo and LST. The energy balance equation is employed to predict the partition between latent and sensible heat flux and to obtain ET. Satellite images such as Landsat or MODIS at different spatial and temporal scales were used to evaluate ET magnitudes before and after large-scale deforestation in the dry forests of Argentina or to assess actual ET across Australia and to identify areas of relevance for the provision of ecosystem services.

Advanced knowledge linking remote sensing, energy balance, and ET estimation has accelerated in the past decades, allowing accessibility of satellite products and information to a broad number of stakeholders including the general public. Although further improvements are needed, new models and applications are rapidly emerging, offering the possibility of multiscale analysis across time and creating new opportunities for ecosystem service studies. The implementation of this high-tech information in basic and applied areas of knowledge such as health, environmental, and education sciences can help us make better decisions that contribute to a better balance between environment and human welfare.

Acknowledgments

The authors thank F. Teste and the two anonymous reviewers who kindly provided comments on an earlier version of the manuscript, and J. Straschnoy for editing work. V. Marchesini thanks the Image Processing Laboratory of the University of Valencia for its hospitality during her remote sensing internship.

References

Aguilera, M. O., D. F. Steinaker, and M. R. Demaría. 2003. Runoff and soil loss in undisturbed and roller-seeded shrublands on semiarid Argentina. *Journal of Range Management* 56:227–233.

Allen, R. G., L. S. Pereira, D. Raes, and M. Smith. 1998. *Evapotranspiración del cultivo: guías para la determinación de los requerimientos de agua de los cultivos.* [Crop evapotranspiration - Guidelines for computing crop water requirements]. Irrigation and drainage paper 56, Rome: FAO.

Allen, R. G., W. O. Pruitt, J. L. Wright, et al. 2006b. A recommendation on standardized surface resistance for hourly calculation of reference ETo by the FAO56 Penman–Monteith method. *Agricultural Water Management* 81:1–22.

Anderson, M., and W. Kustas. 2008. Thermal remote sensing of drought and evapotranspiration. *EOS Transactions American Geophysical Union* 89:233–234.

Anderson, M. C., J. M. Norman, W. P. Kustas, R. Houborg, P. J. Starks, and N. Agam. 2008. A thermal-based remote sensing technique for routine mapping of land-surface carbon, water and energy fluxes from field to regional scales. *Remote Sensing of Environment* 112:4227–4241.

Baldocchi, D. D. 2003. Assessing the eddy covariance technique for evaluating carbon dioxide exchange rates of ecosystems: Past, present and future. *Global Change Biology* 9:479–492.

Barrett-Lennard, E. G. 2003. The interaction between waterlogging and salinity in higher plants: Causes, consequences and implications. *Plant and Soil* 253:35–54.

Beltrán-Przekurat, A., R. A. Pielke Sr., D. P. C. Peters, K. A. Snyder, and A. Rango. 2008. Modeling the effects of historical vegetation change on near-surface atmosphere in the northern Chihuahuan Desert. *Journal of Arid Environments* 72:1897–1910.

Blanco, L. J., C. A. Ferrando, F. N. Biurrum, et al. 2005. Vegetation response to roller chopping and buffelgrass seeding in Argentina. *Rangeland Ecology and Management* 58:219–224.

Blaney, H. F., and W. D. Criddle. 1950. Determining water requirements in irrigated areas from climatological and irrigation data. U.S. Department of Agriculture, SCS-TP 96, 44 pp. Washington, D.C.

Blight, G. E. 2002. Measuring evaporation from soil surfaces for environmental and geotechnical purposes. *Water SA* 28:381–394.

Brest, C. L., and S. Goward. 1987. Deriving surface albedo measurements from narrowband satellite data. *International Journal of Remote Sensing* 8:351–367.

Brutsaert, W., and M. B. Parlange. 1998. Hydrologic cycle explains the evaporation paradox. *Nature* 396:30.

Burgess, S. S. O., M. A. Adams, N. C. Turner, et al. 2001. An improved heat pulse method to measure low and reverse rates of sap flow in woody plants. *Tree Physiology* 21:589–598.

Caparrini, F., F. Castelli, and D. Entekhabi. 2003. Mapping of land-atmosphere heat fluxes and surface parameters with remote sensing data. *Boundary-Layer Meteorology* 107:605–633.

Carlson, T. N., W. J. Capehart, and R. R. Gillies. 1995. A new look at the simplified method for remote sensing of daily evapotranspiration. *Remote Sensing of Environment* 54:161–167.

Carlson, T. N., E. M. Perry, and T. T. Shmugge. 1990. Remote estimation of soil moisture availability and fractional vegetation cover for agricultural fields. *Agricultural and Forest Meteorology* 52:45–69.

Carpenter, S. R., and C. Folke. 2006. Ecology for transformation. *Trends in Ecology and Evolution* 21:309–315.

Chapin, F. S., P. Matson, and H. A. Mooney. 2002. Terrestrial water and energy balance. In *Principles of terrestrial ecosystem ecology*, eds. F. S. Chapin, P. Matson, and H. A. Mooney, 71–96. New York: Springer-Verlag.

Chen, S., J. Chen, L. Guanghui, et al. 2009. Energy balance and partition in Inner Mongolia steppe ecosystems with different land use types. *Agricultural and Forest Meteorology* 149:1800–1809.

Cleugh, H., R. Leuning, Q. Mu, and S. Running. 2007. Regional evaporation estimates from flux tower and MODIS satellite data. *Remote Sensing of Environment* 106:285–304.

Contreras, S., E. G. Jobbágy, P. E. Villagra, M. D. Nosetto, and J. Puigdefábregas. 2011. Remote sensing estimates of supplementary water consumption by arid ecosystems of central Argentina. *Journal of Hydrology* 397:10–22.

Daily, G. 1997. *Nature's services: Societal dependence on natural ecosystems.* Washington, DC: Island Press.

Davin, E. L., N. de Noblet-Ducoudré, and P. Friedlingstein. 2007. Impact of land cover change on surface climate: Relevance of the radiative forcing concept. *Geophysical Research Letters* 34:13702.

Dirmeyer, P. A., and J. Shukla. 1994. Albedo as a modulator of climate response to tropical deforestation. *Journal of Geophysical Research* 99:20863–20877.

Doody, T. M., and R. G. Benyon. 2011. Direct measurement of groundwater uptake through tree roots in a cave. *Ecohydrology* 4:644–649.

El-Nahry, A. H., and A. Y. Hammad. 2009. Assessment of salinity effects and vegetation stress, west of Suez Canal, Egypt, using remote sensing techniques. *Journal of Applied Sciences Research* 5:316–322.

FAO. 2010. Global Forest Resources Assessment. FAO Forestry Paper 163. Rome: FAO

Feidas, H., G. Kokolatos, A. Negri, M. Manyin, N. Chrysoulakis, and Y. Kamarianakis. 2009. Validation of an infrared-based satellite algorithm to estimate accumulated rainfall over the Mediterranean basin. *Theoretical and Applied Climatology* 95:91–109.

Fernandez-Illescas, C. P., A. Porporato, F. Laio, and I. Rodriguez-Iturbe. 2001. The eco-hydrological role of soil texture in a water-limited ecosystem. *Water Resources Research* 37:2863–2872.

Forster, P., V. Ramaswamy, P. Artaxo, et al. 2007. Changes in atmospheric constituents and in radiative forcing. In Climate change 2007: The physical science basis. Contribution of Working Group I to the Fourth Assessment Report of the Intergovernmental Panel on Climate Change [Solomon, S., D. Qin, M. Manning, Z. Chen, M. Marquis, K.B. Averyt, M. Tignor and H.L. Miller (eds.)]. Cambridge University Press, Cambridge, United Kingdom and New York, NY, USA.

Gasparri, N. I., and R. Grau. 2009. Deforestation and fragmentation of Chaco dry forest in NW Argentina (1972–2007). *Forest Ecology and Management* 258:913–921.

Guerschman, J. P., A. I. J. M. Van Dijk, G. Mattersdorf, et al. 2009. Scaling of potential evapotranspiration with MODIS data reproduces flux observations and catchment water balance observations across Australia. *Journal of Hydrology* 369:107–119.

Hamilton, L. S. 2008. *Forests and water: A thematic study prepared in the framework of the Global Forest Resources Assessment 2005*. Rome: FAO.

Hansen, M. C., S. V. Stehman, and P. V. Potapov. 2010. Quantification of global gross forest cover loss. *Proceedings of the National Academy of Sciences of the United States of America* 107:8650–8655.

Houspanossian, J., M. Nosetto, and E. G. Jobbágy. 2012. Radiation budget changes with dry forest clearing in temperate Argentina. *Global Change Biology* 19:1211–1222.

Hurtado, E., and J. A. Sobrino. 2000. Daily net radiation estimated from air temperature and NOAA-AVHRR data: A case study for the Iberian Peninsula. *International Journal of Remote Sensing* 22:1521–1533.

Jackson, R. D., R. J. Reginato, and S. B. Idso. 1977. Wheat canopy temperature: A practical tool for evaluating water requirements. *Water Resources Research* 13:651–656.

Jung, M., M. Reichstein, P. Ciais, et al. 2010. Recent decline in the global land evapotranspiration trend due to limited moisture supply. *Nature* 467:951–954.

Kalma, J., T. McVicar, and M. McCabe. 2008. Estimating land surface evaporation: A review of methods using remotely sensed surface temperature data. *Surveys in Geophysics* 29:421–469.

Kelliher, F. M., R. Leuning, and E. D. Schulze. 1993. Evaporation and canopy characteristics of coniferous forests and grasslands. *Oecologia* 95 153–163.

Kerr, J. T., and M. Ostrovsky. 2003. From space to species: Ecological applications for remote sensing. *Trends in Ecology and Evolution* 18:299–305.

Kinzig, A. P., C. Perrings, F. S. Chapin, et al. 2011. Paying for ecosystem services—Promise and peril. *Science* 334:603–604.

Kremen, C., and R. Ostfeld. 2005. A call to ecologists: Measuring, analysing and managing ecosystem services. *Frontiers in Ecology and the Environment* 3:540–548.

Lambers, H. 2003. Dryland salinity: A key environmental issue in southern Australia. *Plant and Soil* 257:5–7.

Lee, X. 2010. Forest and climate: A warming paradox. *Science* 328:1479.

Lee, X., M. L. Goulden, D. Y. Hollinger, et al. 2011. Observed increase in local cooling effect of deforestation at higher latitudes. *Nature* 479:384–387.

Liang, S. 2000. Narrowband to broadband conversions of land surface albedo: I. Algorithms. *Remote Sensing of Environment* 76:213–238.

MA (Millennium Ecosystem Assessment). 2005. *Ecosystems and human well-being: desertification synthesis*. Washington, DC: Island Press.

Malhi, Y., J. T. Roberts, R. A. Betts, T. J. Killeen, W. Li, and C. A. Nobre. 2008. Climate change, deforestation, and the fate of the Amazon. *Science* 319:169–172.

Marchesini, V. A., R. J. Fernández, and E. G. Jobbágy. 2012. Salt leaching leads to drier soils in disturbed semiarid woodlands of central Argentina. *Oecologia* 171:1003–1012.

Miao, H., S. Chen, J. Chen, et al. 2009. Cultivation and grazing altered evapotranspiration and dynamics in Inner Mongolia steppes. *Agricultural and Forest Meteorology* 149:1810–1819.

Miles, L., A. Newton, R. DeFries, et al. 2006. A global overview of the conservation status of tropical dry forest. *Journal of Biogeography* 33:491–505.

Moore, G. W., and J. L. Heilman. 2011. Proposed principles governing how vegetation changes affect transpiration. *Ecohydrology* 4:351–358.

Mu, Q., F. A. Heinsch, M. Zhao, and S. W. Running. 2007. Development of a global evapotranspiration algorithm based on MODIS and global meteorology data. *Remote Sensing of Environment* 111:519–536.

Mu, Q., M. Zhao, and S. W. Running. 2011. Improvements to a MODIS global terrestrial evapotranspiration algorithm. *Remote Sensing of Environment* 115:1781–1800.

Norman, J. M., W. P. Kustas, and K. S. Humes. 1995. Source approach for estimating soil and vegetation energy fluxes in observations of directional radiometric surface temperature. *Agricultural and Forest Meteorology* 77:263–293.

Nosetto, M. D., E. G. Jobbágy, A. B. Brizuela, and R. B. Jackson. 2012. The hydrologic consequences of land cover change in central Argentina. *Agriculture, Ecosystems and Environment* 154:2–11.

Nosetto, M. D., E. G. Jobbágy, and J. M. Paruelo. 2005. Land-use change and water losses: The case of grassland afforestation across a soil textural gradient in central Argentina. *Global Change Biology* 11:1101–1117.

Paruelo, J. M., M. R. Aguiar, and R. A. Golluscio. 1991. Evaporation estimates in arid environments: An evaluation of some methods for the Patagonian steppe. *Agricultural and Forest Meteorology.* 55:127–132.

Reynolds, J. F., P. R. Kemp, K. Ogle, and R. J. Fernández. 2004. Precipitation pulses, soil water and plant responses: Modifying the 'pulse-reserve' paradigm for deserts of North America. *Oecologia* 141:194–210.

Ritchie, J. C., T. J. Schmugge, A. Rango, and F. R. Schiebe. 2000. Remote sensing applications for monitoring semiarid grasslands at the Sevilleta LTER, New Mexico, 1969–1971. In *IEEE 2000 International Geoscience and Remote Sensing Symposium Proceedings.* Honolulu, HI: International Association of Hydrological Sciences.

Roerink, G. J., Z. Su, and M. Menenti. 2000. S-SEBI: A simple remote sensing algorithm to estimate the surface energy balance. *Physics and Chemistry of the Earth Part B* 25:147–157.

Saunders, R. W. 1990. The determination of broad band surface albedo from AVHRR visible and near-infrared radiances. *International Journal of Remote Sensing* 11:49–67.

Seiler, C., and A. F. Moene. 2011. Estimating actual evapotranspiration from satellite and meteorological data in central Bolivia. *Earth Interactions* 15:1–24.

Sobrino, J. A., M. Gómez, J. C. Jiménez-Muñoz, A. Olioso, and G. Chehbouni. 2005. A simple algorithm to estimate evapotranspiration from DAIS data: Application to the DAISEX campaigns. *Journal of Hydrology* 315:117–125.

Sobrino, J. A., J. C. Jiménez, and L. Paolini. 2004. Land surface temperature retrieval from Landsat TM 5. *Remote Sensing of Environment* 90:434–440.

Su, Z. 2002. The Surface Energy Balance System (SEBS) for estimation of turbulent heat fluxes. *Hydrology and Earth System Sciences* 6:85–99.

Timmermans, W. J., J. C. Jimenez-Munoz, V. Hidalgo, et al. 2011. Estimation of the spatially distributed surface energy budget for AgriSAR 2006, Part I: Remote sensing model intercomparison. *IEEE Journal of Selected Topics in Applied Earth Observations and Remote Sensing* 4:465–481.

Wang, L., P. D'Odorico, J. P. Evans, et al. 2012. Dryland ecohydrology and climate change: Critical issues and technical advances. *Hydrology and Earth System Sciences* 9:4777–4825.

Wilcox, B. P. 2002. Shrub control and streamflow on rangelands: A process based viewpoint. *Journal of Range Management* 55:318–326.

Yebra, M., A. Van Dijk, R. Leuning, A. Huete, and J. P. Guerschman. 2013. Evaluation of optical remote sensing to estimate actual evapotranspiration and canopy conductance. *Remote Sensing of Environment* 129:250–261.

19

Urban Heat Island Effect

J. A. Sobrino, R. Oltra-Carrió, and G. Sòria

University of València, Spain

CONTENTS

19.1 Introduction 417
19.2 Evaluation of the UHI Effect 420
    19.2.1 DESIREX Experimental Campaign: The Madrid Case Study 421
    19.2.2 Thermopolis Experimental Campaign: The Athens
         Case Study 423
19.3 Requirements of a Suitable Remote Sensing Sensor to Analyze
    the UHI Effect 424
    19.3.1 Spatial Resolution 424
    19.3.2 Overpass Time 427
19.4 Assessment of Different Procedures to Retrieve the Land Surface
    Emissivity over Urban Areas 429
    19.4.1 Comparative Analysis of the Land Surface Emissivity
         Products 430
         19.4.1.1 Comparison in Terms of Land Surface Emissivity 430
         19.4.1.2 Comparison in Terms of Land Surface Temperature ... 432
19.5 Conclusions 433
Acknowledgments 434
References 434

19.1 Introduction

In 2011, the percentage of the total world population living in urban areas was 51%, according to the Population Reference Bureau's world population data sheet (PRB 2011). Even though countries define urban in many different ways, from population centers of 100 or more dwellings to only the population living in national and provincial capitals, one thing is clear: the world is being urbanized. Ecosystems regulate the climate through biophysical processes that mediate energy and water balances at the land surface (West et al. 2011) (see Chapter 16). Consequently, the replacement of natural surfaces by artificial ones with different thermal properties, will affect the ecosystem: a natural ecosystem is transformed into an urban or human-dominated ecosystem. Thus, the growth of a city strongly influences the change of local climate. As said by Oke (1987), in some cases, these changes are intentioned to improve the atmospheric environment for specific human uses, for example to protect from the wind or from the rain or to obtain the warm environment needed to grow vegetables in greenhouses. In other cases, the modifications occur inadvertently due to changes in the cover land or by direct atmospheric contamination by pollutants. In any case, with regard to urbanization, both cases are related; a house is constructed to provide a pleasant environment for its inhabitants and at the same time it changes the outside wind and thermal environments.

The urban heat island (UHI) phenomenon is one example of local climate change (e.g. Oke 1981). This effect is characterized by the heating of urban zones in comparison with nonurbanized surroundings. Therefore, the UHI is defined as the difference between the air temperature (AT) within the city and the AT of its surroundings (Equation 19.1) (Morris et al. 2001). The estimation of the UHI can be an indicator of the changes taking place in the ecosystem. The fact that urban ecosystems are often warmer than other ecosystems has an influence on the fauna, altering the species composition of communities (Grimm et al. 2008), and on the flora. For example, the impact of urban areas can alter the vegetation phenology, such as when beginning of flowering is advanced (Roetzer et al. 2000).

UHI can be defined for different layers of the urban atmosphere and for various surfaces. Atmospheric heat islands may be defined for the urban canopy layer (UCL), that layer of the urban atmosphere extending upward from the surface to approximately mean building height, and the urban boundary layer (UBL), the layer above the UCL.

$$UHI = AT_{Urban} - AT_{Rural} \qquad (19.1)$$

Studies by Oke (1982) and Oke et al. (1991) suggesting some causes of the UHI have been adapted in Voogt (2002) and are explained in the following text. One of these causes is the surface geometry. Due to the urban canyon geometry, the effective surface in urban regions is increased and solar radiation is trapped by multiple reflections, which leads to warming. Moreover, the closely spaced buildings reduce the sky view factor and reduce radiative heat loss. The buildings also serve as shelter elements reducing convective heat loss from the surface and near-surface air. Other contributing factors are the surface thermal properties and conditions. Urban materials are better stores of heat than natural ones because they have higher heat capacity and larger thermal admittance. In addition, man-made materials in the city are insulating and waterproof. They reduce evaporation and channelize more energy into sensible heat (rather than into latent heat) that can warm the air. We also should consider the anthropogenic heat, which is the heat released by urban energy use in buildings and vehicles and from humans. The last cause attributed to the UHI effect is the urban greenhouse effect. The polluted, humid, and warmer urban atmosphere emits more thermal radiation downward toward the city surface. UHI can provide positive and negative effects for cities, depending on the latitude, the climatic area, and the time of the year.

For example, the relation between energy consumption and UHI is well defined. Cities in cold climates can anticipate energy savings for space heating. Conversely, cities in hot climates will face extra costs for air conditioning, which may lead to a positive feedback process because conventional air-conditioning systems can increase the UHI effect and might even exacerbate climate change itself (WHO 2003).

One of the most important impacts of the UHI is its influence on human health. High temperatures are related to human health complaints; for instance, the exposure to high temperatures during nighttime may increase the episodes of insomnia. Human diseases and UHI are so connected that even the World Health Organization (WHO 2003) reports them. It classifies the phenomenon as a land use impact over the climate at local and regional scales and defines the thermal stress and air pollution as having a major impact on climate change and health issues. The two problems, thermal stress and air pollution, are associated with urban ecosystems and are strongly linked because hot weather could amplify the production of noxious photochemical smog. So the most vulnerable populations within heat-sensitive regions are urban populations. However, the increase of the urban temperature may have positive effects. For example, in Martens (1998), the effect of a change in temperature on mortality is studied combined with projections of changes in climate conditions of 20 cities. The research found that, for most of the cities included, climate change was likely to lead to a reduction in mortality rates due to decreasing winter mortality, mainly because of cardiovascular mortality in elderly people.

A question might reasonably arise regarding the role of the UHI in the current global climate scenario. The Fourth Assessment Report of the Intergovernmental Panel on Climate Change (IPCC 2007) refers to the global impact (also called *global warming* in the popular literature) of any change in climate over time due to natural variability or as a result of human activity. The report warns about the continuation of the anthropogenic warming throughout centuries even if greenhouse gas (GHG) emissions were to be reduced sufficiently for GHG concentrations to stabilize, due to the time scales associated with climate processes and feedbacks. Urban areas are the sites of most GHG emissions; nevertheless, UHIs themselves are not responsible for global change. As already noted, they are a local climate modification. And despite the expected warming of the cities due to the global change, it is unlikely that the intensity of the heat island will increase because the temperature gradient between city and country is expected to remain similar (Voogt 2002). Nevertheless, some authors, such as Oke (1997), talk about the UHI as an analog for global warming, suggesting that the climate modifications that occur within urban areas may be used as an analog for examining changes anticipated from global climate change.

After all the information exposed, it is obvious that the study of the urban ecosystem is of prime importance. However, quantifying the effects of the urbanization process is not an easy task because normally we do not have a database of pre-urban measurements. Instead, it is common to compare the data from the center of a city with those from rural stations in the surrounding area. Such urban–rural comparisons are at best only an approximation of the urban modification. Moreover, the meteorological network inside a city is not always as complete as desirable, and stations are not always evenly distributed spatially within the city. Consequently, some large areas may remain without coverage. In this sense, the remote sensing data are a powerful tool to study the urban environment, solving previous difficulties. In fact, some works highlight the importance of thermal remote sensing data to study the urban environment and to achieve a more friendly and comfortable ambiance for its inhabitants (Johnson et al. 2009).

When the UHI is monitored with remote sensing data, we must consider surface urban heat island (SUHI; Voogt and Oke 2003) because the parameter studied is no longer the AT but the land surface temperature (LST; Equation 19.2).

$$SUHI = LST_{Urban} - LST_{Rural} \tag{19.2}$$

19.2 Evaluation of the UHI Effect

There are several examples of experimental campaigns that aim to study urban climates, and in most of the cases to quantify the UHI of different urban zones. One strategy to achieve this purpose is to record AT, humidity, and wind speed measures using fixed stations or mobile transects (Oke 1973; Voogt and Oke 1998; Fernández et al. 2004; Najjar et al. 2004; Jasche and Rezende 2007; Masson et al. 2008). Many other studies have applied remote sensing technology using satellite images (Voogt and Oke 2003; Hartz et al. 2006; Li et al. 2009; Stathopoulou et al. 2009) or airborne sensor imagery (Voogt and Oke 1997; Lagouarde et al. 2004; Masson et al. 2008) to describe the behavior of urban surfaces. In this section, we introduce two experimental campaigns founded by the European Space Agency (ESA) in the framework of its Earth Observation Programs. The first one is the Dual-use European Security IR Experiment (DESIREX) carried out in Madrid in 2008, and the second one is the Thermopolis campaign developed in Athens in 2009. In both cases, ground and atmospheric data were collected as well as spaceborne and airborne imagery. So both effects, SUHI and UHI, can be evaluated and correlations between them can be established. The airborne imagery was recorded by the Airborne Hyperspectral Scanner (AHS), property of the Instituto Nacional de Técnica Aeroespacial—Spain (INTA).

19.2.1 DESIREX Experimental Campaign: The Madrid Case Study

The AHS sensor is an airborne imaging 80-band radiometer with 10 bands in the thermal infrared (TIR) range (effective wavelengths of 8.18, 8.66, 9.15, 9.60, 10.07, 10.59, 11.18, 11.78, 12.35, and 12.93 μm). During the DESIREX experimental campaign (Sobrino et al. 2013), the AHS flights took place on different days during June and July. Thirty AHS images were recorded and divided into two different patterns, one from northwest to southeast (named from now on as flight P01) and the other one from south to north (named from now on as flight P02). Both overpasses crossed over the city center, covering the same area with an extension of approximately 17 km². A description of each flight is given in Table 19.1. AHS imagery was atmospherically corrected, and LST values were retrieved applying the Temperature and Emissivity Separation (TES) algorithm (Gillespie et al. 1998).

Figure 19.1 displays the UHI and the SUHI effects. For the SUHI phenomenon, the values are calculated for both overpasses of the AHS. The LST_{urban} in Equation 19.2 is considered as the average LST of the urban area in the AHS images, and the LST_{rural} is the average LST of the area, which results when the urban zone is removed. Limits between urban and rural zones are established according to urban contours and neighborhood limits. Even though each overpass includes different urban and rural zones, the behavior of both plots is similar. The AT data were recorded in fixed points of measurements placed inside the UBL over the buildings, using masts of 3 m of height. Moreover, a rural station was placed outside the city, over a surface covered by green

TABLE 19.1

Airborne Hyperspectral Scanner Flights Carried Out
in DESIREX 2008 Campaign

DOY	Starting Time (UTC)	Flight ID
177	11:11/11:27	P01/P02
177	22:15/22:31	P01/P02
178	04:12/04:26	P01/P02
180	11:32/11:53	P01/P02
180	21:29/21:44	P01/P02
183	11:21/11:44	P01/P02
183	21:59/22:12	P01/P02
184	04:09/04:26	P01/P02
186	11:16/11:32	P01/P02
186	21.59/22:14	P01/P02

Note: Flights at 4 m of spatial resolution; DOY = day of year.

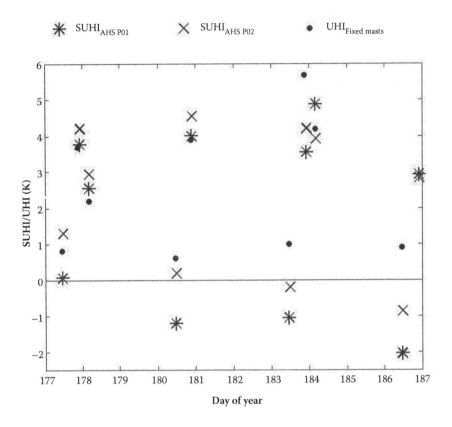

FIGURE 19.1
Surface urban heat island (SUHI) and urban heat island (UHI) effects obtained during DESIREX in the city of Madrid.

grass. The UHI effect (Equation 19.1) was then obtained as the AT difference between the average of AT registered in the urban points and the rural one.

A relevant aspect to highlight from Figure 19.1 is the difference observed between UHI and SUHI values retrieved simultaneously. For noon measurements, the UHI and SUHI have maximum discrepancies; at night, we found reasonable agreement between both effects.

Thus, the use of remote sensing imagery during the night may give the value of the heat island at surface and canopy levels. The same results were achieved by Ben-Dor and Saaroni (1997) when radiometric and air temperatures were compared for the city of Tel Aviv during night flights.

Low SUHI values (sometimes negative values) were obtained for images at noon because the rural surfaces appear warmer than the urban ones. This negative island characterizes some inner city urban areas, such as Madrid, in the middle of the day due to the slow uptake of heat by the city and the large shaded areas from the tall buildings (Voogt 2002). Note that for UHI effect, no negative values are registered at noon, but values around 1 K are measured. So in Madrid, the decrease of the heat island effect at noon is also noticed by the atmospheric data, such as it is also noticed for other cities such as Shanghai (Liu et al. 2007).

Although the SUHI maximum during the DESIREX campaign (5 K) was achieved some hours after the UHI maximum (6 K), they agree within 1 K. According to the 3.2 million inhabitants of Madrid, the expected UHI maximum predicted by Oke (1973) is 9 K, which is 3 K over the obtained UHI maximum obtained during this campaign experience.

Summarizing, UHI intensity is greatest at night and UHI disappears by day or the city is cooler than the rural environs, which is the same behavior that Arnfield (2003) observed in his review of two decades of urban climate research.

19.2.2 Thermopolis Experimental Campaign: The Athens Case Study

TIR airborne measurements were performed over the city of Athens. The AHS instrument was used with the intention of studying the UHI phenomenon over the city from June 18, 2009, to July 24, 2009. Figure 19.2 plots

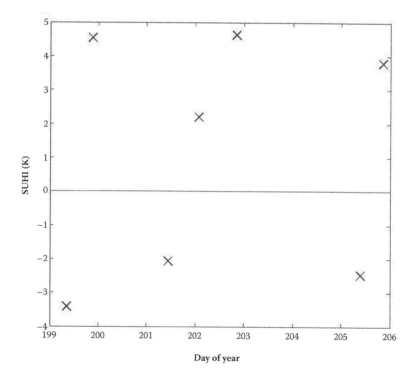

FIGURE 19.2
Surface urban heat island (SUHI) effect obtained during Thermopolis in the city of Athens.

the SUHI values obtained after the atmospheric correction of the imagery and the retrieval of the LST from TES. As in the DESIREX campaign, in the Thermopolis campaign, a cold island (negative UHI effect) is detected at daytime, while the positive UHI phenomenon takes place at night.

19.3 Requirements of a Suitable Remote Sensing Sensor to Analyze the UHI Effect

There are several studies in the literature that report on the SUHI effect using different spaceborne platforms, for example, Advanced Spaceborne Thermal Emission and Reflection Radiometer (ASTER; Hartz et al. 2006; Lu and Weng 2006), Advanced Very High Resolution Radiometer (AVHRR; Streutker 2003), Moderate Resolution Imaging Spectroradiometer (MODIS; Pu et al. 2006), and Landsat (Weng et al. 2004; Yuan and Bauer 2007; Rajasekar and Weng 2009). However, existing satellite remote sensing capabilities (visitation time, spatial and spectral resolutions) are not the best way to properly monitor the SUHI effect.

With this aim in mind, the main objective of this section is to give recommendations on the optimal characteristics in terms of spatial resolution and the time of acquisition that a satellite must possess to properly monitor the SUHI effect. To achieve this purpose, imagery from the DESIREX campaign is considered (Sobrino et al. 2012a).

19.3.1 Spatial Resolution

To analyze the impact of the spatial resolution to properly monitor the SUHI phenomenon inside a city, the AHS images have been aggregated. The aggregation process was applied to the geometrically corrected at-sensor radiance images by averaging all the pixel values that contribute to the output pixel. This exercise has been done using the resampling pixel aggregate tool of the ©ENVI application. This tool does not take into account any influence on the neighboring pixels. Atmospheric effects have been corrected and LST retrieved.

Figure 19.3 shows the LST images generated by the aggregation process. There is a clear loss of information when the spatial resolution decreases. This is especially evident in the 1000 m aggregation image, where the thermal structure of the different districts cannot be distinguished. This indicates that the (at 1 km spatial resolution) onboard satellites have limited ability to resolve the thermal structure of the SUHI effect in big cities; this will, of course, be more critical for small cities. This finding is corroborated quantitatively in Table 19.2, which reports the standard deviation of LST for the urban zone (the area inside the limit drawn over the 4-m resolution image in Figure 19.3) at the different aggregations. We can observe the strong

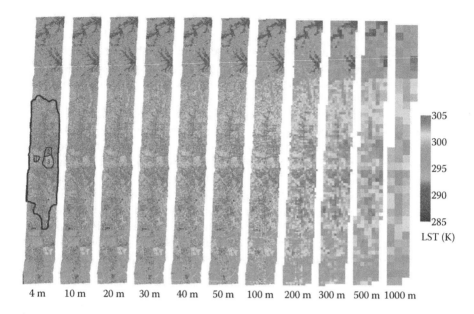

FIGURE 19.3 (See color insert.)
Land surface temperature (LST) obtained from Airborne Hyperspectral Scanner 4-m resolution image and images obtained from the aggregation process described at 10, 20, 30, 40, 50, 100, 200, 300, 500, and 1000 m. The black outer line on the left panel indicates the boundary of Madrid city. Inner lines correspond with the three city districts cited in the text.

TABLE 19.2

Standard Deviation of the LST inside the Urban Area and SUHI Effect from the Aggregated Images

Spatial Resolution (m)	σ (K)	SUHI (K)
4	4.4	4.56
10	3.3	4.56
20	2.8	4.55
30	2.5	4.54
40	2.4	4.54
50	2.2	4.53
100	1.7	4.49
200	1.4	4.49
300	1.2	4.51
500	1.0	4.40
1000	0.8	4.25

Note: LST = land surface temperature; SUHI = surface urban heat island.

variation of 3.6 K (from 4.4 K for the original 4-m resolution image to 0.8 K for the aggregated to 1 km). Table 19.2 also shows the SUHI effect measured for each aggregated image. We can observe that the SUHI hardly decreases with the resolution; a difference of only 0.3 K is obtained between the SUHI at 4-m spatial resolution and the one at 1 km. Therefore, further work must be done to find out the proper spatial resolution.

Usually, SUHI effect in the literature is defined as an average value that represents the difference between the mean surface temperature of a city and the temperature of a surrounding rural area. This is in part because of the low spatial resolution of the remote sensing images used. However, the thermal comfort can vary substantially among different areas inside the same city (Toy et al. 2007). To study this effect, three representative districts have been selected (Figure 19.3). The first one, downtown, is characterized by narrow streets and small buildings with red tiles on top. The second one is characterized by wide streets and tall buildings with insulating materials over the roofs; and the third one is almost covered by a garden. Thermal structures, such as streets or small gardens, can be clearly detected at 10-m and 50-m spatial resolution. For the 100-m resolution image, these patterns become mixed, but we can still differentiate between some structures inside the district. For 500 m and 1 km of spatial resolution, the neighborhoods appear homogeneous and the heterogeneity between them is lost. The parameter chosen as evidence of the variation in the thermal effects inside the city is the maximum SUHI (SUHIM), obtained from the difference between the maximum LST inside each district and the LST of the rural zone. When the SUHIM is presented as a function of the spatial resolution (Figure 19.4),

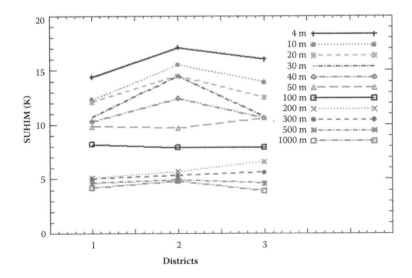

FIGURE 19.4
Maximum SUHI (SUHIM) for the three districts considered at different spatial resolutions.

some factors can be observed: (1) SUHIM is higher than 10 K for the different districts, and also differences in SUHIM between districts can be observed from LST images of 10–40 m of spatial resolution; (2) spatial resolution of 50 m (SUHIM near 10 K for the different districts) can be considered as the critical spatial resolution; and (3) resolutions lower than 50 m present lower SUHIM values (from 8 K for 100 m to 4 K for 1000 m).

19.3.2 Overpass Time

At the moment of the satellite overpass, certain conditions must be fulfilled to achieve an appropriate observation to detect the SUHI phenomenon. The geometry of observation should have as little influence as possible in the recorded data, so the values registered on different days will be more easily compared. In this respect, we have observed that, during the night, the urban surface anisotropy—caused by the different surfaces seen by a sensor because of the 3D structure of the urban scenes—decreases. This statement is proven when the area shared by AHS overpasses P01 and P02 is analyzed because each flight has a different geometry of observation; they are observing different surfaces even when looking at the same area. On the one hand, for daytime flights, different histograms of LST are obtained from each overpass. On the other hand, during nighttime, when there are no direct irradiated surfaces, both histograms are almost the same. The other condition to be imposed is that the UHI and the SUHI should not be too different, or at least the UHI should be simply and accurately obtained from SUHI. The reason for this condition is that the atmospheric effect, especially in the UCL, is what directly influences the comfort of the inhabitants. In Figure 19.1, the urban heat island has been analyzed at the atmospheric and at the surface level, and we have found that both phenomena have a similar value during night. Therefore, nighttime is shaping up as the best option for the satellite overpass. A more in-depth study based on data recorded on the fixed points of measurements during the campaign can be carried out.

To analyze the impact of the overpass time of a satellite in the SUHI estimation, we present an hourly comparison between the air and the surface temperatures measured simultaneously in the fixed points of measurements mentioned in the previous section.

If we analyze the hourly evolution of the LST minus the AT, we find for all sites that, during nighttime, differences are close to 0 K. Moreover, the minimum temperature difference is registered for the sites with lower housing density. And the values are lower between 20:00 and 6:00 UTC for all masts.

Figure 19.5 shows an analysis of the correlations between LST and AT temperatures, which shows the hourly evolution of the correlation coefficient and the error obtained when a linear correlation is assumed between both temperatures. The values plotted are time averaged, for example, for a time equal to four in the abscissa axis; the values plotted are measured as the

(a)

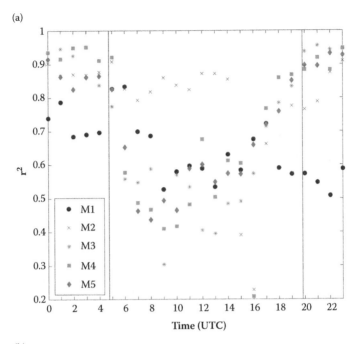

(b)

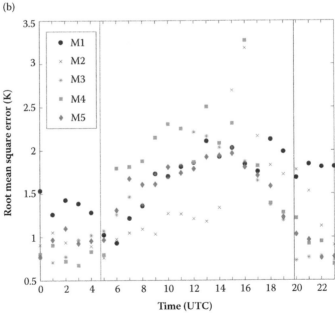

FIGURE 19.5
Daily evolution of (a) the correlation coefficient and (b) estimated error in the determination of
air temperature for every linear fit between air temperature and land surface temperature, and
for each measurement site (M1, M2, M3, M4, and M5).

average of the values registered from 4:00 h UTC to 4:55 h UTC every day during the campaign. According to the plot legend, M1 corresponds to the station outside the city, placed over a green grass surface. M2, M3, M4, and M5 are placed over man-made surfaces at the top of the buildings. M2 is situated in an area of low housing density. M3 and M5 are placed in a medium housing density area and M4 in a high density one. The correlation coefficients are lower and errors are higher for daytime than for nighttime. Higher correlation coefficients are obtained from 21 h to 5 h UTC for the masts located over artificial surfaces (Figure 19.5a). Moreover, the root mean square error (RMSE) for all masts is approximately 1 K around sunrise (Figure 19.5b). Therefore, it is clear that an overpass time immediately before sunrise is the best choice to estimate an SUHI effect that can be compared with the UHI effect obtained from air temperature data.

19.4 Assessment of Different Procedures to Retrieve the Land Surface Emissivity over Urban Areas

Measurements of parameters from thermal remote sensors, such as the LST, strongly depend on the characteristics of the surface, especially its emissivity and geometric form (Voogt and Oke 2003). In fact, an inaccuracy of 1% in emissivity estimation can cause an error of up to 0.78 K in LST (Van de Griend and Owe 1993). The correct retrieval of the land surface emissivity (LSE) of urban surfaces is of utmost importance because a city is very heterogeneous and because the derived LST is the key to studying urban phenomena such as the UHI.

The surface emissivity is a measure of the inherent efficiency of the surface in transforming the energy accumulated into radiant energy. The term of emissivity is introduced to characterize the emission of a surface taking as a reference a black body with the same surface temperature (Equation 19.3)

$$\varepsilon = L(T)/B(T) \tag{19.3}$$

where $L(T)$ is the radiance of the surface at a temperature T and $B(T)$ (Equation 19.4) is the radiance of a black body at surface temperature:

$$B(T) = \frac{c_1}{\lambda^5 \left[\exp\left(\frac{c_2}{\lambda T}\right) - 1 \right]} \tag{19.4}$$

in which $c_1 = 1.191 \times 10^{-8}$ W/(m^2 sr cm^{-4}), $c_2 = 1.439$ cm K and λ is the wavelength in centimeters.

In this section, three methodologies for retrieval of LSE are applied in the framework of the DESIREX campaign (Oltra-Carrió et al. 2012). The methodologies are the NDVI Threshold Method (NDVI™) (Sobrino et al. 2008a), the TES (Gillespie et al. 1998), and the Temperature Independent Spectral Indices (TISI) algorithm (Becker and Li 1990). Each algorithm selected uses one type of separation method to solve the problem of the combined effects of the LST and the LSE. First, the NDVI uses auxiliary information derived from a vegetation index. We can find several examples of NDVI application in the literature, even for urban surfaces (e.g., Rigo et al. 2006; Stathopoulou et al. 2007) where a correlation coefficient (r^2) of 0.79 and an RMSE of 0.010 were found when the emissivities retrieved with the NDVI method were compared with published values. Second, the TES uses the TIR spectral contrast. It relies on an empirical relation between the range of observed TIR emissivities and their minimum value. The TES was designed for the ASTER. Based on numerical simulation, TES should be able to recover temperatures within about ±1.5 K and emissivities within about ±0.015 in homogeneous areas (Gillespie et al. 1998). Hence, in almost all the papers where ASTER imagery is used, the TES algorithm is applied for both LSE and LST retrieval. Two studies over urban areas are Tiangco et al. (2008) and Nichol et al. (2009). Finally, the TISI algorithm uses the middle-infrared (MIR) surface reflectance as a key to solve the emissivity/temperature separation problem. In Dash et al. (2005), the TISI method is applied to the AVHRR, while in Petitcolin and Vermote (2002) it is implemented for the MODIS over southern Africa, where emissivity and surface temperature were derived within ±0.01 and ±1 K, respectively. None of the three algorithms considers the 3D structure of the city, so we will not analyze the contribution that a neighboring surface can introduce in the radiation balance of a particular material.

Finally, the proper accuracy of each method is tested using each LSE map to retrieve the LST by applying a split window (SW) algorithm (Sobrino et al. 2008b).

19.4.1 Comparative Analysis of the Land Surface Emissivity Products

19.4.1.1 Comparison in Terms of Land Surface Emissivity

In order to test the LSE retrieval methodology in urban areas, we selected three different urban surfaces: asphalt of a large avenue inside the city, asphalt shingle (referred to just as shingle from here on), and tile. The last two surfaces are the most common roofing materials used in Madrid. The *in situ* emissivity of the asphalt and the shingle was measured by applying the TES algorithm to the *in situ* data registered with a multiband radiometer. The tile spectrum was extracted from the ASTER spectral library (http://speclib.jpl.nasa.gov); details on sample preparation and sample measurements can be found in Baldridge et al. (2009). Notice that

TABLE 19.3

Statistical Values from the Validation Activity of the Retrieved Land Surface Emissivity Products

	NDVI	TES	TISI
r^2	0.250	0.640	0.730
Bias-mean	0.024	0.015	−0.008
σ	0.051	0.036	0.029
Bias-median	0.009	−0.003	−0.007
MAD	0.033	0.025	0.023
RMSE	0.056	0.039	0.030

Note: TES = Temperature and Emissivity Separation; TISI = Temperature Independent Spectral Indices; MAD = median absolute deviation; RMSE = root mean square error.

no metallic surfaces have been included in the study because they do not follow the empirical relation between the minimum emissivity and the spectral contrast required for the TES implementation (Sobrino et al. 2012b). Table 19.3 shows the statistical results of the comparison between the AHS and ground truth estimations of LSE. Classic statistical parameters, such as the mean difference between the AHS EST and the ground truth EST (bias-mean) and the standard deviation (σ), are shown. Also robust statistical parameters, which are less affected by anomalous values than the classic ones, are presented, such as the median of the differences between the AHS EST and the ground truth EST (bias-median) and the median absolute deviation (MAD). The RMSE for each validation process is also given.

From our analysis, we can extract some information. The LSE ground truth values go from 0.775 at 9.154 μm for the shingle to 0.977 at 8.695 μm for the asphalt. Nevertheless, the range of emissivities retrieved with the three algorithms is not so broad. For the TISI, the range is between 0.817 and 0.974, while for the TES it is between 0.881 and 0.969. For the NDVI, the range is the narrowest, between 0.939 and 0.972. Moreover, the NDVI and the TES do not reproduce properly low values of LSE. In fact, the TES gives good values over 0.93. Notice also that NDVI does not distinguish between different urban materials. The statistical indicators for the NDVI (Table 19.3) showed low correlations between both retrievals, that is, r^2 of 0.25 and RMSE of 0.06. These results were expected because this methodology is based on a vegetation index, and it would not be the best way to characterize man-made materials.

Regarding the TES and the TISI algorithms, the correlation (r^2) is improved: 0.64 and 0.73, respectively. For the TES, the bias-mean and bias-median values are acceptable: −0.015 and 0.003, respectively. And for the TISI, these values are 0.008 and 0.007, respectively. Nevertheless, the RMSE obtained in both cases is bigger than the accuracy estimated for each algorithm in previous studies.

19.4.1.2 Comparison in Terms of Land Surface Temperature

An LST map was obtained for each flight of the AHS by applying the SW algorithm (Sobrino et al. 2008b).

Table 19.4 shows statistical values of the validation of the LST. The study has been divided into two groups. First, taking into account four validation targets, including two natural surfaces (green grass and bare soil) and two man-made surfaces (two different roofs inside the city) for all the nighttime images, there are a total of 18 validation values. Second, taking into account just the man-made surfaces and also for all the nighttime images, this gives a total of eight validation values. The LST retrieved with the NDVI emissivity presents the highest dispersion of the data (the highest MAD and σ values), as well as the highest bias and RMSE values in both groups of study (when considering 18 points and 8 points of validation, respectively). When natural and man-made surfaces are studied, the LST retrieved with the TES and the TISI emissivities present similar behavior, with equal values of MAD and very similar values of σ. However, regarding artificial urban surfaces, the LST retrieved with the TES emissivity presents a higher bias than the TISI one but lower dispersion of the data and also higher linear correlation between the airborne and the *in situ* LST. Focusing on the RMSE, we can observe that when all the surfaces are analyzed, the TES and the TISI present similar values (1.7 and 1.6 K, respectively). Considering just man-made surfaces, the RMSE is the same for both algorithms (2 K). However, in both cases, the RMSE for the TES is more in agreement with its theoretical accuracy than the TISI.

Notice that the TISI algorithm requires more spectral information (TIR bands and at least one band in the MIR region) and more temporal information (one night and one day image) than the other two methods. However, the

TABLE 19.4

Land Surface Temperature Products Retrieved with Split Window Algorithm from the Airborne Hyperspectral Scanner Imagery—Validation Activity Statistical Values

	18 Validation Points: Green Grass, Bare Soil, Two Man-Made Surfaces			Eight Validation Points: Two Man-Made Surfaces		
	NDVI	**TES**	**TISI**	**NDVI**	**TES**	**TISI**
r^2	0.86	0.92	0.90	0.83	0.88	0.67
Bias-mean (K)	−1.50	−1.00	−0.70	−2.10	−1.60	−0.58
σ (K)	1.70	1.30	1.40	1.90	1.20	1.90
Bias-median (K)	−1.20	−1.10	−0.50	−1.90	−1.60	0.29
MAD (K)	1.40	1.10	1.10	1.70	0.90	1.40
RMSE (K)	2.30	1.70	1.60	2.90	2.00	2.00

Note: Land surface temperature computed from the NDVI, the TES, and the TISI. TES = Temperature and Emissivity Separation; TISI = Temperature Independent Spectral Indices; MAD = median absolute deviation; RMSE = root mean square error.

TES algorithm can be applied with only five TIR bands and only one image. In addition, this algorithm has the advantage that the LST can be retrieved at the same time as the LSE. In conclusion, our results show that the TES emissivity seems to be the algorithm that best reproduces the LST over an urban area, without requiring a high temporal resolution of the sensor.

19.5 Conclusions

The growth of a city involves the replacement of natural surfaces by manmade ones and transforms rural ecosystems into urban ones. These changes have direct consequences on the local climate as thermal conditions and gusts of wind are modified. One example of the environmental modification is the UHI. In this study, we have used real data to introduce the concept of the UHI. The DESIREX 2008 experimental campaign provided an excellent opportunity for collecting ground-based measurements and also for validating different algorithms for the retrieval of biophysical parameters from remote sensing data of interest in urban areas. SUHI and UHI effects were measured from LST maps and AT data. The results showed a similar trend for both effects, higher values at night and lower values at noon. Thus, three main conclusions can be extracted from the results: (1) at noon, a negative heat island is registered at the surface level, while a low effect around 1 K is obtained at UBL; (2) the maximum values for SUHI and UHI differ by 1 K, reaching a top value of 5 and 6 K, respectively; nevertheless, the maximum value for the UHI is recorded some hours before the SUHI maximum occurred; and (3) the differences between SUHI and UHI results are minimum during night and maximum during day, which demonstrates that the use of remote sensing imagery during night may give the value of the heat island at the surface and boundary levels.

In addition, we show the characteristics (in terms of spatial resolution and satellite overpass time) that a spaceborne sensor has to have to properly monitor the SUHI effect at the district level in a big city. The results show that spatial resolutions greater than 50 m are needed to properly estimate the SUHI effect at the district level. Spatial resolutions lower than 50 m underestimate the effect and do not distinguish between the different zones inside the city. Note that higher spatial resolutions, which may offer higher amounts of information, involve a lower frequency of revisit time and a lower swath at nadir. Besides this, it has been shown that an overpass time immediately before sunrise is the best choice to estimate the SUHI effect that can be compared with the UHI effect obtained from air temperature data. At present, no spaceborne thermal sensors satisfy these spatial and time conditions at the same time, so these conclusions must be considered for future space missions planning.

Finally, we compare three algorithms (NDVI, TES, and TISI) for the estimation of LSE over urban areas using the AHS sensor. Results show the unsuitability of using the NDVI because it presents the highest errors, and it does not distinguish artificial surfaces with different compositions, while the TES and the TISI permit distinguishing them. For the TES and the TISI, RMSE values of 0.04 and 0.03, respectively, are obtained. When comparing LST estimates, we still observe better agreements with *in situ* values for the TISI and the TES algorithms than for the NDVI. The RMSE computed for the TISI is 1 K higher than its accuracy when only man-made surfaces are observed. For the TES, this difference is just 0.5 K, which corresponds to the requirement for many applications over urban areas, such as the monitoring of the UHI effect. In addition, the TISI algorithm requires more spectral information (TIR bands and at least one band in the MIR region) and temporal information (one night and one day image) than the other two methods. However, the TES algorithm can be applied with only five TIR bands and only one image. In addition, this algorithm has the advantage that the LST can be retrieved at the same time as the LSE. In conclusion, our results show that the TES emissivity is the algorithm that, without the requirement of high temporal resolution of the sensor, best reproduces the LST over an urban area.

Acknowledgments

The authors thank the ESA (DESIREX 2008, project 21717/08/I-LG and THERMOPOLIS 2009 Experimental Campaign, 22693/09/I-EC), and all the people and groups that participated in both experimental campaigns as well as the Ministerio de Ciencia e Innovación (CEOS-SPAIN, project AYA2011-29334-C02-01) for its financial support. During this work, R. Oltra-Carrió received a grant V Segles from the Universitat de València.

References

Arnfield, A. J. 2003. Two decades of urban climate research: A review of turbulence, exchanges of energy and water, and the urban heat island. *International Journal of Climatology* 23:1–26.

Baldridge, A. M., S. J. Hook, C. I. Grove, and G. Rivera. 2009. The ASTER spectral library version 2.0. *Remote Sensing of Environment* 113:711–715.

Becker, F., and Z. L. Li. 1990. Temperature-independent spectral indices in thermal infrared bands. *Remote Sensing of Environment* 32:17–33.

Ben-Dor, E., and H. Saaroni. 1997. Airborne video thermal radiometry as a tool for monitoring microscale structures of the urban heat island. *International Journal of Remote Sensing* 18:3039–3053.

Dash, P., F. M. Göttsche, F. S. Olesen, and H. Fischer. 2005. Separating surface emissivity and temperature using two-channel spectral indices and emissivity composites and comparison with a vegetation fraction method. *Remote Sensing of Environment* 96:1–17.

Fernández, F., J. P. Montávez, J. F. González-Rouco, and F. Valero. 2004. Relación entre la estructura espacial de la isla térmica y la morfología urbana de Madrid [Relationship between the spatial distribution of the urban heat island and the urban pattern of Madrid]. *El Clima entre el Mar y la Montaña* 4:641–650.

Gillespie, A., S. Rokugawa, T. Matsunaga, et al. 1998. A temperature and emissivity separation algorithm for Advanced Spaceborne Thermal Emission and Reflection Radiometer (ASTER) images. *IEEE Transactions on Geoscience and Remote Sensing* 36:1113–1126.

Grimm, N. B., S. H. Faeth, N. E. Golubiewski, et al. 2008. Global change and the ecology of cities. *Science* 319:756–760.

Hartz, D. A., L. Prashad, B. C. Hedquist, J. Golden, and A. J. Brazel. 2006. Linking satellite images and hand-held infrared thermography to observed neighborhood climate conditions. *Remote Sensing of Environment* 104:190–200.

IPCC (Intergovernmental Panel on Climate Change). 2007. Fourth assessment report: Climate change. Geneva, Switzerland: Intergovernmental Panel on Climate Change.

Jasche, A., and T. Rezende. 2007. Detection of the urban heat-island effect from a surface mobile platform. *Revista de Teledetección* 27:59–70.

Johnson, D. P., J. S. Wilson, and G. C. Luber. 2009. Socioeconomic indicators of heat-related health risk supplemented with remotely sensed data. *International Journal of Health Geographics* 8:57.

Lagouarde, J. P., P. Moreau, M. Irvine, et al. 2004. Airborne experimental measurements of the angular variations in surface temperature over urban areas: Case study of Marseille (France). *Remote Sensing of Environment* 93:443–462.

Li, J. J., X. R. Wang, X. J. Wang, W. C. Ma, and H. Zhang. 2009. Remote sensing evaluation of urban heat island and its spatial pattern of the Shanghai metropolitan area, China. *Ecological Complexity* 6:413–420.

Liu, W., C. Ji, J. Zhong, X. Jiang, and Z. Zheng. 2007. Temporal characteristics of the Beijing urban heat island. *Theoretical and Applied Climatology* 87:213–221.

Lu, D., and Q. Weng. 2006. Spectral mixture analysis of ASTER images for examining the relationship between urban thermal features and biophysical descriptors in Indianapolis, Indiana, USA. *Remote Sensing of Environment* 104:157–167.

Martens, W. J. M. 1998. Climate change, thermal stress and mortality changes. *Social Science and Medicine* 46:331–344.

Masson, V., L. Gomes, G. Pigeon, et al. 2008. The Canopy and Aerosol Particles Interactions in Toulouse Urban Layer (CAPITOUL) experiment. *Meteorology and Atmospheric Physics* 102:135–157.

Morris, M., E. P. McClain, and N. Plummer. 2001. Quantification of the influences of wind and cloud on the nocturnal urban heat island of a large city. *Journal of Applied Meteorology* 40:169–182.

Najjar, G., P. P. Kastendeuch, M. P. Stoll, et al. 2004. Le projet Reclus. Télédétection, rayonnement et bilan d'énergie en climatologie urbaine à Strasbourg. *La Météorologie* 46:44–50.

Nichol, J. E., W. Y. Fung, K. S. Lam, and M. S. Wong. 2009. Urban heat island diagnosis using ASTER satellite images and "in situ" air temperature. *Atmospheric Research* 94:276–284.

Oke, T. R. 1973. City size and the urban heat island. *Atmospheric Environment* 7:769–779.

Oke, T. R. 1981. Canyon geometry and the nocturnal urban heat island: Comparison of scale model and field observations. *Journal of Climatology* 1:237–254.

Oke, T. R. 1982. The energetic basis of the urban heat-island. *Quarterly Journal of the Royal Meteorological Society* 108:1–24.

Oke, T. R. 1987. *Boundary layer climates*. London: Routledge.

Oke, T. R. 1997. Urban climates and global environmental change. In *Applied climatology: Principles and practice*, eds. R. D. Thompson and A. Perry, 273–287. London: Routledge.

Oke, T. R., G. T. Johnson, D. G. Steyn, and I. D. Watson. 1991. Simulation of surface urban heat islands under ideal conditions at night. Part 2: Diagnosis of causation. *Boundary-Layer Meteorology* 56:339–358.

Oltra-Carrió, R., J. A. Sobrino, B. Franchand, and F. Nerry. 2012. Land surface emissivity retrieval from airborne sensor over urban areas. *Remote Sensing of Environment* 123:298–305.

Petitcolin, F., and E. Vermote. 2002. Land surface reflectance, emissivity and temperature from MODIS middle and thermal infrared data. *Remote Sensing of Environment* 83:112–134.

PRB (Population Reference Bureau). 2011. *2011 world population data sheet*. Technical Report. Washington, DC: Population Reference Bureau.

Pu, R., P. Gong, R. Michishita, and T. Sasagawa. 2006. Assessment of multi-resolution and multi-sensor data for urban surface temperature retrieval. *Remote Sensing of Environment* 104:211–225.

Rajasekar, U., and Q. H. Weng. 2009. Spatio-temporal modelling and analysis of urban heat islands by using Landsat TM and ETM plus imagery. *International Journal of Remote Sensing* 30:3531–3548.

Rigo, G., E. Parlow, and D. Oesch. 2006. Validation of satellite observed thermal emission with in-situ measurements over an urban surface. *Remote Sensing of Environment* 104:201–210.

Roetzer, T., M. Wittenzeller, H. Haeckel, and J. Nekovar. 2000. Phenology in central Europe—Differences and trends of spring phenophases in urban and rural areas. *International Journal of Biometeorology* 44:60–66.

Sobrino, J. A., J. C. Jimenez-Munoz, G. Soria, et al. 2008a. Land surface emissivity retrieval from different VNIR and TIR sensors. *IEEE Transactions on Geoscience and Remote Sensing* 46:316–327.

Sobrino, J. A., J. C. Jiménez-Muñoz, G. Sòria, et al. 2008b. Thermal remote sensing in the framework of the SEN2FLEX project: Field measurements, airborne data and applications. *International Journal of Remote Sensing* 29:4961–4991.

Sobrino, J. A., R. Oltra-Carrió, J. C. Jiménez-Muñoz, et al. 2012b. Emissivity mapping over urban areas using a classification-based approach: Application to the Dual-use European Security IR Experiment (DESIREX). *International Journal of Applied Earth Observation and Geoinformation* 18:141–147.

Sobrino, J. A., R. Oltra-Carrió, G. Sòria, R. Bianchi, and M. Paganini. 2012a. Impact of spatial resolution and satellite overpass time on evaluation of the surface urban heat island effects. *Remote Sensing of Environment* 117:50–56.

Sobrino, J. A., R. Oltra-Carrió, G. Sòria, et al. 2013. Evaluation of the surface urban heat island effect in the city of Madrid by thermal remote sensing. *International Journal of Remote Sensing* 34:3177–3192.

Stathopoulou, M., C. Cartalis, and M. Petrakis. 2007. Integrating CORINE land cover data and Landsat TM for surface emissivity definition: Application to the urban area of Athens, Greece. *International Journal of Remote Sensing* 28:3291–3304.

Stathopoulou, M., A. Synnefa, C. Caralis, et al. 2009. A surface heat island study of Athens using high-resolution satellite imagery and measurements of the optical and thermal properties of commonly used building and paving materials. *International Journal of Sustainable Energy* 28:59–76.

Streutker, D. R. 2003. Satellite-measured growth of the urban heat island of Houston, Texas. *Remote Sensing of Environment* 85:282–289.

Tiangco, M., A. M. F. Lagmay, and J. Argete. 2008. ASTER-based study of the nighttime urban heat island effect in metro Manila. *International Journal of Remote Sensing* 29:2799–2818.

Toy, S., S. Yilmaz, and H. Yilmaz. 2007. Determination of bioclimatic comfort in three different land uses in the city of Erzurum, Turkey. *Building and Environment* 42:1315–1318.

Van de Griend, A. A., and M. Owe. 1993. On the relationship between thermal emissivity and the normalized difference vegetation index for natural surfaces. *International Journal of Remote Sensing* 14:1119–1131.

Voogt, J. A. 2002. Urban heat island. In *Encyclopedia of global environmental change*, ed. T. Munn. Chichester: Wiley.

Voogt, J. A., and T. R. Oke. 1997. Complete urban surface temperatures. *Journal of Applied Meteorology* 36:1117–1132.

Voogt, J. A., and T. R. Oke. 1998. Radiometric temperatures of urban canyon walls obtained from vehicle traverses. *Theoretical and Applied Climatology* 60:199–217.

Voogt, J. A., and T. R. Oke. 2003. Thermal remote sensing of urban climates. *Remote Sensing of Environment* 86:370–384.

Weng, Q. H., D. S. Lu, and J. Schubring. 2004. Estimation of land surface temperature-vegetation abundance relationship for urban heat island studies. *Remote Sensing of Environment* 89:467–483.

West, P. C., G. T. Narisma, C. C. Barford, C. J. Kucharik, and J. A. Foley. 2011. An alternative approach for quantifying climate regulation by ecosystems. *Frontiers in Ecology and the Environment* 9:126–133.

WHO (World Health Organization). 2003. *Climate change and human health: Risk and responses.* Geneva, Switzerland: World Health Organization.

Yuan, F., and M. E. Bauer. 2007. Comparison of impervious surface area and normalized difference vegetation index as indicators of surface urban heat island effects in Landsat imagery. *Remote Sensing of Environment* 106:375–386.

Section VI

Other Dimensions of Ecosystem Services

20

Multidimensional Approaches in Ecosystem Services Assessment

A. J. Castro Martínez
University of Oklahoma, Oklahoma; University of Almería, Spain

M. García-Llorente
Carlos III University of Madrid, Spain; Autonomous University of Madrid, Spain

B. Martín-López, I. Palomo, and I. Iniesta-Arandia
Autonomous University of Madrid, Spain

CONTENTS

20.1 The Need for a Multidimensional and Interdisciplinary
Framework for Ecosystem Services Assessment.................................442
20.2 The Supply Side of Ecosystem Services....................................444
    20.2.1 Service-Providing Units.....................................444
    20.2.2 Biophysical Indicators for Evaluating Ecosystem Services......445
    20.2.3 Going Spatial: From Early Approaches to Current
    Toolboxes..445
20.3 The Demand Side of Ecosystem Services...................................450
    20.3.1 Ecosystem Services Beneficiaries.............................450
    20.3.2 Sociocultural Valuation....................................451
    20.3.3 Economic Valuation.......................................453
20.4 Discussion and Future Steps: Toward Hybrid Methodologies and
New Concepts 458
20.5 Conclusions..459
Acknowledgments...460
References...461

20.1 The Need for a Multidimensional and Interdisciplinary Framework for Ecosystem Services Assessment

Sustainability science, or the science that focuses on human–nature relationships (MA 2005; Perrings 2007; Perrings et al. 2011), is increasing in research forums particularly through the application of the ecosystem service concept in environmental conservation and management (Seppelt et al. 2011; Burkhard et al. 2012a). Over the past two decades, the ecosystem service concept has gained importance among scientists, managers, and policy-makers worldwide as a way to communicate societal dependence on ecological life support systems integrating both the natural and social science perspectives (Bastian et al. 2012). Many international initiatives, such as the Millennium Ecosystem Assessment (MA), The Economics of Ecosystems and Biodiversity (TEEB), and the Intergovernmental Platform on Biodiversity and Ecosystem Services (IPBES) (Carpenter et al. 2009; de Groot et al. 2010; Seppelt et al. 2011; Burkhard et al. 2012a), have developed interdisciplinary frameworks to tackle the different value dimensions in which ecosystems benefit society and, therefore, make the ecosystem service concept operational.

Despite the progress that has been made, many challenges still remain to integrate the ecosystem service concept into an operational framework (de Groot et al. 2010). Seppelt et al. (2011) recently reviewed the ecosystem service field to provide guidance that enhances the applications of the concept and the credibility of results. Despite the increasing number of publications that present innovative ideas and complementary insights from various perspectives, there is growing uncertainty with respect to the appropriate methodologies and techniques used that limits the comparability and the applicability of the studies (Seppelt et al. 2012). Therefore, one of the major challenges to address is developing a comprehensive framework that integrates the multidimensional value of ecosystem services (i.e., biophysical, sociocultural, and economic) (Lamarque et al. 2011; Chan et al. 2012).

Many authors have noted the importance of identifying the ecosystem's capacity to provide services (supply side) and their social demand (demand side), highlighting that the status of an ecosystem service is influenced not only by the ecosystem's properties but also by societal needs (Paetzold et al. 2010; Syrbe and Walz 2012; Burkhard et al. 2012b). On the supply side, ecosystems and biodiversity are experiencing serious degradation with regard to their capacity to supply services. At the same time, the demand for certain ecosystem services is rapidly increasing as populations and standards of living increase (Liu et al. 2010). Burkhard et al. (2012a) defined the supply side as the capacity of a particular area to provide a specific bundle of ecosystem services within a given time period, and the demand side as the sum of all ecosystem services currently consumed, used, or valued in a particular area over a given time period.

In this sense, remote sensing methodologies and techniques have been mostly used in the past two decades for quantifying and mapping the supply side of ecosystem services, in particular quantifying and map provisioning (e.g., timber or food production) and regulating services (e.g., air quality, climate, extreme events, waste treatment, erosion, and soil fertility) (Ayanu et al. 2012). Remotely sensed information has been used most often as a proxy for biophysical variables (e.g., biomass; see Chapter 5) that in turn are used as proxies for a particular ecosystem service (e.g., carbon storage). Ayanu et al. (2012) recently provided a revision of relevant remote sensing systems, sensors, and methods applicable in quantifying the supplies and demands of provisioning and regulating services. Their results showed that the quantification of services through Earth observation techniques used either regression models (by linking remotely sensed information to a limited number of *in situ* observations) or land use/land cover classifications (see Chapter 10) that are subsequently linked to ecosystem services.

Here, we present a conceptual framework for ecosystem service assessment (Figure 20.1) that links the service-providing units (SPUs) and the ecosystem service beneficiaries (ESBs) concepts and how they could be

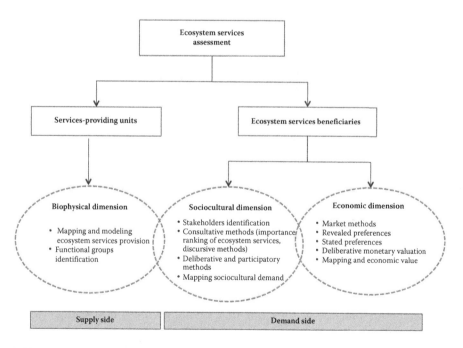

FIGURE 20.1
Conceptual framework showing the links among services-providing units and ecosystem services beneficiaries and how they could be explored through the different dimensions in the ecosystem services assessment.

explored from both the supply side and demand side, considering the multidimensional nature of ecosystem service assessment. Within this framework, we explore which remote sensing methods and techniques are currently available for the assessment of both the supply and the demand side of ecosystem service.

20.2 The Supply Side of Ecosystem Services

20.2.1 Service-Providing Units

Several concepts have been developed to operationalize the specific contributions of ecosystems to the delivery of ecosystem services. One of them is the SPU concept, developed as a method to link species populations with service delivery, primarily at small scales (Luck et al. 2003) but extensible to larger scales. The SPU is defined here as the collection of individuals from a given species and their characteristics necessary to deliver an ecosystem service at the level desired by beneficiaries. This idea was further extended to other levels of organization and scales of analysis. In fact, Luck et al. (2009) suggested the use of social–ecological landscape units as service providers where the capacity of landscapes to provide services would be related to their structural, functional, and social attributes within a socioecological context. There is growing consensus that, among all biodiversity components, it is the functional diversity of the component that mostly ensures the delivery of ecosystem services, particularly regulating services (Díaz et al. 2006; Cardinale et al. 2012; Alcaraz et al. 2013).

Remote sensing may provide useful tools for understanding the SPU concept and its application into regional and global assessments of ecosystem services. Spectral information may help to map land units that are homogeneous in terms of the ecosystem services that they provide. This could be the case with the ecosystem functional types (EFTs) approach. As plant species can be grouped into plant functional types, ecosystems can be grouped into EFTs (Paruelo et al. 2001). EFTs represent groups of ecosystems that share functional characteristics in relation to their exchanges of matter and energy between the biota and the physical environment (Paruelo et al. 2001; Alcaraz-Segura et al. 2006). In other words, EFTs are homogeneous patches of the land that exchange mass and energy with the atmosphere in a common way and, therefore, they can be interpreted as landscape units with the same capacity to provide ecosystem services derived from the ecological process used in their classification (e.g., carbon gains). In this sense, further studies exploring the capacity of EFTs to deliver ecosystem services could be a major breakthrough for Earth observation of ecosystem services.

20.2.2 Biophysical Indicators for Evaluating Ecosystem Services

Several biophysical indicators have been used for assessing ecosystem services. The type of indicator depends on the information available for each case study and the amount of resources willing to be used for evaluating ecosystem services. Some studies have grouped the most common indicators used (Maes et al. 2011; Burkhard et al. 2012b). Table 20.1 presents some of the indicators most often used to assess ecosystem service supply, and it indicates some indicators that might be used for ecosystem service demands. We also provide studies where remote sensing has been or may be used to derive ecosystem service indicators on both the supply and the demand sides.

Due to the urgent need for understanding and mapping the spatial and temporal heterogeneity of ecosystem services, and for maximizing conservation objectives (Polasky et al. 2008; Nelson et al. 2009), remote sensing can be very useful for obtaining biophysical indicators in the supply side of ecosystem service assessment. Approaches from remote sensing can be used for ecosystem service mapping by direct or indirect monitoring and in combination with ecosystem models (Feng et al. 2010). Ecosystem services directly monitored such as habitat for species (see Chapter 3), carbon fixation (see Chapter 6), water provision (see Chapter 12), and climate regulation (see Chapters 17 and 20) require information on aspects of vegetation and water. Indirectly monitored services, such as soil-based services, use surrogate information such as soil status or canopy reflectance. Finally, some services such as flood regulation or soil erosion can be monitored and implemented in ecosystem models through inputs derived from remotely sensed spatial explicit data. Table 20.2 shows some remote sensing products linked to ecosystem services (Feng et al. 2010).

20.2.3 Going Spatial: From Early Approaches to Current Toolboxes

Ecosystem services maps seem to be among the most useful approaches to integrate the ecosystem service concept into ecosystems management (Balvanera et al. 2001; Daily and Matson 2008). They allow the identification of highly valuable areas for conservation (Chan et al. 2006; Naidoo and Rickets 2006) or the identification of ecosystem service supply and demand, which might help in achieving a sustainable use of ecosystem services (Kroll et al. 2012). From the very first ecosystem services maps (Eade and Moran 1996; Costanza et al. 1997), to the current ecosystem service models or toolboxes (e.g., InVEST or ARIES; see Box 20.1), we have witnessed an increase in the different methodologies and purposes for mapping ecosystem services. Nowadays, new tools for mapping services have emerged to support landscape management (see POLYSCAPE in Box 20.1) (Jackson et al. 2013). Mapping works started focusing on the supply side (Burkhard et al. 2012b), but participative methods are being increasingly used to also include the demand side (Bryan et al. 2011; Palomo et al. 2012).

TABLE 20.1

List of Biophysical Indicators and Proxies Used for Evaluating Ecosystem Services Supply and Demand

Ecosystem Services	Biophysical Indicators for the Supply Side	Studies from Remote Sensing	Social Indicators for the Demand Side	Studies from Remote Sensing
Provisioning services				
Food provision	Crop yield production	Doraiswamy et al. (2003)	Crop yield consumption	Stephen et al. (2013)
Food provision	Number of grazing heads		Livestock consumed	
Food provision	Fish population sizes	Chassot et al. (2010)	Number of fish or biomass captured or consumed	Stuart et al. (2006)
Biotic materials	Timber stock	Clementel et al. (2012)	Timber production used	
Water provision	Water availability (i.e., precipitation minus evapotranspiration)	Bahadur (2011)	Water consumed	
Medicinal resources	Number of species from which natural medicines have been derived		Number of drugs using natural compounds	
Genetic pool	Number of crop varieties and livestock breed varieties living in a region/surface	Heller et al. (2012)	Number of crop varieties and livestock breed variety used in a region	
Raw materials	Raw materials existing in nature/surface	Mengzhi (2009)	Quantity extracted, used, or bought/surface/time	
Regulating services				
Air quality regulation	Atmospheric cleaning capacity in ton of pollutants removed		Illnesses avoided related to air contamination	
Water quality regulation	Biomass of nutrients removed by aquatic ecosystems (i.e., N, P)		Illnesses/avoided related to contaminated water or unhealthy water consumption	Lobitz et al. (2000)

Climate regulation	Total amount of carbon (or methane) sequestered-stored	Myeong et al. (2006)	Avoided number of climate refugees, people affected by climate change
Moderation of extreme events	Natural elements or number of species dampening extreme events (flood, storms, avalanches)		Avoided damages caused by flood, storms, avalanches
Erosion protection	Soil erosion rate, erosion related variables (slope, precipitation, vegetation cover, etc.)	Vrieling (2006)	Mass of soil removed from water reservoirs
Pollination	Abundance and species richness of wild pollinators	Schulp and Alkemade (2011)	Benefits for crop production or biodiversity maintenance due to pollination
Cultural services			
Aesthetic enjoyment	View shed extent, iconic elements views		Users of scenic routes
Recreation and tourism	Number of trails or natural areas available for tourism	Nichol and Wong (2005)	Number of visitors
Recreation and tourism (recreational hunting)	Animal population sizes, regeneration rate of species	Ropert-Coudert and Wilson (2005)	Individuals hunted
Intellectual and experiential (scientific knowledge)	Abundance of landscape features or species with scientific value	Mertes (2002)	Number of researchers working with environmental assets

Source: From de Groot, R. S., et al., *Ecological Complexity*, 7, 260–272, 2010; Maes, J., et al. *A spatial assessment of ecosystem services in Europe: Methods, case studies and policy analysis-phase 1*, PEER Report No 3, Partnership for European Environmental Research, Ispra, Italy, 2011; and Burkhard, B., et al., *Ecological Indicators*, 21, 17–29, 2012.

TABLE 20.2

Ecosystem Services Quantified by Remote Sensing Data

Proxy	Ecosystem Services	Spectral Index or Techniques	Sensor Types	References
Carbon storage/ sequestration by different land uses/land cover	Climate regulation; air quality regulation	Advanced Very High Resolution Radiometer (AVHRR) for mapping land covers	LANDSAT	Konarska et al. (2002)
Animal and plant species richness and abundance	Habitat for species; biodiversity conservation	Normalized Difference Vegetation Index (NDVI)—landscape heterogeneity; Light Detection and Ranging (LiDAR)—canopy structure	MODIS	Gould (2000); Balvanera et al. (2006); Carlson et al. (2007)
Carbon uptake through ecosystem carbon gains	Climate regulation	Normalized Difference Vegetation Index (NDVI); net primary productivity (NPP); Enhanced Vegetation Index (EVI); gross primary productivity (GPP); Leaf Area Index (LAI)	MODIS	Gianelle and Vescovo (2007); Olofsson et al. (2008); Gianelle et al. (2009)
Water cycle (green and blue water)	Hydrological regulation; erosion protection; flood control	Normalized Difference Vegetation Index (NDVI)—surface parameters; Leaf Area Index (LAI)—surface parameters	SWIM WEPP LASCAM SEB	Krysanova et al. (2007); Liu and Li (2008); Minacapilli et al. (2009); Williams et al. (2010)
Properties of soil types	Climate regulation; soil fertility	Normalized Difference Vegetation Index (NDVI)—proportion of bare soil; normalized difference wetness index (NDWI)—soil saturation; soil color index (SCI)—soil saturation; LiDAR—soil roughness measurement	LANDSAT MODIS	Lobell et al. (2009); Kheir et al. (2010)

Note: Table offers most common proxies, spectral index or remotely sensed techniques, sensors types used. SWIM = Soil and Water Integrated Model; WEPP = Water Erosion Prediction Project model; LASCAM = Large Scale Catchment Model; SEB = Surface energy balance; LANDSAT = Land-Use Satellite (http://landsat.gsfc.nasa.gov); MODIS = Moderate Resolution Imaging Spectroradiometer (http://modis.gsfc.nasa.gov).

BOX 20.1   MAIN TOOLBOXES FOR ECOSYSTEM SERVICES MAPPING

InVEST: INTEGRATED VALUATION OF ENVIRONMENTAL SERVICES AND TRADEOFFS

InVEST is a family of tools designed by the Natural Capital Project that allow mapping of ecosystem services. It is designed to inform decisions about natural resource management and provides an effective tool for evaluating trade-offs among ecosystem services by estimating the amount and value of ecosystem services that are provided on the current landscape or under future scenarios. InVEST models are spatially explicit, using maps as information sources and producing maps as outputs. InVEST returns results in either biophysical (e.g., tons of carbon stored) or economic terms (e.g., net present value of that sequestered carbon). InVEST allows mapping several ecosystem services These are (1) lands and waters—biodiversity, carbon, hydropower, water purification, reservoir sedimentation, managed timber production, crop pollination; and (2) oceans and coasts—wave energy, coastal vulnerability, marine fish aquaculture, aesthetic quality, overlap analysis (fisheries recreation, habitat risk assessment). The different tools run as an extension of ArcMAP and are available from the Natural Capital Project Web site (http://www.naturalcapitalproject.org/InVEST.html). There is a forum in which users can communicate and exchange experiences.

LAND USE/LAND COVER: REMOTELY SENSING INFORMATION

Land use/land cover has been widely used as a proxy for the quantification and mapping of ecosystem services by assigning ecosystem service values to the different land use/cover types. Remote sensing provides useful data for land use/land cover classification. The classification techniques are based on statistical analysis to obtain discrete classes, and its accuracy depends on the set training areas. Therefore, resulting land use/land cover classification depends on properties of the remote sensing data available. Thus, the accuracy of ecosystem services quantification depends on the accuracy of the land use/land cover classification (Ayanu et al. 2012).

ARIES: ARTIFICIAL INTELLIGENCE FOR ECOSYSTEM SERVICES

ARIES is a web-based technology developed by various institutions (including the University of Vermont) that allows rapid ecosystem service assessment and valuation. It provides an intelligent modeling platform capable of composing complex ecosystem services models from a collection of models specified by the user. It can map the locations and quantity of potential provision of ecosystem services (sources), their human beneficiaries (users), and any biophysical features that can deplete service flows (sinks). Because many models written in the ARIES modeling language are based on a probabilistic, Bayesian approach, they are able to explicitly convey uncertainty about their inputs to their outputs and are capable of operating even in data-scarce conditions where deterministic models cannot run. In the version 1.0 beta release, ecosystem services maps are available for seven areas of the world including the following ecosystem services: water supply, subsistence fisheries, carbon, flood and sediment regulation, coastal protection, aesthetic, and recreation. All information is available at the project Web site (http://ariesonline.org).

POLYSCAPE: EMERGING NEW TOOLS

POLYSCAPE is a geographic information system (GIS) framework designed to explore spatially explicit synergies and trade-offs among ecosystem services to support landscape management. POLYSCAPE currently includes algorithms to explore the impacts of land cover change on flood risk, habitat connectivity, erosion and associated sediment delivery to receptors, carbon sequestration, and agricultural productivity.

Remote sensing provides valuable input data for ecosystem service models or toolboxes commonly used to simulate ecosystem services. Ecosystem service models provide an explicit connection between the ecosystem services to be quantified and the remotely sensed parameters. For instance, the Integrated Valuation of Environmental Services and Trade-offs (InVEST) toolbox quantifies and maps ecosystem services using spatially explicit information on land cover, evapotranspiration, precipitation, and topography that can be derived from remote sensing data. Box 20.1 provides information on the applicability of remote sensing data in characterizing and mapping land use and land covers within InVEST.

20.3 The Demand Side of Ecosystem Services

20.3.1 Ecosystem Services Beneficiaries

Currently, the majority of ecosystem service studies do not explicitly include the preferences and values of different ESBs (Menzel and Teng 2010; Seppelt et al. 2011). However, there is a range of stakeholders with different priorities regarding which ecosystem services are most important for their well-being (McMichael et al. 2003; Díaz et al. 2011) and, consequently, they should be included in the ecosystem service assessment (Egoh et al. 2007). Detaching ecosystem services from their perceived value, as is often currently practiced, implies that these services can be defined without including the values given by those who benefit from them. This approach is not compatible with the ecosystem service definition. Ecosystem services include the direct or indirect contributions of ecosystems to human well-being; thus, including the importance of human preferences in their assessment is a necessary requirement (de Groot et al. 2010; EME 2011).

Recent contributions in Mediterranean protected areas identify three common categories of stakeholders' profiles that should be included in the assessments: tourist population, managers and environmental professionals, and local population or residents with different relationships with the ecosystem services (Martín-López et al. 2007; Castro et al. 2011; García-Llorente et al. 2011a, 2011b). Locals mainly benefit from and enjoy provisioning (related to agropecuarian activities) and cultural services associated with a sense of place and cultural heritage. Professionals are focused on regulating services and those cultural services related to knowledge systems and protected areas (i.e., environmental education, scientific knowledge, or nature tourism). Tourists usually indicate their preferences for cultural services related to recreational activities and aesthetic values linked to urban demands of biophilic stimuli. Understanding this diversity of views improves the analysis of potential social conflicts and the understanding of the ecosystem service trade-offs. On the one hand, the identification and characterization of the different ESBs is a first step to their latter engagement in participatory processes for designing or promoting environmental management policies toward a desirable future (Baker and Landers 2004; Reed 2008; Palomo et al. 2011). On the other hand, the inclusion of different ESBs profiles promotes the combination of different knowledge sources, that is, the experimental (local ecological knowledge) and experiential (technical or scientific knowledge) (García-Llorente et al. 2011b).

20.3.2 Sociocultural Valuation

The number of studies using the sociocultural perspective in service valuation is very limited, and the techniques used have not been as formalized as in the economic assessments (explained here). However, this approach is increasingly gaining attention as a means to value cultural services, address nonmaterial benefits (Chan et al. 2006, 2012), and to reflect a plurality of service values. The sociocultural valuation of ecosystem services includes noneconomic methods to analyze human preferences toward ecosystem service demand, use, enjoyment, and value in which moral, ethical, historical, or social aspects play an important role. Understanding human preferences, attitudes toward nature, and behavioral intentions requires analysis of psychological, historical, and ethical factors in addition to economic approaches (Spash et al. 2009).

In sociocultural valuation, we recognize that ecosystems and their biodiversity provide services related to non-use values, such as the satisfaction of conserving biodiversity, local identity, or local ecological knowledge. The same applies when we want to explore the inherent intrinsic value of species and ecosystems, where the utility function used in the economic dimension does not cover all human motivations (Chan et al. 2006). As Kumar and Kumar (2008) stated, there are issues that transcend the domain of the particular logic of economic choices and, therefore, the different dimensions of human well-being, such as social relationships, health, security, and freedom of choice and action (MA 2003). These cannot be addressed using economic techniques (Wegner and Pascual 2011).

The particular sociocultural values attached to biodiversity and ecosystem services can be explored using qualitative and quantitative techniques that involve direct and indirect consultative methods. Direct consultative methods include techniques that explore individual perceptions and collective preference methods. On the one hand, techniques analyzing individual perceptions of ecosystem service importance or use, apply ranking or rating of preferred ecosystem services through surveys and the use of scales (e.g., Likert scales). In the ranking technique, the respondents usually decide the most important ecosystem services from a panel of existing services in a given ecosystem (e.g., Castro et al. 2011). In the rating technique, respondents rate each service independently (Agbenyega et al. 2009), often using some kind of visual aid (Calvet-Mir et al. 2012). On the other hand, collective preferences, such as discourse-based analysis (Wilson and Howarth 2002), are based on the assumption that the valuation of public goods (e.g., most ecosystem services) should result from a process of free and open public debate incorporating social equity issues and not from the aggregation of individual perceptions. In this set of techniques, a small group of individuals (usually more than 2 but no more than 20) debates and reaches a consensus-based value of ecosystem services.

In indirect consultative methods, respondents are asked to name notions or terms to describe a particular ecosystem; later, the researcher associates these notions with the defined ecosystem service types (e.g., Quetier et al. 2010). In addition, those expressed views could be complemented with the views, ideas, and language included explicitly in communication media (books, articles, laws, conservation programs, webs, etc.). This could be done using content analysis techniques—suitable tools to assess ESB views and values attached to biodiversity and ecosystem service importance (Xenarios and Tziritis 2007; Webb and Raffaelli 2008). Q-methodology is also a promising way to identify ecosystem service values. This technique is focused on how ESBs understand and feel about environmental issues (Sandbrook et al. 2011), with the possibility of exploring ecosystem service priorities and trade-offs depending on the ESBs profile. Other indirect approaches would consist of the use of social preferences toward landscapes, considering landscape as a representation of the capacity of ecosystems to provide ecosystem services to society, and *a posteriori* identification of ecosystem services using the Delphi method (e.g., García-Llorente et al. 2012).

20.3.3 Economic Valuation

In the past decades, the economic dimension has gained the highest importance in both the academic and the political arenas (Gómez-Baggethun et al. 2010). One of the first steps toward its mainstream occurred in the 1990s with the publications of de Groot (1992), Daily (1997), or the research conducted by Costanza et al. (1997) in which 17 ecosystem services for 16 biomes in the world were valued with an average estimation of US$33 trillion per year. (Meanwhile, the global gross national product total was around US$18 trillion per year.) Consequently, nature "production" was estimated to be 1.8 times more than the human "production." Since then, the predominance of ecosystem service economics could be seen by the rise in the number of scientific articles (e.g., Loomis et al. 2000; Kontogianni et al. 2010), and by different global projects such as TEEB, which quantifies the cost of biodiversity loss and ecosystem service degradation at an international scale (TEEB 2010). This approach has been broadly applied to value the ecosystem services provided by different ecosystems such as forests (e.g., Croitoru 2007; Zandersen and Tol 2009), wetlands (e.g., Woodward and Wui 2001; Brander et al. 2006), marine ecosystems (e.g., Turpie et al. 2003; Ressurreição et al. 2011), or by species (Losey and Vaughan 2006; García-Llorente et al. 2011b).

Some of the principal explanations for the predominance of this dimension are given next. First, considering that biodiversity continues to decline (Burkhard et al. 2012b), there is a call to look for the instrumental arguments to support biodiversity conservation in addition to acknowledging its intrinsic value, a vision traditionally supported by conservationists. It is argued that as far as intrinsic values are not measurable, they tend to be ignored in the decision making (Bateman et al. 2002). Following this line of argument, there is currently a growing enthusiasm for the challenge of giving visibility to the ecosystem services provided by biodiversity that are invisible to the markets (positive externalities) because of their lack of price (but not value). This is the case for a large number of regulating services (e.g., pollination, erosion control, or water regulation) and cultural services (e.g., aesthetic values, local ecological knowledge, or the satisfaction of conserving species) (Rodríguez et al. 2005; Gee and Burkhard 2010; Vejre et al. 2010). Policy-makers are particularly sensitive to the cost of actions, and a majority of land use policies are based on economic studies using a cost–benefit analysis that measures the benefits and costs of a policy measure (Balmford et al. 2011). This argument is based on the idea that the more complete the information in economic terms is regarding the contribution of ecosystems and biodiversity to human well-being, the higher the decision-making success (de Groot 2006). The economics of ecosystem services has proven helpful in demonstrating the importance of those ecosystem services without market price, such as (for example) the contribution of pest control and pollination to agricultural production (Gallai et al. 2009) to set priorities between conservation strategies (Martín-López et al. 2007; García-Llorente et al. 2011b), or to analyze the trade-offs or synergies between different biodiversity and ecosystem management options.

Several case studies around the world have demonstrated the importance of maintaining different natural ecosystems not only for their ecological value but also in economic terms, instead of converting them into intensive uses. This has been the case with studies showing the importance of the tropical forests in Cameroon or the wetland ecosystems in Canada (Balmford et al. 2002), the economic importance of sustainable management in wetlands (Birol et al. 2006), coral reef management strategies (Hicks et al. 2009), or the ecosystem services provided by Mediterranean protected areas (Martín-López et al. 2011). In this sense, considering the context of global change and the mainstream economic thinking, the economic valuation of ecosystem services has created a pragmatic and common language between scientists and policy-makers, and a forum of discussion around the idea of how to make explicit the human dependence on ecosystems and their biodiversity.

Within environmental economics, ecosystem services are considered as positive externalities that could be measured through the total economic value (TEV) framework. Different service categories have different types of values attached to them that could be aggregated and isolated for analysis (Pearce and Turner 1990). The TEV is composed of use and non-use values (Figure 20.2). Use values are related to the direct or indirect contributions we receive from ecosystems; non-use values are related to moral or ethical considerations of maintaining biodiversity and its ecosystem services independent of their use values. At the same time, use values are composed of direct use values, indirect use values, and option values. Direct use values usually have an expression in markets and result from the direct human use of ecosystems and their biodiversity, no matter if this is consumptive or extractive (e.g., timber or freshwater) or nonconsumptive or nonextractive (e.g., nature tourism). Meanwhile, indirect use values are generally not reflected in conventional markets and are derived from ecological processes and regulating services (e.g., water purification by aquatic plants or the role of mangrove ecosystems in erosion control or erosion mitigation). Finally, the option value is related to the importance of maintaining a flow of ecosystem services in the future and by definition is associated with any ecosystem service category. Non-use values could be split into existence values related to the satisfaction of conserving ecosystems and their biodiversity even though we will not enjoy or use them; that is, which ones represent a cultural service or philanthropic values related with the satisfaction of knowing that future generations will have access to ecosystem services (bequest value), and the satisfaction of knowing that other people have access to ecosystem

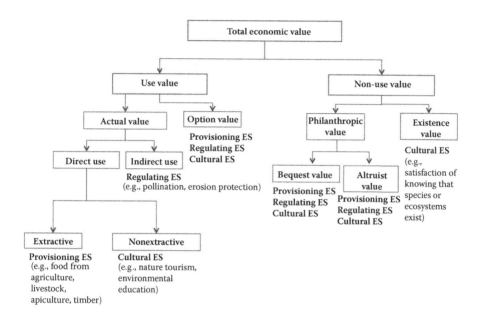

FIGURE 20.2

Representation of the total economic value, its value types, and the ecosystem services categories related to them. ES = ecosystem service.

services (altruist value). Thus, the option value, and bequest and altruist values could be related to all ecosystem services categories (see Bateman et al. 2002; Martín-López et al. 2009; TEEB 2010 for more detail).

In order to assess each value type, a number of different approaches have been designed within environmental economics. These approaches could be classified into three main categories: direct markets, revealed preferences, and stated preferences. The challenge now is to know which valuation method can best fit different biodiversity value types and their ecosystem services with attention to the context and the particular policy purpose (de Groot et al. 2010). In a review of previous works, we provide some guidelines about how to decide which valuation methods are best suited for measuring different valuation types and service categories.

Direct markets use price as a reflection of value and then use data from actual markets to estimate direct use values. These include: (1) market prices used for those provisioning services sold in markets such as the commodities obtained from agriculture or from forest services (such as timber or nontimber forest products); Production function (2) is used to estimate how much a particular ecosystem service that does not have a market price contributes to the delivery of another service that is sold in markets (e.g., the contribution of pollination in apiculture or agricultural production); and (3) the cost approach estimates the expense that would be incurred if ecosystem service contributions needed to recreated through artificial markets using the estimation of avoided cost or replacement cost (TEEB 2010). Both production function and cost approaches are usually used to estimate indirectly the value of regulating services (e.g., how much wetland is saved by providing protection to the coastline against storms and floods) (Figure 20.3).

Revealed preferences estimate the value of a given service without market price through the observation of substitute markets related to the service. The two main techniques are: (1) travel cost (used to estimate recreational services such as nature tourism in a given natural area), which is based on the idea that the cost to arrive to the particular area should be at least equal to the utility obtained (e.g., Shrestha et al. 2002; Martín-López et al. 2009); and (2) hedonic pricing—where a market commodity, usually a property, is described in terms of several attributes including an environmental one (e.g., size, neighborhood, but also the possibility to see an aesthetically pleasant landscape from the window). Then, estimating a demand function for the property, we could infer the value of a change in the environmental attribute (e.g., with landscape views or without them) (e.g., Lansford and Jones 1995; Geoghegan et al. 1997). Travel cost and hedonic pricing are mainly used to estimate the indirect use value of cultural services related to recreational activities (e.g., nature tourism, recreational fishing, recreational hunting, and landscape aesthetic values). However, hedonic pricing

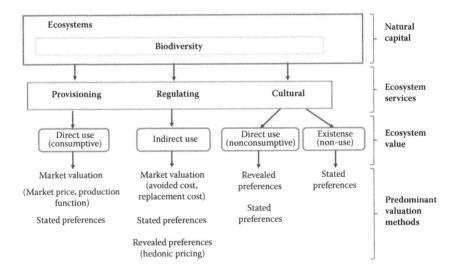

FIGURE 20.3
Graphical representation of the natural capital, the ecosystem services categories delivered, the economic values related to them, and the most commonly used evaluation methods for measuring them.

could be also used to value some regulating services—for example, to estimate the implicit price for air quality service in the price of a property (Figure 20.3).

Finally, stated preference techniques use surveys to create hypothetical markets using to calculate the value of ecosystem services related with both use and non-use values and that could be applied to all ecosystem service categories (Figure 20.3). The three main methods are: (1) the contingent valuation method that elicits public preferences by directly asking people how much they would be willing to pay (or accept) for a change in the quantity or quality of a given ecosystem service in a hypothetical market (Mitchell and Carson 1989). This technique has been one of the most widely used (e.g., Jorgensen et al. 2001; Gürlük 2006; García-Llorente et al. 2011a); (2) choice modeling or conjoint analysis (with the different possibilities of choice experiments, choice ranking, and choice rating) elicits public preferences by asking respondents to choose their preferred option from a series of alternatives of choice sets, each described in terms of different attributes and levels related to the ecosystem services or different environmental plans (e.g., Hanley et al. 2003; Westerberg et al. 2010; Zander and Straton 2010); (3) finally, deliberative monetary methods constitute a hybrid method in which stated preferences techniques are applied in small groups in order to facilitate a participation process (e.g., Zografos and Howarth 2008; Kenter et al. 2011). For an overview of the different valuation techniques, see Bateman et al. (2002), Chee (2004), TEEB (2010), and Turner et al. (2010).

All of these valuation methods present advantages and disadvantages related to information and methodological misspecification, strategic responses, equity problems, unfamiliarity, or sequencing effects, which still need to be improved (Carson et al. 2001; Barkmann et al. 2008; Schläpfer 2008; Turner et al. 2010). However, some of these shortcomings are also related to the inherent limitations of the economic framework itself. Based on neoclassical economics, these techniques assume a utilitarian framework in which individuals in a society are assumed to have rational preferences and try to maximize their profit, advantage, or benefit, and that social interest is an aggregation of individual interests (Dequech 2007; García-Llorente et al. 2011a). Furthermore, the valuation of biodiversity and its ecosystem services differs from the valuation of other goods because of the influence of moral, ethical, or psychological motivations (Hanley and Milne 1996). These values cannot and should not be fully translated into economic terms and have to be complemented or approached using other tools such as sociocultural analysis (TEEB 2010). However, thinking in terms of ecological values, economic valuation dominance implies a risk of being a complexity blinder, hiding the importance of ecosystem processes in the provision of ecosystem services (Norgaard 2010; Sagoff 2011) and having a counterproductive effect on the problem being addressed by favoring the creation of markets that commoditize certain ecosystem services (Gómez-Baggethun and Ruiz-Pérez 2011).

Again, this leads to the necessity of combining or using alternatives to economic valuations with social and biophysical ones.

20.4 Discussion and Future Steps: Toward Hybrid Methodologies and New Concepts

Recent developments in the field include hybrid methodologies that combine nonmonetary and monetary methods or multidimensional methods. One of them is the application of mapping tools in the exercises. On the one hand, this could improve the design and description of multiple economic exercises (Troy and Wilson 2006; Balmford et al. 2011). This is the case of hedonic pricing, travel cost, or choice modeling (e.g., Geoghegan et al. 1997; Brouwer et al. 2010), where GISs permit integration of environmental data and of ecological complexity in the economic exercise. In addition, mapping tools permit us to analyze an ecosystem for the provision of ecosystem services in a sustainable way having also taken into account the social value and demand of those services (Sherrouse et al. 2011; Kroll et al. 2012). The resultant maps provide easily interpretable information that facilitates the analysis of different management options that obtain spatially explicit economic data (Bateman and Jones 2003; Goldstein et al. 2012).

Other examples within the realm of stated preference methods would include the choice modeling without including the monetary or cost attribute (Sayadi et al. 2005) or looking for other options in the payment vehicle that are expressed in terms of willingness to give up time to labors related to biodiversity conservation or to ecosystem service maintenance instead of a conventional willingness to pay (Higuera et al. 2013). This option has been supported by the different ESB profiles; in this way, we avoid the equity problem and also we avoid the assignation of monetary value to things that are considered to be incommensurate with economic values (García-Llorente et al. 2011a). As Cowling et al. (2008) stated "because money is the most commonly used interchangeable commodity, valuation in monetary terms may send the message that a service is more easily replaced by human-manufactured providers than it actually is" (p. 9485). Therefore, the analysis of human values of ecosystem services should be tackled from both the economic (combining monetary and nonmonetary methods) and sociocultural perspectives.

The spatially explicit mapping of ecosystem services available in current toolboxes (e.g., InVEST or ARIES) is based on intensive data requirements, which restricts their application to local scales due to the high costs of surveying the data (Ayanu et al. 2012). In contrast, remote sensing provides data for quantifying and mapping ecosystem services at comparatively low costs, and it offers the possibility of frequent and standardized observations for monitoring (e.g., quantification and mapping of carbon storage at regional or global scales). Currently, most of the lack of knowledge in ecosystem services science relates to quantification and mapping in a spatially explicit way. Hence, remote sensing has the potential to address the basic issues of spatial quantification of ecosystem services. Most of such potential is related to provisioning and regulating ecosystem services such as carbon storage or sequestration (Konarska et al. 2002) or erosion protection (Krysanova et al. 2007). Nowadays, though, the sociocultural and economic dimensions of ecosystem services assessments may also make use of remote sensing products that offer a cost-effective technique for collecting spatial information about services delivered. We only found one study—by Lobitz et al. (2000)—that used remotely sensed indicators of the supply and demand sides to indirectly measure the effects of an infectious disease.

Furthermore, advances in the field have introduced new concepts such as *service benefiting areas* (referring to the demand of ecosystem services from certain areas) and *service connecting areas* (referring to the areas that link ecosystem service supply and demand) (Syrbe and Walz 2012). In this sense, the spatially and explicit quantification of SPUs and service benefiting areas and the importance of linking service providers with ESBs can be considered as opportunities to develop better valuation scenarios for biodiversity conservation (Luck et al. 2003, 2009) and to promote stronger and well-informed valuation applications.

20.5 Conclusions

The assessment of ecosystem services has gained increasing importance among scientists worldwide including the natural, the social, and the economic science perspectives. However, despite the academic progress, many challenges remain to integrate the ecosystem service concept into an operational framework that can be useful for decision making. This chapter presents a review of the current status of assessment of ecosystem services based on the interdisciplinary nature of the methodologies and techniques used today. In particular, we pay attention to the multidimensional nature of ecosystem services (i.e., biophysical, sociocultural, and economic), and we assess them from the supply side as well as the demand side.

Regarding ecosystem service mapping, we provided a list that included some of the common indicators for assessment ecosystem service supply (e.g., precipitation minus evapotranspiration for freshwater provision), and we noted some that might be used to assess ecosystem service demand (e.g., water consumed per surface and time). We also described the main existing toolboxes for mapping ecosystem services (e.g., for example the InVEST toolbox designed by the Natural Capital Project). We also identified the most common indicators derived from remote sensing for services mapping from local to global scale, such as the NDVI index for the quantification of the net primary productivity at multitemporal and spatial scales.

The use of remote sensing data has been mostly utilized for the quantification and mapping of the supply side of provisioning and regulating ecosystem services. Usually, these studies simply characterize the ecological process, such as net primary productivity, without linking it with the potential benefits associated with it, such as food production or climate regulation. Our study reveals the necessity for using a standardized nomenclature of ecosystem services that not only focuses on the ecological process or function but also on the subsequent ecosystem services that the general public perceives as beneficial. Our study also shows that there is still a lack of use of remote sensing data for the assessment of the demand side of ecosystem services. This is probably due to the difficulty in tracking sociocultural aspects, such as social preferences or perceptions, from remotely sensed data. Therefore, further research is needed to provide guidelines that assist in linking remote sensing information with the quantification and mapping of ecosystem services from the supply side with the benefits that society perceives (i.e., the demand side).

Despite the limited number of studies using sociocultural valuation, we described some procedures, including using nonmonetary methods to analyze human preferences. We analyzed some of the reasons why the economic dimension has gained the highest importance in the academic and the political arenas, and we provided some guidelines within the TEV framework about how to decide which valuation method is better for measuring different value types and service categories. In conclusion, we recommend the inclusion of ecosystem service maps evaluated from the biophysical, the sociocultural, and the economic perspectives by combining monetary and nonmonetary methods.

Acknowledgments

The authors thank the three anonymous reviewers for their useful comments. Funding for this project was supported by a grant from the Andalusian Government Department of Environment (GLOCHARID Project) and by the Biodiversity Foundation through the Spanish Millennium Ecosystem Assessment project (http://www.ecomilenio.es/). Support for Antonio J. Castro is also acknowledged from the Centro Andaluz para la Evaluación y Seguimiento del Cambio Global (CAESCG). Support for M. García-Llorente BESAFE Project (Biodiversity and Ecosystem

Services: Arguments for our Future Environment; www.besafe-project. net) funded under the Seventh Framework Programme of the European Commission (Contract No. 282743).

References

Agbenyega, O., P. J. Burgess, M. Cook, and J. Morris. 2009. Application of an ecosystem function framework to perceptions of community woodlands. *Land Use Policy* 26:551–557.

Alcaraz-Segura, D., J. M. Paruelo, and J. Cabello. 2006. Identification of current ecosystem functional types in the Iberian Peninsula. *Global Ecology and Biogeography* 15:200–212.

Alcaraz-Segura, D., J. M. Paruelo, H. E. Epstein, and J. Cabello. 2013. Environmental and human controls of ecosystem functional diversity in temperate South America. *Remote Sensing* 5:127–154.

Ayanu, Y. Z., C. Conrad, T. Nauss, et al. 2012. Quantifying and mapping ecosystem services supplies and demands: A review of remote sensing applications. *Environmental Science and Technology* 46:8529–8541.

Bahadur, K. C. 2011. Assessing strategic water availability using remote sensing, GIS and a spatial water budget model: Case study of the Upper Ing Basin, Thailand. *Hydrological Sciences Journal* 56:994–1014.

Baker, J. P., and D. H. Landers. 2004. Alternative-futures analysis for the Willamette River Basin, Oregon. *Ecological Application* 14:311–312.

Balmford, A., J. Birch, R. Bradbury, et al. 2011. *Measuring and monitoring ecosystem services at the site scale*. Cambridge, UK: Cambridge Conservation Initiative and BirdLife International.

Balmford, A., A. Bruner, P. Cooper, et al. 2002. Economic reasons for conserving wild nature. *Science* 297:950–953.

Balvanera, P., G. C. Daily, P. R. Ehrlich, et al. 2001. Conserving biodiversity and ecosystem services. *Science* 291:2047.

Balvanera, P., A. B. Pfisterer, N. Buchmann, et al. 2006. Quantifying the evidence for biodiversity effects on ecosystem functioning and services. *Ecology Letters* 9:1146–1156.

Barkmann, J., K. Glenk, A. Keil, et al. 2008. Confronting unfamiliarity with ecosystem functions: The case for an ecosystem service approach to environmental valuation with stated preference methods. *Ecological Economics* 65:48–62.

Bastian, O., D. Haase, and K. Grunewald. 2012. Ecosystem properties, potentials and services—The EPPS conceptual framework and an urban application example. *Ecological Indicators* 21:7–16.

Bateman, I. J., and A. P. Jones. 2003. Contrasting conventional with multi-level modelling approaches to meta-analysis: An illustration using UK woodland recreation values. *Land Economics* 2:235–258.

Bateman, I. J., A. P. Jones, A. A. Lovett, I. Lake, and B. H. Day. 2002. Applying geographical information systems (GIS) to environmental and resource economics. *Environmental and Resource Economics* 1–2:219–269.

Birol, E., K. Karousakis, and P. Koundouri. 2006. Using economic methods to inform water resource management policies: A survey and critical appraisal of available methods and an application. *Science of the Total Environment* 365:105–122.

Brander, L., R. Florax, and J. E. Vermaat. 2006. The empirics of wetland valuation: A comprehensive summary and a meta-analysis of the literature. *Environmental and Resource Economics* 33:223–250.

Brouwer, R., T. Dekker, J. Rolfe, and J. Windle. 2010. Choice certainty and consistency in repeated choice experiments. *Environmental and Resource Economics* 46:93–109.

Bryan, B. A., C. M. Raymond, N. D. Crossman, and D. King. 2011. Comparing spatially explicit ecological and social values for natural areas to identify effective conservation strategies. *Conservation Biology* 25:172–181.

Burkhard, B., R. de Groot, R. Costanza, R. Seppelt, S. E. Jorgensen, and M. Potschin. 2012b. Solutions for sustaining natural capital and ecosystem services. *Ecological Indicators* 21:1–6.

Burkhard, B., F. Kroll, S. Nedkov, and F. Müller. 2012a. Mapping ecosystem service supply, demand and budgets. *Ecological Indicators* 21:17–29.

Calvet-Mir, L., E. Gómez-Baggethun, and V. Reyes-García. 2012. Beyond food production: Ecosystem services provided by home gardens. A case study in Vall Fosca, Catalan Pyrenees, northeastern Spain. *Ecological Economics* 74:153–160.

Cardinale, B. J., J. E. Duffy, A. Gonzalez, et al. 2012. Biodiversity loss and its impact on humanity. *Nature* 486:59–67.

Carlson, K. M., G. P. Asner, R. F. Hughes, et al. 2007. Hyperspectral remote sensing of canopy biodiversity in Hawaiian lowland rainforests. *Ecosystems* 4:536–549.

Carpenter, S. R., H. A. Mooney, J. Agard, et al. 2009. Science for managing ecosystem services? Beyond the Millennium Ecosystem Assessment. *Proceedings of the National Academy of Sciences of the United States of America* 106:1305–1312.

Carson, D., A. Gilmore, C. Perry, and K. Gronhaug. 2001. *Qualitative marketing research.* London: Sage.

Castro, A. J., B. Martín-López, D. García-Llorente, P. A. Aguilera, E. López, and J. Cabello. 2011. Social preferences regarding the delivery of ecosystem services in a semiarid Mediterranean region. *Journal of Arid Environments* 75:1201–1208.

Chan, K. M. A., T. Satterfield, and J. Goldstein. 2012. Rethinking ecosystem services to better address and navigate cultural values. *Ecological Economics* 74:8–18.

Chan, K. M. A., M. R. Shaw, D. R. Cameron, E. C. Underwood, and G. C. Daily. 2006. Conservation planning for ES. *PLoS Biology* 4:2138–2152.

Chassot, E., S. Bonhommeau, N. K. Dulvy, et al. 2010. Global marine primary production constrains fisheries catches. *Ecological Letters* 13:495–505.

Chee, Y. E. 2004. An ecological perspective on the valuation of ecosystem services. *Biological Conservation* 120:549–565.

Clementel, F., G. Colle, C. Farruggia, et al. 2012. Estimating forest timber volume by means of "low cost" LiDAR data. *Italian Journal of Remote Sensing* 1:125–140.

Costanza, R., R. d'Arge, R. de Groot, et al. 1997. The value of the world's ecosystem services and natural capital. *Nature* 387:253–260.

Cowling, R. M., B. Egoh, A. T. Knight, et al. 2008. An operational model for mainstreaming ES for implementation. *Proceedings of the National Academy of Sciences of the United States of America* 105:9483–9488.

Croitoru, L. 2007. How much are Mediterranean forests worth? *Forest Policy and Economics* 9:536–545.

Daily, G. 1997. *Nature's services: Societal dependence on natural ecosystems*. Washington, DC: Island Press.

Daily, G. C., and P. Matson. 2008. Ecosystem services: From theory to implementation. *Proceedings of the National Academy of Sciences of the United States of America* 105:9455–9456.

de Groot, R. 2006. Function-analysis and valuation as a tool to assess land use conflicts in planning for sustainable, multi-functional landscapes. *Landscape and Urban Planning* 75:175–186.

de Groot, R. S. 1992. *Functions of nature: Evaluation of nature in environmental planning, management and decision making*. Groningen, The Netherlands: Wolters-Noordhoff B. V.

de Groot, R. S., R. Alkemade, L. Braat, L. Hein, and L. Willemen. 2010. Challenges in integrating the concept of ecosystem services and values in landscape planning, management and decision making. *Ecological Complexity* 7:260–272.

Dequech, D. 2007. Neoclassical, mainstream, orthodox, and heterodox economics. *Journal of Post Keynesian Economics* 30:279–302.

Díaz, S., J. Fargione, F. S. Chapin III, and D. Tilman. 2006. Biodiversity loss threatens human well-being. *PLoS Biology* 4:1300–1305.

Díaz, S., F. Quétier, D. M. Cáceres, et al. 2011. Linking functional diversity and social actor strategies in a framework for interdisciplinary analysis of nature's benefits to society. *Proceedings of the National Academy of Sciences of the United States of America* 3:895–902.

Doraiswamy, P. C., S. Moulin, and P. W. Cook. 2003. Crop yield assessment from remote sensing. *Photogrammetric Engineering and Remote Sensing* 69:665–674.

Eade, J. D. O., and D. Moran. 1996. Spatial economic valuation: Benefits transfer using geographical information systems. *Journal of Environmental Management* 48:97–110.

Egoh, B., M. Rouget, B. Reyers, et al. 2007. Integrating ecosystem services into conservation assessments: A review. *Ecological Economics* 63:714–721.

EME (Spanish Millennium Ecosystem Assessment). 2011. La Evaluación de los Ecosistemas del Milenio de España. Síntesis de resultados. Fundación Biodiversidad. Ministerio de Medio Ambiente, y Medio Rural y Marino, Spain.

Feng, X., F. Bojie, and Y. Yang. 2010. Remote sensing of ecosystem services: An opportunity for spatially explicit assessment. *Chinese Geographical Science* 20:522.

Gallai, N., J. M. Salles, J. Settele, and B. E. Vaissière. 2009. Economic valuation of the vulnerability of world agriculture confronted with pollinator decline. *Ecological Economics* 68:810–821.

García-Llorente, M., B. Martín-López, S. Díaz, and C. Montes. 2011b. Can ecosystem properties be fully translated into service values? An economic valuation of aquatic plant services. *Ecological Applications* 21:3083–3103.

García-Llorente, M., B. Martín-López, and C. Montes. 2011a. Exploring the motivations of protesters in contingent valuation: Insights for conservation policies. *Environmental Science and Policy* 14:76–88.

García-Llorente, M., B. Martín-López, P. A. L. D. Nunes, A. J. Castro, and C. Montes. 2012. A choice experiment study for land use scenarios in semi-arid watersheds environments. *Journal of Arid Environments* 87:219–230.

Gee, K., and B. Burkhard. 2010. Cultural ecosystem services in the context of offshore wind farming: A case study from the west coast of Schleswig-Holstein. *Ecological Complexity* 7:349–358.

Geoghegan, J., L. A. Wainger, and N. E. Bockstael. 1997. Spatial landscape indices in a hedonic framework and ecological economics analysis using GIS. *Ecological Economics* 23:251–264.

Gianelle, D., and L. Vescovo. 2007. Determination of green herbage ratio in grasslands using spectral reflectance. Methods and ground measurements. *International Journal of Remote Sensing* 5:931–942.

Gianelle, D., L. Vescovo, B. Marcolla, et al. 2009. Ecosystem carbon fluxes and canopy spectral reflectance of a mountain meadow. *International Journal of Remote Sensing* 2:435–449.

Goldstein, J. H., G. Caldarone, T. K. Duarte, et al. 2012. Integrating ecosystem-service tradeoffs into land-use decisions. *Proceedings of the National Academy of Sciences of the United States of America* 1:6.

Gómez-Baggethun, E., R. de Groot, P. L. Lomas, and C. Montes. 2010. The history of ecosystem services in economic theory and practice: From early notions to markets and payment schemes. *Ecological Economics* 69:1209–1218.

Gómez-Baggethun, E., and M. Ruiz-Pérez. 2011. Economic valuation and the commodification of ecosystem services. *Progress in Physical Geography* 5:613–628.

Gould, W. 2000. Remote sensing of vegetation, plant species richness and regional biodiversity hotspots. *Ecological Applications* 10:1861–1870.

Gürlük, S. 2006. The estimation of ecosystem services' value in the region of Misi rural development project: Results from a contingent valuation survey. *Forest Policy and Economics* 9:209–218.

Hanley, N., and J. Milne. 1996. Ethical beliefs and behaviour in contingent valuation surveys. *Journal of Environmental Planning and Management* 2:255–272.

Hanley, N., F. Schläpfer, and J. Spurgeon. 2003. Aggregating the benefits of environmental improvements: Distance-decay functions for use and non-use values. *Journal of Environmental Management* 3:297–304.

Heller, E., J. M. Rhemtulla, S. Lele, et al. 2012. Global croplands and their water use for food security in the twenty-first century. *Photogrammetric Engineering and Remote Sensing* 78:815–827.

Hicks, C. C., T. R. McClanahan, J. E. Cinner, and J. M. Hills. 2009. Trade-offs in values assigned to ecological goods and services associated with different coral reef management strategies. *Ecology and Society* 14:10.

Higuera, D., B. Martín-López, and A. Sánchez-Jabba. 2013. Social preferences towards ecosystem services provided by cloud forests in the neotropics: Implications for conservation strategies. *Regional Environmental Change.* doi:10.1007/s10113-012-0379-1.

Jackson, B., P. Timothy, F. Sinclair, et al. 2013. Polyscape: A GIS mapping framework providing efficient and spatially explicit landscape-scale valuation of multiple ecosystem services. *Landscape and Urban Planning* 112:74–88.

Jorgensen, B. S., M. A. Wilson, and T. A. Haberlein. 2001. Fairness in the contingent valuation of environmental public goods: Attitude towards paying for environmental improvement at two levels of scope. *Ecological Economics* 36:133–148.

Kenter, J. O., T. Hyde, M. Christie, and I. Fazey. 2011. The importance of deliberation in valuing ecosystem services in developing countries—Evidence from the Solomon Islands. *Global Environmental Change* 21:505–521.

Kheir, R. B., M. H. Greve, P. K. Bocher, et al. 2010. Predictive mapping of soil organic carbon in wet cultivated lands using classification-tree based models: The case study of Denmark. *Journal of Environmental Management* 91:1150–1160.

Konarska, K. M., P. C. Sutton, and M. Castellon. 2002. Evaluating scale dependence of ecosystem service valuation: A comparison of NOAA-AVHRR and Landsat TM datasets. *Ecological Economics* 41:491–507.

Kontogianni, A., G. W. Luck, and M. Skourtos. 2010. Valuing ecosystem services on the basis of service-providing units: A potential approach to address the "endpoint problem" and improve stated preference methods. *Ecological Economics* 7:1479–1487.

Kroll, F., F. Müller, D. Haase, and N. Fohrer. 2012. Rural-urban gradient analysis of ecosystem services supply and demand dynamics. *Land Use Policy* 29:521–535.

Krysanova, V., F. Hattermann, and F. Wechsung. 2007. Implications of complexity and uncertainty for integrated modeling and impact assessment in river basins. *Environmental Modelling and Software* 22:701–709.

Kumar, M., and P. Kumar. 2008. Valuation of the ecosystem services: A psycho-cultural perspective. *Ecological Economics* 64:808–819.

Lamarque, P., F. Quétier, and S. Lavorel. 2011. The diversity of the ecosystem services concept and its implications for their assessment and management. *Comptes Rendus Biologies* 334:441–449.

Lansford, N. H., and L. L. Jones. 1995. Recreational and aesthetic value of water using hedonic price analysis. *Journal of Agriculture Resource Economics* 2:341–355.

Liu, S., C. Robert, F. Stephen, and T. Austin. 2010. Valuing ecosystem services. *Annals of the New York Academy of Sciences* 1185:54–78.

Liu, X., and J. Li. 2008. Application of SCS model in estimation of runoff from small watershed in Loess Plateau of China. *Chinese Geographical Science* 18:235–241.

Lobell, D. B., S. M. Lesch, D. L. Corwin, et al. 2009. Regional-scale assessment of soil salinity in the Red River Valley using multi-year MODIS EVI and NDVI. *Journal of Environmental Quality* 1:35–41.

Lobitz, B., L. Beck, A. Huq, et al. 2000. Climate and infectious disease: Use of remote sensing for detection of *Vibrio cholerae* by indirect measurement. *Proceedings of the National Academy of Sciences of the United States of America* 4:1438–1443.

Loomis, J., P. Kent, L. Strange, L. Fausch, and A. Covich. 2000. Measuring the total economic value of restoring ecosystem services in an impaired river basin: Results from a contingent valuation survey. *Ecological Economics* 33:103–117.

Losey, J. E., and M. Vaughan. 2006. The economic value of ecological services provided by insects. *BioScience* 56:311–323.

Luck, G. W., G. C. Daily, and P. R. Ehrlich. 2003. Population diversity and ecosystem services. *Trends in Ecology and Evolution* 18:331–336.

Luck, G. W., R. Harrington, P. A. Harrison, et al. 2009. Quantifying the contribution of organisms to the provision of ecosystem services. *BioScience* 59:223–235.

MA (Millennium Ecosystem Assessment). 2003. *Ecosystems and human well-being: A Framework for Assessment.* Washington, D. C.: Island Press and World Resources Institute.

MA (Millenium Ecosystem Assessment). 2005. *Ecosystems and human well-being: Current states and trends.* Washington, DC: World Resources Institute.

Maes, J., L. Braat, K. Jax, et al. 2011. *A spatial assessment of ecosystem services in Europe: Methods, case studies and policy analysis-phase 1.* PEER Report No 3. Ispra, Italy: Partnership for European Environmental Research.

Martín-López, B., M. García-Llorente, I. Palomo, and C. Montes. 2011. The conservation against development paradigm in protected areas: Valuation of ecosystem services in the Doñana social–ecological system (southwestern Spain). *Ecological Economics* 70:1481–1491.

Martín-López, B., E. Gómez-Baggethun, P. L. Lomas, and C. Montes. 2009. Effects of spatial and temporal scales on cultural services valuation areas. *Journal of Environmental Management* 2:1050–1059.

Martín-López, B., C. Montes, and J. Benayas. 2007. The non-economic motives behind the willingness to pay for biodiversity conservation. *Biological Conservation* 139:67–82.

McMichael, A. J., D. Campbell-Lendrum, C. F. Corvalan, et al. 2003. *Climate change and human health: Risks and responses.* Geneva: World Health Organization.

Mengzhi, D. 2009. Study on tobacco spatial agglomeration pattern based on remote sensing and GIS methods in Henan province, China. *Geoscience and Remote Sensing Symposium IEEE International, IGARSS 2009* 2:646–649.

Menzel, S., and J. Teng. 2010. Ecosystem services as a stakeholder-driven concept for conservation science. *Conservation Biology* 3:907–909.

Mertes, L. A. K. 2002. Satellite remote sensing for detailed landslide inventories using change detection and image fusion. *Freshwater Biology* 47:799–816.

Minacapilli, M., C. Agnese, F. Blanda, et al. 2009. Estimation of actual evapotranspiration of Mediterranean perennial crops by means of remote-sensing based surface energy balance models. *Hydrology and Earth System Sciences* 13:1061–1074.

Mitchell, R., and R. Carson. 1989. *Using surveys to value public goods: The contingent valuation method.* Washington, DC: Resources for the Future.

Myeong, S., D. J. Nowak, and M. J. Duggin. 2006. A temporal analysis of urban forest carbon storage using remote sensing. *Remote Sensing of Environment* 101:277–282.

Naidoo, R., and T. H. Ricketts. 2006. Mapping the economic costs and benefits of conservation. *PLoS Biology* 11:360.

Nelson, G. C., M. W. Rosegrant, J. Koo, et al. 2009. *Climate change: Impact on agriculture and costs of adaptation.* Washington, DC: IFPRI Food Policy Report.

Nichol, J., and M. S. Wong. 2005. Satellite remote sensing for detailed landslide inventories using change detection and image fusion. *International Journal of Remote Sensing* 26:1913–1926.

Norgaard, R. B. 2010. Ecosystem services: From eye-opening metaphor to complexity blinder. *Ecological Economics* 69:1219–1227.

Olofsson, P., F. Lagergren, A. Lindroth, et al. 2008. Towards operational remote sensing of forest carbon balance across northern Europe. *Biogeosciences* 5:817–832.

Paetzold, A., P. H. Warren, and L. L. Maltby. 2010. A framework for assessing ecological quality based on ecosystem services. *Ecological Complexity* 7:273–281.

Palomo, I., B. Martín-López, C. López-Santiago, and C. Montes. 2011. Participatory scenario planning for natural protected areas management under the ecosystem services framework: The Doñana social–ecological system, SW Spain. *Ecology and Society* 16:23.

Palomo, I., B. Martín-López, M. Potschin, R. Haines-Young, and C. Montes. 2012. National Parks, buffer zones and surrounding landscape: Mapping ecosystem services flows. *Ecosystem Services Journal* 4:104–116. doi:10.1016/j.ecoser.2012.09.001.

Paruelo, J. M., E. G. Jobbágy, and O. E. Sala. 2001. Current distribution of ecosystem functional types in temperate South America. *Ecosystems* 4:683–698.

Pearce, D. W., and R. K. Turner. 1990. *Economics of natural resources and the environment.* Hemel Hempstead and London: Harvester Wheatsheaf.

Perrings, C. 2007. Future challenges. *Proceedings of the National Academy of Sciences of the United States of America* 104:15179–15180.

Perrings, C., A. Duraiappah, A. Larigauderie, and H. Mooney. 2011. The biodiversity and ecosystem services science-policy interface. *Science* 331:17–19.

Polasky, S., E. Nelson, J. Camm, et al. 2008. Where to put things? Spatial land management to sustain biodiversity and economic returns. *Biological Conservation* 141:1505–1524.

Quetier, F., F. Rivoal, P. Marty, J. de Chazal, W. Thuiller, and S. Lavorel. 2010. Social representations of an alpine grassland landscape and socio-political discourses on rural development. *Regional Environmental Change* 10:119–130.

Reed, M. S. 2008. Stakeholder participation for environmental management: A literature review. *Biological Conservation* 141:2417–2431.

Ressurreição, A., J. Gibbons, T. Ponce Dentinho, M. Kaiser, R. S. Santos, and G. Edwards-Jones. 2011. Economic valuation of species loss in the open sea. *Ecological Economics* 4:729–739.

Rodríguez, J. P., T. D. Beard, J. Agard Jr., et al. 2005. Interactions among ecosystem services. In *Ecosystems and human well-being: Scenarios. Volume 2. Working group, Millennium Ecosystem Assessment. Findings of the scenarios*, eds. S. R. Carpenter, P. L. Pingali, E. M. Bennett, and M. B. Zurek, 431–448. Washington, DC: Island Press.

Ropert-Coudert, Y., and R. P. Wilson. 2005. Trends and perspectives in animal-attached remote-sensing. *Frontiers in Ecology and the Environment* 3:437–444.

Sagoff, M. 2011. The quantification and valuation of ecosystem services. *Ecological Economics* 70:497–502.

Sandbrook, C., I. R. Scales, V. Ivan, et al. 2011. Value plurality among conservation professionals. *Conservation Biology* 25:285–294.

Sayadi, S., M. C. G. Roa, and J. C. Requena. 2005. Ranking versus scale rating in conjoint analysis: Evaluating landscapes in mountainous regions in southeastern Spain. *Ecological Economics* 55:539–550.

Schläpfer, F. 2008. Contingent valuation: A new perspective. *Ecological Economics* 64:729–740.

Schulp, C. J. E., and R. Alkemade. 2011. Consequences of uncertainty in global-scale land cover maps for mapping ecosystem functions: An analysis of pollination efficiency. *Remote Sensing* 3:2057–2075.

Seppelt, R., C. F. Dormann, F. V. Eppink, et al. 2011. A quantitative review of ecosystem service studies: Approaches, shortcomings and the road ahead. *Journal of Applied Ecology* 48:630–636.

Seppelt, R., B. Fath, B. Burkhard, et al. 2012. Form follows function? Proposing a blueprint for ecosystem service assessments based on reviews and case studies. *Ecological Indicators* 21:145–154.

Sherrouse, B. C., J. M. Clement, and D. J. Semmens. 2011. A GIS application for assessing, mapping, and quantifying the social values of ecosystem services. *Applied Geography* 31:748–760.

Shrestha, R. K., A. F. Seidl, and A. S. Moraes. 2002. Value of recreational fishing in the Brazilian Pantanal: A travel cost analysis using count data models. *Ecological Economics* 42:289–299.

Spash, C. L., K. Urama, R. Burton, et al. 2009. Motives behind willingness to pay for improving biodiversity in a water ecosystem: Economics, ethics and social psychology. *Ecological Economics* 4:955–964.

Stephen, B., M. Walter, and A. Ivan. 2013. Remote sensing in agricultural livestock welfare monitoring: Practical considerations. In *Wireless sensor networks & ecological monitoring*, eds. S. Mukhopadhyay and J. A. Jiang, 179–193. Heidelberg, Berlin: Springer-Verlag.

Stuart, N., T. Barratt, and C. Place. 2006. Classifying the neotropical savannas of Belize using remote sensing and ground survey. *Journal of Biogeography* 33:476–490.

Syrbe, R. U., and U. Walz. 2012. Spatial indicators for the assessment of ecosystem services: Providing, benefiting and connecting areas and landscape metrics. *Ecological Indicators* 21:80–88.

TEEB (The Economics of Ecosystems and Biodiversity). 2010. *The economics of ecosystems and biodiversity: Ecological and economic foundations.* London: Earthscan.

Troy, A., and M. A. Wilson. 2006. Mapping ecosystem services: Practical challenges and opportunities in linking GIS and value transfer. *Ecological Economics* 60:435–449.

Turner, R. K., S. Morse-Jones, and B. Fisher. 2010. Ecosystem valuation: A sequential decision support system and quality assessment issues. *Annals of the New York Academy of Science* 1185:79–101.

Turpie, J. K., B. J. Heydenrych, and S. J. Lamberth. 2003. Economic value of terrestrial and marine biodiversity in the Cape Floristic region: Implications for defining effective and socially optimal conservation strategies. *Biological Conservation* 112:233–251.

Vejre, H., S. Jensen, F. Jellesmark, et al. 2010. Demonstrating the importance of intangible ecosystem services from peri-urban landscapes. *Ecological Complexity* 7:338–348.

Vrieling, A. 2006. Satellite remote sensing for water erosion assessment: A review. *Catena* 65:2–18.

Webb, T. J., and D. Raffaelli. 2008. Conversations in conservation: Revealing and dealing with language differences in environmental management. *Journal of Applied Ecology* 45:1198–1204.

Wegner, G., and U. Pascual. 2011. Cost-benefit analysis in the context of ecosystem services for human well-being: A multidisciplinary critique. *Global Environmental Change* 21:492–504.

Westerberg, V. H., L. Robert, and S. B. Olsen. 2010. To restore or not? A valuation of social and ecological functions of the Marais des Baux wetland in southern France. *Ecological Economics* 12:2383–2393.

Williams, J. D., S. Dun, D. S. Robertson, et al. 2010. WEPP simulations of dryland cropping systems in small drainages of northeastern Oregon. *Journal of Soil and Water Conservation* 1:22–23.

Wilson, M. A., and R. B. Howarth. 2002. Distributional fairness and ecosystem service valuation. *Ecological Economics* 41:421–429.

Woodward, R. T., and Y. S. Wui. 2001. The economic value of wetland services: A meta-analysis. *Ecological Economics* 37:257–270.

Xenarios, S., and I. Tziritis. 2007. Improving pluralism in multi criteria decision aid approach through focus group technique and content analysis. *Ecological Economics* 62:692–703.

Zander, K. K., and A. Straton. 2010. An economic assessment of the value of tropical river ecosystem services: Heterogeneous preferences among Aboriginal and non-Aboriginal Australians. *Ecological Economics* 69:2417–2426.

Zandersen, M., and R. S. J. Tol. 2009. A meta-analysis of forest recreation values in Europe. *Journal of Forest Economics* 15:109–130.

Zografos, C., and R. B. Howarth. 2008. Towards a deliberative ecological economics. In *Deliberative ecological economics*, eds. C. Zografos and R. B. Howarth, 1–20. Delhi: Oxford University Press.

Index

A

aboveground biomass (AGB)
  estimations, 27–30
  measurement of, 19
aboveground net primary production
  (ANPP), 25, 27, 88, 93, 96–97,
  107, 285
aboveground net primary production
  (ANPP) estimation
  RUE estimation and, 94
  through remote sensing, 92–93
  through successive biomass harvests,
  89–92
absorbed light, 41
absorbed photosynthetically active
  radiation (APAR), 24, 26, 50
acequias, 339–340
active sensors, for biomass estimations,
  27–30
ADvanced Earth Observing Satellite II
  (ADEOS-2), 48
Advanced Microwave Scanning
  Radiometer Earth Observing
  System (AMSR-E), 72, 307
Advanced Scatterometer (ASCAT), 247
Advanced Spaceborne Thermal
  Emission and Reflection
  Radiometer (ASTER), 251, 424
advanced synthetic aperture radar
  (ASAR), 74
Advanced Very High Resolution
  Radiometer (AVHRR), 64–66,
  68, 70–71, 160, 336, 401, 424, 470
aerosol optical depth (AOD), 128
aerosols, 128, 353, 380
  in Amazon tropical forest, 139–142
AGB, see aboveground biomass (AGB)
agricultural areas, 207
agricultural expansion, 25, 127–128, 219
Agro-IBIS dynamic global vegetation
  model, 6
Airborne Visible/Infrared Spectrometer
  (AVIRIS), 336

airborn sensors, to monitor snow cover,
  336–337
albedo, 380–382, 407–410
albedo estimation, 389–393, 404–407
Amazon tropical forest
  aerosols in, 139–142
  biomass burning emission
  estimation, 125–144
  biomass distribution in, 129
  trace gases emission estimation in,
  139–142
animal ecology, 165–171
animal species
  ecology and conservation of,
  168–171
  population dynamics, 169–170
  remote detection of, 157
annual aboveground NPP (AANPP), 69
ANPP, see aboveground net primary
  production (ANPP)
AOD, see aerosol optical depth (AOD)
APAR estimation, RUE estimation and,
  94–95
Arctic tundra
  carbon cycling pools and processes
  in, 63–77
  "greening" of, 65, 77
  land-atmosphere exchange of carbon
  in, 71–73
  remote sensing of biomass and
  primary production in, 65–68
  soil carbon processes in, 73–74
Argentina, large-scale deforestation in,
  407–410
Artificial Intelligence for Ecosystem
  Services (ARIES), 6, 268,
  449–450
ASTER, see Advanced Spaceborne
  Thermal Emission and
  Reflection Radiometer (ASTER)
Athens thermopolis experimental
  campaign, 423–424
atmosphere, water in, 235, 245

B

bare soil, 207
BDRoute, 207
bedload erosion, 267
beneficiaries, of ecosystem services,
        450–451
Bidirectional Reflectant Distribution
        Function (BRDF) model, 381
biodiversity, 8, 25, 126, 202
    assessment, of national park
        networks, 185
    assessment and monitoring, 151–172
    carbon gains and, 185–186, 192–193
    economics of, 180, 442, 453–458
    ecosystem functioning and, 165–168
    ecosystem services and, 180
    reduction of, 151–152
    spatial resolution requirements for
        study of, 153–157, 472
biogeochemical models, 262
biomass, 24
    ANPP estimation through harvest
        of, 89–92
    distribution in Amazon tropical
        forest, 129
    estimation, 19
    estimations, 27–30
    harvesting, 106
    remote sensing of, in Arctic tundra,
        65–68
    remote sensing of, in boreal forests,
        68–71
    stock of, 23
    trace gases emission estimation,
        139–142
biomass burning emission estimation,
        125–144
    aerosols, 139–142
    CCATT-BRAMS model, 134–135
    emission model assessment, 142–143
    field data and inventory comparison,
        135–137
    FRE distribution, 137–139
    FRP integration, 130–134
    materials and methods, 129–137
    methods of, 128
    results, 137–143
    thermal anomalies detections, 129–130

BIOME-BGC, 68
biophysical variables, 6–7
    changes in, in regional climate
        model, 363–366
    from EFTs, 360–363
    for evaluating ecosystem services,
        445, 446–448
    hydrologic, 251
Boreal Ecosystem Atmosphere Study
        (BOREAS), 69
boreal forests
    "browning" of, 65, 77
    land-atmosphere exchange of carbon
        in, 71–73
    remote sensing of biomass and
        primary production in, 68–71
    soil carbon processes in, 73–74
Bowen ratio energy balance
        (BREB), 315
BRDF, *see* Bidirectional Reflectant
        Distribution Function (BRDF)
        model
BREB, *see* Bowen ratio energy balance
        (BREB)
buffer capacity, 26–27

C

Calypso satellite, 235
canopy reflectance simulation model,
        384–385
carbon (C), 18
    budgets, 106
    land-atmosphere exchange of, 71–73
    released by wildfires, 30–31, 75–76
    soil carbon processes, 73–74
carbon cycle, 7–8
    key processes, 18–19
    models, 262
    monitoring of, 23–31
    observation, at northern high
        latitudes, 63–77
    processes, 63–65
carbon dioxide (CO_2), 18, 52
    assimilation, 50
    fixation, 41, 47, 181
    uptake, 49, 64
carbon dynamics, 17–31
carbon gains, 25–27

congruence between functional
    diversity and, 185–186, 192–193
estimation, using LUE models,
    105–115
in national park networks, 192–193
qualification of, 182–185
carbon monoxide (CO), 52, 140–141
carbon-related ESs, scale issues in
    evaluation of, 22–23
carbon sequestration, 18–19, 40, 126
carbon sinks, 64
carbon storage, 23, 40, 63–64
carbon uptake, estimation of, 39–55
carotenoids, 45, 46
CASI-1500, *see* Compact Airborne
    Spectrographic Imager
    (CASI-1500)
catchment scale analysis, 201–221
    classification of RALC, 208–212
    datasets, 205–208
    discussion, 218–220
    introduction to, 202–204
    modeling relationships between
        land cover and stream ecology,
        212–216
    results, 216–218
    study area, 204–205
cattle, 87, 91
cheatgrass, 155
Chinese research, 4
chlorophyll, in leaf pigments, 42–44
chlorophyll fluorescence, 41, 49–54, 55
    advancements in, 51–52
    basics, origins, and characteristics of,
        49–51
    final comments on, 54
    future of, 52, 54
    remote sensing tools, 52–53
chlorophyll indices, 43, 44
CHRIS (Compact High Resolution
    Imaging Spectrometer), 48
CHRIS/PROBA, 46
CICES, *see* Common International
    Classification of Ecosystem
    Services (CICES)
citizen science, 333
CLC, 208, 214
climate change, 18, 127, 232, 250
climate regulation, 40

climate regulation services, 351–374
    case study for drought episode,
        366–372
    discussion and conclusions, 373–374
    ecosystem-climate feedbacks,
        353–354
    identification of EFTs, 357–370
    introduction to, 352–356
    modeling of ecosystem-atmosphere
        interactions, 354–356
    in regional modeling, 363–372
climate regulation value (CRV) index,
    352–353
clouds, 235, 237–238
Cloudsat satellite, 235
colored dissolved organic matter
    (CDOM), 274
combustion factor, 139
Common International Classification of
    Ecosystem Services (CICES),
    127, 330
Compact Airborne Spectrographic
    Imager (CASI-1500), 51
conservation
    ecosystem functioning and, 165–171
    human well-being and, 180
    national parks networks and, 179–196
    of wildlife populations, 168–171
conservation biogeography, 159–161
CORINE (Coordination of Information
    on the Environment), 159–160,
    161, 208, 217, 251
Coupled Chemistry-Aerosol-Tracer
    Transport Model coupled
    to Brazilian Regional
    Atmospheric Modeling System
    (CCATT-BRAMS), 133–135, 139
    biomass burned estimation and,
        134–135
cover fraction, 265–266
crops
    chlorophyll content, 42–44
    irrigated, 295–298
    LUE in, 111
CRV, *see* climate regulation value (CRV)
    index
cultural services, 5, 202, 298, 304, 319,
    340–341
cyrosphere, water in, 245–246

D

data
  flux, 55
  point cloud, 29–30
  sources, 4
  spectral, 24
Data Assimilation Research Testbed
      (DART), 270–271
date of maximum EVI (DMAX), 184, 187,
      357–360
decomposition, 73–74
deforestation, 25, 127
  impact of large-scale, 407–410
delta coastal areas, 267–268
DEM, *see* digital elevation model (DEM)
DESIREX experimental campaign,
      421–423, 424, 430
digital elevation model (DEM), 334
disturbances, 21–22
DMAX, *see* date of maximum EVI
      (DMAX)
double crops, 25
droughts, 247, 249, 366–372
dynamic vegetation models, 355–356

E

Earth Explorer 8, 53
Earth observation (EO), for species
      diversity assessment and
      monitoring, 151–172
Earth Observing-1 (EO-1), 48
Earth Radiation Budget Experiment
      (ERBE), 380
Earth Radiation Budget Satellite
      (ERBS), 380
ecological change, 232
ecological quality ration (EQR), 213
ecology, of wildlife populations, 168–171
economic valuation, of ecosystem
      services, 453–458
ecosystem-atmosphere interactions,
      modeling, 354–356
ecosystem-climate feedbacks, 353–354
ecosystem functional types (EFTs), 181,
      356, 444
  assessment, of national park
      networks, 185
  biophysical properties from, 360–363

  definition of, 357–359
  identification of, 182–185, 357–370
  land service parameterization of,
      360–361
  satellite data record, 357
  of southern South America, 359–360
  spatial patterns of, 186–189
  U.S. Geological Survey vs. EFT-
      derived biophysical properties,
      361–363
ecosystem functioning, 18
  measuring, 165–171
  species diversity and, 165–168
ecosystems, buffer capacity of, 26–27
ecosystem services
  beneficiaries, 450–451
  biodiversity and, 8, 180
  biophysical indicators for, 445,
      446–448
  carbon cycle and, 7–8
  carbon dynamics and, 17–31
  cascade, 126
  classification of, 19–21, 127, 330
  climate regulation, 351–374
  concept of, 127
  cultural services, 5, 202, 298, 304, 319
  defined, 3, 19
  degradation of, 180
  demand side of, 450–458
  economics of, 180, 442, 453–458
  economic valuation, 453–458
  energy balance, 379–394, 399–411
  energy balance and, 9
  estimation of photosynthetic stress
      for terrestrial, 39–55
  evapotranspiration and, 401–407
  final, 19–21
  hydrological, 229–254, 261–276
  intermediate, 19–21, 23–31
  mapping, 6, 9, 445, 449–450
  modeling, 6
  monitoring of, 23–31
  mountain, 329–344
  multidimensional approaches to
      assessment, 441–460
  payments for, 127
  provisioning services, 126, 337–338
  provision of, 21–23
  quantification of, 262–268, 330–332

regulating services, 126, 202, 338–339
remote sensing, 399–411
remote sensing of, 3–7, 17–18
in river ecosystems, 201–222
role of, 379–380
scale issues, 22–23
scientific progress in field of, 181
sociocultural valuation, 451–453
supply side of, 444–450
supporting, 202
tradeoffs among, 21–22
uses of, 17–18
water cycle and, 8
energy, released by wildfires, 30–31
energy balance, 9
    canopy reflectance simulation model,
        384–385
    ecosystem, 38–82
    ecosystem services related to,
        379–394
    evapotranspiration and, 399–411
    introduction to, 379–383
    simulated case studies, 385–387
    simulated reflectance, 387–393
    wetlands reflected energy, 383–393
energy budgets, 265
Enhanced Vegetation Index (EVI), 24,
        25, 43, 72, 106–107, 167–168, 181,
        184, 289, 290, 292–295, 404
Environmental Mapping and Analysis
        Program (EnMAP), 47, 48
environmental variability, 170–171
ENVISAT, *see* European Environmental
        Satellite (ENVISAT)
EQR, *see* ecological quality
        ration (EQR)
ERBE, *see* Earth Radiation Budget
        Experiment (ERBE)
ERBS, *see* Earth Radiation Budget
        Satellite (ERBS)
erosion control, 23, 25
ESA, *see* European Space Agency (ESA)
ET, *see* evapotranspiration (ET)
European Environmental Satellite
        (ENVISAT), 53, 74
European Space Agency (ESA), 307
evapotranspiration (ET), 6, 265,
        379–380
    ecosystem services and, 401–407

energy balance and, 399–411
estimation of, 312–317, 402–404
mapping, using thermal satellite
        imagery, 314–315
remote sensing of, 402–404
sensors and satellites to measure,
        241–242
EVI, *see* Enhanced Vegetation Index
        (EVI)
EVI dynamic, 292–295
evop, 285

F

fAPAR, *see* fraction of the absorbed
        photosynthetically active
        radiation (fAPAR)
field-based observations, 4
final ESs, 19–21
fire radiative energy (FRE), 30, 128,
        137–139, 141
fire radiative power (FRP), 30–31, 128,
        130–134, 140
fires, 24, 30–31
    biomass burning emission
        estimation, 125–144
    carbon emissions from, 75–76
fire thermal anomaly (FTA)
        algorithm, 30
floods, 24, 247, 249
fluorescence, 49–54, 55
FLuorescence EXplorer (FLEX), 52, 53
fluorescence imaging spectrometer
        (FIS), 52
flux data, 55
FluxNet database, 316–317
flux towers, 55
food, 40
Food and Agricultural Organization
        (FAO), 251
forage growth rate, *see* forage
        production
forage production
    monitoring of, 87–100
    remote sensing, 98–99
forests, 43
    biomass estimation, 28–30
    boreal, 65, 68–73
    large-scale deforestation of, 407–410

fraction of the absorbed photosynthetically active radiation (fAPAR), 24, 43, 68, 94–95, 98, 99
Fraunhofer lines, 51–52
FRE, *see* fire radiative energy (FRE)
FRE distribution, 137–139
FRP, *see* fire radiative power (FRP)
FRP integration, 130–134
FTA, *see* fire thermal anomaly (FTA)
functional diversity
    assessment, 185
    congruence between carbon gains and, 185–186, 192–193

G

GCPM, *see* Geostationary Carbon Process Mapper (GCPM)
GDEM, *see* Global Digital Elevation Model (GDEM)
GEO BON, *see* Group on Earth Observations Biodiversity Observation Network (GEO BON)
geographic information system (GIS), 203
Geographic Information Systems (GIS), 311
GEOSS, *see* Global Earth Observation System of Systems (GEOSS)
Geostationary Carbon Process Mapper (GCPM), 52, 53
Geostationary Fourier Transform Spectrometer (GeoFTS), 53
Geostationary Operational Environmental Satellite (GOES), 130, 133, 137–139
geostationary orbit (GEO), 54
GIS, *see* Geographic Information Systems (GIS)
Global Digital Elevation Model (GDEM), 251
Global Earth Observation System of Systems (GEOSS), 4
GLobal Imager (GLI), 48
Global Land Cover 2000 (GLC2000), 161–163
Global System for Monitoring Ecosystem Service Change, 3–4

global warming, *see* climate change
GlobCover, 163
GOCE, *see* Gravity Field and Steady-State Ocean Circulation Explorer (GOCE)
GOES, *see* Geostationary Operational Environmental Satellite (GOES)
GOSAT (Greenhouse gases Observing SATellite), 53
GRACE, 248
grasslands, 40, 220
Gravity Field and Steady-State Ocean Circulation Explorer (GOCE), 248
grazing, 26
green biomass, assessment of, 40
green economy, 232
greenhouse gases, 30
greenhouse gases (GHGs), 419–420
greenness anomalies, 283–299
    climate and satellite data for estimation of, 288–290
    defined, 288
gross primary productivity (GPP), 40–44, 64, 68, 76
ground water, 247–248
groundwater
    dependence gradient, 292–295
    detecting ecosystem reliance on, 283–299
    impact on vegetation dynamics, 291–292
    remote sensing, 247–248
    sensors and satellites to measure, 244
groundwater-dependent ecosystems (GDEs), 283–299
    introduction to, 284–286
    methods of study, 286–292
    results of study, 292–298
Group on Earth Observations, 3–4
Group on Earth Observations Biodiversity Observation Network (GEO BON), 4
Gudalfeo study case, 267–268

H

heat loss, 41
herbaceous/shrub vegetation, 207

HRU, *see* hydrologic response units (HRU)
HSI, *see* HyperSpectral Imager (HSI)
human activities, 21–22, 203, 379
human-nature relationships, 442
human well-being, 18, 40, 180
hybrid methodologies, 458–459
hydrological ecosystem services, 339–340
  assessment, 261–276
  drivers and pressures of, 249–250
  provision of, 230–231, 234–249
  remote sensing, 229–254, 261–276
  society and, 230–232
  strains on, 232
  water cycle and, 232–234
hydrologic biophysiographic variables, 251
hydrologic cycle, 263; *see also* water cycle
hydrologic modeling, 262–264
  ecosystem services quantification and, 262–268
  integrating remote sensing data with, 250–253, 261–276, 311–312
  remote sensing and, 268–272
hydrologic response units (HRU), 264
hydrologic-state variables, 251–252
Hyperion, 46, 48
hyperspatial imagery, 155–157
HyperSpectral Imager (HSI), 48
Hyperspectral Infrared Imager (HyspIRI), 48
hyperspectral remote sensing, 44, 156–157
HyspIRI, 47

I

ice, 240–241, 245–246
IGBP-DISCover, 161
IKONOS, 250
impact functions, 21–22
incidental sunlight, 41
indicators, selection of, 23–24
INMET, *see* National Institute of Meteorology (INMET)
Integrated Valuation of Ecosystem Services and Tradeoffs (InVEST), 6, 266–268, 449

interferometric SAR (InSAR), 28–29
interferometry techniques, 28–29
Intergovernmental Panel on Climate Change (IPCC), 419
Intergovernmental Platform on Biodiversity and Ecosystem Services (IPBES), 442
Intergovernmental Science-Policy Platform on Biodiversity and Ecosystem Services, 3
intermediate ESs, 19–21
intermediate services, monitoring of, 23–31
International Geosphere–Biosphere Programme dataset, 161
International Union for Conservation of Nature (IUCN), 151
invasive species, 153, 155
irrigated crops, 295–298
IUCN, *see* International Union for Conservation of Nature (IUCN)
IUCN Red List, 163

K

Kruskal-Wallis tests, 112–113, 114

L

land-atmosphere exchange of carbon
  remote sensing of, in Arctic tundra, 71–73
  remote sensing of, in boreal forests, 71–73
land clearing, 25, 27
land cover, 127, 157–165
  biological responses and, 219–220
  common, 207
  for ecosystem services mapping, 449
  effect, on stream ecology, 217–218
  fraction, 265–266
  global land cover projects, 161–165
  LUE estimates across types of, 110–112
  mapping, 158–159
  modeling relationship between stream ecology integrity and, 212–216

Riparian Area Land Cover (RALC),
203–222
spatial indicators, 213–215
Land Cover database, 208
land cover spatial indicator (LCSI),
212–213
land function dynamics assessment, 6
Land Parameter Retrieval Model
(LPRM), 247
Landsat satellites, 55, 72, 74, 75, 155, 158,
171, 265–266, 275, 335–336, 401
land service parameterization, of EFTs,
360–361
landslides, 249
land surface dynamics, 249
land surface emissivity (LSE), 430–432
land surface emissivity, over urban
areas, 429–433
land surface temperature (LST), 168, 403,
429–430, 432–433
land use, 127
changes in, 157, 232
for ecosystem services mapping, 449
land use planning, 17
Laser Imaging Detection and Ranging
(LIDAR), 336
LCSI, *see* land cover spatial indicator
(LCSI)
Leaf Area Index (LAI), 43, 264
leaf pigments, 42–49
chlorophyll content, 42–44
cycles, 44–47
LET, 402, 404
LIDAR, 19, 28, 29–30, *see* Laser Imaging
Detection and Ranging
(LIDAR)
light detection and ranging (LIDAR),
156, 235
light use efficiency (LUE), 40, 43–44, 45,
46, 47, 49, 55, 107
light use efficiency (LUE) estimation,
105–115
across organization levels and land
cover types, 110–112
materials and methods, 107–108
results, 108–114
time interval of, 112–114
l'Indice Biologique Global Normalisé
(IBGN), 208, 212–213

livestock systems, remote sensing for,
87–99
Living Plant Index, 152
LSE, *see* land surface emissivity (LSE)
LST, 404–407, *see* land surface
temperature (LST)
LUE, *see* light use efficiency (LUE)

M

Madrid DESIREX experimental
campaign, 421–423
maintenance services, of snow cover,
338–340
maize, 43
MAP-EVI regional function, 292
mass balance equation, 18
Mato Grasso State, 141
maximum difference water index
(MDWI), 248
maximum SUHI (SUHIM), 426–427
mean annual precipitation (MAP), 27,
288–289
MEdium Resolution Imaging
Spectrometer (MERIS), 51,
53, 247
MERIS, *see* Medium Resolution
Imaging Spectrometer
(MERIS)
MERIS Terrestrial Chlorophyll Index
(MTCI), 43
METEOSAT, 250
methane (CH4), 52
production, 73–74
micrometeorological tower sites,
316–317
microwave radiation (MR), 235, 245–246,
306–307
microwave sensors, 337
middle-infrared (MIR) radiance,
131, 246
Millennium Ecosystem Assessment,
3, 18, 19, 232–233, 442
Moderate Resolution Imaging
Spectroradiometer
(MODIS), 6, 43, 46, 48, 66,
72, 75, 129–130, 153, 168, 182,
184, 235, 265, 334–335, 381,
404, 424

FRE distribution, 137–139
FRP integration, 130–134
national parks networks, 186–196
MODIS Land Cover Product Collection 5 (MCD12Q1), 163
Monte Desert, 286–288
Monteith's model, 24, 25, 106–107
Morocco
national parks networks, 183–184
spatial patterns of EFTs in, 186–189
mountain ecosystem services, 329–344
cultural services, 340–341
maintenance services, 338–340
provisioning services, 337–338
regulating services, 338–339
Sierra Nevada Biosphere Reserve, 341–343
MR, *see* microwave radiation (MR)
multidimensional approaches, 441–460
need for, 442–444
multisensor techniques, 235

N

NASA QuickScat satellite, 74
National Centers for Environmental Prediction/National Center for Atmospheric Research (NCEP/NCAR), 365
National Institute of Meteorology (INMET), 142
national parks networks
carbon gains in, 192–193
congruence between functional diversity and carbon gains in, 185–186, 192–193
ecosystem services assessment of, 179–196
functional diversity assessment of, 185
Morocco, 183–184
Portugal, 183
representativeness and rarity of, 189–192
Spain, 183
National Polar-orbiting Partnership (NPP), 55
national statistics, 4
nature, as series of connected systems, 263

NDII, *see* normalized difference infrared index (NDII)
NDVI, *see* Normalized Difference Vegetation Index (NDVI)
near-infrared (NIR), 43, 64, 205, 235
net ecosystem exchange (NEE), 18, 71–73, 77, 106
net ecosystem production (NEP), 18, 77
net primary production (NPP), 23, 24, 26, 40, 64, 68, 70–71, 76, 106, 165, 166–167, 181
estimations, 24–27
normalized difference infrared index (NDII), 248
Normalized Difference Snow Index (NDSI), 334
Normalized Difference Vegetation Index (NDVI), 24–26, 43, 64–68, 70–72, 94–96, 106–107, 161, 166–169, 181, 211–212, 253, 265, 285, 309–310, 357–358
normalized difference water index (NDWI), 248
northern high latitudes, remote sensing of carbon cyling processes at, 63–77
numerical simulation models, 4

O

object-based image analysis (OBIA), 209–211, 218–219, 220–221
optical remote sensing, 19
Orbiting Carbon Observatory-2 (OCO-2), 52, 53

P

PAR, *see* photosynthetically active radiation (PAR)
particulate matter, 30
pasture renewal, 127
peatlands, 43
PEM, *see* production efficiency models (PEMs)
pest controls, 127
photochemical reflectance index (PRI), 25, 44–47, 49
photosynthesis, 18, 40, 379, 401

photosynthetically active radiation
        (PAR), 24, 69, 89, 97, 106–107, 167
photosynthetic efficiency, 50–51
photosynthetic stress
    chlorophyll fluorescence and, 49–54
    estimation of, 39–55
    leaf pigments and, 42–49
phytomass, 67, 469
planetary boundary layer (PBL), 135
point cloud data, 29–30
pollution sinks, 219
POLYSCAPE, 6, 450
Portugal
    national parks networks, 183, 186–196
    spatial patterns of EFTs in, 186–189
precipitation, 235, 238–239, 245, 248–249,
        288–289
Precipitation Estimation from Remotely
        Sensed Information using
        Artificial Neural Networks
        (PERSIANN-CCS), 245
PRI, *see* photochemical reflectance index
        (PRI)
primary production
    aboveground net primary production
        (ANPP), 25, 27, 88, 93, 96–97, 107
    gross, 40–44, 64, 68, 76
    net, 23, 24, 26, 40, 64, 68, 76, 106, 165,
        166–167, 181
    remote sensing of, in Arctic tundra,
        65–68
    remote sensing of, in boreal forests,
        68–71
PROBA, 48
process equation, 18
production efficiency models (PEMs),
        40, 68, 70
production functions, 19–21
Program for Deforestation Assessment
        in the Brazilian Legal
        Amazonia (PRODES), 142
provisioning services, 126
    by snow cover, 337–338
pyranometer, 380

Q

QuickBird data, 155–156, 250

R

RADAR, *see* radio detection and ranging
        (RADAR)
radar, 28–29
radiation
    fAPAR, 24, 43, 68, 94–95, 98
    fraction of the absorbed
        photosynthetically active
        radiation (fAPAR), 24, 43, 68
    microwave, 245–246
    photosynthetically active, 24, 69,
        89, 97
    reflected from land surface, 306
    solar, 380–382, 389–393, 403
    visible, 235
radiation use efficiency (RUE), 24–25, 88,
        99–100
    estimation, 92–98
radiative transfer models, 248
radio detection and ranging (RADAR),
        235, 251
radiometric indices, 24, 25, 26
RALC, *see* Riparian Area Land Cover
        (RALC)
regional modeling
    biophysical property changes in,
        363–366
    climate regulation services in,
        363–372
regulating services, 126, 202
    of snow cover, 338–339
remote sensing, 4
    ANPP estimation through, 92–93
    applied to SWAT, 252–253
    of carbon cycling processes, 63–77
    of carbon emissions from fire, 75–76
    of ecosystem services, 3–7, 17–18,
        399–411
    to estimate ET, 402–404
    forage monitoring system, 98–99
    of hydrological ecosystem services,
        229–254
    hydrological modeling and, 250–253,
        261–276
    hydrologic modeling and, 268–272
    hyperspectral, 44, 156–157
    of land-atmosphere exchange of
        carbon, 71–73

LIDAR, 19, 28, 29–30
for livestock systems, 87–99
of national parks networks, 179–196
optical, 19
of primary production, in Arctic,
65–68
of primary production, in boreal
forests, 68–71
of RUE, 24–25
of soil carbon processes, 73–74
of species diversity, 151–172
of surface soil moisture, 303–319
synthetic aperture radar (SAR),
19, 28–29
of urban heat island effect, 424–429
of vegetation biomass, in Arctic, 65–68
of vegetation biomass, in boreal
forests, 68–71
water quality monitoring and, 272–275
respiration, 18
Riparian Area Land Cover (RALC),
203–222
classification automation and
accuracy assessment, 211–212
classification of, 208–209
map, 216–217, 475
mapping over broad territories,
218–219
OBIA designed for, 209–211, 218–219
riparian vegetation, influence of, on
river ecology, 201–221
riparian zone, 202–203
river ecosystem
functional processes, 203
influence of riparian vegetation on,
201–221
riparian zone, 202–203
services provided by, 202
river ecosystems, 201–222
root mean square error (RMSE),
384–385, 429
root zone, 265
RUE estimation, 93–98
Rules Dam, 267

S

SAILHFlood, 384–385
saltcedar, 155

satellite images, 5–7, 24, 47, 48, 181
for detecting ecosystem reliance on
groundwater, 283–299
hyperspatial imagery, 155–157
to measure snow cover, 334–336
to measure water cycle elements,
236–244
satellite sensors, monitoring ecosystem
services with, 3–7
scale issues, in evaluation of carbon-
related ESs, 22–23
SCanning Imaging Absorption
spectroMeter for
Atmospheric CHartography
(SCIAMACHY), 53
scanning multichannel microwave
radiometer (SMMR), 73
Scopus database, 4
SDMs, *see* species distribution models
(SDMs)
seasonality, 25, 27
SEBAL, *see* Surface Energy Balance
Algorithm for Land (SEBAL)
sediment loads, 267
seminatural bare soil, 207
Sentinel, 55
service-providing units (SPUs), 444
SEVIRI, *see* Spinning Enhanced Visible
and Infrared Imager (SEVIRI)
Shuttle Radar Topography Mission
(SRTM), 251
Sierra Nevada Biosphere Reserve,
341–343
Simplified Surface Energy Balance
Index (S-SEBI), 403–404
SMA, *see* spectral mixture
analysis (SMA)
SMAP, *see* Soil Moisture Active and
Passive (SMAP)
SMMR, *see* scanning multichannel
microwave radiometer (SMMR)
SMOS, *see* Soil Moisture and Ocean
Salinity (SMOS)
SNOTEL, 333
snow, 240–241, 245–246
snow cover
airborne sensors for, 336–337
case study of monitoring, 341–343
cultural services by, 340–341

maintenance services, 338–340
maps, 271
monitoring of, 330–332
monitoring services provided by,
    332–337
oblique photographs, 333–334
provisioning services by, 337–338
regulating services, 338–339
satellite measurements, 334–336
in situ measurements, 332–334
snowpack, 329–344
snow surveys, 332–333
snow telemetry, 333
sociocultural valuation, 451–453
soil
    classification by spectral mixture
        analysis, 308–309
    moisture, 243, 247
    seminatural bare soil, 207
    surface moisture monitoring, 303–319
    surface soil layer, 265
soil and ground water, 247–248
Soil and Water Assessment Tool
    (SWAT), 252–253, 264
soil carbon processes
    remote sensing of, in Arctic tundra,
        73–74
    remote sensing of, in boreal forests,
        73–74
Soil Moisture Active and Passive
    (SMAP), 305
Soil Moisture and Ocean Salinity
    (SMOS), 305
soil respiration, 73–74
solar radiation, 265, 403
    efficiency models, 105–120
    estimation, 389–393, 404–407
solar radiation reflection (albedo),
    380–382
South America
    EFTs of southern, 359–360
    weather systems, 129
soybean, 43
Spain
    Madrid DESIREX experimental
        campaign, 421–423
    national parks networks, 183, 186–196
    Sierra Nevada Biosphere Reserve,
        341–343

spatial patterns of EFTs in, 186–189
spatial scale, 22–23
Special Sensor Microwave/Imager
    (SSM/I), 73
species
    distribution modeling, 159–161
    finding from space, 153–157
    forecasting fate of, 164
    global assessment, 161–165
    invasive, 153, 155
    loss, 151–152
    niches, 157–165
species distribution models (SDMs),
    159–161
species diversity; *see also* biodiversity
    assessment and monitoring, 151–172
    congruence between carbon gains
        and, 192–193
    ecosystem functioning and, 165–171
    land cover and, 157–165
species-energy hypothesis, 165–168
spectral data, 24
spectral mixture analysis (SMA), soil
    state classification by, 308–309
spectral vegetation indices, 106
Spinning Enhanced Visible and
        Infrared Imager (SEVIRI), 235
SPUs, *see* service-providing
        units (SPUs)
SRTM, *see* Shuttle Radar Topography
        Mission (SRTM)
SSM/I, *see* Special Sensor Microwave/
        Imager (SSM/I)
stress factors, 21–22
Suomi National Polar-orbiting
        Partnership (NPP), 55
supporting services, 202
surface albedo, 380–382
Surface Energy Balance Algorithm for
        Land (SEBAL), 252
surface soil layer, 265
surface soil moisture
    estimating evapotranspiration for
        valuation of, 312–317
    introduction to, 304–305
    monitoring by remote sensing,
        303–319
surface temperature, 168, 403, 407–410,
        432–433

surface urban heat island (SUHI),
421–423, 426–427
surface water, 242, 246–247
SWAT, *see* Soil and Water Assessment
Tool (SWAT)
synthetic aperture radar (SAR), 19, 28–29
System of Economic and Environmental
Accounts (SEEA), 126

T

TEEB, *see* The Economics of Ecosystems
and Biodiversity (TEEB)
TEM, *see* terrestrial ecosystem model
(TEM)
temperature, 41; *see also* surface
temperature
Temperature and Emissivity Separation
(TES) algorithm, 421
temporal dynamics, 283–299
temporal scale, 22–23
TERRA-AQUA, 48
terrestrial ecosystem model (TEM), 68
terrestrial ecosystem services,
estimation of photosynthetic
stress for, 39–55
terrestrial vegetation models, 262
TES, *see* Temperature and Emissivity
Separation (TES) algorithm
TEV, *see* total economic value (TEV)
framework
The Economics of Ecosystems and
Biodiversity (TEEB), 180, 442
Thermal And Near-infrared Sensor for
carbon Observation–Fourier
Transform Spectrometer
(TANSO–FTS/GOSAT), 51
thermal anomalies detections, 129–130
thermal infrared (TIR), 421
thermal satellite imagery, mapping
evapotranspiration using,
314–315
Therman And Near-infrared Sensor for
carbon Observation-Fourier
Transform Spectometer
(TANSO-FTS), 53
thermopolis experimental campaign,
423–424
timber, 126

total economic value (TEV) framework,
454–455
trace gases, 128
trace gases emission estimation, in
Amazon tropical forest,
139–142
tree vegetation, 207
tropical forests
biomass burning emission
estimation, 125–144
LUE in, 111
Tropical Rainfall Measuring Mission
(TRMM), 235, 262
tropical savannas, 111

U

UCL, *see* urban canopy layer (UCL)
UHI, *see* urban heat island (UHI) effect
urban areas, 207, 220
urban canopy layer (UCL), 418
urban heat island (UHI) effect, 417–434
Athens case study, 423–424
evaluation of, 420–424
introduction to, 417–420
land surface emissivity, 429–433
Madrid case study, 421–423
overpass time for, 427–429
remote sensing of, 424–429
spatial resolution for, 424–427
urbanization, 219
U.S. Geological Survey, 361–363

V

Valuation Policy of Acre Environmental
Forest Asset (PVAAFA),
142–143
vapor pressure deficit (VPD), 41
vegetation, 24
dynamic vegetation models, 355–356
herbaceous/shrub, 207
interannual variability of, 363
light use efficiency (LUE) by, 105–120
soil moisture stress on, 312–317
tree, 207
vegetation biomass
remote sensing of, in Arctic tundra,
65–68

remote sensing of, in boreal forests, 68–71
vegetation dynamics, impact of groundwater on, 291–292
vegetation indices (VIs), 24–26, 43
  Enhanced Vegetation Index (EVI), 24, 25, 43, 72, 106–107, 167–168, 181, 184, 289, 290, 292–295, 404
  Normalized Difference Vegetation Index (NDVI), 24–26, 64–68, 70–72, 94–96, 106–107, 161, 166–169, 181, 211–212, 253, 265, 285, 309–310, 357–358
  satellite-derived enhanced, 182–185
  spectral, 106
vegetation water content (VWC), 248
vegetative cover, 265
very high spatial resolution (VHSR) mapping, 203–204, 205
VIs, *see* vegetation indices (VIs)
visible and infrared regions (VNIR), 307
visible radiation (VIS), 235
VPD, *see* vapor pressure deficit (VPD)
VSWIR, 48

W

waste processing, 126
water, 126
  atmospheric, 235, 245
  availability, 230, 232
  budgets, 265
  classification of, 207
  in cyrosphere, 245–246
  damage mitigation, 234, 248–249
  demand for, 232
  detecting ecosystem reliance on groundwater, 283–299
  essential role of, 230
  groundwater, 244, 247–248, 283–299
  management of, 232
  quality, monitoring of, 246–247, 272–275
  scarcity, 230
  soil and ground, 247–248
  storage capacity, 230
  surface, 242, 246–247
  vegetation, 248

water band index (WBI), 248
water cycle, 8, 232–234, 261–262
  hydrological services and, 232–234
  sensors and satellites to measure elements of, 236–244
Water Framework Directive (WFD), 208
water quality monitoring, remote sensing and, 272–275
water regulation, 25, 126
Watershed integrated Management for Mediterranean environments (WiMMed), 266, 271
water supply, 235, 245–248
water vapor, 235, 236–237, 265
WBI, *see* water band index (WBI)
Weather Research and Forecasting (WRF), 365
wetlands, 295–298, 382–383
wetlands reflected energy, 383–393
  canopy reflectance simulation model, 384–385
  simulated case studies, 385–387
  simulated reflectance, 387–393
WFD, *see* Water Framework Directive (WFD)
Wildfire Automated Biomass Burning Algorithm (WFABBA), 130
wildfires, 30–31
  biomass burning emission estimation, 125–144
  carbon emissions from, 75–76
wildlife populations
  assessment and monitoring, 151–172
  conservation of, 168–171
  ecology of, 168–171
WiMMED, *see* Watershed integrated Management for Mediterranean environments (WiMMed)
woodlands, 126, 295–298
WRF, *see* Weather Research and Forecasting (WRF)

X

xanthophyll cycle, 44, 45, 46